MW00843385

STUDENT'S SOLUTIONS MANUAL

BEVERLY FUSFIELD

BASIC COLLEGE MATHEMATICS
TENTH EDITION

Margaret L. Lial
American River College

Stanley A. Salzman
American River College

Diana L. Hestwood
Minneapolis Community and Technical College

 Pearson

The author and publisher of this book have used their best efforts in preparing this book. These efforts include the development, research, and testing of the theories and programs to determine their effectiveness. The author and publisher make no warranty of any kind, expressed or implied, with regard to these programs or the documentation contained in this book. The author and publisher shall not be liable in any event for incidental or consequential damages in connection with, or arising out of, the furnishing, performance, or use of these programs.

Reproduced by Pearson from electronic files supplied by the author.

Copyright © 2018, 2014, 2010 Pearson Education, Inc.
Publishing as Pearson, 330 Hudson Street, NY NY 10013

All rights reserved. No part of this publication may be reproduced, stored in a retrieval system, or transmitted, in any form or by any means, electronic, mechanical, photocopying, recording, or otherwise, without the prior written permission of the publisher. Printed in the United States of America.
    2 2020

 Pearson

ISBN-13: 978-0-13-447411-3
ISBN-10: 0-13-447411-2

Table of Contents

Copyright © 2018 Pearson Education, Inc.

Copyright © 2018 Pearson Education, Inc.

Copyright © 2018 Pearson Education, Inc.

Copyright © 2018 Pearson Education, Inc.

CHAPTER 1 WHOLE NUMBERS

1.1 Reading and Writing Whole Numbers

1.1 Margin Exercises

1. **(a)** 342

   4 is in the *tens* place.

   (b) 714

   4 is in the *ones* place.

   (c) 479

   4 is in the *hundreds* place.

2. The place value of each digit:

   (a) 14,218

   1: ten-thousands

   4: thousands

   2: hundreds

   1: tens

   8: ones

   (b) 460,329

   4: hundred-thousands

   6: ten-thousands

   0: thousands

   3: hundreds

   2: tens

   9: ones

3. The digits in each period (group) of 3,251,609,328:

   (a) 3 is in the billions period.

   (b) 251 is in the millions period.

   (c) 609 is in the thousands period.

   (d) 328 is in the ones period.

4. **(a)** 18 is eighteen.

   (b) 36 is thirty-six.

   (c) 418 is four hundred eighteen.

   (d) 902 is nine hundred two.

5. **(a)** 3104 is three <u>thousand</u>, one <u>hundred</u> four.

   (b) 95,372 is ninety-five <u>thousand</u>, three hundred seventy-two.

   (c) 100,075,002 is one hundred million, seventy-five thousand, two.

   (d) 11,022,040,000 is eleven billion, twenty-two million, forty thousand.

6. **(a)** One thousand, four hundred thirty-seven is 1437.

   (b) Nine hundred seventy-one thousand, six is 971,006.

   (c) Eighty-two million, three hundred twenty-five is 82,000,325.

7. **(a)** The U.S. population in the 2015 column is 325 million and is written in digits as 32<u>5</u>,000,000.

   (b) The estimated U.S. population in the 2020 column is 341 million and is written in digits as 341,000,000.

   (c) The household income in 2010 was $46,326 and is written in words as forty-six thousand, three hundred twenty-six dollars.

   (d) The estimated average yearly salary in 2020 is $36,169 and is written in words as thirty-six thousand, one hundred sixty-nine dollars.

1.1 Section Exercises

1. The digit in the hundreds place in the whole number 3<u>0</u>65 is 0. **(c)**

3. <u>18</u>,<u>0</u>15; ten-thousands: 1; hundreds: 0

5. 7,6<u>2</u>8,<u>5</u>92,183; millions: 8; thousands: 2

7. The digits in the thousands period in the whole number 552,<u>687</u>,318 are 687.

9. <u>3</u>,<u>561</u>,<u>435</u>; millions: 3; thousands: 561; ones: 435

11. Evidence suggests that this is true. It is common to count using fingers.

13. 23,115 is twenty-three thousand, one hundred fifteen. The statement is *false* (no "and").

15. 346,009 is three hundred forty-six <u>thousand</u>, <u>nine</u>.

17. 25,756,665 is twenty-five million, seven hundred fifty-six thousand, six hundred sixty-five.

19. Sixty-three thousand, one hundred sixty-three is 6<u>3</u>,1<u>63</u>.

Copyright © 2018 Pearson Education, Inc.

21. Ten million, two hundred twenty-three is 10,000,223.

23. Seventy-nine million, six hundred eighty thousand in digits is 79,680,000.

25. Fifty million, fifty-one thousand, five hundred seven in digits is 50,051,507.

27. Fifty-four million, seven hundred fifty thousand in digits is 54,750,000.

29. Eight hundred trillion, six hundred twenty-one million, twenty thousand, two hundred fifteen in digits is 800,000,621,020,215.

31. The least used method of transportation is public transportation. 6,069,589 in words is six million, sixty-nine thousand, five hundred eighty-nine.

33. The number of people who walk to work or work at home is 7,894,911. In words, 7,894,911 is written as seven million, eight hundred ninety-four thousand, nine hundred eleven.

1.2 Adding Whole Numbers

1.2 Margin Exercises

1. (a) $2 + 6 = \underline{8}$; $6 + \underline{2} = \underline{8}$

   (b) $9 + 5 = 14$; $5 + 9 = 14$

   (c) $4 + 7 = 11$; $7 + 4 = 11$

2. (a)
$$
\begin{array}{r}
3 \\
8 \quad 3 + 8 = 11 \\
5 \quad 11 + 5 = 16 \\
4 \quad 16 + 4 = 20 \\
+6 \quad 20 + 6 = 26 \\
\hline
26
\end{array}
$$

   (b)
$$
\begin{array}{r}
5 \\
6 \quad 5 + 6 = 11 \\
3 \quad 11 + 3 = 14 \\
2 \quad 14 + 2 = 16 \\
+4 \quad 16 + 4 = 20 \\
\hline
20
\end{array}
$$

   (c)
$$
\begin{array}{r}
9 \\
6 \quad 9 + 6 = 15 \\
8 \quad 15 + 8 = 23 \\
7 \quad 23 + 7 = 30 \\
+3 \quad 30 + 3 = 33 \\
\hline
33
\end{array}
$$

   (d)
$$
\begin{array}{r}
3 \\
8 \quad 3 + 8 = 11 \\
6 \quad 11 + 6 = 17 \\
4 \quad 17 + 4 = 21 \\
+8 \quad 21 + 8 = 29 \\
\hline
29
\end{array}
$$

3. (a)
$$
\begin{array}{r}
26 \\
+73 \\
\hline
99
\end{array}
$$
   (b)
$$
\begin{array}{r}
534 \\
+265 \\
\hline
799
\end{array}
$$

   (c)
$$
\begin{array}{r}
42{,}305 \\
+11{,}563 \\
\hline
53{,}868
\end{array}
$$

4. (a)
$$
\begin{array}{r}
\overset{1}{6}6 \\
+27 \\
\hline
93
\end{array}
$$
   6 ones and 7 ones = 13 ones

   (b)
$$
\begin{array}{r}
\overset{1}{5}8 \\
+33 \\
\hline
91
\end{array}
$$
   8 ones and 3 ones = 11 ones

   (c)
$$
\begin{array}{r}
\overset{1}{5}6 \\
+37 \\
\hline
93
\end{array}
$$
   6 ones and 7 ones = 13 ones

   (d)
$$
\begin{array}{r}
\overset{1}{3}4 \\
+49 \\
\hline
83
\end{array}
$$
   4 ones and 9 ones = 13 ones

5. (a)
$$
\begin{array}{r}
\overset{1\,2}{1}62 \\
4\,271 \\
372 \\
+8\,976 \\
\hline
13{,}\underline{78}1
\end{array}
$$

   (b)
$$
\begin{array}{r}
\overset{2\,21}{7\,8}21 \\
435 \\
72 \\
305 \\
+1\,693 \\
\hline
10{,}326
\end{array}
$$

6. (a)
$$
\begin{array}{r}
816 \\
363 \\
17 \\
2 \\
5 \\
+7654 \\
\hline
8857
\end{array}
$$

Copyright © 2018 Pearson Education, Inc.

(b)

```
   15,829
      765
       78
       15
        9
        7
 + 13,179
 ─────────
   29,882
```

7. The shortest route from Lake Buena Vista to Conway is as follows:

4	*Lake Buena Vista to Resort Area*
6	*Resort Area to Pine Castle*
3	*Pine Castle to Belle Isle*
+ 6	*Belle Isle to Conway*
19	*miles*

8. The next shortest route from Orlando to Clear Lake is as follows:

5	*Orlando to Pine Hills*
8	*Pine Hills to Altamonte Springs*
5	*Altamonte Springs to Casselberry*
6	*Casselberry to Bertha*
7	*Bertha to Winter Park*
+ 7	*Winter Park to Clear Lake*
38	*miles*

9.

```
    22
   526
   297
   526
 + 297
 ──────
  1646 feet
```

The amount of fencing needed is 1646 feet, which is the perimeter of (distance around) the project.

10. (a)

```
    63
     4
     9
 + 28
 ──────
   104  correct
```

(b)

```
   927
   395
    64
 + 251
 ──────
  1637  correct
```

(c)

```
     79
    218
      7
 + 639
 ──────
    943  (953 is incorrect.)
```

(d)

```
   21,892
   11,746
 + 43,925
 ─────────
   77,563  (79,563 is incorrect.)
```

1.2 Section Exercises

1.

```
    43
 + 54
 ─────
    97
```

3.

```
    56
 + 33
 ─────
    89
```

5.

```
   317
 + 572
 ──────
   889
```

7.

```
   318
   151
 + 420
 ──────
   889
```

9.

```
  6310
   252
 + 1223
 ───────
  7785
```

11. Line up the numbers in columns. Then start at the right and add the ones digits. Add the ten digits next, and, finally, the hundreds digits.

```
   932
    44
 + 613
 ──────
  1589  correct
```

13. Line up:

```
    1251
    4311
 + 2114
 ───────
   7676  (7686 is incorrect.)
```

15. Line up:

```
   12,142
   43,201
 + 23,103
 ─────────
   78,446
```

Copyright © 2018 Pearson Education, Inc.

17. Line up:
$$
\begin{array}{r}
3213 \\
+\ 5715 \\
\hline
8928
\end{array}
$$

19. Line up:
$$
\begin{array}{r}
38{,}204 \\
+\ 21{,}020 \\
\hline
59{,}224
\end{array}
$$

21.
$$
\begin{array}{r}
\overset{1}{8}7 \\
+\ 63 \\
\hline
150 \text{ correct}
\end{array}
$$

23.
$$
\begin{array}{r}
\overset{1}{8}6 \\
+\ 69 \\
\hline
155 \text{ correct}
\end{array}
$$

25.
$$
\begin{array}{r}
\overset{1}{4}7 \\
+\ 74 \\
\hline
121 \ (111 \text{ is incorrect.})
\end{array}
$$

27.
$$
\begin{array}{r}
\overset{1}{6}7 \\
+\ 78 \\
\hline
145
\end{array}
$$

29.
$$
\begin{array}{r}
\overset{1}{7}3 \\
+\ 29 \\
\hline
102
\end{array}
$$

31.
$$
\begin{array}{r}
7\overset{1}{4}6 \\
+\ 905 \\
\hline
1651
\end{array}
$$

33.
$$
\begin{array}{r}
3\overset{1}{0}6 \\
+\ 848 \\
\hline
1154
\end{array}
$$

35.
$$
\begin{array}{r}
2\overset{11}{7}8 \\
+\ 135 \\
\hline
413
\end{array}
$$

37.
$$
\begin{array}{r}
9\overset{1}{2}8 \\
+\ 843 \\
\hline
1771
\end{array}
$$

39.
$$
\begin{array}{r}
5\overset{11}{2}6 \\
+\ 884 \\
\hline
1410
\end{array}
$$

41.
$$
\begin{array}{r}
3\overset{1}{5}\overset{1}{7}4 \\
+\ 2817 \\
\hline
6391
\end{array}
$$

43.
$$
\begin{array}{r}
7\overset{111}{8}96 \\
+\ 3\,728 \\
\hline
11{,}624
\end{array}
$$

45.
$$
\begin{array}{r}
9\overset{111}{6}25 \\
+\ 7\,986 \\
\hline
17{,}611
\end{array}
$$

47.
$$
\begin{array}{r}
9\,0\overset{22}{5}6 \\
78 \\
6\,089 \\
+\ \ \ 731 \\
\hline
15{,}954
\end{array}
$$

49. *Step 1*: Add the digits in the ones column.

Step 2: Add the digits in the tens column, including the regrouped 2.

Step 3: Add the hundreds column, including the regrouped 1.

Step 4: Add the thousands column, including the regrouped 1.

$$
\begin{array}{r}
\overset{112}{}18 \\
708 \\
9\,286 \\
+\ \ \ 636 \\
\hline
10{,}648
\end{array}
$$

51.
$$
\begin{array}{r}
\overset{1\,11}{4}22 \\
6\,074 \\
435 \\
+\ 8\,663 \\
\hline
15{,}594
\end{array}
$$

53.
$$
\begin{array}{r}
\overset{1\ \ 1}{3}21 \\
9\,603 \\
8 \\
21 \\
+\ 1\,604 \\
\hline
11{,}557
\end{array}
$$

55.
$$
\begin{array}{r}
2\,1\overset{112}{0}9 \\
63 \\
16 \\
3 \\
+\ 9\,887 \\
\hline
12{,}078
\end{array}
$$

Copyright © 2018 Pearson Education, Inc.

57.
$$\begin{array}{r} \overset{2\,3\,3}{553} \\ 97 \\ 2772 \\ 437 \\ 63 \\ +\ 328 \\ \hline 4250 \end{array}$$

59.
$$\begin{array}{r} \overset{2\,1\,2}{413} \\ 85 \\ 9\,919 \\ 602 \\ 31 \\ +\ 1\,218 \\ \hline 12,268 \end{array}$$

61. Add up to check addition.

$$\begin{array}{r} \underline{2091} \\ 832 \\ 468 \\ +\ 791 \\ \hline 2091 \text{ correct} \end{array}$$

63. Add up to check addition.

$$\begin{array}{r} \underline{1857} \\ 326 \\ 852 \\ +\ 679 \\ \hline 1857 \text{ correct} \end{array}$$

65. Add up to check addition.

$$\begin{array}{r} \underline{5420} \\ 4713 \\ 28 \\ 615 \\ +\ 64 \\ \hline 5420 \text{ correct} \end{array}$$

67. Add up to check addition.

$$\begin{array}{r} \underline{11,577} \\ 678 \\ 7\,952 \\ 56 \\ 718 \\ +\ 2\,173 \\ \hline 11,377 \text{ incorrect; should be 11,577} \end{array}$$

69. Add up to check addition.

$$\begin{array}{r} \underline{14,332} \\ 4714 \\ 27 \\ 77 \\ 8878 \\ +\ 636 \\ \hline 14,332 \text{ correct} \end{array}$$

71. Changing the order in which two numbers are added does not change the sum. You can add two numbers from bottom to top when checking addition.

73. The shortest route between Southtown and Rena is through Thomasville.

$$\begin{array}{rl} 21 & \textit{Southtown to Thomasville} \\ +\ 12 & \textit{Thomasville to Rena} \\ \hline 33 & \textit{miles} \end{array}$$

75. The shortest route between Thomasville and Murphy is through Rena and Austin.

$$\begin{array}{rl} 12 & \textit{Thomasville to Rena} \\ 15 & \textit{Rena to Austin} \\ +\ 11 & \textit{Austin to Murphy} \\ \hline 38 & \textit{miles} \end{array}$$

77. To find the total amount raised, add the amounts raised at each event.

$$\begin{array}{rl} \$3\,482 & \textit{flea market} \\ +\ 12,860 & \textit{annual auction} \\ \hline \$16,342 & \textit{total amount raised} \end{array}$$

79.
$$\begin{array}{rl} 413 & \textit{women} \\ +\ 286 & \textit{men} \\ \hline 699 & \textit{total people} \end{array}$$

81.
$$\begin{array}{rl} 13,786 & \textit{on-campus day} \\ 3\,497 & \textit{on-campus night} \\ +\ 2\,874 & \textit{on-line} \\ \hline 20,157 & \textit{total students} \end{array}$$

83. To find the perimeter, add the lengths of the 4 sides in any order.

$$\begin{array}{rl} 325 & \\ 160 & \\ 325 & \\ +\ 160 & \\ \hline 970 & \textit{feet} \end{array}$$

970 feet is the total distance around the lot.

Copyright © 2018 Pearson Education, Inc.

85. To find the perimeter, add the lengths of the 3 sides in any order.

$$\begin{array}{r} 30 \\ 24 \\ + 18 \\ \hline 72 \end{array} \text{ feet}$$

Martin will need 72 feet of lumber.

87. The largest four-digit number possible, using the digits 4, 1, 9, and 2 each once, will begin with the largest digit in the thousands place and the remaining digits will descend in size until the smallest digit is in the ones place. Therefore, the largest four-digit number possible is 9421.

88. The smallest four-digit number possible, using the digits 4, 1, 9, and 2 each once, will begin with the smallest digit in the thousands place and the remaining digits will ascend in size until the largest digit is in the ones place. Therefore, the smallest four-digit number possible is 1249.

89. The largest five-digit number possible, using the digits 6, 2, and 7 at least once, will use the largest digit in the ten-thousands, thousands, and hundreds place and the remaining digits will descend in size until the smallest digit is in the ones place. Therefore, the largest five-digit number possible is 77,762.

90. The smallest five-digit number possible, using the digits 6, 2, and 7 at least once, will use the smallest digit in the ten-thousands, thousands, and hundreds place and the remaining digits will ascend in size until the largest digit is in the ones place. Therefore, the smallest five-digit number possible is 22,267.

91. The largest seven-digit number possible is 9,994,433.

92. The smallest seven-digit number possible is 3,334,499.

93. To write the largest seven-digit number possible, using the digits 4, 3, and 9, and using each digit at least twice, write the largest digits on the left, and use the smaller digits as you move right.

94. To write the smallest seven-digit number possible, using the digits 4, 3, and 9, and using each digit at least twice, write the smallest digits on the left, and use the larger digits as you move right.

1.3 Subtracting Whole Numbers

1.3 Margin Exercises

1. **(a)** $8 + 2 = 10$:

$10 - \underline{2} = 8$ or $10 - \underline{8} = 2$

(b) $7 + 4 = 11$:

$11 - 7 = 4$ or $11 - 4 = 7$

(c) $15 + 22 = 37$:

$37 - 15 = 22$ or $37 - 22 = 15$

(d) $23 + 55 = 78$:

$78 - 55 = 23$ or $78 - 23 = 55$

2. **(a)** $7 - 5 = 2$:

$7 = 5 + \underline{2}$ or $7 = 2 + 5$

(b) $9 - 4 = 5$:

$9 = 4 + 5$ or $9 = 5 + 4$

(c) $21 - 15 = 6$:

$21 = 15 + 6$ or $21 = 6 + 15$

(d) $58 - 42 = 16$:

$58 = 42 + 16$ or $58 = 16 + 42$

3. **(a)** $\begin{array}{r} 74 \\ - 43 \\ \hline 31 \end{array}$ 4 ones $-$ 3 ones $=$ 1 one
7 tens $-$ 4 tens $=$ 3 tens

(b) $\begin{array}{r} 68 \\ - 24 \\ \hline 44 \end{array}$ 8 ones $-$ 4 ones $=$ 4 ones
6 tens $-$ 2 tens $=$ 4 tens

(c) $\begin{array}{r} 429 \\ - 318 \\ \hline 111 \end{array}$

(d) $\begin{array}{r} 3927 \\ - 2614 \\ \hline 1313 \end{array}$

4. **(a)** $\begin{array}{r} 76 \\ - 45 \\ \hline 31 \end{array}$ *Subtraction problem* $\begin{array}{r} 45 \\ + \underline{31} \\ \hline \underline{76} \end{array}$ *Addition problem*

Match: 31 is correct.

(b) $\begin{array}{r} 53 \\ - 22 \\ \hline 21 \end{array}$ *Subtraction problem* $\begin{array}{r} 22 \\ + 21 \\ \hline 43 \end{array}$ *Addition problem*

Not a match: 21 is incorrect.

(*continued*)

Copyright © 2018 Pearson Education, Inc.

Rework.

$$\begin{array}{r} 53 \\ -22 \\ \hline 31 \end{array} \quad \text{Subtraction} \atop \text{problem} \qquad \begin{array}{r} 22 \\ +31 \\ \hline 53 \end{array} \quad \text{Addition} \atop \text{problem}$$

Match: 31 is correct.

(c)
$$\begin{array}{r} 374 \\ -251 \\ \hline 113 \end{array} \quad \text{Subtraction} \atop \text{problem} \qquad \begin{array}{r} 251 \\ +113 \\ \hline 364 \end{array} \quad \text{Addition} \atop \text{problem}$$

Not a match: 113 is incorrect.

Rework.

$$\begin{array}{r} 374 \\ -251 \\ \hline 123 \end{array} \quad \text{Subtraction} \atop \text{problem} \qquad \begin{array}{r} 251 \\ +123 \\ \hline 374 \end{array} \quad \text{Addition} \atop \text{problem}$$

Match: 123 is correct.

(d)
$$\begin{array}{r} 7531 \\ -4301 \\ \hline 3230 \end{array} \quad \text{Subtraction} \atop \text{problem}$$

$$\begin{array}{r} 4301 \\ +3230 \\ \hline 7531 \end{array} \quad \text{Addition} \atop \text{problem}$$

Match: 3230 is correct.

5. (a)
$$\begin{array}{r} \overset{4\,18}{\cancel{5}\,\cancel{8}} \\ -1\,9 \\ \hline 3\,9 \end{array}$$

(b)
$$\begin{array}{r} \overset{7\,16}{\cancel{8}\,\cancel{6}} \\ -3\,8 \\ \hline 4\,8 \end{array}$$

(c)
$$\begin{array}{r} \overset{3\,11}{\cancel{4}\,1} \\ -2\,7 \\ \hline 1\,4 \end{array}$$

(d)
$$\begin{array}{r} 8\overset{5\,13}{\cancel{6}\,\cancel{3}} \\ -4\,7 \\ \hline 8\,1\,6 \end{array}$$

(e)
$$\begin{array}{r} 7\overset{5\,12}{\cancel{6}\,\cancel{2}} \\ -1\,5\,7 \\ \hline 6\,0\,5 \end{array}$$

6. (a)
$$\begin{array}{r} \overset{8\,12}{9\,\cancel{2}\,7} \\ -4\,3 \\ \hline \underline{8\,8\,4} \end{array}$$

(b)
$$\begin{array}{r} \overset{5\,16\,15}{\cancel{6}\,\cancel{7}\,\cancel{5}} \\ -8\,6 \\ \hline 5\,8\,9 \end{array}$$

(c)
$$\begin{array}{r} \overset{3\,16\,17}{\cancel{4}\,\cancel{7}\,\cancel{7}} \\ -3\,8\,9 \\ \hline 8\,8 \end{array}$$

(d)
$$\begin{array}{r} \overset{0\,13\,10\,17}{1\,\cancel{4}\,\cancel{1}\,\cancel{7}} \\ -9\,8\,8 \\ \hline 4\,2\,9 \end{array}$$

(e)
$$\begin{array}{r} \overset{7\,16\,13}{8\,\cancel{7}\,\cancel{3}\,9} \\ -3\,8\,9\,2 \\ \hline 4\,8\,4\,7 \end{array}$$

7. (a)
$$\begin{array}{r} \overset{9}{\overset{1\,\cancel{10}\,16}{\cancel{2}\,\cancel{0}\,\cancel{6}}} \\ -1\,7\,7 \\ \hline 2\,9 \end{array}$$

(b)
$$\begin{array}{r} \overset{9}{\overset{6\,\cancel{10}\,13}{\cancel{7}\,\cancel{0}\,\cancel{3}}} \\ -4\,1\,5 \\ \hline 2\,8\,8 \end{array}$$

(c)
$$\begin{array}{r} \overset{9}{\overset{6\,\cancel{10}\,12}{\cancel{7}\,\cancel{0}\,\cancel{2}\,4}} \\ -2\,6\,3\,2 \\ \hline 4\,3\,9\,2 \end{array}$$

8. (a)
$$\begin{array}{r} \overset{9}{\overset{2\,\cancel{10}\,18}{\cancel{3}\,\cancel{0}\,\cancel{8}}} \\ -1\,5\,9 \\ \hline \underline{1\,4}\,9 \end{array}$$

(b)
$$\begin{array}{r} 5\overset{6\,10}{\cancel{7}\,\cancel{0}} \\ -3\,6\,8 \\ \hline 2\,0\,2 \end{array}$$

(c)
$$\begin{array}{r} \overset{0\,14\,16\,10}{1\,\cancel{5}\,\cancel{7}\,\cancel{0}} \\ -9\,8\,3 \\ \hline 5\,8\,7 \end{array}$$

(d)
$$\begin{array}{r} \overset{99}{\overset{6\,\cancel{10}\,\cancel{10}\,11}{\cancel{7}\,\cancel{0}\,\cancel{0}\,\cancel{1}}} \\ -5\,1\,9\,3 \\ \hline 1\,8\,0\,8 \end{array}$$

(e)
$$\begin{array}{r} \overset{99}{\overset{3\,\cancel{10}\,\cancel{10}\,10}{\cancel{4}\,\cancel{0}\,\cancel{0}\,\cancel{0}}} \\ -1\,7\,8\,2 \\ \hline 2\,2\,1\,8 \end{array}$$

Copyright © 2018 Pearson Education, Inc.

9. **(a)**

$$
\begin{array}{r}
357 \\
-\ 168 \\
\hline
189
\end{array}
\qquad
\begin{array}{r}
168 \\
+\ 189 \\
\hline
357
\end{array}
$$

Match: 189 is correct.

(b)

$$
\begin{array}{r}
570 \\
-\ 328 \\
\hline
252
\end{array}
\qquad
\begin{array}{r}
328 \\
+\ 252 \\
\hline
580
\end{array}
$$

Not a match: 252 is incorrect.

Rework.

$$
\begin{array}{r}
5\ \overset{6}{7}\ \overset{10}{\cancel{0}} \\
-\ 3\ 2\ 8 \\
\hline
2\ 4\ 2
\end{array}
\qquad
\begin{array}{r}
328 \\
+\ 242 \\
\hline
570
\end{array}
$$

Match: 242 is correct.

(c)

$$
\begin{array}{r}
14,726 \\
-\ 8\,839 \\
\hline
5\,887
\end{array}
\qquad
\begin{array}{r}
8839 \\
+\ 5887 \\
\hline
14,726
\end{array}
$$

Match: 5887 is correct.

10. **(a)**

$$
\begin{array}{r}
\$7\,1,9\,0\,0 \\
-\ 4\,0,2\,0\,0 \\
\hline
\$3\,1,7\,0\,0
\end{array}
\quad
\begin{array}{l}
\textit{Associate of Arts degree} \\
\textit{Not a high school graduate}
\end{array}
$$

(b)

$$
\begin{array}{r}
\overset{0\ 9\ 9}{\$1\ 0\ 0},\overset{10}{\cancel{0}}\,0\,0 \\
-\ 7\,1,9\,0\,0 \\
\hline
\$2\,8,1\,0\,0
\end{array}
\quad
\begin{array}{l}
\textit{Bachelor's degree} \\
\textit{Associate of Arts degree} \\
\textit{More earnings}
\end{array}
$$

1.3 Section Exercises

1.

$$
\begin{array}{r}
48 \\
-\ 32 \\
\hline
16
\end{array}
\qquad
\textit{Check:}
\qquad
\begin{array}{r}
16 \\
+\ 32 \\
\hline
48
\end{array}
$$

3.

$$
\begin{array}{r}
86 \\
-\ 53 \\
\hline
33
\end{array}
\qquad
\textit{Check:}
\qquad
\begin{array}{r}
53 \\
+\ 33 \\
\hline
86
\end{array}
$$

5.

$$
\begin{array}{r}
77 \\
-\ 60 \\
\hline
17
\end{array}
\qquad
\textit{Check:}
\qquad
\begin{array}{r}
60 \\
+\ 17 \\
\hline
77
\end{array}
$$

7.

$$
\begin{array}{r}
335 \\
-\ 122 \\
\hline
213
\end{array}
\qquad
\textit{Check:}
\qquad
\begin{array}{r}
213 \\
+\ 122 \\
\hline
335
\end{array}
$$

9.

$$
\begin{array}{r}
552 \\
-\ 451 \\
\hline
101
\end{array}
\qquad
\textit{Check:}
\qquad
\begin{array}{r}
451 \\
+\ 101 \\
\hline
552
\end{array}
$$

11.

$$
\begin{array}{r}
7352 \\
-\ 241 \\
\hline
7111
\end{array}
\qquad
\textit{Check:}
\qquad
\begin{array}{r}
241 \\
+\ 7111 \\
\hline
7352
\end{array}
$$

13.

$$
\begin{array}{r}
5546 \\
-\ 2134 \\
\hline
3412
\end{array}
\qquad
\textit{Check:}
\qquad
\begin{array}{r}
3412 \\
+\ 2134 \\
\hline
5546
\end{array}
$$

15.

$$
\begin{array}{r}
6259 \\
-\ 4148 \\
\hline
2111
\end{array}
\qquad
\textit{Check:}
\qquad
\begin{array}{r}
4148 \\
+\ 2111 \\
\hline
6259
\end{array}
$$

17.

$$
\begin{array}{r}
24,392 \\
-\ 11,232 \\
\hline
13,160
\end{array}
\qquad
\textit{Check:}
\qquad
\begin{array}{r}
13,160 \\
+\ 11,232 \\
\hline
24,392
\end{array}
$$

19.

$$
\begin{array}{r}
46,253 \\
-\ 5\,143 \\
\hline
41,110
\end{array}
\qquad
\textit{Check:}
\qquad
\begin{array}{r}
41,110 \\
+\ 5\,143 \\
\hline
46,253
\end{array}
$$

21.

$$
\begin{array}{r}
54 \\
-\ 42 \\
\hline
12
\end{array}
\qquad
\textit{Check:}
\qquad
\begin{array}{r}
42 \\
+\ 12 \\
\hline
54
\end{array}
\quad \text{correct}
$$

23.

$$
\begin{array}{r}
89 \\
-\ 27 \\
\hline
63
\end{array}
\qquad
\textit{Check:}
\qquad
\begin{array}{r}
63 \\
+\ 27 \\
\hline
90
\end{array}
\quad \text{incorrect}
$$

Rework.

$$
\begin{array}{r}
89 \\
-\ 27 \\
\hline
62
\end{array}
\qquad
\textit{Check:}
\qquad
\begin{array}{r}
62 \\
+\ 27 \\
\hline
89
\end{array}
\quad \text{correct}
$$

25.

$$
\begin{array}{r}
382 \\
-\ 261 \\
\hline
131
\end{array}
\qquad
\textit{Check:}
\qquad
\begin{array}{r}
261 \\
+\ 131 \\
\hline
392
\end{array}
\quad \text{incorrect}
$$

Rework.

$$
\begin{array}{r}
382 \\
-\ 261 \\
\hline
121
\end{array}
\qquad
\textit{Check:}
\qquad
\begin{array}{r}
261 \\
+\ 121 \\
\hline
382
\end{array}
\quad \text{correct}
$$

27.

$$
\begin{array}{r}
4683 \\
-\ 3542 \\
\hline
1141
\end{array}
\qquad
\textit{Check:}
\qquad
\begin{array}{r}
3542 \\
+\ 1141 \\
\hline
4683
\end{array}
\quad \text{correct}
$$

29.

$$
\begin{array}{r}
8643 \\
-\ 1421 \\
\hline
7212
\end{array}
\qquad
\textit{Check:}
\qquad
\begin{array}{r}
7212 \\
+\ 1421 \\
\hline
8633
\end{array}
\quad \text{incorrect}
$$

Rework.

$$
\begin{array}{r}
8643 \\
-\ 1421 \\
\hline
7222
\end{array}
\qquad
\textit{Check:}
\qquad
\begin{array}{r}
7222 \\
+\ 1421 \\
\hline
8643
\end{array}
\quad \text{correct}
$$

Copyright © 2018 Pearson Education, Inc.

31. In subtraction, regrouping is necessary when the digit in the <u>minuend</u> has less value than the digit in the subtrahend, which is directly <u>below</u> it.

33. In the ones column, 5 is less than 7, so in order to subtract, regroup 1 ten as 10 ones. Then subtract 7 ones from 15 ones in the ones column. Finally, subtract 3 tens from 6 tens in the tens column.

$$
\begin{array}{r}
{}^{6\ 15}\!\!\not7\,\not5 \\
-\ 3\,7 \\
\hline
3\,8
\end{array}
$$

35.
$$
\begin{array}{r}
{}^{8\ 14}\!\!\not9\,\not4 \\
-\ 4\,9 \\
\hline
4\,5
\end{array}
$$

37.
$$
\begin{array}{r}
{}^{4\ 17}\!\!\not5\,\not7 \\
-\ 3\,8 \\
\hline
1\,9
\end{array}
$$

39.
$$
\begin{array}{r}
{}^{7\ 12}\!\!\not8\,\not2\,8 \\
-\ 5\,4\,7 \\
\hline
2\,8\,1
\end{array}
$$

41.
$$
\begin{array}{r}
{}^{6\ 11}\!\!\not7\,\not7\,1 \\
-\ 2\,5\,2 \\
\hline
5\,1\,9
\end{array}
$$

43.
$$
\begin{array}{r}
{}^{4\ 12\ 18}\!\!\not7\,\not5\,\not3\,\not8 \\
-\ \ \ \ 4\,7\,9 \\
\hline
7\,0\,5\,9
\end{array}
$$

45.
$$
\begin{array}{r}
{}^{8\ 17\ 18}\!\!\not9\,\not9\,\not8\,\not8 \\
-\ 2\,3\,9\,9 \\
\hline
7\,5\,8\,9
\end{array}
$$

47.
$$
\begin{array}{r}
{}^{7\ 10}\!\!\not8\,\not0 \\
-\ 7\,3 \\
\hline
7
\end{array}
$$

49.
$$
\begin{array}{r}
{}^{9}\\
{}^{2\ 10\ 18}\!\!\not3\,\not0\,\not8 \\
-\ 2\,8\,9 \\
\hline
1\,9
\end{array}
$$

51.
$$
\begin{array}{r}
{}^{3\ 10\ \ 3\ 11}\!\!\not4\,\not0\,\not4\,\not1 \\
-\ 1\,2\,0\,8 \\
\hline
2\,8\,3\,3
\end{array}
$$

53.
$$
\begin{array}{r}
{}^{8\ 12\ 10}\!\!\not9\,\not3\,\not0\,5 \\
-\ 1\,5\,3\,0 \\
\hline
7\,7\,7\,5
\end{array}
$$

55.
$$
\begin{array}{r}
{}^{7\ 10}\!\!1\,5\,\not8\,\not0 \\
-\ 1\,0\,7\,7 \\
\hline
5\,0\,3
\end{array}
$$

57.
$$
\begin{array}{r}
{}^{9}\\
{}^{1\ 10\ 10}\!\!\not2\,\not0\,\not0\,6 \\
-\ 1\,8\,5\,0 \\
\hline
1\,5\,6
\end{array}
$$

59.
$$
\begin{array}{r}
{}^{1\ 13\ 10}\!\!8\,\not2\,\not4\,\not0 \\
-\ 6\,0\,5\,6 \\
\hline
2\,1\,8\,4
\end{array}
$$

61.
$$
\begin{array}{r}
{}^{9}\\
{}^{7\ 14\ 10\ 13}\!\!\not8\,\not5\,\not0\,\not3 \\
-\ 2\,8\,1\,6 \\
\hline
5\,6\,8\,7
\end{array}
$$

63.
$$
\begin{array}{r}
{}^{9}\\
{}^{7\ 10\ \ 6\ 10\ 15}\!\!\not8\,\not0,\not7\,\not0\,\not5 \\
-\ 6\,1,6\,6\,7 \\
\hline
1\,9,0\,3\,8
\end{array}
$$

65.
$$
\begin{array}{r}
{}^{9\ \ 9}\\
{}^{5\ 10\ 10\ 10}\!\!\not6\not6,\not0\,\not0\,\not0 \\
-\ 3\,4,4\,4\,4 \\
\hline
3\,1,5\,5\,6
\end{array}
$$

67.
$$
\begin{array}{r}
{}^{9\ \ 9}\\
{}^{1\ 10\ 10\ 17\ 10}\!\!\not2\,\not0,\not0\,\not8\,\not0 \\
-\ 1\,3,4\,9\,6 \\
\hline
6\,5\,8\,4
\end{array}
$$

69. When checking the accuracy of an answer to an addition problem, you can use <u>subtraction</u>.

71.
$$
\begin{array}{r}
9428 \\
-\ 4509 \\
\hline
4919
\end{array}
\qquad
\begin{array}{r}
4509 \\
+\ 4919 \\
\hline
9428
\end{array}
\ \text{correct}
$$

73.
$$
\begin{array}{r}
2548 \\
-\ 2278 \\
\hline
270
\end{array}
\qquad
\begin{array}{r}
2278 \\
+\ 270 \\
\hline
2548
\end{array}
\ \text{correct}
$$

75.
$$
\begin{array}{r}
93,758 \\
-\ 52,869 \\
\hline
40,889
\end{array}
\qquad
\begin{array}{r}
52,869 \\
+\ 40,889 \\
\hline
93,758
\end{array}
\ \text{correct}
$$

77.
$$
\begin{array}{r}
36,778 \\
-\ 17,405 \\
\hline
19,373
\end{array}
\qquad
\begin{array}{r}
19,373 \\
+\ 17,405 \\
\hline
36,778
\end{array}
\ \text{correct}
$$

Copyright © 2018 Pearson Education, Inc.

79. Possible answers are:

$3 + 2 = 5$ could be changed to $5 - 2 = 3$ or $5 - 3 = 2$.

$6 - 4 = 2$ could be changed to $2 + 4 = 6$ or $4 + 2 = 6$.

81. To find how many fewer calories a woman burns, subtract the number of calories a woman burns from the number of calories a man burns.

$$
\begin{array}{rl}
187 & \textit{calories man burns} \\
- 140 & \textit{calories woman burns} \\
\hline
47 & \textit{fewer calories woman burned}
\end{array}
$$

A woman burns 47 fewer calories.

83.
$$
\begin{array}{rl}
612 & \textit{new record} \\
- 543 & \textit{old record} \\
\hline
69 & \textit{difference in numbers}
\end{array}
$$

There are 69 more tornadoes in the new record.

85.
$$
\begin{array}{rl}
746 & \textit{height tower top is above the water} \\
- 500 & \textit{height tower top is above the roadway} \\
\hline
246 & \textit{height roadway is above the water}
\end{array}
$$

The roadway is 246 feet above the water.

87.
$$
\begin{array}{rl}
6970 & \textit{jobs six years ago} \\
- 3700 & \textit{jobs that remain} \\
\hline
3270 & \textit{jobs eliminated}
\end{array}
$$

There were 3270 jobs eliminated.

89.
$$
\begin{array}{rl}
14{,}608 & \textit{flags manufactured} \\
- 5\,069 & \textit{flags sold} \\
\hline
9\,539 & \textit{flags remaining}
\end{array}
$$

There are 9539 flags that remain unsold.

91.
$$
\begin{array}{rl}
\$913 & \textit{monthly housing expense} \\
- 650 & \textit{monthly rent} \\
\hline
\$263 & \textit{increase in monthly housing expense}
\end{array}
$$

They will pay $263 more per month if they buy a house.

93.
$$
\begin{array}{rl}
5299 & \textit{distance from NYC to Buenos Aires} \\
- 5158 & \textit{distance from Los Angeles to Dublin} \\
\hline
141 & \textit{difference}
\end{array}
$$

The first trip is 141 miles longer than the second.

95.
$$
\begin{array}{rl}
264{,}311 & \textit{total hip replacement patients} \\
- 125{,}423 & \textit{patients 65 years or older} \\
\hline
138{,}888 & \textit{difference}
\end{array}
$$

This year, there were 138,888 hip replacement procedures were done in the United States on patients under age 65.

97.
$$
\begin{array}{rl}
540 & \textit{calories in a Big Mac} \\
- 230 & \textit{calories in a 6-inch Veggie Delite} \\
\hline
310 & \textit{fewer calories in the Veggie Delite}
\end{array}
$$

$$
\begin{array}{rl}
29 & \textit{grams of fat in a Big Mac} \\
- 3 & \textit{grams of fat in a 6-inch Veggie Delite} \\
\hline
26 & \textit{fewer grams of fat in the Veggie Delite}
\end{array}
$$

99.
$$
\begin{array}{rl}
320 & \textit{calories in Roasted Chicken} \\
5 & \textit{calories in 2 tsp. of mustard} \\
+ 45 & \textit{calories in 1 tsp. of olive oil} \\
\hline
370 & \textit{total calories}
\end{array}
$$

$$
\begin{array}{rl}
5 & \textit{grams of fat in Roasted Chicken} \\
0 & \textit{grams of fat in 2 tsp. of mustard} \\
+ 5 & \textit{grams of fat in 1 tsp. of olive oil} \\
\hline
10 & \textit{total grams of fat}
\end{array}
$$

1.4 Multiplying Whole Numbers

1.4 Margin Exercises

1. **(a)**
$$
\begin{array}{rl}
8 & \textit{factor} \\
\times 5 & \textit{factor} \\
\hline
40 & \textit{product}
\end{array}
$$

(b)
$$
\begin{array}{rl}
6 & \textit{factor} \\
\times 4 & \textit{factor} \\
\hline
24 & \textit{product}
\end{array}
$$

(c)
$$
\begin{array}{rl}
7 & \textit{factor} \\
\times 6 & \textit{factor} \\
\hline
42 & \textit{product}
\end{array}
$$

(d)
$$
\begin{array}{rl}
3 & \textit{factor} \\
\times 9 & \textit{factor} \\
\hline
27 & \textit{product}
\end{array}
$$

2. **(a)** $7 \times 4 = 28$

(b) $0 \times 9 = 0$

(c) $8(5) = 40$

(d) $6 \cdot 5 = 30$

(e) $(1)(8) = 8$

3. **(a)** $3 \times 2 \times 5 = (3 \times 2) \times 5 = 6 \times 5 = 30$

(b) $4 \cdot 7 \cdot 1 = (4 \cdot 7) \cdot 1 = 28 \cdot 1 = 28$

(c) $(8)(3)(0) = 24(0) = 0$

4. **(a)**
$$
\begin{array}{r}
\overset{1}{5}3 \\
\times 5 \\
\hline
\underline{265}
\end{array}
$$

Copyright © 2018 Pearson Education, Inc.

(b) $\begin{array}{r} 79 \\ \times\ 0 \\ \hline 0 \end{array}$

(c) $\begin{array}{r} {}^{46} \\ 758 \\ \times\ 8 \\ \hline 6064 \end{array}$

(d) $\begin{array}{r} {}^{52} \\ 2831 \\ \times\ 7 \\ \hline 19{,}817 \end{array}$

(e) $\begin{array}{r} {}^{513} \\ 4714 \\ \times\ 8 \\ \hline 37{,}712 \end{array}$

5. **(a)** $63 \times 10 = 63\underline{0}$ *Attach* 0.

   (b) $305 \times 100 = 30{,}500$ *Attach* 00.

   (c) $714 \times 1000 = 714{,}000$ *Attach* 000.

6. **(a)** 17×50

   $\begin{array}{r} 17 \\ \times\ 5 \\ \hline 85 \end{array}$ $17 \times 50 = 85\underline{0}$ *Attach* 0.

   (b) 73×400

   $\begin{array}{r} 73 \\ \times\ 4 \\ \hline 292 \end{array}$ $73 \times 400 = 29{,}200$ *Attach* 00.

   (c) 180×30

   $\begin{array}{r} 18 \\ \times\ 3 \\ \hline 54 \end{array}$ $180 \times 30 = 5400$ *Attach* 00.

   (d) 4200×80

   $\begin{array}{r} 42 \\ \times\ 8 \\ \hline 336 \end{array}$ $4200 \times 80 = 336{,}000$ *Attach* 000.

   (e) 800×600

   $\begin{array}{r} 8 \\ \times\ 6 \\ \hline 48 \end{array}$ $800 \times 600 = 480{,}000$ *Attach* 0000.

7. **(a)** $\begin{array}{r} 35 \\ \times\ 54 \\ \hline 140 \\ 175 \quad Add \\ \hline \underline{18}90 \end{array}$

(b) $\begin{array}{r} 76 \\ \times\ 49 \\ \hline 684 \\ 304 \quad Add \\ \hline 3724 \end{array}$

8. **(a)** $\begin{array}{r} 52 \\ \times\ 16 \\ \hline 312 \\ 52 \\ \hline 832 \end{array}$ $\begin{array}{l} \leftarrow 6 \times 52 \\ \leftarrow 1 \times 52 \end{array}$

   (b) $\begin{array}{r} 81 \\ \times\ 49 \\ \hline 729 \\ 324 \\ \hline 3969 \end{array}$ $\begin{array}{l} \leftarrow 9 \times 81 \\ \leftarrow 4 \times 81 \end{array}$

   (c) $\begin{array}{r} 234 \\ \times\ 73 \\ \hline 702 \\ 1638 \\ \hline 17{,}082 \end{array}$ $\begin{array}{l} \leftarrow 3 \times 234 \\ \leftarrow 7 \times 234 \end{array}$

   (d) $\begin{array}{r} 835 \\ \times\ 189 \\ \hline 7515 \\ 6680 \\ 835 \\ \hline 157{,}815 \end{array}$ $\begin{array}{l} \leftarrow 9 \times 835 \\ \leftarrow 8 \times 835 \\ \leftarrow 1 \times 835 \end{array}$

9. **(a)** $\begin{array}{r} 28 \\ \times\ 60 \\ \hline \underline{00} \\ 168 \\ \hline \underline{1680} \end{array}$

   (b) $\begin{array}{r} 728 \\ \times\ 50 \\ \hline 36{,}400 \end{array}$

   (c) $\begin{array}{r} 562 \\ \times\ 109 \\ \hline 5058 \\ 5620 \\ \hline 61{,}258 \end{array}$ $1 \times 562 = 562$ *Insert* 0

   (d) $\begin{array}{r} 3526 \\ \times\ 6002 \\ \hline 7052 \\ 2115600 \\ \hline 21{,}163{,}052 \end{array}$ $6 \times 3526 = 21{,}156$ *Insert* 00

Copyright © 2018 Pearson Education, Inc.

10. (a)
$$
\begin{array}{r}
314 \\
\times\ \ 14 \\
\hline
1256 \\
314\ \ \ \\
\hline
4396
\end{array}
$$

The total cost of 314 garden sprayers is \$4396.

(b)
$$
\begin{array}{r}
139 \\
\times\ \ 64 \\
\hline
556 \\
834\ \ \ \\
\hline
8896
\end{array}
$$

The total cost of 64 tires is \$8896.

(c)
$$
\begin{array}{r}
34{,}300 \\
\times\ \ \ \ 12 \\
\hline
68600 \\
34300\ \ \ \\
\hline
411{,}600
\end{array}
$$

The total cost of 12 delivery vans is \$411,600.

1.4 Section Exercises

1. When multiplying two factors,, if you multiply by the larger number first, rather than the smaller number first, the product will always be <u>the same</u>.

3. When you multiply any number by zero, the answer is always <u>zero</u>.

5. $2 \times 6 \times 2 = (2 \times 6) \times 2 = 12 \times 2 = 24$
or $2 \times (6 \times 2) = 2 \times 12 = 24$.

7. $7 \cdot 8 \cdot 0 = (7 \cdot 8) \cdot 0 = 56 \cdot 0 = 0$
or $7 \cdot (8 \cdot 0) = 7 \cdot 0 = 0$.

The product of any number and 0 is 0.

9. $4 \cdot 1 \cdot 6 = (4 \cdot 1) \cdot 6 = 4 \cdot 6 = 24$
or $4 \cdot (1 \cdot 6) = 4 \cdot 6 = 24$.

11. $\underbrace{(4)\,(5)}\,(2) \qquad$ or $\quad (4)\,\underbrace{(5)\,(2)}$
$\qquad 20\,(2) = 40 \qquad\qquad (4)\ 10 = 40$

13. Two factors may be multiplied in any order to get the same answer. The commutative properties of addition and multiplication are similar since with either you may add or multiply two numbers in any order.

15.
$$
\begin{array}{r}
\overset{3}{3}5 \\
\times\ 6 \\
\hline
210
\end{array}
$$

$6 \cdot 5 = 30$ Write 0, carry 3 tens.
$6 \cdot 3 = 18$ Add 3 to get 21. Write 21.

17.
$$
\begin{array}{r}
\overset{2}{3}4 \\
\times\ 7 \\
\hline
238
\end{array}
$$

$7 \cdot 4 = 28$ Write 8, carry 2 tens.
$7 \cdot 3 = 21$ Add 2 to get 23. Write 23.

19.
$$
\begin{array}{r}
\overset{21}{6}42 \\
\times\ \ 5 \\
\hline
3210
\end{array}
$$

$5 \cdot 2 = 10$ Write 0, carry 1 ten.
$5 \cdot 4 = 20$ Add 1 to get 21. Write 1, carry 2.
$5 \cdot 6 = 30$ Add 2 to get 32. Write 32.

21.
$$
\begin{array}{r}
6\overset{1}{2}4 \\
\times\ \ 3 \\
\hline
1872
\end{array}
$$

$3 \cdot 4 = 12$ Write 2, carry 1 ten.
$3 \cdot 2 = 6$ Add 1 to get 7. Write 7.
$3 \cdot 6 = 18$ Write 18.

23.
$$
\begin{array}{r}
2\overset{21}{1}53 \\
\times\ \ 4 \\
\hline
8612
\end{array}
$$

$4 \cdot 3 = 12$ Write 2, carry 1 ten.
$4 \cdot 5 = 20$ Add 1 to get 21. Write 1, carry 2.
$4 \cdot 1 = 4$ Add 2 to get 6. Write 6.
$4 \cdot 2 = 8$ Write 8.

25.
$$
\begin{array}{r}
2\overset{2}{5}21 \\
\times\ \ \ 4 \\
\hline
10{,}084
\end{array}
$$

$4 \cdot 1 = 4$ Write 4.
$4 \cdot 2 = 8$ Write 8.
$4 \cdot 5 = 20$ Write 0, carry 2.
$4 \cdot 2 = 8$ Add 2 to get 10. Write 10.

27.
$$
\begin{array}{r}
2\overset{44}{5}61 \\
\times\ \ \ 8 \\
\hline
20{,}488
\end{array}
$$

$8 \cdot 1 = 8$ Write 8.
$8 \cdot 6 = 48$ Write 8, carry 4.
$8 \cdot 5 = 40$ Add 4 to get 44. Write 4, carry 4.
$8 \cdot 2 = 16$ Add 4 to get 20. Write 20.

Copyright © 2018 Pearson Education, Inc.

29.
$$
\begin{array}{r}
\overset{4\,6\,1}{36{,}921} \\
\times\ \ \ \ 7 \\
\hline
258{,}447
\end{array}
$$

$7 \cdot 1 = 7$ Write 7.
$7 \cdot 2 = 14$ Write 4, carry 1.
$7 \cdot 9 = 63$ Add 1 to get 64. Write 4, carry 6.
$7 \cdot 6 = 42$ Add 6 to get 48. Write 8, carry 4.
$7 \cdot 3 = 21$ Add 4 to get 25. Write 25.

31. You can use multiples of 10 to multiply
86×200. First, multiply <u>86</u> $\times 2$ to get <u>172</u>.
Then, attach <u>two</u> zeros to the right of this number
for a final answer of <u>17,200</u>.

33.
$$
\begin{array}{r}
80 \\
\times\ 6 \\
\hline
\\
\end{array}
\qquad
\begin{array}{r}
8 \\
\times\ 6 \\
\hline
48
\end{array}
\qquad
\begin{array}{r}
80 \\
\times\ 6 \\
\hline
480
\end{array}
\ \ \textit{Attach } 0.
$$

35.
$$
\begin{array}{r}
740 \\
\times\ 3 \\
\hline
\\
\end{array}
\qquad
\begin{array}{r}
74 \\
\times\ 3 \\
\hline
222
\end{array}
\qquad
\begin{array}{r}
740 \\
\times\ 3 \\
\hline
2220
\end{array}
\ \ \textit{Attach } 0.
$$

37.
$$
\begin{array}{r}
600 \\
\times\ 6 \\
\hline
\\
\end{array}
\qquad
\begin{array}{r}
6 \\
\times\ 6 \\
\hline
36
\end{array}
\qquad
\begin{array}{r}
600 \\
\times\ 6 \\
\hline
3600
\end{array}
\ \ \textit{Attach } 00.
$$

39.
$$
\begin{array}{r}
125 \\
\times\ 30 \\
\hline
\\
\end{array}
\qquad
\begin{array}{r}
125 \\
\times\ 3 \\
\hline
375
\end{array}
\qquad
\begin{array}{r}
125 \\
\times\ 30 \\
\hline
3750
\end{array}
\ \ \textit{Attach } 0.
$$

41.
$$
\begin{array}{r}
1635 \\
\times\ 40 \\
\hline
\\
\end{array}
\qquad
\begin{array}{r}
1635 \\
\times\ 4 \\
\hline
6540
\end{array}
\qquad
\begin{array}{r}
1635 \\
\times\ 40 \\
\hline
65{,}400
\end{array}
\ \ \textit{Attach } 0.
$$

43.
$$
\begin{array}{r}
900 \\
\times\ 300 \\
\hline
\\
\end{array}
\qquad
\begin{array}{r}
9 \\
\times\ 3 \\
\hline
27
\end{array}
\qquad
\begin{array}{r}
900 \\
\times\ 300 \\
\hline
270{,}000
\end{array}
\ \ \textit{Attach } 0000.
$$

45.
$$
\begin{array}{r}
43{,}000 \\
\times\ 2\,000 \\
\hline
\\
\end{array}
\qquad
\begin{array}{r}
43 \\
\times\ 2 \\
\hline
86
\end{array}
\qquad
\begin{array}{r}
43{,}000 \\
\times\ 2\,000 \\
\hline
86{,}000{,}000
\end{array}
\ \ \textit{Attach } 000000.
$$

47. $970 \cdot 50$
$$
\begin{array}{r}
97 \\
\times\ 5 \\
\hline
485
\end{array}
$$

$970 \cdot 50 = 48{,}500$ *Attach* 00.

49. $800 \cdot 900$
$$
\begin{array}{r}
8 \\
\times\ 9 \\
\hline
72
\end{array}
$$

$800 \cdot 900 = 720{,}000$ *Attach* 0000.

51. $9700 \cdot 200$
$$
\begin{array}{r}
97 \\
\times\ 2 \\
\hline
194
\end{array}
$$

$9700 \cdot 200 = 1{,}940{,}000$ *Attach* 0000.

53.
$$
\begin{array}{r}
28 \\
\times\ 17 \\
\hline
196 \\
28\ \ \ \\
\hline
476
\end{array}
\quad
\begin{array}{l}
\leftarrow 7 \times 28 \\
\leftarrow 1 \times 28 \\
\\
\end{array}
$$

55.
$$
\begin{array}{r}
75 \\
\times\ 32 \\
\hline
150 \\
225\ \ \ \\
\hline
2400
\end{array}
\quad
\begin{array}{l}
\leftarrow 2 \times 75 \\
\leftarrow 3 \times 75 \\
\\
\end{array}
$$

57.
$$
\begin{array}{r}
83 \\
\times\ 45 \\
\hline
415 \\
332\ \ \ \\
\hline
3735
\end{array}
\quad
\begin{array}{l}
\leftarrow 5 \times 83 \\
\leftarrow 4 \times 83 \\
\\
\end{array}
$$

59. $(58)(41)$
$$
\begin{array}{r}
58 \\
\times\ 41 \\
\hline
58 \\
232\ \ \ \\
\hline
2378
\end{array}
\quad
\begin{array}{l}
\leftarrow 1 \times 58 \\
\leftarrow 4 \times 58 \\
\\
\end{array}
$$

61. $(67)(92)$
$$
\begin{array}{r}
67 \\
\times\ 92 \\
\hline
134 \\
603\ \ \ \\
\hline
6164
\end{array}
\quad
\begin{array}{l}
\leftarrow 2 \times 67 \\
\leftarrow 9 \times 67 \\
\\
\end{array}
$$

63. $(28)(564)$
$$
\begin{array}{r}
564 \\
\times\ 28 \\
\hline
4512 \\
1128\ \ \ \\
\hline
15{,}792
\end{array}
\quad
\begin{array}{l}
\leftarrow 8 \times 564 \\
\leftarrow 2 \times 564 \\
\\
\end{array}
$$

65. $(619)(35)$
$$
\begin{array}{r}
619 \\
\times\ 35 \\
\hline
3095 \\
1857\ \ \ \\
\hline
21{,}665
\end{array}
\quad
\begin{array}{l}
\leftarrow 5 \times 619 \\
\leftarrow 3 \times 619 \\
\\
\end{array}
$$

67. $(55)(286)$
$$
\begin{array}{r}
286 \\
\times\ 55 \\
\hline
1430 \\
1430\ \ \ \\
\hline
15{,}730
\end{array}
\quad
\begin{array}{l}
\leftarrow 5 \times 286 \\
\leftarrow 5 \times 286 \\
\\
\end{array}
$$

Copyright © 2018 Pearson Education, Inc.

69.
```
      735
   ×  112
    1470    ← 2 × 735
     735    ← 1 × 735
     735    ← 1 × 735
   82,320
```

71.
```
      538
   ×  342
    1076    ← 2 × 538
    2152    ← 4 × 538
    1614    ← 3 × 538
  183,996
```

73. First multiply 9352 by 4. Then multiply 9352 by 6, making sure to line up the tens. Then multiply 9352 by 2, making sure to line up the hundreds. Then add the partial products.

```
       9352
     ×  264
      37408    ← 4 × 9352
      56112    ← 6 × 9352
      18704    ← 2 × 9352
    2,468,928
```

75.
```
      215
   ×  307
    1505    ← 7 × 215
    6450    ← 30 × 215
   66,005
```

77.
```
      428
   ×  201
     428    ← 1 × 428
    8560    ← 20 × 428
   86,028
```

79.
```
      6310
    × 3078
     50480    ← 8 × 6310
     44170    ← 7 × 6310
    189300    ← 30 × 6310
  19,422,180
```

81.
```
      2195
    × 1038
     17560    ← 8 × 2195
     6585     ← 3 × 2195
     21950    ← 10 × 2195
   2,278,410
```

83. To multiply by 10, 100, or 1000, just attach one, two, or three zeros, respectively, to the number you are multiplying and that's your answer.

85. *Approach* To find the number of balls purchased, multiply the number of cartons (300) by the number of balls per carton (10).

```
    300    cartons
  ×  10    balls per carton
   3000    balls
```

3000 balls were purchased.

87.
```
      269    cost per night
   ×   12    nights
      538
      269
     3228    cost
```

The cost of a 12-night stay is $3228.

89.
```
      365    days per year
   ×   66    gallons per day
     2190
     2190
   24,090    gallons per year
```

The average person in the United States uses 24,090 gallons of water in one year.

91. 75 first-aid kits at $8 per kit

```
     75
   ×  8
    600
```

The total cost is $600.

93. 65 rebuilt alternators at $24 per alternator

```
      65
    ×  24
     260
     130
    1560
```

The total cost is $1560.

95. 206 laptop computers at $548 per computer

```
      548
    ×  206
     3288
    10960
   112,888
```

The total cost is $112,888.

Copyright © 2018 Pearson Education, Inc.

97. $21 \cdot 43 \cdot 56 = (21 \cdot 43) \cdot 56$
$$= 903 \cdot 56$$
$$= 50{,}568$$

99.
$$
\begin{array}{r}
450 \\
\times\ 85 \\
\hline
2250 \\
3600 \\
\hline
38{,}250
\end{array}
$$
trees per acre
acres

trees

38,250 trees are needed to plant 85 acres.

101.
$$
\begin{array}{r}
8{,}391{,}881 \\
-\ 645{,}169 \\
\hline
7{,}746{,}712
\end{array}
$$
population of New York City
population of Boston
more people

7,746,712 more people live in New York City than in Boston.

103. Multiply the number of flat-screen televisions purchased (12) by the price per television ($970). Then multiply the number of speaker systems purchased (8) by the price per speaker system ($315).

$$
\begin{array}{r}
970 \\
\times\ 12 \\
\hline
11{,}640
\end{array}
$$
cost per television
number of televisions
total cost of televisions

$$
\begin{array}{r}
315 \\
\times\ 8 \\
\hline
2520
\end{array}
$$
cost per speaker system
number of speaker systems
total cost of speaker systems

The total cost is the sum of these values:
$$\$11{,}640 + \$2520 = \$14{,}160$$

105. (a) $189 + 263 = 452$

(b) $263 + 189 = 452$

106. commutative

107. (a) $(65 + 81) + 135 = 146 + 135 = 281$

(b) $65 + (81 + 135) = 65 + 216 = 281$

108. associative

109. (a) $220 \times 72 = 15{,}840$

(b) $72 \times 220 = 15{,}840$

110. commutative

111. (a) $(26 \times 18) \times 14 = 468 \times 14 = 6552$

(b) $26(18 \times 14) = 26 \times 252 = 6552$

112. associative

113. No. Some examples are

1. $7 - 5 = 2$, but $5 - 7$ does not equal 2.

2. $12 - 6 = 6$, but $6 - 12$ does not equal 6.

3. $(8 - 2) - 5 = 1$, but $8 - (2 - 5)$ does not equal 1.

114. No. Some examples are

1. $10 \div 2 = 5$, but $2 \div 10$ does not equal 5.

2. $(16 \div 8) \div 2 = 1$, but $16 \div (8 \div 2)$ does not equal 1.

1.5 Dividing Whole Numbers

1.5 Margin Exercises

1. **(a)** $24 \div 6 = 4$: $6\overline{)24}$ with quotient 4 or $\dfrac{24}{6} = 4$

(b) $9\overline{)36}$ with quotient 4: $36 \div 9 = 4$ or $\dfrac{36}{9} = 4$

(c) $\dfrac{42}{6} = 7$: $42 \div 6 = 7$ or $6\overline{)42}$ with quotient 7

2. **(a)** $15 \div 3 = 5$
dividend: 15; divisor: 3; quotient: 5

(b) $18 \div 6 = 3$
dividend: 18; divisor: 6; quotient: 3

(c) $\dfrac{28}{7} = 4$

dividend: 28; divisor: 7; quotient: 4

(d) $9\overline{)27}$ with quotient 3
dividend: 27; divisor: 9; quotient: 3

3. **(a)** $0 \div 5 = 0$

(b) $\dfrac{0}{9} = 0$

(c) $\dfrac{0}{24} = 0$

(d) $37\overline{)0}$ with quotient 0

4. **(a)** $5\overline{)15}$ with quotient 3; $5 \cdot 3 = \underline{15}$ or $3 \cdot 5 = \underline{15}$

(b) $\dfrac{32}{4} = 8$; $4 \cdot 8 = 32$ or $8 \cdot 4 = 32$

(c) $45 \div 9 = 5$; $9 \cdot 5 = 45$ or $5 \cdot 9 = 45$

Copyright © 2018 Pearson Education, Inc.

5. (a) $\dfrac{4}{0}$; undefined

(b) $\dfrac{0}{4} = 0$

(c) $0\overline{)36}$; undefined

(d) $36\overline{)\overset{0}{0}}$

(e) $100 \div 0$; undefined

(f) $0 \div 100 = 0$

6. (a) $8 \div 8 = 1$

(b) $15\overline{)\overset{1}{15}}$

(c) $\dfrac{37}{37} = 1$

7. (a) $9 \div 1 = 9$

(b) $1\overline{)\overset{18}{18}}$

(c) $\dfrac{43}{1} = 43$

8. (a) $2\overline{)\overset{12}{24}}$ $\dfrac{2}{2} = 1, \ \dfrac{4}{2} = 2$

(b) $3\overline{)\overset{31}{93}}$ $\dfrac{9}{3} = 3, \ \dfrac{3}{3} = 1$

(c) $4\overline{)\overset{22}{88}}$ $\dfrac{8}{4} = 2, \ \dfrac{8}{4} = 2$

(d) $2\overline{)\overset{312}{624}}$ $\dfrac{6}{2} = 3, \ \dfrac{2}{2} = 1, \ \dfrac{4}{2} = 2$

9. (a) $2\overline{)\overset{62 \ \mathbf{R}1}{125}}$ $\dfrac{12}{2} = 6, \ \dfrac{5}{2} = 2\,\mathbf{R}1$

(b) $3\overline{)\overset{71 \ \mathbf{R}2}{215}}$ $\dfrac{21}{3} = 7, \ \dfrac{5}{3} = 1\,\mathbf{R}2$

(c) $4\overline{)5\,{}^1 3\,{}^1 8}\,{}^{1\ 3\ 4\ \mathbf{R}2}$ $\dfrac{5}{4} = 1\,\mathbf{R}1, \ \dfrac{13}{4} = 3\,\mathbf{R}1,$

$\dfrac{18}{4} = 4\,\mathbf{R}2$

(d) $\dfrac{819}{5}$

$5\overline{)8\,{}^3 1\,{}^1 9}\,{}^{1\ 6\ 3\ \mathbf{R}4}$ $\dfrac{8}{5} = 1\,\mathbf{R}3, \ \dfrac{31}{5} = 6\,\mathbf{R}1,$

$\dfrac{19}{5} = 3\,\mathbf{R}4$

10. (a) $4\overline{)5\,{}^1 3\,{}^1 0}\,{}^{1\ 3\ 2\ \mathbf{R}\underline{2}}$

(b) $7\overline{)5\,1\,{}^2 5}\,{}^{7\ 3\ \mathbf{R}4}$

(c) $3\overline{)1\,8\,8\,{}^2 5}\,{}^{6\,2\ 8\ \mathbf{R}1}$

(d) $6\overline{)1\,4\,{}^2 1\,{}^3 5}\,{}^{2\ 3\ 5\ \mathbf{R}5}$

11. (a) $2\overline{)65}\,{}^{32\ \mathbf{R}1}$

divisor $\times$ *quotient* $+$ *remainder* $=$ *dividend*

$\downarrow \qquad\quad \downarrow \qquad\qquad \downarrow \qquad\qquad \downarrow$

$\ 2 \quad \times \quad 32 \quad + \quad\ 1$

$\qquad\quad \underline{64} \qquad\quad + \quad \underline{1} \quad = \quad 65$

The answer is correct.

(b) $7\overline{)586}\,{}^{83\ \mathbf{R}4}$

divisor $\times$ *quotient* $+$ *remainder* $=$ *dividend*

$\downarrow \qquad\quad \downarrow \qquad\qquad \downarrow \qquad\qquad \downarrow$

$\ 7 \quad \times \quad 83 \quad + \quad\ 4$

$\qquad 581 \qquad\quad + \quad\ 4 \quad = \quad 585$

$\qquad\qquad\qquad\qquad\qquad\qquad\qquad\quad \uparrow$

$\qquad\qquad\qquad\qquad\qquad\qquad\quad$ *incorrect*

Rework.

$7\overline{)5\,8\,{}^2 6}\,{}^{8\ 3\ \mathbf{R}5}$ *Now check:*

divisor $\times$ *quotient* $+$ *remainder* $=$ *dividend*

$\downarrow \qquad\quad \downarrow \qquad\qquad \downarrow \qquad\qquad \downarrow$

$\ 7 \quad \times \quad 83 \quad + \quad\ 5$

$\qquad 581 \qquad\quad + \quad\ 5 \quad = \quad 586$

$\qquad\qquad\qquad\qquad\qquad\qquad\qquad\quad \uparrow$

$\qquad\qquad\qquad\qquad\qquad\qquad\quad$ *correct*

The correct answer is 83 $\mathbf{R}5$.

Copyright © 2018 Pearson Education, Inc.

(c) $3\overline{)1223}$ 407 **R2**

divisor $\times$ *quotient* $+$ *remainder* $=$ *dividend*

$\quad\downarrow\qquad\quad\downarrow\qquad\quad\downarrow\qquad\qquad\downarrow$

$\quad 3\quad\times\quad 407\quad+\quad 2$

$\qquad\qquad\quad 1221\qquad+\quad 2\quad=\quad 1223$

The answer is correct.

(d) $5\overline{)2383}$ 476 **R3**

divisor $\times$ *quotient* $+$ *remainder* $=$ *dividend*

$\quad\downarrow\qquad\quad\downarrow\qquad\quad\downarrow\qquad\qquad\downarrow$

$\quad 5\quad\times\quad 476\quad+\quad 3$

$\qquad\qquad\quad 2380\qquad+\quad 3\quad=\quad 2383$

The answer is correct.

12. **(a)** 258: ends in 8, divisible by 2

(b) 307: ends in 7, not divisible by 2

(c) 4216: ends in 6, divisible by 2

(d) 73,000: ends in 0, divisible by 2

13. **(a)** 743: $7 + 4 + 3 = 14$
14 is *not* divisible by 3.
So, 743 is *not* divisible by 3.

(b) 5325: $5 + 3 + 2 + 5 = \underline{15}$
$\underline{15}$ is divisible by 3.
So, 5325 is divisible by $\underline{3}$.

(c) 374,214: $3 + 7 + 4 + 2 + 1 + 4 = 21$
21 is divisible by 3.
So, 374,214 is divisible by 3.

(d) 205,633: $2 + 0 + 5 + 6 + 3 + 3 = 19$
19 is *not* divisible by 3.
So, 205,633 is *not* divisible by 3.

14. **(a)** 180: ends in 0, divisible by 5

(b) 635: ends in 5, divisible by 5

(c) 8364: does not end in 0 or 5, not divisible by 5

(d) 206,105: ends in 5, divisible by 5

15. **(a)** 270: ends in 0, divisible by 10

(b) 495: does not end in 0, not divisible by 10

(c) 5030: ends in 0, divisible by 10

(d) 14,380: ends in 0, divisible by 10

1.5 Section Exercises

1. Three common symbols used to indicate multiplication are

$$\times \quad ; \quad () () \quad ; \quad \text{and} \quad \cdot \,(\text{dot}).$$

3. $24 \div 4 = 6 :\quad 4\overline{)24}^{\,6}\quad$ or $\quad \dfrac{24}{4} = 6$

5. $\dfrac{45}{9} = 5 :\quad 9\overline{)45}^{\,5}\quad$ or $\quad 45 \div 9 = 5$

7. $2\overline{)16}^{\,8} :\quad 16 \div 2 = 8\quad$ or $\quad \dfrac{16}{2} = 8$

9. When a number is divided by 1, the answer is always the $\underline{\text{number}}$ itself.

11. $9 \div 9 = 1$

Any nonzero number divided by itself is 1.

13. $\dfrac{14}{2} = 7$

15. $22 \div 0$; When 0 is the divisor, write $\underline{\text{undefined}}$ as the answer.

17. $\dfrac{24}{1} = 24$

19. $15\overline{)0}^{\,0}$; When dividing 0 by a nonzero number, the answer is $\underline{0}$.

21. $0\overline{)43}$ is undefined

23. 8670 ends in 0, so it is divisible by 2, 5, and 10. The sum of its digits, 21, is divisible by 3, so 8670 is divisible by 3.

25. 9,221,784 ends in 4, so it is divisible by 2, but not divisible by 5 or 10. The sum of its digits, 33, is divisible by 3, so 9,221,784 is divisible by 3.

27. $3\overline{)7\,^{1}5}^{\,2\ 5}$ *Check:* $\begin{array}{r} 25 \\ \times\ 3 \\ \hline 75 \end{array}$

29. $7\overline{)12\,^{5}6}^{\,1\ 8}$ *Check:* $\begin{array}{r} 18 \\ \times\ 7 \\ \hline 126 \end{array}$

31. $4\overline{)1216}^{\,3\,0\,4}$ *Check:* $\begin{array}{r} 304 \\ \times\ 4 \\ \hline 1216 \end{array}$

Copyright © 2018 Pearson Education, Inc.

33.
$$\begin{array}{r} 6\;2\;7\ \mathbf{R}1 \\ 4\overline{)2\,5\,{}^{1}0\,{}^{2}9} \end{array}$$

Check: $(4 \times 627) + 1 = 2508 + 1 = 2509$

35.
$$\begin{array}{r} 1\;5\;2\;2\ \mathbf{R}5 \\ 6\overline{)9\,{}^{3}1\,{}^{1}3\,{}^{1}7} \end{array}$$

Check: $(6 \times 1522) + 5 = 9132 + 5 = 9137$

37.
$$\begin{array}{r} 3\,0\,9 \\ 6\overline{)1\,8\,5\,4} \end{array}$$

Check: $6 \times 309 = 1854$

39. $12{,}020 \div 4$
$$\begin{array}{r} 3\,0\,0\,5 \\ 4\overline{)1\,2{,}0\,2\,0} \end{array}$$

Check: $4 \times 3005 = 12{,}020$

41. $30{,}036 \div 6$
$$\begin{array}{r} 5\,0\,0\,6 \\ 6\overline{)3\,0{,}0\,3\,6} \end{array}$$

Check: $6 \times 5006 = 30{,}036$

43. $2434 \div 3$
$$\begin{array}{r} 8\,1\,1\ \mathbf{R}1 \\ 3\overline{)2\,4\,3\,4} \end{array}$$

Check: $3 \times 811 + 1 = 2433 + 1 = 2434$

45. $12{,}947 \div 5$
$$\begin{array}{r} 2\;5\;8\;9\ \mathbf{R}2 \\ 5\overline{)1\,2{,}\,{}^{2}9\,{}^{4}4\,{}^{4}7} \end{array}$$

Check: $5 \times 2589 + 2 = 12{,}945 + 2 = 12{,}947$

47. $\dfrac{21{,}040}{8}$
$$\begin{array}{r} 2\;6\;3\,0 \\ 8\overline{)2\,1{,}\,{}^{5}0\,{}^{2}4\,0} \end{array}$$

Check: $8 \times 2630 = 21{,}040$

49. $\dfrac{74{,}751}{6}$
$$\begin{array}{r} 1\;2{,}\,4\;5\;8\ \mathbf{R}3 \\ 6\overline{)7\,{}^{1}4{,}\,{}^{2}7\,{}^{3}5\,{}^{5}1} \end{array}$$

Check:
$(6 \times 12{,}458) + 3 = 74{,}748 + 3 = 74{,}751$

51. $\dfrac{71{,}776}{7}$
$$\begin{array}{r} 1\,0{,}\;2\;5\;3\ \mathbf{R}5 \\ 7\overline{)7\,1{,}\,{}^{1}7\,{}^{3}7\,{}^{2}6} \end{array}$$

Check:
$(7 \times 10{,}253) + 5 = 71{,}771 + 5 = 71{,}776$

53. $\dfrac{128{,}645}{7}$
$$\begin{array}{r} 1\;8{,}\,3\;7\;7\ \mathbf{R}6 \\ 7\overline{)1\,2\,{}^{5}8{,}\,{}^{2}6\,{}^{5}4\,{}^{5}5} \end{array}$$

Check:
$(7 \times 18{,}377) + 6 = 128{,}639 + 6 = 128{,}645$

55.
$$\begin{array}{r} 3\,7\,5\ \mathbf{R}2 \\ 5\overline{)1\,8\,7\,7} \end{array}$$

Check: $(5 \times 375) + 2 = 1875 + 2 = \underline{1877}$
correct

57.
$$\begin{array}{r} 1\,9\,0\,8\ \mathbf{R}2 \\ 3\overline{)5\,7\,2\,5} \end{array}$$

Check: $(3 \times 1908) + 2 = 5724 + 2 = 5726$
incorrect

Rework.

$$\begin{array}{r} 1\;9\,0\,8\ \mathbf{R}1 \\ 3\overline{)5\,{}^{2}7\,2\,{}^{2}5} \end{array}$$

Check: $3 \times 1908 + 1 = 5724 + 1 = 5725$

59.
$$\begin{array}{r} 6\,5\,0\ \mathbf{R}2 \\ 7\overline{)4\,6\,9\,2} \end{array}$$

Check: $7 \times 650 + 2 = 4550 + 2 = 4552$
incorrect

Rework.

$$\begin{array}{r} 6\;7\,0\ \mathbf{R}2 \\ 7\overline{)4\,6\,{}^{4}9\,2} \end{array}$$

Check: $7 \times 670 + 2 = 4690 + 2 = 4692$
correct

61.
$$\begin{array}{r} 3\;5\,6\,8\ \mathbf{R}2 \\ 6\overline{)2\,1{,}4\,0\,9} \end{array}$$

Check: $6 \times 3568 + 2 = 21{,}408 + 2 = 21{,}410$
incorrect

Rework.

$$\begin{array}{r} 3\;\;5\;6\;8\ \mathbf{R}1 \\ 6\overline{)2\,1{,}\,{}^{3}4\,{}^{4}0\,{}^{4}9} \end{array}$$

Check: $(6 \times 3568) + 1 = 21{,}408 + 1 = 21{,}409$

63.
$$\begin{array}{r} 2\,0\,0\,2\ \mathbf{R}3 \\ 8\overline{)1\,6{,}0\,1\,9} \end{array}$$

Check: $(8 \times 2002) + 3 = 16{,}016 + 3 = 16{,}019$
correct

65.
$$\begin{array}{r} 1\,1{,}5\,2\,3\ \mathbf{R}2 \\ 6\overline{)6\,9{,}1\,4\,0} \end{array}$$

Check:
$(6 \times 11{,}523) + 2 = 69{,}138 + 2 = 69{,}140$
correct

67.
$$\begin{array}{r} 9\,6\,2\,8\ \mathbf{R}7 \\ 9\overline{)8\,6{,}6\,5\,5} \end{array}$$

Check: $(9 \times 9628) + 7 = 86{,}652 + 7 = 86{,}659$
Does not match dividend, incorrect

Rework.

$$\begin{array}{r} 9\;\;6\,2\,8\ \mathbf{R}3 \\ 9\overline{)8\,6{,}\,{}^{5}6\,{}^{2}5\,{}^{7}5} \end{array}$$

Check: $(9 \times 9628) + 3 = 86{,}652 + 3 = 86{,}655$

Copyright © 2018 Pearson Education, Inc.

69.
$$8\overline{)222{,}576} \quad 27{,}822$$

Check: $8 \times 27{,}822 = 222{,}576$ *correct*

71. Forgetting to add the remainder is a common error. The check should be $6 \times 497 + 5 = 2982 + 5 = 2987$. Then the answer checks.

73. To find the number of tables that can be set, divide the number of napkins (2624) by the number of napkins it takes to set each table (8).

$$8\overline{)26\,{}^2 2\,{}^6 4} \quad 3\ 2\ 8$$

328 tables can be set.

75.
$$8\overline{)76{,}\,{}^4 800} \quad 9\ 600$$

9600 drumsticks are produced each hour.

77.
$$9\overline{)43\,{}^7 6{,}\,{}^4 500} \quad 4\ 8{,}500$$

Each employee received $48,500.

79.
$$4\overline{)6\,{}^2 6\,{}^2 0} \quad 1\ 6\ 5$$

There are 165 locations in this area.

81.
$$6\overline{)6{,}8\,{}^2 2\,{}^4 5{,}\,{}^3 000} \quad 1{,}1\ 3\ 7{,}500 \quad \text{(or use a calculator)}$$

Each person received $1,137,500.

83.
$$8\overline{)6{,}9\,{}^5 0\,{}^2 0{,}\,{}^4 000} \quad 8\ 6\ 2{,}500$$

The number of blueberries picked each hour was 862,500.

85. 60 ends in 0, so it is divisible by 2, 5, and 10. The sum of its digits, 6, is divisible by 3, so 60 is divisible by 3.

87. 92 ends in 2, so it is divisible by 2, but not divisible by 5 or 10. The sum of its digits, 11, is not divisible by 3, so 92 is not divisible by 3.

89. 445 ends in 5, so it is divisible by 5, but not divisible by 2 or 10. The sum of its digits, 13, is not divisible by 3, so 445 is not divisible by 3.

91. 903 ends in 3, so it is not divisible by 2, 5, or 10. The sum of its digits, 12, is divisible by 3, so 903 is divisible by 3.

93. 5166 ends in 6, so it is divisible by 2, but not divisible by 5 or 10. The sum of its digits, 18, is divisible by 3, so 5166 is divisible by 3.

95. 21,763 ends in 3, so it is not divisible by 2, 5, or 10. The sum of its digits, 19, is not divisible by 3, so 21,763 is not divisible by 3.

1.6 Long Division

1.6 Margin Exercises

1. (a)
$$\begin{array}{r} 8\ 2 \\ 28\overline{)2\ 2\ 9\ 6} \\ 2\ 2\ 4 \quad \leftarrow 8 \times 28 \\ \hline 5\ 6 \\ 5\ 6 \quad \leftarrow 2 \times 28 \\ \hline 0 \end{array}$$

(b)
$$\begin{array}{r} 6\ 4 \\ 16\overline{)1\ 0\ 2\ 4} \\ 9\ 6 \quad \leftarrow 6 \times 16 \\ \hline 6\ 4 \\ 6\ 4 \quad \leftarrow 4 \times 16 \\ \hline 0 \end{array}$$

(c)
$$\begin{array}{r} 1\ 4\ 4 \\ 61\overline{)8\ 7\ 8\ 4} \\ 6\ 1 \quad \leftarrow 1 \times 61 \\ \hline 2\ 6\ 8 \\ 2\ 4\ 4 \quad \leftarrow 4 \times 61 \\ \hline 2\ 4\ 4 \\ 2\ 4\ 4 \quad \leftarrow 4 \times 61 \\ \hline 0 \end{array}$$

(d)
$$\begin{array}{r} 2\ 9 \\ 93\overline{)2\ 6\ 9\ 7} \\ 1\ 8\ 6 \quad \leftarrow 2 \times 93 \\ \hline 8\ 3\ 7 \\ 8\ 3\ 7 \quad \leftarrow 9 \times 93 \\ \hline 0 \end{array}$$

2. (a)
$$\begin{array}{r} 5\ 6 \\ 24\overline{)1\ 3\ 4\ 4} \\ 1\ 2\ 0 \quad \leftarrow 5 \times 24 \\ \hline 1\ 4\ 4 \\ 1\ 4\ 4 \quad \leftarrow 6 \times 24 \\ \hline 0 \end{array}$$

(b)
$$\begin{array}{r} 6\ 2\ \textbf{R}8 \\ 72\overline{)4\ 4\ 7\ 2} \\ 4\ 3\ 2 \quad \leftarrow 6 \times 72 \\ \hline 1\ 5\ 2 \\ 1\ 4\ 4 \quad \leftarrow 2 \times 72 \\ \hline 8 \end{array}$$

Copyright © 2018 Pearson Education, Inc.

(c)
$$
\begin{array}{r}
8\;3 \quad \mathbf{R}21 \\
65\,\overline{)5\;4\;1\;6} \\
5\;2\;0 \quad \leftarrow 8\times 65 \\
\overline{2\;1\;6} \\
1\;9\;5 \quad \leftarrow 3\times 65 \\
\overline{2\;1}
\end{array}
$$

(d)
$$
\begin{array}{r}
7\;4 \quad \mathbf{R}63 \\
89\,\overline{)6\;6\;4\;9} \\
6\;2\;3 \quad \leftarrow 7\times 89 \\
\overline{4\;1\;9} \\
3\;5\;6 \quad \leftarrow 4\times 89 \\
\overline{6\;3}
\end{array}
$$

3.   **(a)**
$$
\begin{array}{r}
1\;0\;7 \quad \mathbf{R}\underline{4} \\
17\,\overline{)1\;8\;2\;3} \\
1\;7 \quad\quad \leftarrow 1\times 17 \\
\overline{1\;2\;3} \\
1\;1\;9 \quad \leftarrow 7\times 17 \\
\overline{\underline{4}}
\end{array}
$$

(b)
$$
\begin{array}{r}
2\;0\;8 \quad \mathbf{R}7 \\
23\,\overline{)4\;7\;9\;1} \\
4\;6 \quad\quad \leftarrow 2\times 23 \\
\overline{1\;9\;1} \\
1\;8\;4 \quad \leftarrow 8\times 23 \\
\overline{7}
\end{array}
$$

(c)
$$
\begin{array}{r}
4\;0\;8 \quad \mathbf{R}21 \\
39\,\overline{)1\;5,9\;3\;3} \\
1\;5\;6 \quad\quad \leftarrow 4\times 39 \\
\overline{3\;3\;3} \\
3\;1\;2 \quad \leftarrow 8\times 39 \\
\overline{2\;1}
\end{array}
$$

(d)
$$
\begin{array}{r}
3\;0\;0 \quad \mathbf{R}62 \\
78\,\overline{)2\;3,4\;6\;2} \\
2\;3\;4 \quad\quad \leftarrow 3\times 78 \\
\overline{6\;2}
\end{array}
$$

4.   **(a)** $70 \div 10 = 7$

One zero is dropped.

(b) $2600 \div 100 = 26$

Two zeros are dropped.

(c) $505{,}000 \div 1000 = 505$

Three zeros are dropped.

5.   **(a)** $50\,\overline{)6250}$

Drop 1 zero from the divisor and the dividend.

$$
\begin{array}{r}
1\;2\;5 \\
5\,\overline{)6\;2\;5} \\
5 \\
\overline{1\;2} \\
1\;0 \\
\overline{2\;5} \\
2\;5 \\
\overline{0}
\end{array}
$$

The quotient is 125.

(b) $130\,\overline{)131{,}040}$

Drop 1 zero from the divisor and the dividend.

$$
\begin{array}{r}
1\;0\;0\;8 \\
13\,\overline{)1\;3,1\;0\;4} \\
1\;3 \\
\overline{1\;0\;4} \\
1\;0\;4 \\
\overline{0}
\end{array}
$$

The quotient is 1008.

(c) $3400\,\overline{)190{,}400}$

Drop 2 zeros from the divisor and the dividend.

$$
\begin{array}{r}
5\;6 \\
34\,\overline{)1\;9\;0\;4} \\
1\;7\;0 \\
\overline{2\;0\;4} \\
2\;0\;4 \\
\overline{0}
\end{array}
$$

The quotient is 56.

6.   **(a)**
$$
\begin{array}{r}
3\;8 \\
16\,\overline{)6\;0\;8} \\
4\;8 \\
\overline{1\;2\;8} \\
1\;2\;8 \\
\overline{0}
\end{array}
\qquad
\begin{array}{r}
38 \\
\times\;16 \\
\overline{2\;2\;8} \\
3\;8 \\
\overline{6\;0\;8} \quad \leftarrow correct
\end{array}
$$

Multiply the quotient and the divisor.

(b)
$$
\begin{array}{r}
4\;2 \quad \mathbf{R}178 \\
426\,\overline{)1\;9,1\;7\;0} \\
1\;7\;0\;4 \\
\overline{1\;1\;3\;0} \\
9\;5\;2 \\
\overline{1\;7\;8}
\end{array}
\qquad
\begin{array}{r}
426 \\
\times\;\;42 \\
\overline{8\;5\;2} \\
1\;7\;0\;4 \\
\overline{1\;7,8\;9\;2} \\
+\;\;1\;7\;8 \\
\overline{1\;8,0\;7\;0}
\end{array}
$$

The result does not match the dividend.

(continued)

Copyright © 2018 Pearson Education, Inc.

Rework.

$$
\begin{array}{r}
45 \\
426\overline{)19,170} \\
1\,7\,0\,4 \\
\hline
2\,1\,3\,0 \\
2\,1\,3\,0 \\
\hline
0
\end{array}
$$

A check shows that 45 is correct.

(c)

$$
\begin{array}{r}
57\,\mathbf{R}18 \\
514\overline{)29,316} \\
2\,5\,7\,0 \\
\hline
3\,6\,1\,6 \\
3\,5\,9\,8 \\
\hline
1\,8
\end{array}
$$

$$
\begin{array}{r}
5\,1\,4 \\
\times\quad 5\,7 \\
\hline
3\,5\,9\,8 \\
25\,7\,0 \\
\hline
29,298 \\
+\quad 1\,8 \\
\hline
29,316
\end{array}
$$

Multiply the quotient and the divisor.

Add the remainder

← *correct*

1.6 Section Exercises

1.
$$
\begin{array}{r}
5 \\
50\overline{)2650}
\end{array}
$$
5; 53; 530

5 goes over the 5, because $\frac{265}{50}$ is about 5. The answer must then be a two-digit number or 53.

3.
$$
\begin{array}{r}
2 \\
18\overline{)4500}
\end{array}
$$
2; 25; 250

2 goes over the 5, because $\frac{45}{18}$ is about 2. The answer must then be a three-digit number or 250.

5.
$$
\begin{array}{r}
1 \\
86\overline{)10,327}
\end{array}
$$
12; 120 **R**7; 1200

1 goes over the 3, because $\frac{103}{86}$ is about 1. The answer must then be a three-digit number or 120 **R**7.

7.
$$
\begin{array}{r}
1 \\
26\overline{)28,735}
\end{array}
$$
11; 110; 1105 **R**5

1 goes over the 8, because $\frac{28}{26}$ is about 1. The answer must then be a four-digit number or 1105 **R**5.

9.
$$
\begin{array}{r}
7 \\
21\overline{)149,826}
\end{array}
$$
71; 713; 7134 **R**12

7 goes over the 9, because $\frac{149}{21}$ is about 7. The answer must then be a four-digit number or 7134 **R**12.

11.
$$
\begin{array}{r}
9 \\
523\overline{)470,800}
\end{array}
$$
9 **R**100; 90 **R**100; 900 **R**100

9 goes over the 8, because $\frac{4708}{523}$ is about 9. The answer must then be a three-digit number or 900 **R**100.

13. Use 2 as a trial divisor, since 18 is closer to 20 than to 10.

$$
\frac{131}{2} = 6 \text{ with 1 left over}
$$

Because $6 \times 18 = 108$ and $131 - 108 = 23$, which is greater than the divisor, use 7 instead. To find the next digit in the quotient, use 2 as a trial divisor again.

$$
\frac{5}{2} = 2 \text{ with 1 left over}
$$

Because $2 \times 18 = 36$ and $59 - 36 = 23$, which is greater than the divisor, use 3 instead.

$$
\begin{array}{r}
73\,\mathbf{R}5 \\
18\overline{)1319} \\
126 \\
\hline
59 \\
54 \\
\hline
5
\end{array}
\qquad
\begin{array}{r}
\textbf{Check:} \\
73 \\
\times\ 18 \\
\hline
584 \\
73 \\
\hline
1314 \\
+\ \ 5 \\
\hline
1319\ \checkmark\ \textit{matches}
\end{array}
$$

15.
$$
\begin{array}{r}
476\,\mathbf{R}15 \\
23\overline{)10,963} \\
9\,2 \\
\hline
1\,7\,6 \\
1\,6\,1 \\
\hline
1\,5\,3 \\
1\,3\,8 \\
\hline
1\,5
\end{array}
\qquad
\begin{array}{r}
\textbf{Check:} \\
476 \\
\times\ 23 \\
\hline
1428 \\
952 \\
\hline
10,948 \\
+\ \ 15 \\
\hline
10,963
\end{array}
$$

17.
$$
\begin{array}{r}
2407\,\mathbf{R}1 \\
26\overline{)62,583} \\
5\,2 \\
\hline
1\,0\,5 \\
1\,0\,4 \\
\hline
1\,8\,3 \\
1\,8\,2 \\
\hline
1
\end{array}
\qquad
\begin{array}{r}
\textbf{Check:} \\
2407 \\
\times\ 26 \\
\hline
14442 \\
4814 \\
\hline
62,582 \\
+\ \ 1 \\
\hline
62,583
\end{array}
$$

Copyright © 2018 Pearson Education, Inc.

19.

```
         1 1 4 6  R15   Check:      1146
   74│8 4, 8 1 9                  ×   74
      7 4                          4584
     ─────                         8022
      1 0 8                       ──────
        7 4                       84,804
       ────                       +   15
        3 4 1                     ──────
        2 9 6                     84,819
       ─────
          4 5 9
          4 4 4
         ─────
            1 5
```

21.

```
          3 3 3 1  R82   Check:      3331
   153│5 0 9, 7 2 5               ×   153
       4 5 9                        9993
      ──────                       16655
       5 0 7                        3331
       4 5 9                       ───────
      ──────                       509,643
        4 8 2                      +    82
        4 5 9                      ───────
       ──────                      509,725
          2 3 5
          1 5 3
         ──────
            8 2
```

23.

```
          8 5 0   Check:      850
   420│3 5 7, 0 0 0          ×   420
       3 3 6 0              ───────
      ──────                 17000
        2 1 0 0              3400
        2 1 0 0             ───────
       ───────              357,000
              0
```

25.

```
       1 0 1  R4   Check:      101
   35│3 5 4 9               ×   35
                            ─────
                             505
                             303
                            ─────
                             3535
                            +    4
                            ─────
                             3539  incorrect
```

Rework.

```
       1 0 1  R14   Check:      101
   35│3 5 4 9                ×   35
      3 5                    ─────
     ─────                    505
      0 4 9                   303
        3 5                  ─────
       ────                   3535
        1 4                  +   14
                            ─────
                             3549  ✓ matches
```

The correct answer is 101 **R**14.

27.

```
          6 5 8  R9   Check:      658
   28│1 8, 4 2 4               ×   28
                              ──────
                               5 264
                               13 16
                              ──────
                               18,424
                               +    9
                              ──────
                               18,433  incorrect
```

Rework.

```
          6 5 8
   28│1 8, 4 2 4
      1 6 8
     ──────
       1 6 2
       1 4 0
      ──────
         2 2 4
         2 2 4
        ──────
              0
```

The correct answer is 658.

29.

```
              6 2  R3   Check:      614
   614│3 8, 0 6 8               ×   62
                              ──────
                               1 228
                               36 84
                              ──────
                               38,068
                               +    3
                              ──────
                               38,071  incorrect
```

Rework.

```
              6 2
   614│3 8, 0 6 8
       3 6 8 4
      ──────
        1 2 2 8
        1 2 2 8
       ──────
              0
```

The correct answer is 62.

31. Divide. (We can drop two zeros.)

```
          4 5
   19│8 5 5
      7 6
     ────
       9 5
       9 5
      ────
         0
```

She drives this route 45 weeks each year.

Copyright © 2018 Pearson Education, Inc.

33. First, add the number of wall and table clocks.

$$
\begin{array}{r}
272 \ \textit{wall clocks} \\
+ \ 308 \ \textit{table clocks} \\
\hline
580 \ \textit{wall and table sum}
\end{array}
$$

Subtract the sum from the total number of clocks.

$$
\begin{array}{r}
636 \ \textit{total number of clocks} \\
- \ 580 \ \textit{wall and table sum} \\
\hline
56 \ \textit{standing floor clocks}
\end{array}
$$

He worked on 56 floor clocks this year.

We could also arrange our work as follows:

total	−	wall	−	table	=	floor
clocks		clocks		clocks		clocks
636	−	272	−	308	=	56

35. Divide. (or use a calculator)

$$
\begin{array}{r}
3\,5\,5 \\
96\overline{)3\,4,0\,8\,0} \\
2\,8\,8 \\
\hline
5\,2\,8 \\
4\,8\,0 \\
\hline
4\,8\,0 \\
4\,8\,0 \\
\hline
0
\end{array}
$$

Judy's monthly payment is $355.

37. If one ring is sold each minute, then 60 rings are sold each hour.

$$
\begin{array}{r}
60 \ \textit{rings per hour} \\
\times \ \ 24 \ \textit{hours per day} \\
\hline
1440 \ \textit{rings per day} \\
\times \ \ \ 30 \ \textit{days} \\
\hline
43{,}200 \ \textit{rings}
\end{array}
$$

There are 43,200 diamond rings sold in 30 days.

39. **(a)** Divide the number of Big Macs eaten by 39 years.

$$
\begin{array}{r}
6\,4\,8 \\
39\overline{)2\,5,2\,7\,2} \\
2\,3\,4 \\
\hline
1\,8\,7 \\
1\,5\,6 \\
\hline
3\,1\,2 \\
3\,1\,2 \\
\hline
0
\end{array}
$$

He has eaten an average of 648 Big Macs a year for the last 39 years.

(b) 2 Big Macs eaten each day would be $2 \times 365 = 730$. He has eaten less than 2 Big Macs each day.

41. $\dfrac{\$0}{3} = \$0$ Each person will receive $0.

42. When 0 is divided by any nonzero number, the result is 0.

43. $8 \div 0$ is undefined.

44. It is impossible. If you have 6 cookies, it is not possible to divide them among zero people.

45. **(a)** $14 \div 1 = 14$

(b)
$$
\begin{array}{r}
1\,7 \\
1\overline{)1\,7}
\end{array}
$$

(c) $\dfrac{38}{1} = 38$

46. Yes. Some examples are $18 \cdot 1 = 18$; $26 \cdot 1 = 26$; and $43 \cdot 1 = 43$.

47. **(a)** $32{,}000 \div 10 = 3200$

Drop 1 zero from the dividend and 1 zero from the divisor.

(b) $32{,}000 \div 100 = 320$

Drop 2 zeros from the dividend and 2 zeros from the divisor.

(c) $32{,}000 \div 1000 = 32$

Drop 3 zeros from the dividend and 3 zeros from the divisor.

48. Drop the same number of zeros from the dividend that appear in the divisor. The result is the quotient. With a divisor of 10, drop one zero; with 100, drop two zeros; with 1000, drop three zeros.

Summary Exercises *Whole Numbers Computation*

1. 6<u>3</u>1,5<u>4</u>8; ten-thousands: 3; tens: 4

3. 9,<u>1</u>81,57<u>6</u>,423; hundred-millions: 1; thousands: 6

5. 425,208,733 is four hundred twenty-five million, two hundred eight thousand, seven hundred thirty-three.

7. $166 + 739$
$$
\begin{array}{r}
\overset{1}{}\overset{1}{} \\
166 \\
+ \ 739 \\
\hline
905
\end{array}
$$

Copyright © 2018 Pearson Education, Inc.

9.
$$
\begin{array}{r}
{}^{8}7{}^{1}9{}^{8}8 \\
-\,3\,8\,9 \\
\hline
4\,0\,9
\end{array}
$$

11. $75 + 81{,}579 + 506 + 4$

$$
\begin{array}{r}
1\;1{}^{1}2 \\
7\,5 \\
8\,1{,}5\,7\,9 \\
5\,0\,6 \\
+\quad\quad 4 \\
\hline
8\,2{,}1\,6\,4
\end{array}
$$

13.
$$
\begin{array}{r}
{}^{4}5{}^{14}5{,}{}^{9}0{}^{9}0{}^{10}0 \\
-\,1\,7{,}3\,2\,6 \\
\hline
3\,7{,}6\,7\,4
\end{array}
$$

15. $56 \times 10 = 560$ *Attach* 0.

17.
$$
\begin{array}{cc}
\begin{array}{r} 5\,0\,0 \\ \times\,7\,0\,0 \\ \hline \end{array} &
\begin{array}{r} 5 \\ \times\,7 \\ \hline 3\,5 \end{array} &
\begin{array}{r} 5\,0\,0 \\ \times\,7\,0\,0 \\ \hline 3\,5\,0{,}0\,0\,0 \end{array}
\end{array}
$$
  Attach 0000.

19. One million, two hundred thirty-eight thousand, two hundred one is written as 1,238,201.

21. $8 \div 8 = 1$

23. $\dfrac{12}{0}$ is undefined.

25. $7 \times 8 = 56$

27. $8 \cdot 4 \cdot 3 = (8 \cdot 4) \cdot 3 = 32 \cdot 3 = 96$

29.
$$
\begin{array}{r}
2\;7\;5\;0 \;\textbf{R2} \\
3\,\overline{)8\,{}^{2}2\,{}^{1}5\,2}
\end{array}
$$

Check: $3 \times 2750 + 2 = 8250 + 2 = 8252$

31.
$$
\begin{array}{r}
6\,5 \\
\times\;\;5\,2 \\
\hline
1\,3\,0 \\
3\,2\,5 \\
\hline
3\,3\,8\,0
\end{array}
$$

33. $(28)(72)$
$$
\begin{array}{r}
7\,2 \\
\times\;\;2\,8 \\
\hline
5\,7\,6 \quad \leftarrow 8 \times 72 \\
1\,4\,4 \quad \leftarrow 2 \times 72 \\
\hline
2\,0\,1\,6
\end{array}
$$

35.
$$
\begin{array}{r}
7\,8 \\
25\,\overline{)1\,9\,5\,0} \\
1\,7\,5 \quad \leftarrow 7 \times 25 \\
\hline
2\,0\,0 \\
2\,0\,0 \quad \leftarrow 8 \times 25 \\
\hline
0
\end{array}
$$

37.
$$
\begin{array}{r}
3\,6\,0\,2 \\
\times\,5\,0\,0\,8 \\
\hline
2\,8\,8\,1\,6 \\
1\,8\,0\,1\,0\,0\,0 \\
\hline
1\,8{,}0\,3\,8{,}8\,1\,6
\end{array}
$$

39. $630\,\overline{)32{,}760}$ Drop 1 zero.
$$
\begin{array}{r}
5\,2 \\
63\,\overline{)3\,2\,7\,6} \\
3\,1\,5 \quad \leftarrow 5 \times 63 \\
\hline
1\,2\,6 \\
1\,2\,6 \quad \leftarrow 2 \times 63 \\
\hline
0
\end{array}
$$

41.
$$
\begin{array}{r}
5\;\;7\;\;3\;\textbf{R3} \\
8\,\overline{)4\,5\,{}^{5}8\,{}^{2}7}
\end{array}
$$

Check: $8 \times 573 + 3 = 4584 + 3 = 4587$

43.
$$
\begin{array}{r}
6\,6\,2 \\
\times\;3\,1\,5 \\
\hline
3\,3\,1\,0 \\
6\,6\,2 \\
1\,9\,8\,6 \\
\hline
2\,0\,8{,}5\,3\,0
\end{array}
$$

45. From the table, there were 56,307 workers. This number in words is fifty-six thousand, three hundred seven.

47.
$$
\begin{array}{r}
{}^{4}5{}^{10}0\,4 \\
-\,2\,5\,2 \\
\hline
2\,5\,2
\end{array}
$$

252 million more tons will pass through the canal.

49.
$$
\begin{array}{r}
{}^{9} \\
2\,5\,{}^{5}\cancel{6}\,{}^{10}0\,{}^{10}0 \\
-\quad\quad 6 \\
\hline
2\,5\,5\,9\,4
\end{array}
$$

There were twenty-five thousand, five hundred ninety-four more lives lost.

1.7   Rounding Whole Numbers

1.7   Margin Exercises

1.   **(a)** 373 (nearest ten)
Underline the tens place: 3$\underline{7}$3
373 is closer to 3$\underline{7}$0.

     (b) 1482 (nearest thousand)
Underline the thousands place: $\underline{1}$482
1482 is closer to $\underline{1}$000.

Copyright © 2018 Pearson Education, Inc.

(c) 89,512 (nearest hundred)
Underline the hundreds place: 89,512
89,512 is closer to 89,500.

2. (a) 62
Underline the tens place. Next digit is 4 or less.
Change the digit to the right of the underlined
place to zero. Leave 6 as 6.
62 rounded to the nearest ten is 60.

(b) 94
Underline the tens place. Next digit is 4 or less.
Change the digit to the right of the underlined
place to zero.
94 rounded to the nearest ten is 90.

(c) 134
Underline the tens place. Next digit is 4 or less.
Change the digit to the right of the underlined
place to zero.
134 rounded to the nearest ten is 130.

(d) 7543
Underline the tens place. Next digit is 4 or less.
Change the digit to the right of the underlined
place to zero.
7543 rounded to the nearest ten is 7540.

3. (a) 3683 Next digit is 5 or more.
Change 3 to 4. All digits to the right of the
underlined place are changed to zero.
3683 rounded to the nearest thousand is 4000.

(b) 6502 Next digit is 5 or more.
Change 6 to 7. All digits to the right of the
underlined place are changed to zero.
6502 rounded to the nearest thousand is 7000.

(c) 84,621 Next digit is 5 or more.
Change 4 to 5. All digits to the right of the
underlined place are changed to zero.
84,621 rounded to the nearest thousand is 85,000.

(d) 55,960 Next digit is 5 or more.
Change 5 to 6. All digits to the right of the
underlined place are changed to zero.
55,960 rounded to the nearest thousand is 56,000.

4. (a) 3458 to the nearest ten
3458 Next digit is 5 or more.
Change 5 to 6. All digits to the right of the
underlined place are changed to zeros.
3458 rounded to the nearest ten is 3460.

(b) 6448 to the nearest hundred
6448 Next digit is 4 or less.
Leave 4 as 4. All digits to the right of the
underlined place are changed to zeros.
6448 rounded to the nearest hundred is 6400.

(c) 73,077 to the nearest hundred
73,077 Next digit is 5 or more.
Change 0 to 1. All digits to the right of the
underlined place are changed to zeros.
73,077 rounded to the nearest hundred is 73,100.

(d) 85,972 to the nearest hundred
85,972 Next digit is 5 or more.
Change 9 to 10. Write zero and carry the 1.
All digits to the right of the underlined place are
changed to zeros.
85,972 rounded to the nearest hundred is 86,000.

5. (a) 14,598 to the nearest ten-thousand
14,598
The ten-thousands place does not change because
the digit to the right is 4 or less. All digits to the
right of the underlined place are changed to
zeros.
14,598 rounded to the nearest ten-thousand is
10,000.

(b) 724,518,715 to the nearest million
724,518,715
Change 4 to 5 because the digit to the right is 5 or
more.
All digits to the right of the underlined place are
changed to zeros.
724,518,715 rounded to the nearest million is
725,000,000.

6. (a) **To the nearest ten:** 549
Next digit is 5 or more.
Tens place $(4 + 1 = 5)$ changes.
549 rounded to the nearest ten is 550.

To the nearest hundred: 549
Next digit is 4 or less.
Hundreds place stays the same.
549 rounded to the nearest hundred is 500.

(b) **To the nearest ten:** 458
Next digit is 5 or more.
Tens place $(5 + 1 = 6)$ changes.
458 rounded to the nearest ten is 460.

To the nearest hundred: 458
Next digit is 5 or more.
Hundreds place $(4 + 1 = 5)$ changes. All digits
to the right of the underlined place are changed to
zeros.
458 rounded to the nearest hundred is 500.

Copyright © 2018 Pearson Education, Inc.

(c) To the nearest ten: 93<u>0</u>8
Next digit is 5 or more.
Tens place $(0 + 1 = 1)$ changes.
9308 rounded to the nearest ten is 9310.

To the nearest hundred: 9<u>3</u>08
Next digit is 4 or less.
Hundreds place stays the same.
9308 rounded to the nearest hundred is 9300.

7. **(a) To the nearest ten:** 40<u>7</u>8
Next digit is 5 or more.
Tens place $(7 + 1 = 8)$ changes.
4078 rounded to the nearest ten is 4080.

To the nearest hundred: 4<u>0</u>78
Next digit is 5 or more.
Hundreds place $(0 + 1 = 1)$ changes.
4078 rounded to the nearest hundred is 4100.

To the nearest thousand: <u>4</u>078
Next digit is 4 or less.
Thousands place stays the same.
4078 rounded to the nearest thousand is 4000.

(b) To the nearest ten: 46,3<u>6</u>4
Next digit is 4 or less.
Tens place stays the same.
46,364 rounded to the nearest ten is 46,360.

To the nearest hundred: 46,<u>3</u>64
Next digit is 5 or more.
Hundreds place $(3 + 1 = 4)$ changes.
46,364 rounded to the nearest hundred is 46,400.

To the nearest thousand: 4<u>6</u>,364
Next digit is 4 or less.
Thousands place stays the same.
46,364 rounded to the nearest thousand is 46,000.

(c) To the nearest ten: 268,3<u>2</u>8
Next digit is 5 or more.
Tens place $(2 + 1 = 3)$ changes.
268,328 rounded to the nearest ten is 268,330.

To the nearest hundred: 268,<u>3</u>28
Next digit is 4 or less.
Hundreds place stays the same.
268,328 rounded to the nearest hundred is 268,300.

To the nearest thousand: 26<u>8</u>,328
Next digit is 4 or less.
Thousands place stays the same.
268,328 rounded to the nearest thousand is 268,000.

8. **(a)**

16	20	rounded to the nearest ten
74	70	
58	60	
+ 31	+ 30	
	180	estimated answer

(b)

53	50	rounded to the nearest ten
− 19	− 20	
	30	estimated answer

(c)

46	5<u>0</u>	rounded to the nearest ten
× 74	× 7<u>0</u>	
	35<u>00</u>	estimated answer

9. **(a)**

358	400	rounded to the nearest
743	700	hundred
822	800	
+ 978	+ 1000	
	2900	estimated answer

(b)

842	800	rounded to the nearest
− 475	− 500	hundred
	300	estimated answer

(c)

723	700	rounded to the nearest
× 478	× 500	hundred
	350,000	estimated answer

10. **(a)**

36	40	first digit rounded; all
3852	4000	others changed to zero
749	7<u>00</u>	
+ 5474	+ <u>5</u>000	
	9740	estimated answer

(b)

2583	3000	first digit rounded; all
− 765	− 800	others changed to zero
	2200	estimated answer

(c)

648	600	first digit rounded; all
× 67	× 70	others changed to zero
	42,000	estimated answer

1.7 Section Exercises

1. 624 to the nearest <u>ten</u> is 620.

3. 86,813 to the nearest <u>hundred</u> is 86,800.

5. 78,499 to the nearest <u>thousand</u> is 78,000.

7. 12,987 to the nearest <u>ten-thousand</u> is 10,000.

Copyright © 2018 Pearson Education, Inc.

9. 855 rounded to the nearest ten: 860

85<u>5</u> Underline the 5 in the tens place. Next digit is 5 or more, so add one to the 5 in the tens place. All digits to the right of the underlined place change to zero.

11. 6771 rounded to the nearest hundred: 6800

6<u>7</u>71 Next digit is 5 or more. Hundreds place changes (7 + 1 = 8). All digits to the right of the underlined place change to zero.

13. 28,472 rounded to the nearest hundred: 28,500

28,<u>4</u>72 Next digit is 5 or more. Hundreds place changes (4 + 1 = 5). All digits to the right of the underlined place change to zero.

15. 5996 rounded to the nearest hundred: 6000

5<u>9</u>96 Next digit is 5 or more. Hundreds place changes (9 + 1 = 10). Write 0 and carry 1. All digits to the right of the underlined place change to zero.

17. 15,758 rounded to the nearest thousand: 16,000

1<u>5</u>,758 Next digit is 5 or more. Thousands place changes (5 + 1 = 6). All digits to the right of the underlined place change to zero.

19. 7,760,058,721 rounded to the nearest billion: 8,000,000,000

<u>7</u>,760,058,721 Next digit is 5 or more. Billions place changes (7 + 1 = 8). All digits to the right of the underlined place change to zero.

21. 595,008 rounded to the nearest ten-thousand: 600,000

5<u>9</u>5,008 Next digit is 5 or more. Ten-thousands place changes (9 + 1 = 10). Write 0 and carry 1. All digits to the right of the underlined place change to zero.

23. 4,860,220 rounded to the nearest million: 5,000,000

<u>4</u>,860,220 Next digit is 5 or more. Millions place changes (4 + 1 = 5). All digits to the right of the underlined place change to zero.

25. **To the nearest ten:** 44<u>7</u>6

Underline the 7 in the tens place. Next digit is 5 or more, so add 1 to the 7. All digits to the right of the underlined place are changed to zero. **4480**

To the nearest hundred: 4<u>4</u>76

Underline the 4 in the hundreds place. Next digit is 5 or more, so add 1 to the 4. All digits to the right of the underlined place are changed to zero. **4500**

To the nearest thousand: <u>4</u>476

Underline the 4 in the thousands place. Next digit is 4 or less, so leave 4 as 4 in the thousands place. All digits to the right of the underlined place are changed to zero. **4000**

27. **To the nearest ten:** 33<u>7</u>4

Next digit is 4 or less. All digits to the right of the underlined place are changed to zero. Leave 7 as 7. **3370**

To the nearest hundred: 3<u>3</u>74

Next digit is 5 or more. All digits to the right of the underlined place are changed to zero. Add 1 to 3. **3400**

To the nearest thousand: <u>3</u>374

Next digit is 4 or less. All digits to the right of the underlined place are changed to zero. Leave 3 as 3. **3000**

29. **To the nearest ten:** 60<u>4</u>8

Next digit is 5 or more. All digits to the right of the underlined place are changed to zero. Add 1 to 4. **6050**

To the nearest hundred: 6<u>0</u>48

Next digit is 4 or less. All digits to the right of the underlined place are changed to zero. Leave 0 as 0. **6000**

To the nearest thousand: <u>6</u>048

Next digit is 4 or less. All digits to the right of the underlined place are changed to zero. Leave 6 as 6. **6000**

31. **To the nearest ten:** 53<u>4</u>3

Next digit is 4 or less. All digits to the right of the underlined place are changed to zero. Leave 4 as 4. **5340**

To the nearest hundred: 5<u>3</u>43

Next digit is 4 or less. All digits to the right of the underlined place are changed to zero. Leave 3 as 3. **5300**

(continued)

Copyright © 2018 Pearson Education, Inc.

To the nearest thousand: 5343

Next digit is 4 or less. All digits to the right of the underlined place are changed to zero. Leave 5 as 5. **5000**

33.   **To the nearest ten:** 19,539

Next digit is 5 or more. All digits to the right of the underlined place are changed to zero. Add 1 to 3. **19,540**

To the nearest hundred: 19,539

Next digit is 4 or less. All digits to the right of the underlined place are changed to zero. Leave 5 as 5. **19,500**

To the nearest thousand: 19,539

Next digit is 5 or more. All digits to the right of the underlined place are changed to zero. Add 1 to 9. Write 0 and carry 1. **20,000**

35.   **To the nearest ten:** 26,292

Next digit is 4 or less. All digits to the right of the underlined place are changed to zero. Leave 9 as 9. **26,290**

To the nearest hundred: 26,292

Next digit is 5 or more. All digits to the right of the underlined place are changed to zero. Add 1 to 2. **26,300**

To the nearest thousand: 26,292

Next digit is 4 or less. All digits to the right of the underlined place are changed to zero. Leave 6 as 6. **26,000**

37.   **To the nearest ten:** 93,706

Next digit is 5 or more. All digits to the right of the underlined place are changed to zero. Add 1 to 0. **93,710**

To the nearest hundred: 93,706

Next digit is 4 or less. All digits to the right of the underlined place are changed to zero. Leave 7 as 7. **93,700**

To the nearest thousand: 93,706

Next digit is 5 or more. All digits to the right of the underlined place are changed to zero. Add 1 to 3. **94,000**

39.   *Step 1* Locate the place to be rounded and underline it.

Step 2 Look only at the next digit to the right. If this digit is 5 or more, increase the underlined digit by 1.

Step 3 Change all digits to the right of the underlined place to zeros.

41.   *Estimate:*   *Exact:*

$$
\begin{array}{rr}
30 & 25 \\
60 & 63 \\
50 & 47 \\
+\,80 & +\,84 \\
\hline
220 & 219
\end{array}
$$

43.   *Estimate:*   *Exact:*

$$
\begin{array}{rr}
80 & 78 \\
-\,40 & -\,43 \\
\hline
40 & 35
\end{array}
$$

45.   *Estimate:*   *Exact:*

$$
\begin{array}{rr}
70 & 67 \\
\times\,30 & \times\,34 \\
\hline
2100 & 268 \\
 & 201 \\
\hline
 & 2278
\end{array}
$$

47.   *Estimate:*   *Exact:*

$$
\begin{array}{rr}
900 & 863 \\
700 & 735 \\
400 & 438 \\
+\,800 & +\,792 \\
\hline
2800 & 2828
\end{array}
$$

49.   *Estimate:*   *Exact:*

$$
\begin{array}{rr}
900 & 883 \\
-\,400 & -\,448 \\
\hline
500 & 435
\end{array}
$$

51.   *Estimate:*   *Exact:*

$$
\begin{array}{rr}
800 & 752 \\
\times\,400 & \times\,375 \\
\hline
320,000 & 3760 \\
 & 5264 \\
 & 2256 \\
\hline
 & 282,000
\end{array}
$$

53.   *Estimate:*   *Exact:*

$$
\begin{array}{rr}
8\,000 & 8\,215 \\
60 & 56 \\
700 & 729 \\
+\,4\,000 & +\,3\,605 \\
\hline
12,760 & 12,605
\end{array}
$$

Copyright © 2018 Pearson Education, Inc.

55.

Estimate:	Exact:
700	687
− 500	− 529
200	158

57.

Estimate:	Exact:
900	939
× 30	× 29
27,000	8 4 5 1
	1 8 7 8
	27,231

59. Perhaps the best explanation is that 3492 is closer to 3500 than 3400, but 3492 is closer to 3000 than to 4000.

61. 7̲6,000,000 Next digit is 5 or more, so add 1 to 7. Change all digits to the right of the underlined place to zero. **80, 000, 000 people**

3̲25,000,000 Next digit is 5 or more, so add 1 to 3. Change all digits to the right of the underlined place to zero. **330, 000, 000 people**

63. 34̲8,900 Next digit is 5 or more. Change all digits to the right of the underlined place to zero. Add 1 to 8. **349,000 streets**

34̲8,900 Next digit is 5 or more. All digits to the right of the underlined place are changed to zero. Add 1 to 4. **350,000 streets**

65. To the nearest ten-thousand: 39,8̲36,000

Next digit is 5 or more, so add 1 to 3. Change digits to the right of the underlined place to zeros.

39,840,000 tickets

To the nearest hundred-thousand: 39,8̲36,000

Next digit is 4 or less. Hundred-thousands place doesn't change. All digits to the right of the underlined place are changed to zero.

39,800,000 tickets

To the nearest million: 39̲,836,000

Next digit is 5 or more, so add 1 to 9. Write 0 and carry 1. Change digits to the right of the underlined place to zeros.

40,000,000 tickets

67. To the nearest thousand: 19,265̲,780

Next digit is 5 or more, so add 1 to 5. Change digits to the right of the underlined place to zeros.

19,266,000 players

To the nearest ten-thousand: 19,26̲5,780

Next digit is 5 or more, so add 1 to 6. Change digits to the right of the underlined place to zeros.

19,270,000 players

To the nearest hundred-thousand: 19,2̲65,780

Next digit is 5 or more, so add 1 to 2. Change digits to the right of the underlined place to zeros.

19,300,000 players

69. 71,499 would round to 71,000, so the smallest whole number it could have been is 71,500.

70. 72,500 would round to 73,000, so the largest whole number it could have been is 72,499.

71. 7500 is the smallest possible whole number that will round to 8000 using front end rounding.

72. 8499 is the largest possible whole number that will round to 8000 using front end rounding.

73.

Exact:	Rounding to the nearest ten:
3925	3930
11,243	11,240
15,974	15,970
17,916	17,920
534,883	534,880
2,788,000	2,788,000

74.

Exact:	Front end rounded:
3925	4000
11,243	10,000
15,974	20,000
17,916	20,000
534,883	500,000
2,788,000	3,000,000

75. (a)

Driving	2,788,000	→	3, 000, 000
Brushing teeth	3925	→	− 4,000
			2, 996,000

There were about 2, 996,000 more accidents from driving than from brushing teeth.

(b)

Driving	2,788,000
Brushing teeth	− 3925
	2,784,075

There were 2,784,075 more accidents from driving than from brushing teeth.

Copyright © 2018 Pearson Education, Inc.

(c) The difference between the answers may not be important since the purpose of the estimated answer is to determine whether the exact answer is reasonable.

1.8 Exponents, Roots, and Order of Operations

1.8 Margin Exercises

1. **(a)** 4^2: exponent is 2; base is 4.

   $4^2 = 4 \times 4 = 16$

   (b) 3^4: exponent is 4; base is 3.

   $3^4 = 3 \times 3 \times 3 \times 3 = 81$

   (c) 2^6: exponent is 6; base is 2.

   $2^6 = 2 \times 2 \times 2 \times 2 \times 2 \times 2 = 64$

2. **(a)** Because $2^2 = 4$, $\sqrt{4} = 2$.

   (b) Because $6^2 = 36$, $\sqrt{36} = 6$.

   (c) Because $15^2 = 225$, $\sqrt{225} = 15$.

   (d) Because $1^2 = 1$, $\sqrt{1} = 1$.

3. **(a)** $4 + 5 + 2^2$ *Evaluate exponent.*
   $4 + 5 + 2 \cdot 2$ *Multiply.*
   $4 + 5 + \underline{4}$ *Add from left to right.*
   $9 + \underline{4} = \underline{13}$

   (b) $3^2 + 2^3$ *Evaluate exponent.*
   $3 \cdot 3 + 2 \cdot 2 \cdot 2$ *Multiply from left to right.*
   $9 + 8 = 17$ *Add.*

   (c) $60 \div \sqrt{36} \div 2$ *Square root*
   $60 \div 6 \div 2$ *Divide left to right.*
   $10 \div 2 = 5$ *Divide.*

   (d) $8 + 6(14 \div 2)$ *Work inside parentheses.*
   $8 + 6(7)$ *Multiply.*
   $8 + 42 = 50$ *Add.*

4. **(a)** $12 - 6 + 4^2$ *Evaluate exponent.*
   $12 - 6 + \underline{4} \cdot \underline{4}$ *Multiply.*
   $12 - 6 + 16$ *Subtract left to right.*
   $6 + 16 = \underline{22}$ *Add.*

   (b) $2^3 + 3^2 - (5 \cdot 3)$ *Work inside parentheses.*
   $2^3 + 3^2 - 15$ *Evaluate exponents.*
   $8 + 9 - 15$ *Add left to right.*
   $17 - 15 = 2$ *Subtract.*

(c) $20 \div 2 + (7 - 5)$ *Work inside parentheses.*
$20 \div 2 + 2$ *Divide.*
$10 + 2 = 12$ *Add.*

(d) $15 \cdot \sqrt{9} - 8 \cdot \sqrt{4}$ *Square roots*
$15 \cdot 3 - 8 \cdot 2$ *Multiply.*
$45 - 16 = 29$ *Subtract.*

1.8 Section Exercises

1. 3^2: exponent is 2, base is 3.

3. 5^2: exponent is 2, base is 5.

5. 8^2: exponent is 2, base is 8.

   $8^2 = 8 \cdot 8 = 64$

7. 15^2: exponent is 2, base is 15.

   $15^2 = 15 \cdot 15 = 225$

9. From the table, $4^2 = 16$, so $\sqrt{16} = 4$.

11. From the table, $8^2 = 64$, so $\sqrt{64} = 8$.

13. From the table, $10^2 = 100$, so $\sqrt{100} = 10$.

15. From the table, $12^2 = 144$, so $\sqrt{144} = 12$.

17. *False:* 5^2 means that 5 is used as a factor 2 times, so $5^2 = 5 \cdot 5 = 25$.

19. *False:* 1 raised to any power is 1. Here, $1^3 = 1 \cdot 1 \cdot 1 = 1$.

21. $6^2 = \underline{36}$, so $\sqrt{36} = 6$
    $6 \cdot 6 = \underline{36}$, so $\sqrt{36} = 6$

23. $25^2 = \underline{625}$, so $\sqrt{625} = 25$

25. $100^2 = \underline{10,000}$, so $\sqrt{10,000} = 100$

27. A perfect square is the square of a whole number. The number 25 is the square of 5 because $5 \cdot 5 = 25$.

    The number 50 is not a perfect square. There is no whole number that can be multiplied by itself (squared) to get 50.

29. $3^2 + 8 - 5$ *Exponent*
    $9 + 8 - 5$ *Add.*
    $17 - 5 = 12$ *Subtract.*   **TRUE**

31. $6 + 8 \div 2$ *Divide.*
    $6 + 4 = 10$ *Add.*

    The given statement is *false.* Multiplications and divisions are performed from *left* to *right*, then additions and subtractions.

Copyright © 2018 Pearson Education, Inc.

33. $3^2 + 8 - 5$ *Exponent*
     $9 + 8 - 5$ *Add.*
     $17 - 5 = 12$ *Subtract.*

35. $25 \div 5(8 - 4)$ *Parentheses*
     $25 \div 5(4)$ *Divide.*
     $5(4) = 20$ *Multiply.*

37. $5 \cdot 3^2 + \dfrac{0}{8}$ *Exponent*

     $5 \cdot 9 + \dfrac{0}{8}$ *Multiply.*

     $45 + \dfrac{0}{8}$ *Divide.*

     $45 + 0 = 45$ *Add.*

39. $4 \cdot 1 + 8(9 - 2) + 3$ *Parentheses*
     $4 \cdot 1 + 8 \cdot 7 + 3$ *Multiply.*
     $4 + 56 + 3$ *Add.*
     $60 + 3 = 63$ *Add.*

41. $2^2 \cdot 3^3 + (20 - 15) \cdot 2$ *Parentheses*
     $2^2 \cdot 3^3 + 5 \cdot 2$ *Exponents*
     $4 \cdot 27 + 5 \cdot 2$ *Multiply.*
     $108 + 10 = 118$ *Add.*

43. $5 \cdot \sqrt{36} - 2(4)$ *Square root*
     $5 \cdot 6 - 2(4)$ *Multiply.*
     $30 - 8 = 22$ *Subtract.*

45. $8(2) + 3 \cdot 7 - 7$ *Multiply.*
     $16 + 21 - 7$ *Add.*
     $37 - 7 = 30$ *Subtract.*

47. $2^3 \cdot 3^2 + 3(14 - 4)$ *Parentheses*
     $2^3 \cdot 3^2 + 3(10)$ *Exponents*
     $8 \cdot 9 + 3(10)$ *Multiply.*
     $72 + 30 = 102$ *Add.*

49. $7 + 8 \div 4 + \dfrac{0}{7}$ *Divide.*

     $7 + 2 + \dfrac{0}{7}$ *Divide.*

     $7 + 2 + 0$ *Add.*
     $9 + 0 = 9$ *Add.*

51. $3^2 + 6^2 + (30 - 21) \cdot 2$ *Parentheses*
     $3^2 + 6^2 + 9 \cdot 2$ *Exponents*
     $9 + 36 + 9 \cdot 2$ *Multiply.*
     $9 + 36 + 18$ *Add.*
     $45 + 18 = 63$ *Add.*

53. $7 \cdot \sqrt{81} - 5 \cdot 6$ *Square root*
     $7 \cdot 9 - 5 \cdot 6$ *Multiply.*
     $63 - 30 = 33$ *Subtract.*

55. $8 \cdot 2 + 5(3 \cdot 4) - 6$ *Parentheses*
     $8 \cdot 2 + 5(12) - 6$ *Multiply.*
     $16 + 60 - 6$ *Add.*
     $76 - 6 = 70$ *Subtract.*

57. $4 \cdot \sqrt{49} - 7(5 - 2)$ *Parentheses*
     $4 \cdot \sqrt{49} - 7 \cdot 3$ *Square root*
     $4 \cdot 7 - 7 \cdot 3$ *Multiply.*
     $28 - 21 = 7$ *Subtract.*

59. $5^2 \cdot 2^2 + (8 - 4) \cdot 2$ *Parentheses*
     $5^2 \cdot 2^2 + 4 \cdot 2$ *Exponents*
     $25 \cdot 4 + 4 \cdot 2$ *Multiply.*
     $100 + 8 = 108$ *Add.*

61. $5 + 9 \div 3 + 6 \cdot 3$ *Divide.*
     $5 + 3 + 6 \cdot 3$ *Multiply.*
     $5 + 3 + 18$ *Add.*
     $8 + 18 = 26$ *Add.*

63. $8 \cdot \sqrt{49} - 6(9 - 4)$ *Parentheses*
     $8 \cdot \sqrt{49} - 6(5)$ *Square root*
     $8 \cdot 7 - 6(5)$ *Multiply.*
     $56 - 6(5)$ *Multiply.*
     $56 - 30 = 26$ *Subtract.*

65. $8 + 8 \div 8 + 6 + \dfrac{5}{5}$ *Divide.*

     $8 + 1 + 6 + 1$ *Add.*
     $9 + 6 + 1$ *Add.*
     $15 + 1 = 16$ *Add.*

67. $6 \cdot \sqrt{25} - 7(2)$ *Square root*
     $6 \cdot 5 - 7 \cdot 2$ *Multiply.*
     $30 - 14 = 16$ *Subtract.*

69. $9 \cdot \sqrt{16} - 3 \cdot \sqrt{25}$ *Square roots*
     $9 \cdot 4 - 3 \cdot 5$ *Multiply.*
     $36 - 15 = 21$ *Subtract.*

71. $7 \div 1 \cdot 8 \cdot 2 \div (21 - 5)$ *Parentheses*
     $7 \div 1 \cdot 8 \cdot 2 \div 16$ *Divide.*
     $7 \cdot 8 \cdot 2 \div 16$ *Multiply.*
     $56 \cdot 2 \div 16$ *Multiply.*
     $112 \div 16 = 7$ *Divide.*

Copyright © 2018 Pearson Education, Inc.

73.

$15 \div 3 \cdot 2 \cdot 6 \div (14 - 11)$ *Parentheses*

$15 \div 3 \cdot 2 \cdot 6 \div 3$ *Divide.*

$5 \cdot 2 \cdot 6 \div 3$ *Multiply.*

$10 \cdot 6 \div 3$ *Multiply.*

$60 \div 3 = 20$ *Divide.*

75.

$6 \cdot \sqrt{25} - 4 \cdot \sqrt{16}$ *Square roots*

$6 \cdot 5 - 4 \cdot 4$ *Multiply.*

$30 - 16 = 14$ *Subtract.*

77.

$5 \div 1 \cdot 10 \cdot 4 \div (17 - 9)$ *Parentheses*

$5 \div 1 \cdot 10 \cdot 4 \div 8$ *Divide.*

$5 \cdot 10 \cdot 4 \div 8$ *Multiply.*

$50 \cdot 4 \div 8$ *Multiply.*

$200 \div 8 = 25$ *Divide.*

79.

$2 \cdot (7 - 5) \cdot 6$ *Parentheses*

$2 \cdot \quad 2 \quad \cdot 6$ *Multiply from left to right.*

$4 \cdot 6 = 24$

80.

$4 \cdot (3 + 2) \cdot 2$ *Parentheses*

$4 \cdot \quad 5 \quad \cdot 2$ *Multiply from left to right.*

$20 \cdot 2 = 40$

81.

$(20 - 7) - 2$ *Parentheses*

$13 - 2 = 11$

82.

$15 + (1 \cdot 3)$ *Parentheses*

$15 + 3 = 18$

83.

$8 \cdot 5 - (3 + 2)$ *Parentheses*

$8 \cdot 5 - 5$ *Multiply.*

$40 - 5 = 35$ *Subtract.*

84.

$9 \cdot (3 - 2) \cdot 3$ *Parentheses*

$9 \cdot \quad 1 \quad \cdot 3$ *Multiply from left to right.*

$9 \cdot 3 = 27$

85.

$(4 - 3) \cdot 2 \cdot \sqrt{25} \cdot 5 \cdot 2 + 2$ *Parentheses*

$1 \quad \cdot 2 \cdot \sqrt{25} \cdot 5 \cdot 2 + 2$ *Square root*

$1 \quad \cdot 2 \cdot \quad 5 \cdot 5 \cdot 2 + 2$ *Multiply from left to right.*

$100 + 2 = 102$ *Add.*

86.

$15(5 - 2) \cdot 4$ *Parentheses*

$15 \quad \cdot 3 \quad \cdot 4 = 180$ *Multiply from left to right.*

87.

$2 \cdot \sqrt{36} \cdot 5 - 3 + 4 - 8$ *Square root*

$2 \cdot \quad 6 \cdot 5 - 3 + 4 - 8$ *Multiply from left to right.*

$60 \quad - 3 + 4 - 8 = 53$ *Add and subtract from left to right.*

88.

$\sqrt{49} + (7 - 5) \cdot 2 - 10$ *Square root*

$7 + (7 - 5) \cdot 2 - 10$ *Parentheses*

$7 + \quad 2 \quad \cdot 2 - 10$ *Multiply.*

$7 + \quad 4 \quad - 10 = 43$ *Add and subtract from left to right.*

89.

$8 \cdot 2 + 6 - 5 \cdot \sqrt{16} + 6 - 5$ *Square root*

$8 \cdot 2 + 6 - 5 \cdot \quad 4 + 6 - 5$ *Multiply from left to right.*

$16 + 6 - 20 + 6 - 5 = 3$ *Add and subtract from left to right.*

90.

$35 - \sqrt{25} + 7 \cdot 2 + 3 - 45$ *Square root*

$35 - \quad 5 \quad + 7 \cdot 2 + 3 - 45$ *Multiply.*

$35 - \quad 5 \quad + 14 + 3 - 45 = 2$ *Add and subtract from left to right.*

1.9 Reading Pictographs, Bar Graphs, and Line Graphs

1.9 Margin Exercises

1. **(a)** The Rock and Roll/Rhythm and Blues stamp had the second greatest number (4) of sales since that row has the second most stamp symbols.

(b) There is about one more stamp in the Rock and Roll/Rhythm and Blues row than in the Art of Disney Romance row. Since one stamp represents 20 million stamps, there were about 20,000,000 more Rock and Roll/Rhythm and Blues stamps sold.

2. **(a)** Secret account or card: 27 out of 100

(b) Misrepresented purchased price: 39 out of 100

(c) Hid purchases: 42 out of 100

(d) Hid debt: 26 out of 100

(e) Did not tell credit score: 12 out of 100

3. **(a)** For the year 2050, the dot is very close to the horizontal line marked 4, so the predicted population is about 400,000,000 people in 2050.

(b) About 475,000,000 people in 2075

(c) About 575,000,000 people in 2100

1.9 Section Exercises

1. The number of pictures or symbols for the Home Depot retail stores is 4. Since each symbol represents 500 stores, Home Depot has $4 \cdot 500 = 2000$ stores.

Copyright © 2018 Pearson Education, Inc.

3. There are $9\frac{1}{2}$ symbols, so the number of retail stores for Walmart is

$$9\frac{1}{2} \cdot 500 = (9 \cdot 500) + \frac{1}{2} \cdot 500$$
$$= 4500 + 250 = 4750 \text{ stores.}$$

5. From the pictograph, McDonald's has the greatest number of retail stores. There are 28 symbols, so the number of retail stores for McDonald's is about

$$28 \cdot 500 = 14,000.$$

7. According to the pictograph, Costco has 500 stores and Dollar General has $22 \times 500 = 11,000$ stores. Subtract 500 from 11,000 to find the difference.

$$11,000 - 5000 = 10,500 \text{ fewer stores}$$

Alternatively, Dollar General has 22 symbols while Costco has 1. So, Dollar General has $22 - 1 = 21$ more symbols. This represents $21 \cdot 500 = 10,500$ more stores.

9. According to the directions above the chart, 100 working adults were surveyed.

11. From the bar graph, 9 people out of 100 found their career as a result of training for a job.

13. **(a)** According to the bar graph, the greatest number of people found their job by <u>seeing an ad</u>.

(b) From the bar graph, <u>25 people</u> out of 100 found their career as a result of seeing an ad.

15. According to the bar graph, $18 - 9 = 9$ more people out of 100 found their career as a result of "Studied in school" than "Luck or chance."

17. According to the line graph, the year with the greatest number of installations is <u>2017</u>, with <u>7000 installations</u>.

19. According to the line graph, the increase in the number of installations from 2013 to 2014 is $7000 - 2500 = 4500$.

21. Possible answers are

1. shortage of units to install.
2. lack of qualified workers.
3. poor economy.
4. less demand for solar products.

23.

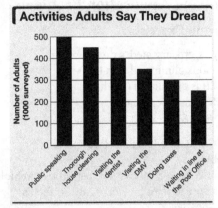

24. There are six categories, so the greatest number of responses that could have been given by each adult in the survey was 6.

25. The least number of responses that could have been given by each adult in the survey was 0.

26. $500 + 350 + 300 + 450 + 250 + 400 = 2250$
There were 2250 responses.

27. $6 \times 1000 = 6000$
The maximum number or responses possible in the survey was 6000.

28. For a sample size of 10,000, multiply the original number of responses by 10. For a sample size of 100,000, multiply the original number of responses by 100. For a sample size of 500, divide the original number of responses by 2.

Activity	Sample Size		
	10,000	100,000	500
Public speaking	5000	50,000	250
Visiting the DMV	3500	35,000	175
Doing taxes	3000	30,000	150
Thorough house cleaning	4500	45,000	225
Waiting in line at Post Office	2500	25,000	125
Visiting the dentist	4000	40,000	200

Copyright © 2018 Pearson Education, Inc.

1.10   Solving Application Problems

1.10   Margin Exercises

1.   **(a)** A grocery clerk's hourly wage:
$1.80 is eliminated (too low)
$180 is eliminated (too high)
$18 is most reasonable
$1.80; $\boxed{\$18}$; $180

     (b) The total length of five sports-utility vehicles:
One vehicle is at least 10 feet, so five vehicles measure at least 50 feet. Only 80 feet is reasonable.
8 feet; 18 feet; $\boxed{80 \text{ feet}}$; 800 feet

     (c) The cost of heart bypass surgery:
$1000; $\boxed{\$100{,}000}$; $10,000,000

2.   **(a)** *Step 1*
The total number of fossils is given, and the number each person receives must be found.

     Step 2
"Divided equally" indicates division should be used.

     Step 3
80 fossils divided equally among 4 people gives an estimate of 20 fossils per person.

     Step 4

$$4\overline{)84} \quad \text{with quotient } 21$$

     Step 5
Each person receives 21 fossils.

     Step 6

$$
\begin{array}{r}
21 \quad \text{\textit{amount received by each person}} \\
\times \ 4 \quad \text{\textit{number of people}} \\
\hline
84 \quad \text{\textit{total fossils; matches}}
\end{array}
$$

     (b) *Step 1*
The total number of children and the number of children assigned to each counselor is given. The number of counselors needed must be found.

     Step 2
"12 children are assigned to *each* camp counselor" indicates division should be used.

     Step 3
If we replace 408 and 12 with 400 and 10, our estimate is $400 \div 10 = 40$.

Step 4

$$
\begin{array}{r}
34 \\
12\overline{)408} \\
\underline{36} \\
48 \\
\underline{48} \\
0
\end{array}
$$

Step 5
There are 34 counselors needed.

Step 6

$$
\begin{array}{r}
34 \quad \text{\textit{counselors}} \\
\times \ 12 \quad \text{\textit{children per counselor}} \\
\hline
408 \quad \text{\textit{total children; matches}}
\end{array}
$$

3.   **(a)** *Step 1*
The number of points on examinations and quizzes is given. The total number of points must be found.

     Step 2
"Her total points" indicates addition should be used.

     Step 3
Rounding each score to the nearest ten gives 90, 80, 80, 100, 20, 10, 20, and 10. The sum of these scores gives an estimate of 410 points.

     Step 4

$$
\begin{array}{r}
\overset{3}{9}2 \\
81 \\
83 \\
98 \\
15 \\
14 \\
15 \\
+ \ 12 \\
\hline
410 \quad \text{\textit{matches estimate}}
\end{array}
$$

     Step 5
Her total is 410 points.

     Step 6
The answer is the same as the estimate, so it is reasonable. Add the numbers again to check.

     (b) *Step 1*
The customer contacts for each day are given. The total number of contacts for the week must be found.

     Step 2
Adding the daily numbers to find the weekly total seems reasonable.

(continued)

Copyright © 2018 Pearson Education, Inc.

Step 3
Rounding each day's contacts to the nearest ten gives 80, 60, 120, 100, and 200. The sum gives an estimate of 560 customer contacts.

Step 4
$$
\begin{array}{r}
22 \\
78 \\
64 \\
118 \\
102 \\
+\,196 \\
\hline
558
\end{array}
$$

Step 5
Sue had 558 customer contacts for the week.

Step 6
The answer is reasonable. A check shows that the answer is correct.

4. **(a)** *Step 1*
The areas in square miles of two states is given. The difference must be found.

Step 2
"Difference" indicates subtraction should be used.

Step 3
Rounding each number of square miles to the nearest thousand gives
$663,000 - 269,000 = 394,000$ square miles.

Step 4
$$
\begin{array}{r}
663,267 \\
-\,268,580 \\
\hline
394,687
\end{array}
$$

Step 5
The difference in the number of square miles is 394,687 square miles.

Step 6
The answer is reasonable.

Check:
$$
\begin{array}{r}
268,580 \\
+\,394,687 \\
\hline
663,267 \quad matches
\end{array}
$$

(b) *Step 1*
We know the beginning balance and the check amount, and need to find the new balance.

Step 2
Subtracting the numbers seems reasonable.

Step 3
Using the values $14,900 and $1200 in place of $14,863 and $1180 gives us
$14,900 - $1200 = $13,700$ as an estimate.

Step 4
$$
\begin{array}{r}
14,863 \\
-\,1\,180 \\
\hline
13,683
\end{array}
$$

Step 5
The amount remaining in the club account is $13,683.

Step 6
The answer is reasonable.
Check:
$$
\begin{array}{r}
13,683 \\
+\,1\,180 \\
\hline
14,863 \quad matches
\end{array}
$$

5. **(a)** *Step 1*
The amount per camera, number of cameras, and total rebate are given. The final cost is to be found.

Step 2
Use multiplication to find the cost for all cameras, and then use subtraction to find the final cost.

Step 3
An estimate would be $320 \cdot \$130 - \$2500$
$= \$41,600 - \$2500 = \$39,100$.

Step 4
$$
\begin{array}{rr}
\$129 & \$41,022 \\
\times\,318 & -\,2\,470 \\
\hline
\$41,022 & \$38,552
\end{array}
$$

Step 5
The final cost of the cameras is $38,552.

Step 6
The answer is close to the estimate, so it is reasonable. Check by adding the rebate to the final cost and then dividing by 318.
$\$38,552 + \$2470 = \$41,022$

$$
\begin{array}{r}
129 \\
318\overline{)41,022} \\
\underline{318} \quad \leftarrow 1 \times 318 \\
922 \\
\underline{636} \quad \leftarrow 2 \times 318 \\
2862 \\
\underline{2862} \quad \leftarrow 9 \times 318 \\
0
\end{array}
$$

$129 matches the amount per camera given in the problem.

Copyright © 2018 Pearson Education, Inc.

(b) *Step 1*
The number of books sold and the number of those returned is given. The profit for each book sold is also given. The total profit on the books sold and not returned must be found.

Step 2
"Remaining" indicates subtraction. The number of books returned should be subtracted from the number sold and that number multiplied by the profit per book.

Step 3
Estimate that 13,000 books are sold, and 1000 of those are returned. A reasonable estimate of the number of books remaining after the returns is $13,000 - 1000 = 12,000$. Each book has a profit of \$6, so $12,000 \times \$6 = \$72,000$.

Step 4

12,628	11,765
$-\ 863$	$\times\quad 6$
11,765	\$70,590

Step 5
The total profit for the books remaining after the returns is \$70,590.

Step 6
The answer is reasonable since it is close to the estimate.

Check:
$$6\overline{)70,590}\quad\begin{array}{r}11,765\\+\ 863\\\hline 12,628\end{array}$$

1.10 Section Exercises

1.   Choice **(c)**, 25 hours, is *not* reasonable since there are only 24 hours in a day.

3.   Choice **(a)**, \$5, is reasonable for the cost of lunch at a fast food restaurant.

5.   *Step 1*
Given the number of five types of sandwiches sold, find the total number of sandwiches sold.

Step 2
Addition seems reasonable since we want a total.

Step 3
The total number of sandwiches should be about $600 + 900 + \underline{1000} + 800 + \underline{2000} = \underline{5300}$ sandwiches.

Step 4
$$\begin{array}{r}322\\602\\935\\1328\\757\\+\ 1586\\\hline 5208\end{array}$$

Step 5
The total number of sandwiches sold is 5208.

Step 6
The answer is reasonably close to the estimate. Check by adding the values again.

7.   *Step 1*
Given two prices for car rentals, find the difference.

Step 2
"How much is saved" indicates subtraction could be used.

Step 3
An estimate is $\$300 - \$200 = \underline{\$100\ saved}$.

Step 4
$$\begin{array}{r}296\\-\ 192\\\hline 104\end{array}$$

Step 5
Paying ahead will save \$104.

Step 6
The answer is reasonably close to the estimate.
Check: $\$104 + \$192 = \$296$

9.   *Step 1*
The number of kits packaged in one hour is given and the total number of kits packaged in 24 hours must be found.

Step 2
Multiply the number of kits packaged in 1 hour by 24.

Step 3
A reasonable answer would be $200 \times 20 = 4000$.

Step 4
$$\begin{array}{r}236\\\times\ 24\\\hline 944\\472\quad\\\hline 5664\end{array}$$

Step 5
5664 kits are packaged in 24 hours.

(continued)

Copyright © 2018 Pearson Education, Inc.

Step 6
The answer is reasonably close to the estimate, which is clearly low.

Check: 2 3 6
             24⟌5 6 6 4

11. *Step 1*
     Find the number of toys each child will receive.

     Step 2
     "Same number of toys to each" indicates division should be used.

     Step 3
     3000 divided by 700 gives an estimate of about 4 toys.

     Step 4
                    4
         657⟌2 6 2 8
              2 6 2 8
                    0

     Step 5
     Each child will receive 4 toys.

     Step 6
     $4 \times 657 = 2628$

13. *Step 1*
     The amount saved per month and the number of months are given. Find the total amount saved.

     Step 2
     Multiply the two given values.

     Step 3
     A reasonable answer would be $30 \times 5 = \$150$.

     Step 4
        34
      $\times$ 5
      ─────
       170

     Step 5
     $170 could be saved.

     Step 6
     The answer is reasonably close to the estimate.

     Check: 3 4
                   5⟌1 7 ^{2}0

15. *Step 1*
     Given two numbers, find the total.

     Step 2
     "Find the total" indicates addition could be used.

Step 3
An estimate is $100,000 + 300,000 = \underline{400,000}$ deaths.

Step 4
    110,070
  + 250,152
  ─────────
    360,222

Step 5
The total number of Union deaths in the Civil War was 360,222.

Step 6
The answer is reasonably close to the estimate.
Check: $360,222 - 250,152 = 110,070$

17. *Step 1*
     Given two numbers, find the difference.

     Step 2
     "How many more" indicates subtraction could be used.

     Step 3
     An estimate is $300,000 - 100,000 = 200,000$ deaths.

     Step 4
         250,152
       − 110,070
       ─────────
         140,082

     Step 5
     There were 140,082 more Union deaths that resulted from disease than from battle.

     Step 6
     The answer is reasonably close to the estimate.
     Check: $140,082 + 110,070 = 250,152$

19. *Step 1*
     Given four numbers, find the total.

     Step 2
     "Find the total" indicates addition could be used.

     Step 3
     An estimate is $100,000 + 300,000 + 90,000 + 200,000 = 690,000$ deaths.

     Step 4
         110,070
         250,152
          94,120
       + 164,300
       ─────────
         618,642

(continued)

Copyright © 2018 Pearson Education, Inc.

Step 5
The total number of deaths in the Civil War was 618,642.

Step 6
The answer is reasonably close to the estimate.
Check: $618,642 - 164,300 - 94,120 - 250,152 = 110,070$

21. *Step 1*
Find her monthly savings.

Step 2
Her monthly take home pay and expenses are given. "Remainder" indicates subtraction may be used.

Step 3
Estimate:
$2000 - $700 - $300 - $400 - $200 - $200
= $200

Step 4

$695	$2240
340	$- 1890$
435	$350
240	
$+ 180$	
1890	

Step 5
Her monthly savings are $350.

Step 6
The answer seems reasonable for the given estimate.

Check: $350 + $1890 = $2240. Re-add the expenses to check the sum of $1890.

23. *Step 1*
The square feet in one acre is given and the square feet in the given number of acres must be found.

Step 2
Multiply the square feet in one acre by the total number of acres.

Step 3
An estimate is $40,000 \times 100 = 4,000,000$ ft^2.

Step 4

$$\begin{array}{r} 43,560 \\ \times\ 138 \\ \hline 348480 \\ 130680 \\ 43560 \\ \hline 6,011,280 \end{array}$$

Step 5
There are 6,011,280 square feet in 138 acres. The answer is reasonable considering the rounding that was done in finding the estimate.

Step 6
Check:
$$138\overline{)6,011,280} \quad \text{43,560}$$

25. *Step 1*
The number and cost of wheelchairs and recorder-players are given and the total cost of all items must be found.

Step 2
Find the cost of all wheelchairs and the cost of all recorder-players. Then add these costs to get the total cost.

Step 3
The cost of the wheelchairs is about $1000 \times 6 = 6000.

The cost of the recorder-players is about $900 \times 20 = $18,000$.

Estimate: $6000 + $18,000 = $24,000

Step 4
cost of wheelchairs:
$$\begin{array}{r} \$1256 \\ \times\quad 6 \\ \hline \$7536 \end{array}$$

cost of recorder-players:
$$\begin{array}{r} \$895 \\ \times\ 15 \\ \hline 4475 \\ 895 \\ \hline \$13,425 \end{array}$$

Step 5
The total cost is $13,425 + $7536 = $20,961.

Step 6
The answer is reasonably close to the estimate. Check by repeating Step 4.

27. Possible answers are

Addition: more; total; gain of
Subtraction: less; loss of; decreased by
Multiplication: twice; of; product
Division: divided by; goes into; per
Equals: is; are

29. *Step 1*
The daily sales figures are given and the weekly total must be found.

Step 2
Add the daily sales figures to get the weekly total.

(continued)

Copyright © 2018 Pearson Education, Inc.

Step 3
Estimate: $2000 + $3000 + $3000 + $2000 +
$4000 + $3000 + $3000 = $20,000

Step 4

Add.
$$
\begin{array}{r}
\overset{3\ 34}{\$2\ 358} \\
3\ 056 \\
2\ 515 \\
1\ 875 \\
3\ 978 \\
3\ 219 \\
+\ 3\ 008 \\
\hline
\$20,009
\end{array}
\begin{array}{l}
Monday \\
Tuesday \\
Wednesday \\
Thursday \\
Friday \\
Saturday \\
Sunday \\
\\
Total\ cost
\end{array}
$$

Step 5
The weekly total is $20,009.

Step 6
The answer is reasonably close to the estimate.
Check by repeating Step 4.

31. *Step 1*
Find the final weight of the car.

Step 2
The second engine weighs more than the first
engine. Find the difference and add it to the
weight of the car.

Step 3
582 and 634 both round to 600, so the difference
is 0 and an estimate of the car's weight is just its
original weight, 2425 pounds.

Step 4

$$
\begin{array}{r}
634 \\
-\ 582 \\
\hline
52
\end{array}
\begin{array}{l}
new\ engine \\
old\ engine \\
\\
difference
\end{array}
\qquad
\begin{array}{r}
2425 \\
+\ 52 \\
\hline
2477
\end{array}
\begin{array}{l}
original\ weight \\
additional\ weight \\
\\
new\ weight
\end{array}
$$

Step 5
The car will weigh 2477 pounds.

Step 6
The answer is reasonable since it's just slightly
more than the original weight.

Check:
$$
\begin{array}{r}
2425 \\
-\ 582 \\
\hline
1843
\end{array}
\qquad
\begin{array}{r}
1843 \\
+\ 634 \\
\hline
2477
\end{array}
$$

33. *Step 1*
The costs of two hotels are given. Find the
amount saved by staying at the less expensive
hotel for seven nights.

Step 2
First subtract the cost of the least expensive hotel
and then multiply this savings by six.

Step 3
Estimate: $100 − $60 = $40
$60 × 6 = $360

Step 4

Subtract.
$$
\begin{array}{r}
\$139 \\
-\ 61 \\
\hline
\$78
\end{array}
\begin{array}{l}
Harrah's\ Lake\ Tahoe \\
Harrah's\ Reno \\
\\
Savings\ for\ 1\ night
\end{array}
$$

Multiply.
$$
\begin{array}{r}
\$78 \\
\times\ 6 \\
\hline
\$468
\end{array}
$$

Step 5
The savings for 6 nights is $468.

Step 6
The answer is reasonably close to the estimate.

Check: Lake Tahoe cost: 6 × $139 = $834
Reno cost: 6 × $61 = $366
Difference: $834 − $366 = $468

35. *Step 1*
Find out how much money each team received
given the amount of money raised, the expenses,
and the number of teams.

Step 2
Subtract the expenses from the amount raised,
then divide the result by the number of teams.

Step 3
Estimate: $8000 − $800 = $7200
$7200 ÷ 20 = $360

Step 4

Subtract.
$$
\begin{array}{r}
\$7588 \\
-\ 838 \\
\hline
\$6750
\end{array}
$$

Divide.
$$
\begin{array}{r}
\$3\ 7\ 5 \\
18\overline{\smash{\big)}\ \$6\ 7\ 5\ 0} \\
\underline{5\ 4} \\
1\ 3\ 5 \\
\underline{1\ 2\ 6} \\
9\ 0 \\
\underline{9\ 0} \\
0
\end{array}
$$

Step 5
Each team received $375.

Step 6
The answer is reasonably close to the estimate.
Check: 18 × $375 = $6750
$6750 + $838 = $7588

Copyright © 2018 Pearson Education, Inc.

37. *Step 1*
The total seating requirement is given along with the information needed to find the number of seats on the main floor. The number of rows of seats in the balcony is given, and the number of seats in each row of the balcony must be found.

Step 2
First, the number of seats on the main floor must be found. Next, the number of seats on the main floor must be subtracted from the total number of seats to find the number of seats in the balcony. Finally, the number of seats in the balcony must be divided by the number of rows of seats in the balcony.

Step 3
Seats on main floor: $30 \times 25 = 750$
Seats in balcony: $1250 - 750 = 500$
Seats in each row in balcony: $500 \div 25 = 20$
Since we didn't round, our estimate matches our exact answer.

Step 4
See the calculations in Step 3.

Step 5
The number of seats in each row of the balcony is 20.

Step 6
The answer matches the estimate, as expected.
Check: Seats in balcony: $20 \times 25 = 500$
                      Seats on main floor: $30 \times 25 = 750$
                      Total seats: $500 + 750 = 1250$

39. *Step 1*
Find the total cost of all Safety and Exterior Options.

Step 2
Add the costs of the options to find the total cost.

Step 3
Estimate:
$300 + $200 + $100 + $70 + $400 + 200
  $= 1270

Step 4
$$\begin{array}{r} \overset{144}{\$\ 299} \\ 185 \\ 99 \\ 67 \\ 395 \\ +\ 209 \\ \hline \$1254 \end{array}$$

Step 5
The total cost is $1254.

Step 6
The answer is reasonably close to the estimate. Check by re-adding the costs.

40. *Step 1*
The total cost of all Interior Options must be found.

Step 2
Add the costs of the options to find the total cost.

Step 3
Estimate:
$300 + $50 + $300 + $1800 + $80 + 400
  $= 2930

Step 4
$$\begin{array}{r} \overset{43}{\$321} \\ 51 \\ 299 \\ 1799 \\ 82 \\ +\ 449 \\ \hline \$3001 \end{array}$$

Step 5
The total cost is $3001.

Step 6
The answer is reasonably close to the estimate. Check by re-adding the costs.

41. *Step 1*
An option package is offered. We must find how much can be saved if the customer buys the option package instead of paying for each option separately.

Step 2
First, add the costs of the options. Then subtract the cost of the option package from this total.

Step 3
Estimate:
$300 + $70 + $400 + $200 = 970
$970 - $800 = 170

Step 4
Add.
$$\begin{array}{rl} \overset{23}{\$299} & \textit{VIP Plus Security System} \\ 67 & \textit{Alloy wheel locks} \\ 395 & \textit{Paint protection} \\ +\ 209 & \textit{Lower body moulding} \\ \hline \$970 & \textit{Total cost} \end{array}$$

(continued)

Copyright © 2018 Pearson Education, Inc.

Subtract.

$$
\begin{array}{rl}
\$970 & \textit{Total cost} \\
-\ 785 & \textit{Option package} \\
\hline
\$185 & \textit{Amount saved}
\end{array}
$$

Step 5
Jill can save $185 by buying the option package.

Step 6
The answer is reasonably close to the estimate.
Check:

$$
\begin{array}{r}
\$785 \\
+\ 185 \\
\hline
\$970
\end{array}
$$

42. *Step 1*
An option package is offered. We must find how much can be saved if the customer buys the option package instead of paying for each option separately.

Step 2
First, add the costs of the options. Then subtract the cost of the option package from this total.

Step 3
Estimate:
$300 + $200 + $400 + $300 + $50 + $400
= $1650
$1650 − $1500 = $150

Step 4
Add.

$$
\begin{array}{rl}
\overset{43}{\$299} & \textit{VIP Plus Security System} \\
185 & \textit{Roof rack crossbars} \\
395 & \textit{Paint protection} \\
321 & \textit{Carpet floor mats} \\
51 & \textit{Cargo nets} \\
+\ 449 & \textit{XM Satellite Radio} \\
\hline
\$1700 & \textit{Total cost}
\end{array}
$$

Subtract.

$$
\begin{array}{rl}
\$1700 & \textit{Total cost} \\
-\ 1495 & \textit{Option package} \\
\hline
\$205 & \textit{Amount saved}
\end{array}
$$

Step 5
Samuel can save $205 by buying the option package.

Step 6
The answer is reasonably close to the estimate.
Check:

$$
\begin{array}{r}
\$1495 \\
+\ 205 \\
\hline
\$1700
\end{array}
$$

Chapter 1 Review Exercises

1. <u>6,573</u>; thousands: 6; ones: 573

2. <u>36,215</u>; thousands: 36; ones: 215

3. <u>105,724</u>; thousands: 105; ones: 724

4. <u>1,768,710,618</u>; billions: 1; millions: 768; thousands: 710; ones: 618

5. 728 is seven hundred twenty-eight.

6. 15,310 is fifteen thousand, three hundred ten.

7. 319,215 is three hundred nineteen thousand, two hundred fifteen.

8. 62,500,005 is sixty-two million, five hundred thousand, five.

9. Ten-thousand, eight is 10,008.

10. Two hundred million, four hundred fifty-five is 200,000,455.

11.
$$
\begin{array}{r}
\overset{1}{7}2 \\
+\ 38 \\
\hline
110
\end{array}
$$

12.
$$
\begin{array}{r}
\overset{1}{5}4 \\
+\ 67 \\
\hline
121
\end{array}
$$

13.
$$
\begin{array}{r}
\overset{1}{8}\overset{1}{0}7 \\
4606 \\
+\ \ \ 51 \\
\hline
5464
\end{array}
$$

14.
$$
\begin{array}{r}
8\ 2\overset{1}{1}5 \\
9 \\
+\ 7\ 433 \\
\hline
15,657
\end{array}
$$

15.
$$
\begin{array}{r}
2\ \overset{11}{1}30 \\
453 \\
8\ 107 \\
+\ \ \ 296 \\
\hline
10,986
\end{array}
$$

16.
$$
\begin{array}{r}
\overset{221}{5}684 \\
218 \\
2960 \\
+\ \ \ 983 \\
\hline
9845
\end{array}
$$

Copyright © 2018 Pearson Education, Inc.

17.
$$\overset{1\,1}{\underset{5}{\,}}\overset{3\,3}{7}32$$
11,069
37
1 595
+ 22,169
40,602

18.
$$\overset{1\ \ 3\,1}{3\ 4}51$$
12,286
43
1 291
+ 32,784
49,855

19.
$$\overset{5\,14}{6\ 4}$$ *Check:* $$\overset{1}{2}8$$
− 2 8   + 36
3 6   64

20.
$$\overset{3\,16}{4\ 6}$$ *Check:* $$\overset{1}{1}9$$
− 1 9   + 27
2 7   46

21.
$$\overset{2\,16\,15}{3\ 7\ 5}$$ *Check:* $$\overset{1\,1}{1}86$$
− 1 8 6   + 189
1 8 9   375

22.
$$\overset{4\,16\,13}{5\ 7\ 3}$$ *Check:* $$\overset{1\,1}{3}89$$
− 3 8 9   + 184
1 8 4   573

23.
$$\overset{6\,13\,10\,16}{7\ 4\ 1\ 6}$$ *Check:* $$\overset{1\,1\,1}{5}67$$
− 5 6 7   + 6849
6 8 4 9   7416

24.
$$\overset{4\,11\,10\,10}{5\ 2\ 1\ 0}$$ *Check:* $$\overset{1\,1\,1}{4}327$$
− 8 8 3   + 883
4 3 2 7   5210

25.
$$\overset{1\,11\,10\,10}{2\ 2\ 1\ 0}$$ *Check:* $$\overset{1\,1\,1}{1}986$$
− 1 9 8 6   + 224
2 2 4   2210

26.
$$\overset{\ \ \ \ 9}{\overset{8\ 16\,10\,14}{9\,9,7\ 0\ 4}}$$ *Check:* $$\overset{1\ 11}{2}5,866$$
− 73,8 3 8   + 73,838
25, 8 6 6   99,704

27.
7
× 7
49

28.
8
× 0
0

29. 8(4) = 32

30. 8(8) = 64

31. (5)(9) = 45

32. (6)(7) = 42

33. 7 · 8 = 56

34. 9 · 9 = 81

35. 5 × 4 × 2
(5 × 4) × 2
20 × 2 = 40

36. 9 × 1 × 5
(9 × 1) × 5
9 × 5 = 45

37. 4 × 4 × 3
(4 × 4) × 3
16 × 3 = 48

38. 2 × 2 × 2
(2 × 2) × 2
4 × 2 = 8

39. (6)(0)(8) = 0   Any number times 0 equals 0.

40. (7)(1)(6)
(7 · 1) · 6
7 · 6 = 42

41. 6 · 1 · 8
(6 · 1) · 8
6 · 8 = 48

42. 7 · 7 · 0 = 0   Any number times 0 equals 0.

43.
$$\overset{2}{2}8$$
× 3
84

44.
$$\overset{4}{4}6$$
× 8
368

45.
$$\overset{7}{5}8$$
× 9
522

46.
9 8
× 1
9 8

Copyright © 2018 Pearson Education, Inc.

47.
$$\begin{array}{r} \overset{2\,4}{6\,2\,5} \\ \times\ \ 8 \\ \hline 5000 \end{array}$$

48.
$$\begin{array}{r} \overset{5\,3}{3\,7\,4} \\ \times\ \ 8 \\ \hline 2992 \end{array}$$

49.
$$\begin{array}{r} \overset{1\,1\,3}{1\,3\,4\,9} \\ \times\ \ \ 4 \\ \hline 5396 \end{array}$$

50.
$$\begin{array}{r} \overset{3\,1}{9\,1\,6\,3} \\ \times\ \ \ 5 \\ \hline 45{,}815 \end{array}$$

51.
$$\begin{array}{r} \overset{1\,1}{7\,4\,5\,6} \\ \times\ \ \ 2 \\ \hline 14{,}912 \end{array}$$

52.
$$\begin{array}{r} \overset{6\,5}{2\,8\,8\,0} \\ \times\ \ \ 7 \\ \hline 20{,}160 \end{array}$$

53.
$$\begin{array}{r} \overset{1}{9}3,\overset{2}{1}0\,5 \\ \times\ \ \ \ 5 \\ \hline 465{,}525 \end{array}$$

54.
$$\begin{array}{r} \overset{1\,6}{2}1,\overset{5\,2}{8}7\,3 \\ \times\ \ \ \ 8 \\ \hline 174{,}984 \end{array}$$

55.
$$\begin{array}{r} 35 \\ \times\ 25 \\ \hline 175 \quad \leftarrow 5 \times 35 \\ 70\ \ \quad \leftarrow 2 \times 35 \\ \hline 875 \end{array}$$

56.
$$\begin{array}{r} 74 \\ \times\ 32 \\ \hline 148 \quad \leftarrow 2 \times 74 \\ 222\ \ \quad \leftarrow 3 \times 74 \\ \hline 2368 \end{array}$$

57.
$$\begin{array}{r} 98 \\ \times\ 12 \\ \hline 196 \quad \leftarrow 2 \times 98 \\ 98\ \ \quad \leftarrow 1 \times 98 \\ \hline 1176 \end{array}$$

58.
$$\begin{array}{r} 68 \\ \times\ 75 \\ \hline 340 \quad \leftarrow 5 \times 68 \\ 476\ \ \quad \leftarrow 7 \times 68 \\ \hline 5100 \end{array}$$

59.
$$\begin{array}{r} 472 \\ \times\ 33 \\ \hline 1416 \quad \leftarrow 3 \times 472 \\ 1416\ \ \quad \\ \hline 15{,}576 \end{array}$$

60.
$$\begin{array}{r} 392 \\ \times\ 77 \\ \hline 2744 \quad \leftarrow 7 \times 392 \\ 2744\ \ \quad \\ \hline 30{,}184 \end{array}$$

61. (use a calculator)
$$\begin{array}{r} 4051 \\ \times\ 219 \\ \hline 887{,}169 \end{array}$$

62. (use a calculator)
$$\begin{array}{r} 1527 \\ \times\ 328 \\ \hline 500{,}856 \end{array}$$

63.
$$\begin{array}{r} \$12 \quad \textit{cost per calculator} \\ \times\ 30 \quad \textit{calculators} \\ \hline \$360 \quad \textit{total cost} \end{array}$$

64.
$$\begin{array}{r} \$14 \quad \textit{cost of subscription} \\ \times\ 76 \quad \textit{subscribers} \\ \hline 84\ \ \quad \\ 98\ \ \ \quad \\ \hline \$1064 \quad \textit{total cost} \end{array}$$

65.
$$\begin{array}{r} 318 \quad \textit{sets} \\ \times\ \$64 \quad \textit{cost per set} \\ \hline 1272\ \ \quad \\ 1908\ \ \ \quad \\ \hline \$20{,}352 \quad \textit{total cost} \end{array}$$

66.
$$\begin{array}{r} 114 \quad \textit{ear plugs} \\ \times\ \$6 \quad \textit{cost per plug} \\ \hline \$684 \quad \textit{total cost} \end{array}$$

67.
$$\begin{array}{r} 280 \\ \times\ 50 \\ \hline \end{array} \qquad \begin{array}{r} 28 \\ \times\ 5 \\ \hline 140 \end{array} \qquad \begin{array}{r} 280 \\ \times\ 50 \\ \hline 14{,}000 \end{array} \textit{Attach } 00.$$

68.
$$\begin{array}{r} 340 \\ \times\ 70 \\ \hline \end{array} \qquad \begin{array}{r} 34 \\ \times\ 7 \\ \hline 238 \end{array} \qquad \begin{array}{r} 340 \\ \times\ 70 \\ \hline 23{,}800 \end{array} \textit{Attach } 00.$$

Copyright © 2018 Pearson Education, Inc.

69.

$$\begin{array}{r} 517 \\ \times\ 400 \\ \hline \end{array} \qquad \begin{array}{r} 517 \\ \times\ 4 \\ \hline 2068 \end{array} \qquad \begin{array}{r} 517 \\ \times\ 400 \\ \hline 206{,}800 \end{array} \quad \textit{Attach } 00.$$

70.

$$\begin{array}{r} 637 \\ \times\ 500 \\ \hline \end{array} \qquad \begin{array}{r} 637 \\ \times\ 5 \\ \hline 3185 \end{array} \qquad \begin{array}{r} 637 \\ \times\ 500 \\ \hline 318{,}500 \end{array} \quad \textit{Attach } 00.$$

71.

$$\begin{array}{r} 16{,}000 \\ \times\ 8000 \\ \hline \end{array} \qquad \begin{array}{r} 16 \\ \times\ 8 \\ \hline 128 \end{array} \qquad \begin{array}{r} 16{,}000 \\ \times\ 8000 \\ \hline 128{,}000{,}000 \end{array} \quad \textit{Attach } 000000.$$

72.

$$\begin{array}{r} 43{,}000 \\ \times\ 2100 \\ \hline \end{array} \qquad \begin{array}{r} 43 \\ \times\ 21 \\ \hline 903 \end{array} \qquad \begin{array}{r} 43{,}000 \\ \times\ 2100 \\ \hline 90{,}300{,}000 \end{array} \quad \textit{Attach } 00000.$$

73. $20 \div 4 = 5$

74. $35 \div 5 = 7$

75. $42 \div 7 = 6$

76. $18 \div 9 = 2$

77. $\dfrac{54}{9} = 6$

78. $\dfrac{36}{9} = 4$

79. $\dfrac{49}{7} = 7$

80. $\dfrac{0}{6} = 0$

81. $\dfrac{148}{0}$ is undefined.

82. $\dfrac{0}{23} = 0$

83. $\dfrac{64}{8} = 8$

84. $\dfrac{81}{9} = 9$

85.
$$4\overline{)328} \;=\; 82 \qquad \textit{Check:}\ \begin{array}{r} 82 \\ \times\ 4 \\ \hline 328 \end{array}$$

86.
$$3\overline{)29\,^24} \;=\; 98 \qquad \textit{Check:}\ \begin{array}{r} 98 \\ \times\ 3 \\ \hline 294 \end{array}$$

87.
$$6\overline{)26{,}\,^25\,^13\,^12} \;=\; 4422 \qquad \textit{Check:}\ \begin{array}{r} 211 \\ 4422 \\ \times\ 6 \\ \hline 26{,}532 \end{array}$$

88.
$$76\overline{)26{,}752} \;=\; 352$$
$$\begin{array}{r} 228 \\ \hline 395 \\ 380 \\ \hline 152 \\ 152 \\ \hline 0 \end{array}$$
$$\textit{Check:}\ \begin{array}{r} 352 \\ \times\ 76 \\ \hline 2112 \\ 2464 \\ \hline 26{,}752 \end{array}$$

89. $2704 \div 18$
$$18\overline{)2704} \;=\; 150 \ \mathbf{R}4$$
$$\begin{array}{r} 18 \\ \hline 90 \\ 90 \\ \hline 04 \\ 00 \\ \hline 4 \end{array}$$
$$\textit{Check:}\ \begin{array}{r} 150 \\ \times\ 18 \\ \hline 1200 \\ 150 \\ \hline 2700 \\ +\ \ 4 \\ \hline 2704 \end{array}$$

90. $15{,}525 \div 125$
$$125\overline{)15{,}525} \;=\; 124 \ \mathbf{R}25$$
$$\begin{array}{r} 125 \\ \hline 302 \\ 250 \\ \hline 525 \\ 500 \\ \hline 25 \end{array}$$
$$\textit{Check:}\ \begin{array}{r} 124 \\ \times\ 125 \\ \hline 620 \\ 248 \\ 124 \\ \hline 15{,}500 \\ +\ \ 25 \\ \hline 15{,}525 \end{array}$$

Copyright © 2018 Pearson Education, Inc.

91. 817 rounded to the nearest ten: 820

8$\underline{1}$7   Next digit is 5 or more. Tens place changes $(1 + 1 = 2)$. The digit to the right of the underlined place changes to zero.

92. 15,208 rounded to the nearest hundred: 15,200

15,$\underline{2}$08   Next digit is 4 or less. Hundreds place does not change. All digits to the right of the underlined place change to zero.

93. 20,643 rounded to the nearest thousand: 21,000

2$\underline{0}$,643   Next digit is 5 or more. Thousands place changes $(0 + 1 = 1)$. All digits to the right of the underlined place change to zero.

94. 67,485 rounded to the nearest ten-thousand: 70,000

$\underline{6}$7,485   Next digit is 5 or more. Ten-thousands place changes $(6 + 1 = 7)$. All digits to the right of the underlined place change to zero.

95. **To the nearest ten:** 348$\underline{7}$

Next digit is 5 or more. Tens place changes $(8 + 1 = 9)$. The digit to the right of the underlined place changes to zero. **3490**

To the nearest hundred: 3$\underline{4}$87

Next digit is 5 or more. Hundreds place changes $(4 + 1 = 5)$. All digits to the right of the underlined place are changed to zero. **3500**

To the nearest thousand: $\underline{3}$487

Next digit is 4 or less. Thousands place does not change. All digits to the right of the underlined place are changed to zero. **3000**

96. **To the nearest ten:** 20,0$\underline{6}$5

Next digit is 5 or more. Tens place changes $(6 + 1 = 7)$. The digit to the right of the underlined place changes to zero. **20,070**

To the nearest hundred: 20,$\underline{0}$65

Next digit is 5 or more. Hundreds place changes $(0 + 1 = 1)$. All digits to the right of the underlined place are changed to zero. **20,100**

To the nearest thousand: 2$\underline{0}$,065

Next digit is 4 or less. Thousands place does not change. All digits to the right of the underlined place are changed to zero. **20,000**

97. **To the nearest ten:** 98,2$\underline{0}$1

Next digit is 4 or less. Tens place does not change. The digit to the right of the underlined place changes to zero. **98,200**

To the nearest hundred: 98,$\underline{2}$01

Next digit is 4 or less. Hundreds place does not change. All digits to the right of the underlined place are changed to zero. **98,200**

To the nearest thousand: 9$\underline{8}$,201

Next digit is 4 or less. Thousands place does not change. All digits to the right of the underlined place are changed to zero. **98,000**

98. **To the nearest ten:** 352,11$\underline{8}$

Next digit is 5 or more. Tens place changes $(1 + 1 = 2)$. The digit to the right of the underlined place changes to zero. **352,120**

To the nearest hundred: 352,$\underline{1}$18

Next digit is 4 or less. Hundreds place does not change. All digits to the right of the underlined place are changed to zero. **352,100**

To the nearest thousand: 352,$\underline{1}$18

Next digit is 4 or less. Thousands place does not change. All digits to the right of the underlined place are changed to zero. **352,000**

99. From the table, $4^2 = 16$, so $\sqrt{16} = 4$.

100. From the table, $7^2 = 49$, so $\sqrt{49} = 7$.

101. From the table, $12^2 = 144$, so $\sqrt{144} = 12$.

102. From the table, $14^2 = 196$, so $\sqrt{196} = 14$.

103. 7^3: exponent is 3; base is 7.
$7^3 = 7 \cdot 7 \cdot 7 = 343$

104. 3^6: exponent is 6; base is 3.
$3^6 = 3 \cdot 3 \cdot 3 \cdot 3 \cdot 3 \cdot 3 = 729$

105. 5^3: exponent is 3; base is 5.
$5^3 = 5 \cdot 5 \cdot 5 = 125$

106. 4^5: exponent is 5; base is 4.
$4^5 = 4 \cdot 4 \cdot 4 \cdot 4 \cdot 4 = 1024$

107. $7^2 - 15$ *Exponent*
49 − 15 = 34   *Subtract.*

108. $6^2 - 10$ *Exponent*
36 − 10 = 26   *Subtract.*

109. $2 \cdot 3^2 \div 2$ *Exponent*
$2 \cdot 9 \div 2$ *Multiply.*
$18 \div 2 = 9$ *Divide.*

Copyright © 2018 Pearson Education, Inc.

110. $9 \div 1 \cdot 2 \cdot 2 \div (11 - 2)$ *Parentheses*
$9 \div 1 \cdot 2 \cdot 2 \div 9$ *Divide.*
$9 \cdot 2 \cdot 2 \div 9$ *Multiply.*
$18 \cdot 2 \div 9$ *Multiply.*
$36 \div 9 = 4$ *Divide.*

111. $\sqrt{9} + 2(3)$ *Square root*
$3 + 2 \cdot 3$ *Multiply.*
$3 + 6 = 9$ *Add.*

112. $6 \cdot \sqrt{16} - 6 \cdot \sqrt{9}$ *Square roots*
$6 \cdot 4 - 6 \cdot 3$ *Multiply.*
$24 - 18 = 6$ *Subtract.*

113. From the bar graph, 8 parents out of 100 nagged their children about washing hands after using the bathroom.

114. From the bar graph, 5 parents out of 100 nagged their children about taking shoes off when coming inside.

115. From the bar graph, the greatest number of parents, 25, nagged their children about keeping bedroom clean.

116. From the bar graph, the least number of parents, 3, nagged their children about hanging up wet bath towels.

117. *Estimate:* 40 million × 365

$$\begin{array}{r} 400 \\ \times\ 40 \\ \hline 16{,}000 \end{array}$$

40 million × 400 = 16,000 million or 16,000,000,000 payments (*Attach* 000 000.)

Exact: 40 million × 365

$$\begin{array}{r} 365 \\ \times\ 40 \\ \hline 14{,}600 \end{array}$$

40 million × 365 = 14,600 million or 14,600,000,000 payments (*Attach* 000 000.)

The bank processes 14,600 million payments in a year.

118. *Step 1*
Find the total revolutions.

Step 2
We know the revolutions per minute and the number of minutes.

Number of revolutions × minutes = total revolutions.

Step 3
An estimate is 1000 × 60 = 60,000 revolutions.

Step 4
1400 × 60 = 84,000 revolutions

Step 5
There were 84,000 revolutions.

Step 6
The answer is reasonably close to the estimate considering the rounding.

Check: 84,000 ÷ 60 = 1400

119. *Step 1*
Find the difference in the populations of California and Texas.

Step 2
Difference indicates subtraction.

Step 3
An estimate is 40,000,000 − 30,000,000 = 10,000,000 people.

Step 4
Exact:

$$\begin{array}{r} 38{,}802{,}500 \\ -\ 26{,}956{,}958 \\ \hline 11{,}845{,}542 \end{array}$$

Step 5
The difference in population is 11,845,542.

(*continued*)

Step 6
The answer is reasonably close to the estimate.
Check:
11,845,542 + 26,956,958 = 38,802,500

120. *Step 1*
Find the difference in the populations of Alaska and Wyoming.

Step 2
Difference indicates subtraction.

Step 3
An estimate is 700,000 − 600,000 = 100,000 people.

Step 4
Exact:

$$\begin{array}{r} 736{,}732 \\ -\ 584{,}153 \\ \hline 152{,}579 \end{array}$$

Step 5
The difference in population is 152,579.

Step 6
The answer is reasonably close to the estimate.
Check: 152,579 + 584,153 = 736,732

Copyright © 2018 Pearson Education, Inc.

121. *Step 1*
Find the total cost to replace her transmission.

Step 2
Add the cost of the transaxle, the labor (multiply the number of hours times the hourly rate), and the sales tax.

Step 3
Estimate:
Transaxle: round $2633 to $3000
Labor: $8 \times \$90 = \$720$
Tax: round $230 to $200
Total: $\$3000 + \$720 + \$200 = \$3920$

Step 4
Exact:
Transaxle: $2633
Labor: $8 \times \$90 = \$720$
Tax: $230
Total: $\$2633 + \$720 + \$230 = \$3583$

Step 5
The total cost is $3583.

Step 6
The answer is reasonably close to the estimate. Check by repeating Step 4.

122. *Step 1*
Find the total cost to rent a truck.

Step 2
Add the cost of the rental and the cost of the mileage (multiply the number of miles times the rate per mile).

Step 3
Estimate:
Rental: round $55 to $60
Mileage: round 89 to 90 and multiply by $2,
    $90 \times \$2 = \$180$
Total: $\$60 + \$180 = \$240$

Step 4
Rental: $55
Mileage: $89 \times \$2 = \$178$
Total: $\$55 + \$178 = \$233$

Step 5
The total cost is $233.

Step 6
The answer is reasonably close to the estimate. Check by repeating Step 4.

123. *Step 1*
Find the total cost to buy the ovens.

Step 2
Multiply the number of baking ovens times the cost of each baking oven and the number of warming ovens times the cost of each warming oven. Add to find the total cost.

Step 3
Estimate:
Baking ovens: $30 \times \$2000 = \$60,000$
Warming ovens: $30 \times \$900 = \$27,000$
Total: $\$60,000 + \$27,000 = \$87,000$

Step 4
Exact:
Baking ovens: $32 \times \$1538 = \$49,216$
Warming ovens: $28 \times \$887 = \$24,836$
Total: $\$49,216 + \$24,836 = \$74,052$

Step 5
The total cost is $74,052.

Step 6
The answer is reasonably close to the estimate. Check by repeating Step 4.

124. *Step 1*
Find the total monthly collections.

Step 2
We know the number of daily customers and the daily rate. We know the number of weekend-only customers and the rate.
Number of customers $\times$ daily rate
$+$ number of customers $\times$ weekend rate
$=$ total collections.

Step 3
Estimate:

$$(60 \times \$20) + (20 \times \$10) = \$1200 + \$200$$
$$= \$1400$$

Step 4
$$(62 \times \$24) + (21 \times \$11) = \$1488 + \$231$$
$$= \$1719$$

Step 5
The total monthly collections are $1719.

Step 6
The answer is reasonably close to the estimate. Check by repeating Step 4.

Copyright © 2018 Pearson Education, Inc.

125. *Step 1*
Find the difference in the amount spent on others and the amount spent on themselves.

Step 2
Difference indicates subtraction.

Step 3
An estimate is $600 - $100 = 500.

Step 4
Exact:
$$\begin{array}{r} \$620 \\ -\ \$107 \\ \hline \$513 \end{array}$$

Step 5
The difference in amount spent is $513.

Step 6
The answer is reasonably close to the estimate.
Check: $513 + $107 = $620

126. *Step 1*
Find the new account balance.

Step 2
We know the amount of the payments and the old balance.

Old balance − payment amounts = new balance.

Step 3
Estimate: $2000 - $500 - $400 = $1100

Step 4
$1924 - $520 - $385 = $1019

Step 5
She has $1019 in her bank account.

Step 6
The answer is reasonably close to the estimate.
Check: $1019 + $520 + $385 = $1924

127. *Step 1*
Find out how many pounds of pork are needed.

Step 2
We know the total number of cans and we must divide that total by 175 since each group of 175 cans requires 1 pound of pork. The number of groups × 1 pound = total pounds.

Step 3
Estimate: $9000 \div 200 = 45$ pounds

Step 4
$$\frac{8750}{175} = 50 \text{ pounds}$$

Step 5
50 pounds of pork are needed.

Step 6
The answer is reasonably close to the estimate.

Check:
$(175 \text{ cans per pound}) \cdot (50 \text{ pounds}) = 8750 \text{ cans}$

128. *Step 1*
Find how many hours it takes to produce all the plates.

Step 2
We know the total number of plates and we know how many are produced each hour.

$$\frac{total\ number}{number\ per\ hour} = total\ hours$$

Step 3
An estimate: $30,000 \div 1000 = 30$ hours

Step 4
$$\frac{32,538}{986} = 33 \text{ hours}$$

Step 5
It will take 33 hours.

Step 6
The answer is reasonably close to the estimate.
Check: $33 \cdot 986 = 32,538$

129. *Step 1*
Find the total number of acres fertilized.

Step 2
We know the total amount of fertilizer and how much each acre needs. Total pounds ÷ pounds needed per acre = total acres.

Step 3
Estimate: $30,000 \div 600 = 50$

Step 4
$$\frac{32,500}{625} = 52 \text{ acres}$$

Step 5
52 acres can be spread with 32,500 pounds of nitrogen sulfate.

Step 6
The answer is reasonably close to the estimate.
Check: $52 \times 625 = 32,500$

Copyright © 2018 Pearson Education, Inc.

130. *Step 1*
Find the number of homes that can be fenced.

Step 2
Divide the number of feet of fencing available by the number of feet needed for each home.

Step 3
Estimate: $6000 \div 200 = 30$ homes

Step 4
$$\frac{5760}{180} = 32 \text{ homes}$$

Step 5
32 homes can be fenced.

Step 6
The answer is reasonably close to the estimate.
Check:
$(180 \text{ feet per home}) \cdot (32 \text{ homes}) = 5760 \text{ feet}$

131. *Step 1*
Find the cost of the program to the company.

Step 2
Multiply the number of employees times the cost per employee.

Step 3
Estimate: $200,000 \times \$60 = \$12,000,000$

Step 4

$$
\begin{array}{r}
1635 \\
\times \quad 6 \\
\hline
9810
\end{array}
$$

$163,500 \times \$60 = \$9,810,000$ *Attach* 000.

Step 5
The cost to the company is $9,810,000.

Step 6
The answer is reasonably close to the estimate.
Check: $\$9,810,000 \div \$60 = 163,500$

132. *Step 1*
Find how many bottles of water are consumed.

Step 2
Multiply the people by the number of bottles per person.

Step 3
Estimate: $300 \times 200,000,000 = 60,000,000,000$

Step 4

$$
\begin{array}{r}
1857 \\
\times \quad 27 \\
\hline
12999 \\
3714 \\
\hline
50139
\end{array}
$$

$270 \times 185,700,000 = \$50,139,000,000$
Attach 000000.

Step 5
50,139,000,000 bottles of water are consumed.

Step 6
The answer is reasonably close to the estimate.
Check:
$50,139,000,000 \div 270 = 185,700,000$

Chapter 1 Mixed Review Exercises

1. $4(83)$
$$
\begin{array}{r}
\overset{1}{8}3 \\
\times \ 4 \\
\hline
332
\end{array}
$$

2. $7(64)$
$$
\begin{array}{r}
\overset{2}{6}4 \\
\times \ 7 \\
\hline
448
\end{array}
$$

3.
$$
\begin{array}{r}
\overset{2}{3}\overset{10}{\cancel{0}}\ 9 \\
- \ \ 5\ 6 \\
\hline
2\ 5\ 3
\end{array}
$$

4.
$$
\begin{array}{r}
\overset{7}{8}\overset{12}{\cancel{2}}\overset{15}{\cancel{3}}\overset{\ }{\cancel{5}} \\
- \ 2\ 4\ 7 \\
\hline
5\ 8\ 8
\end{array}
$$

5.
$$
\begin{array}{r}
\overset{1\ 1}{6}\ 6\ 2 \\
+ \ 3\ 7\ 9 \\
\hline
1\ 0\ 4\ 1
\end{array}
$$

6.
$$
\begin{array}{r}
\overset{1\ 1}{7}\ 8\ 9 \\
+ \ 8\ 7\ 2 \\
\hline
1\ 6\ 6\ 1
\end{array}
$$

7.
$$
\begin{array}{r}
\overset{\ \ 0\ \overset{13}{3}\ 10}{3}8,1\,4\,\cancel{0} \\
- \ \ \ 6\,0\,7\,8 \\
\hline
3\,2,0\,6\,2
\end{array}
$$
Check:
$$
\begin{array}{r}
\overset{1\,1}{3}2,0\,6\,2 \\
+ \ 6\,0\,7\,8 \\
\hline
3\,8,1\,4\,0
\end{array}
$$

8. $\dfrac{9}{0}$; undefined

Copyright © 2018 Pearson Education, Inc.

9. $\dfrac{7}{1} = 7$ $\ (7 \cdot 1 = 7)$

10. $27{,}600 \div 4 = 6900$

$$4\overline{\smash{)}27\,^3 6} \quad \text{Attach 00.}$$
$$\ \ 6\ 9$$

11. $18{,}480 \div 8$

$$8\overline{\smash{)}18{,}\,^2 480}$$
$$\ 2\ 3\ 10$$

12.
$$
\begin{array}{r}
8\,430 \\
\times\ \ \ \ 128 \\
\hline
67\,440 \leftarrow 8 \times 8430 \\
168\,60\ \ \leftarrow 2 \times 8430 \\
843\,0\ \ \ \ \ \\
\hline
1{,}079{,}040\ \ \ \
\end{array}
$$

13.
$$
\begin{array}{r}
\overset{14\ \ 1}{21{,}702} \\
\times\ \ \ \ \ 6 \\
\hline
130{,}212
\end{array}
$$

14. Use a calculator to show that
$$34\,\overline{\smash{)}3\ 6\ 7\ 2} = 108.$$

15. Use a calculator to show that
$$68\,\overline{\smash{)}1\ 4{,}\ 0\ 7\ 6} = 207.$$

16. 376,853 is three hundred seventy-six thousand, eight hundred fifty-three in words.

17. 408,610 is four hundred eight thousand, six hundred ten in words.

18. 8749 rounded to the nearest hundred: 8<u>7</u>49
Next digit is 4 or less. Hundreds place doesn't change. All digits to the right of the underlined place are changed to zero. **8700**

19. 400,503 rounded to the nearest thousand:
40<u>0</u>,503
Next digit is 5 or more. Thousands place changes $(0 + 1 = 1)$. All digits to the right of the underlined place are changed to zero. **401,000**

20. From the table, $8^2 = 64$, so $\sqrt{64} = 8$.

21. From the table, $9^2 = 81$, so $\sqrt{81} = 9$.

22.
$$
\begin{array}{r}
\$3\,0\,8\ \text{\textit{cost per pair}} \\
\times\ \ \ 18\ \text{\textit{pairs}} \\
\hline
2\,4\,6\,4\ \ \ \ \ \\
3\,0\,8\ \ \ \ \ \ \ \\
\hline
\$5\,5\,4\,4\ \text{\textit{total cost}}
\end{array}
$$

23.
$$
\begin{array}{r}
\$3\,7\,0\ \text{\textit{cost per dishwasher}} \\
\times\ \ \ 84\ \text{\textit{dishwashers}} \\
\hline
1\,4\,8\,0\ \ \ \ \ \ \ \\
29\,6\,0\ \ \ \ \ \ \ \ \\
\hline
\$31\,0\,8\,0\ \text{\textit{total cost}}
\end{array}
$$

24.
$$
\begin{array}{r}
2\,0\,8\ \text{\textit{baseball hats}} \\
\times\ \ \$1\,1\ \text{\textit{cost of hat}} \\
\hline
2\,0\,8\ \ \ \ \ \\
2\,0\,8\ \ \ \ \ \ \ \\
\hline
\$22\,8\,8\ \text{\textit{total cost}}
\end{array}
$$

25.
$$
\begin{array}{r}
6\,0\,7\ \text{\textit{boxes of avocados}} \\
\times\ \ \$2\,6\ \text{\textit{cost per box}} \\
\hline
3\,6\,4\,2\ \ \ \ \ \ \\
12\,1\,4\ \ \ \ \ \ \ \\
\hline
\$15{,}7\,8\,2\ \text{\textit{total cost}}
\end{array}
$$

26.
$$
\begin{array}{r}
\overset{1}{5}\,2\ \text{\textit{cards per deck}} \\
\times\ \ \ 9\ \text{\textit{decks}} \\
\hline
4\,6\,8\ \text{\textit{total cards}}
\end{array}
$$

There are 468 cards in nine decks.

27.
$$
\begin{array}{r}
1\,7\,0\,0\ \text{\textit{volunteers}} \\
\times\ \ \ 31\ \text{\textit{pounds per volunteer}} \\
\hline
52{,}7\,0\,0\ \text{\textit{pounds}}
\end{array}
$$

There were 52,700 pounds of fruits and vegetables picked by the volunteers.

28.
$$
\begin{array}{r}
\$380 \\
-\ \$100 \\
\hline
\$280
\end{array}
$$

A "push-type" mower costs $280.

29.
$$
\begin{array}{r}
\overset{7\ 14\ 4\ 10}{\$2\,1\,8{,}4\,5\,0}\ \text{\textit{Amount needed}} \\
-\ \$1\,0\,3{,}8\,1\,5\ \text{\textit{Amount raised}} \\
\hline
\$1\,1\,4{,}6\,3\,5\ \text{\textit{Amount left to be raised}}
\end{array}
$$

$114,635 more needs to be raised.

30. Multiply the number of rentals times the sum of the rental fee and the launch fee.

4-person	$6 \times (\$28 + \$2) = 6(\$30)$	$180
6-person	$15 \times (\$38 + \$2) = 15(\$40)$	600
10-person	$10 \times (\$70 + \$2) = 10(\$72)$	720
12-person	$3 \times (\$75 + \$2) = 3(\$77)$	231
16-person	$2 \times (\$85 + \$2) = 2(\$87)$	$+\ 174$
		$1905

Total receipts were $1905.

Copyright © 2018 Pearson Education, Inc.

Chapter 1 Test   **51**

31. Multiply the number of rentals times the sum of the rental fee and the launch fee.

4-person	$38 \times (\$28 + \$2) = 38(\$30)$	$1\,140
6-person	$73 \times (\$38 + \$2) = 73(\$40)$	$2\,920
10-person	$58 \times (\$70 + \$2) = 58(\$72)$	$4\,176
12-person	$34 \times (\$75 + \$2) = 34(\$77)$	$2\,618
16-person	$18 \times (\$85 + \$2) = 18(\$87)$	$+\,1\,566

$12,420

Total receipts were $12,420.

32.
$$2717 \text{ Burj Khalifa}$$
$$-\,1250 \text{ Empire State Building}$$
$$1467 \text{ difference}$$

The Burj Dubai is 1467 feet taller than the Empire State Building.

33.
$$1776 \text{ 1 WTC in New York}$$
$$-\,1483 \text{ Petronas Towers}$$
$$293 \text{ difference}$$

The 1 WTC in New York will be 293 feet taller than the Petronas Towers.

34. (a)
2717 Burj Khalifa Dubai
1776 1 WTC in New York
1667 Taipel 101
1614 World Financial Center
1483 Petronas Towers
1451 Willis Tower
+ 1250 Empire State Building
11,958 combined height in feet

The combined height of these seven buildings is 11,958 feet.

(b) Two miles is the same as $2 \times 5280 = 10,560$ feet, so the combined height is greater than two miles.

11,958 combined height
− 10,560 feet in two miles
1398 difference

The combined height of the seven buildings is 1398 feet more than a two miles.

35. One yard is the same as three feet, so 100 yards is the same as 300 feet. The Willis Tower is 1451 feet tall, so divide 1451 by 300.

$$300\overline{)1451}$$
quotient 4, 1200, 251

The height of the Willis Tower is equivalent to more than the length of 4 football fields (a little less than 5).

Chapter 1 Test

1. 9205 is nine thousand, two hundred five.

2. 25,065 is twenty-five thousand, sixty-five.

3. Four hundred twenty-six thousand, five is 426,005.

4.
853
66
4022
+ 3589
8530

5.
17,063
7
12
1 505
93,710
+ 333
112,630

6.
9009
− 7964
1045

7.
9075
− 2869
6206

8. $7 \times 6 \times 4 = (7 \times 6) \times 4 = 42 \times 4 = 168$

9. $57 \cdot 3000$
57
× 3
171

$57 \cdot 3000 = 171,000$ Attach 000.

Copyright © 2018 Pearson Education, Inc.

10. 85(19)

```
      85        Check:        8 5
  ×   19              19 ) 1 6 1 5
     765                    1 5 2
      85                      9 5
    1615                      9 5
                                0
```

11.

```
    7381        Check:          7 3 8 1
  ×  603              603 ) 4, 4 5 0, 7 4 3
   22143                    4 2 2 1
  442860                    2 2 9 7
 4,450,743                  1 8 0 9
                            4 8 8 4
                            4 8 2 4
                              6 0 3
                              6 0 3
                                  0
```

12.

```
        7 0 4 7     Check:   7047
  16 ) 1 1 2, 7 5 2        ×   16
      1 1 2               42282
      0 7 5               7047
        6 4              112,752
        1 1 2
        1 1 2
            0
```

13. $\dfrac{835}{0}$ is undefined

14. $19{,}241 \div 42$

```
          4 5 8 R5
  42 ) 1 9, 2 4 1
      1 6 8
        2 4 4
        2 1 0
          3 4 1
          3 3 6
              5
```

15.

```
            1 6 0
  280 ) 4 4, 8 0 0
        2 8 0
        1 6 8 0
        1 6 8 0
            0 0
```

16. 6347 rounded to the nearest ten: 63<u>4</u>7

Next digit is 5 or more. Tens place changes $(4+1=5)$. The digit to the right of the underlined place changes to zero. **6350**

17. 76,489 rounded to the nearest thousand: 7<u>6</u>,489

Next digit is 4 or less. Thousands place does not change. All digits to the right of the underlined place change to zero. **76,000**

18.
$5^2 + 8(2)$ *Exponent*
$25 + 8(2)$ *Multiply.*
$25 + 16 = 41$ *Add.*

19.
$7 \cdot \sqrt{64} - 14 \cdot 2$ *Square root*
$7 \cdot 8 - 14 \cdot 2$ *Multiply.*
$56 - 28 = 28$ *Subtract.*

20. Estimate:
$\$500 + \$500 + \$500 + \$400 - \$800 = \$1100$
Exact: Add the rent collected.

```
       1  11
  $    485
       500
       515
  +    425
     $1925
```

Subtract expenses.

```
        812
  $1 9̶ 2̶ 5
  −  7 8 5
  $1 1 4 0   amount left
```

He has $1140 left.

21. Estimate: $90{,}000 \div 400 = 225$ acres

Exact: Divide the total number of gallons by the number of gallons produced from one acre.

```
           2 3 1
  374 ) 8 6, 3 9 4
        7 4 8
        1 1 5 9
        1 1 2 2
            3 7 4
            3 7 4
                0
```

It would take 231 acres.

22. Estimate:
$\$2000 - \$500 - \$200 - \$200 = \$1100$
Exact: $\$1906 - \$528 - \$195 - \$235 = \$948$

Her new balance is $948.

Copyright © 2018 Pearson Education, Inc.

23. If we multiply the number of identifications each minute, 48, by the number of minutes in an hour, 60, and then multiply that number by the number of hours, 4, we'll get the total number of identifications in 4 hours.

    Estimate: $(50 \times 60 \times 4) + (40 \times 60 \times 3)$
    $= (3000 \times 4) + (2400 \times 3)$
    $= 12{,}000 + 7200 = 19{,}200$ chicks

    Exact: $(48 \times 60 \times 4) + (36 \times 60 \times 3)$
    $= (2880 \times 4) + (2160 \times 3)$
    $= 11{,}520 + 6480 = 18{,}000$ chicks

    The total number of baby chicks identified is 18,000.

24. (1) Locate the place to which you are rounding and underline it.

    (2) Look only at the next digit to the right. If this digit is a 4 or less, do not change the underlined digit. If the digit is a 5 or more, increase the underlined digit by 1.

    (3) Change all digits to the right of the underlined place to zeros.

    Each person's example will vary, but the following two examples illustrate the two general cases.

    (A) 1<u>2</u>4,999 rounds to 120,000

    (B) 1<u>2</u>5,000 rounds to 130,000

25. (1) Read the problem carefully.

    (2) Work out a plan.

    (3) Estimate a reasonable answer.

    (4) Solve the problem.

    (5) State the answer.

    (6) Check your work.

Copyright © 2018 Pearson Education, Inc.

CHAPTER 2   MULTIPLYING AND DIVIDING FRACTIONS

2.1 Basics of Fractions

2.1 Margin Exercises

1. The figure has 4 equal parts.
    Three parts are shaded:
    One part is unshaded: $\frac{1}{4}$

2. An area equal to 8 of the $\frac{1}{7}$ parts is shaded.

    $$\frac{8}{7}$$

3. (a) $\frac{2}{3}$ ← Numerator
           ← Denominator

    (b) $\frac{1}{4}$ ← Numerator
           ← Denominator

    (c) $\frac{8}{5}$ ← Numerator
           ← Denominator

    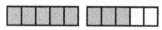

    (d) $\frac{5}{2}$ ← Numerator
           ← Denominator

    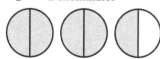

4. (a) Proper fractions: numerator *smaller* than denominator.

    $$\frac{2}{3}, \frac{3}{4}, \frac{1}{3}$$

    (b) Improper fractions: numerator *greater than or equal to* denominator.

    $$\frac{4}{3}, \frac{8}{8}, \frac{3}{1}$$

2.1 Section Exercises

1. $\frac{4}{5}$ ← Numerator
         ← Denominator

3. $\frac{9}{8}$ ← Numerator
         ← Denominator

5. The fraction $\frac{3}{8}$ represents $\underline{3}$ of the $\underline{8}$ equal parts into which a whole is divided.

7. The fraction $\frac{5}{24}$ represents $\underline{5}$ of the $\underline{24}$ equal parts into which a whole is divided.

9. The figure has 4 equal parts.
    Three parts are shaded: $\frac{3}{4}$
    One part is unshaded: $\frac{1}{4}$

11. The figure has 3 equal parts.
    One part is shaded: $\frac{1}{3}$
    Two parts are unshaded: $\frac{2}{3}$

13. Each of the two figures is divided into 5 parts and 7 are shaded: $\frac{7}{5}$
    Three are unshaded: $\frac{3}{5}$

15. Five of the 6 bills have a lifespan of 2 years or greater: $\frac{5}{6}$
    Four of the 6 bills have a lifespan of 4 years or less: $\frac{4}{6}$
    Two of the 6 bills have a lifespan of 9 years: $\frac{2}{6}$

17. There are 25 students, and 8 are hearing impaired.

    $\frac{8}{25}$ ← hearing impaired students (numerator)
         ← total students (denominator)

19. There are 520 rooms. 217 are for guests with pets, and $520 - 217 = 303$ are for guests without pets.

    Fraction for guests without pets: $\frac{303}{520}$

21. Proper fractions: numerator *less* than denominator.

    $$\frac{1}{3}, \frac{5}{8}, \frac{7}{16}$$

    Improper fractions: numerator *greater than or equal to* denominator.

    $$\frac{8}{5}, \frac{6}{6}, \frac{12}{2}$$

23. Proper fractions: numerator *smaller* than denominator.

    $$\frac{3}{4}, \frac{9}{11}, \frac{7}{15}$$

    Improper fractions: numerator *greater than or equal to* denominator.

    $$\frac{3}{2}, \frac{5}{5}, \frac{19}{18}$$

Copyright © 2018 Pearson Education, Inc.

25. **(a)** Three pieces are missing, so $\frac{3}{8}$ of the pie was eaten.

27. Answers will vary. One possibility is

$\frac{3}{4}$ ← Numerator
  ← Denominator

The denominator shows the number of equal parts in the whole and the numerator shows how many of the parts are being considered.

2.2 Mixed Numbers

2.2 Margin Exercises

1. **(a)** The figure shows 1 whole object with 3 equal parts, all shaded, and a second whole with 2 parts shaded, so 5 parts are shaded in all.

$$1\frac{2}{3} = \frac{5}{3}$$

(b) Since each of these diagrams is divided into $\underline{4}$ pieces, the denominator will be $\underline{4}$. The number of pieces shaded is $\underline{9}$.

$$2\frac{1}{4} = \frac{9}{4}$$

2. **(a)** $6\frac{1}{2}$ $6 \cdot \underline{2} = \underline{12}$ Multiply 6 and 2.

$12 + 1 = 13$ Add 1.

$$6\frac{1}{2} = \frac{13}{2}$$

(b) $7\frac{3}{4}$ $7 \cdot 4 = 28$ Multiply 7 and 4.

$28 + 3 = 31$ Add 3.

$$7\frac{3}{4} = \frac{31}{4}$$

(c) $4\frac{7}{8}$ $4 \cdot 8 = 32$ Multiply 4 and 8.

$32 + 7 = 39$ Add 7.

$$4\frac{7}{8} = \frac{39}{8}$$

(d) $8\frac{5}{6}$ $8 \cdot 6 = 48$ Multiply 8 and 6.

$48 + 5 = 53$ Add 5.

$$8\frac{5}{6} = \frac{53}{6}$$

3. **(a)** $\frac{6}{5}$ Divide $\underline{6}$ by $\underline{5}$.

$$\begin{array}{r} 1 \leftarrow \text{Whole number part} \\ 5\overline{)6} \\ \underline{5} \\ 1 \leftarrow \text{Remainder} \end{array}$$

(b) Five pieces are remaining, so $\frac{5}{8}$ of the pie remains.

$$\frac{6}{5} = 1\frac{1}{\underline{5}}$$

(b) $\frac{9}{4}$ Divide 9 by 4.

$$\begin{array}{r} 2 \leftarrow \text{Whole number part} \\ 4\overline{)9} \\ \underline{8} \\ 1 \leftarrow \text{Remainder} \end{array}$$

$$\frac{9}{4} = 2\frac{1}{4}$$

(c) $\frac{35}{5}$ Divide 35 by 5.

$$\begin{array}{r} 7 \leftarrow \text{Whole number part} \\ 5\overline{)3\,5} \\ \underline{3\,5} \\ 0 \leftarrow \text{Remainder} \end{array}$$

$$\frac{35}{5} = 7$$

(d) $\frac{78}{7}$ Divide 78 by 7.

$$\begin{array}{r} 1\,1 \leftarrow \text{Whole number part} \\ 7\overline{)7\,8} \\ \underline{7} \\ 8 \\ \underline{7} \\ 1 \leftarrow \text{Remainder} \end{array}$$

$$\frac{78}{7} = 11\frac{1}{7}$$

2.2 Section Exercises

1. $\frac{12}{12}$ is an improper fraction since the numerator is *greater than or equal to* the denominator. The statement is *true*.

3. $7\frac{2}{5}$ $7 \cdot 5 = 35$ Multiply 7 and 5.

$35 + 2 = 37$ Add 2.

$$7\frac{2}{5} = \frac{37}{5}$$

The mixed number $7\frac{2}{5}$ can be changed to the improper fraction $\frac{37}{5}$, not $\frac{14}{5}$. The statement is *false*.

Copyright © 2018 Pearson Education, Inc.

5. $6\dfrac{1}{2}$ $6 \cdot 2 = 12$ Multiply 6 and 2.

$12 + 1 = 13$ Add 1.

$6\dfrac{1}{2} = \dfrac{13}{2}$

The mixed number $6\dfrac{1}{2}$ can be changed to the improper fraction $\dfrac{13}{2}$, not $\dfrac{12}{2}$. The statement is *false*.

7. $1\dfrac{1}{4}$ $1 \cdot 4 = 4$ Multiply 1 and 4.

$4 + 1 = 5$ Add 1.

$1\dfrac{1}{4} = \dfrac{5}{4}$

9. $4\dfrac{3}{5}$ $4 \cdot 5 = 20$ Multiply 4 and 5.

$20 + 3 = 23$ Add 3.

$4\dfrac{3}{5} = \dfrac{23}{5}$

11. $8\dfrac{1}{2}$ $8 \cdot 2 = 16$ Multiply 8 and 2.

$16 + 1 = 17$ Add 1.

$8\dfrac{1}{2} = \dfrac{17}{2}$

13. $10\dfrac{1}{8}$ $10 \cdot 8 = 80$ Multiply 10 and 8.

$80 + 1 = 81$ Add 1.

$10\dfrac{1}{8} = \dfrac{81}{8}$

15. $10\dfrac{3}{4}$ $10 \cdot 4 = 40$ Multiply 10 and 4.

$40 + 3 = 43$ Add 3.

$10\dfrac{3}{4} = \dfrac{43}{4}$

17. Write $5\dfrac{4}{5}$ as an improper fraction.

$5 \cdot 5 = 25$ Multiply 5 and 5.

$25 + 4 = 29$ Add 4. The numerator is 29.

$5\dfrac{4}{5} = \dfrac{29}{5}$ Use the same denominator.

19. $8\dfrac{3}{5}$ $8 \cdot 5 = 40$ Multiply 8 and 5.

$40 + 3 = 43$ Add 3.

$8\dfrac{3}{5} = \dfrac{43}{5}$

21. $4\dfrac{10}{11}$ $4 \cdot 11 = 44$ Multiply 4 and 11.

$44 + 10 = 54$ Add 10.

$4\dfrac{10}{11} = \dfrac{54}{11}$

23. $32\dfrac{3}{4}$ $32 \cdot 4 = 128$ Multiply 32 and 4.

$128 + 3 = 131$ Add 3.

$32\dfrac{3}{4} = \dfrac{131}{4}$

25. $18\dfrac{5}{12}$ $18 \cdot 12 = 216$ Multiply 18 and 12.

$216 + 5 = 221$ Add 5.

$18\dfrac{5}{12} = \dfrac{221}{12}$

27. $17\dfrac{14}{15}$ $17 \cdot 15 = 255$ Multiply 17 and 15.

$255 + 14 = 269$ Add 14.

$17\dfrac{14}{15} = \dfrac{269}{15}$

29. $7\dfrac{19}{24}$ $7 \cdot 24 = 168$ Multiply 7 and 24.

$168 + 19 = 187$ Add 19.

$7\dfrac{19}{24} = \dfrac{187}{24}$

31. The improper fraction $\dfrac{4}{3}$ can be changed to the mixed number $1\dfrac{1}{3}$, not $1\dfrac{1}{4}$. The statement is *false*.

33. The statement "Some improper fractions can be written as a whole number with no fraction part" is *true*. For example, $\dfrac{6}{2} = 3$.

35. $\dfrac{4}{3}$

$$3\overline{)4} \quad \begin{array}{l} 1 \leftarrow \text{Whole number part} \\ \underline{3} \\ 1 \leftarrow \text{Remainder} \end{array}$$

$\dfrac{4}{3} = 1\dfrac{1}{3}$

37. $\dfrac{9}{4}$

$$4\overline{)9} \quad \begin{array}{l} 2 \leftarrow \text{Whole number part} \\ \underline{8} \\ 1 \leftarrow \text{Remainder} \end{array}$$

$\dfrac{9}{4} = 2\dfrac{1}{4}$

Copyright © 2018 Pearson Education, Inc.

39. $\dfrac{54}{6}$

$$6\overline{)54}\quad\leftarrow\text{Whole number part: }9$$

$$\begin{array}{r}9 \leftarrow \text{Whole number part}\\ 6\,\overline{)\,5\,4}\\ 5\,4\\ \hline 0 \leftarrow \text{Remainder}\end{array}$$

$$\dfrac{54}{6}=9$$

41. $\dfrac{38}{5}$

$$\begin{array}{r}7 \leftarrow \text{Whole number part}\\ 5\,\overline{)\,3\,8}\\ 3\,5\\ \hline 3 \leftarrow \text{Remainder}\end{array}$$

$$\dfrac{38}{5}=7\dfrac{3}{5}$$

43. $\dfrac{63}{4}$

Divide 63 by 4.

$$\begin{array}{r}1\,5 \leftarrow \text{Whole number part}\\ 4\,\overline{)\,6\,3}\\ 4\\ \hline 2\,3\\ 2\,0\\ \hline 3 \leftarrow \text{Remainder}\end{array}$$

The quotient 15 is the whole number part of the mixed number. The remainder 3 is the numerator of the fraction, and the denominator remains as 4.

$$\dfrac{63}{4}=15\dfrac{3}{4}$$

45. $\dfrac{47}{9}$

$$\begin{array}{r}5 \leftarrow \text{Whole number part}\\ 9\,\overline{)\,4\,7}\\ 4\,5\\ \hline 2 \leftarrow \text{Remainder}\end{array}$$

$$\dfrac{47}{9}=5\dfrac{2}{9}$$

47. $\dfrac{65}{8}$

$$\begin{array}{r}8 \leftarrow \text{Whole number part}\\ 8\,\overline{)\,6\,5}\\ 6\,4\\ \hline 1 \leftarrow \text{Remainder}\end{array}$$

$$\dfrac{65}{8}=8\dfrac{1}{8}$$

49. $\dfrac{84}{5}$

$$\begin{array}{r}1\,6 \leftarrow \text{Whole number part}\\ 5\,\overline{)\,8\,4}\\ 5\\ \hline 3\,4\\ 3\,0\\ \hline 4 \leftarrow \text{Remainder}\end{array}$$

$$\dfrac{84}{5}=16\dfrac{4}{5}$$

51. $\dfrac{112}{4}$

$$\begin{array}{r}2\,8 \leftarrow \text{Whole number part}\\ 4\,\overline{)\,1\,1\,2}\\ 8\\ \hline 3\,2\\ 3\,2\\ \hline 0 \leftarrow \text{Remainder}\end{array}$$

$$\dfrac{112}{4}=28$$

53. $\dfrac{183}{7}$

$$\begin{array}{r}2\,6 \leftarrow \text{Whole number part}\\ 7\,\overline{)\,1\,8\,3}\\ 1\,4\\ \hline 4\,3\\ 4\,2\\ \hline 1 \leftarrow \text{Remainder}\end{array}$$

$$\dfrac{183}{7}=26\dfrac{1}{7}$$

55. Multiply the denominator by the whole number and add the numerator. The result becomes the new numerator, which is placed over the original denominator.

$$2\dfrac{1}{2}\qquad (2\cdot 2)+1=5\qquad \dfrac{5}{2}$$

57. $250\dfrac{1}{2}\quad 250\cdot 2=500$

$$500+1=501$$

$$250\dfrac{1}{2}=\dfrac{501}{2}$$

59. $333\dfrac{1}{3}\quad 333\cdot 3=999$

$$999+1=1000$$

$$333\dfrac{1}{3}=\dfrac{1000}{3}$$

Copyright © 2018 Pearson Education, Inc.

61. Write $522\frac{3}{8}$ as an improper fraction.

$$522 \cdot 8 = 4176$$

$$4176 + 3 = 4179 \qquad \text{Add 3. The numerator is } 4179.$$

$$522\frac{3}{8} = \frac{4179}{8} \qquad \text{Use the same denominator.}$$

63. $\dfrac{617}{4}$

$$
\begin{array}{r}
1\ 5\ 4 \quad \leftarrow \text{Whole number part} \\
4\overline{)6\ 1\ 7} \\
\underline{4} \\
2\ 1 \\
\underline{2\ 0} \\
1\ 7 \\
\underline{1\ 6} \\
1 \quad \leftarrow \text{Remainder}
\end{array}
$$

$$\frac{617}{4} = 154\frac{1}{4}$$

65. The commands used will vary. The following is from a TI-83 Plus:

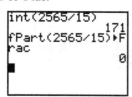

$$\frac{2565}{15} = 171$$

67. The commands used will vary. The following is from a TI-83 Plus:

```
int(3917/32)
           122
fPart(3917/32)▶F
rac
         13/32
```

$$\frac{3917}{32} = 122\frac{13}{32}$$

Note: You can use the following procedure on any calculator. Divide 3917 by 32 to get 122.40625. Subtract 122. Multiply by 32 to get 13. The mixed number is $122\frac{13}{32}$.

69. The following fractions are proper fractions.

$$\frac{2}{3}, \frac{4}{5}, \frac{3}{4}, \frac{7}{10}$$

70. (a) The proper fractions in Exercise 69 are the ones where the <u>numerator</u> is less than the <u>denominator</u>.

(b) $\dfrac{2}{3}$;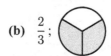

$\dfrac{4}{5}$;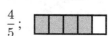

$\dfrac{3}{4}$;

$\dfrac{7}{10}$;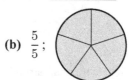

(c) The proper fractions in Exercise 69 are all <u>less</u> than 1.

71. The following fractions are improper fractions.

$$\frac{5}{5}, \frac{10}{3}, \frac{6}{5}$$

72. (a) The improper fractions in Exercise 71 are the ones where the <u>numerator</u> is equal to or greater than the <u>denominator</u>.

(b) $\dfrac{5}{5}$;

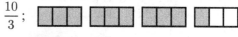

$\dfrac{10}{3}$;

$\dfrac{6}{5}$;

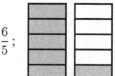

(c) The improper fractions in Exercise 71 are all equal to or <u>greater</u> than 1.

73. The following fractions can be written as whole or mixed numbers.

$$\frac{5}{3}$$

$$
\begin{array}{r}
1 \quad \leftarrow \text{Whole number part} \\
3\overline{)5} \\
\underline{3} \\
2 \quad \leftarrow \text{Remainder}
\end{array}
$$

$$\frac{5}{3} = 1\frac{2}{3}$$

(continued)

Copyright © 2018 Pearson Education, Inc.

$$\frac{7}{7}$$

$$\begin{array}{r} 1 \\ 7\overline{\smash{\big)}\,7} \\ \underline{7} \\ 0 \end{array}$$ ← Whole number part

← Remainder

$$\frac{7}{7} = 1$$

$$\frac{11}{6}$$

$$\begin{array}{r} 1 \\ 6\overline{\smash{\big)}\,11} \\ \underline{6} \\ 5 \end{array}$$ ← Whole number part

← Remainder

$$\frac{11}{6} = 1\frac{5}{6}$$

74. **(a)** The fractions that can be written as whole or mixed numbers in Exercise 73 are <u>improper</u> fractions, and their value is always <u>greater than or equal to</u> 1.

(b) $\frac{5}{3} = 1\frac{2}{3}$;

$\frac{7}{7} = 1$;

$\frac{11}{6} = 1\frac{5}{6}$;

2.3   Factors

2.3   Margin Exercises

1. **(a)** Factorizations of 18:
$1 \cdot 18 = 18$ $2 \cdot 9 = 18$ $3 \cdot 6 = 18$
The factors of 18 are 1, 2, <u>3</u>, 6, <u>9</u>, and 18.

(b) Factorizations of 16:
$1 \cdot 16 = 16$ $2 \cdot 8 = 16$ $4 \cdot 4 = 16$
The factors of 16 are 1, 2, 4, 8, and 16.

(c) Factorizations of 80:
$1 \cdot 80 = 80$ $2 \cdot 40 = 80$ $4 \cdot 20 = 80$
$5 \cdot 16 = 80$ $8 \cdot 10 = 80$
The factors of 80 are 1, 2, 4, 5, 8, 10, 16, 20, 40, and 80.

2. 4, 7, 9, 13, 17, 19, 29, 33
7, 13, 17, 19, and 29 are prime because they are divisible only by themselves and 1.

3. 2, 4, 5, 6, 8, 10, 11, 13, 19, 21, 27, 28, 33, 36, 42
2, 5, 11, 13, and 19 each have no factor other than themselves or 1.
4, 6, 8, 10, 28, 36, and 42 each have a factor of 2.
21, 27, and 33 have a factor of 3.
So 4, 6, 8, 10, 21, 27, 28, 33, 36, and 42 are composite.

4. **(a)** $8 \div 2 = 4$
$4 \div 2 = 2$ *prime*
$8 = 2 \cdot \underline{2} \cdot \underline{2}$

(b) $28 \div 2 = 14$
$14 \div 2 = 7$ *prime*
$28 = 2 \cdot \underline{2} \cdot \underline{7}$

(c) $18 \div 2 = 9$
$9 \div 3 = 3$ *prime*
$18 = 2 \cdot 3 \cdot 3$

(d) $40 \div 2 = 20$
$20 \div 2 = 10$
$10 \div 2 = 5$ *prime*
$40 = 2 \cdot 2 \cdot 2 \cdot 5$

5. **(a)** This division is done from the "bottom-up."

$$\begin{array}{r} 1 \\ 3\overline{\smash{\big)}\,3} \end{array}$$ *Quotient is 1.*   Divide 3 by 3.
$$3\overline{\smash{\big)}\,9}$$ Divide 9 by 3.
$$2\overline{\smash{\big)}\,18}$$ Divide 18 by 2.
$$2\overline{\smash{\big)}\,36}$$ Divide 36 by 2.

This division is done from the "top-down."

$$\begin{array}{r} 18 \\ 2\overline{\smash{\big)}\,36} \end{array}$$ Divide 36 by 2.
$$\begin{array}{r} 9 \\ 2\overline{\smash{\big)}\,18} \end{array}$$ Divide 18 by 2.
$$\begin{array}{r} 3 \\ 3\overline{\smash{\big)}\,9} \end{array}$$ Divide 9 by 3.
$$\begin{array}{r} 1 \\ 3\overline{\smash{\big)}\,3} \end{array}$$ Divide 3 by 3.
Quotient is 1.

Either method is correct and yields the prime factorization as follows:

$$36 = 2 \cdot 2 \cdot 3 \cdot \underline{3} = 2^2 \cdot \underline{3^2}$$

Copyright © 2018 Pearson Education, Inc.

(b)

$2\overline{)54}$ Divide 54 by 2.

$3\overline{)27}$ Divide 27 by 3.

$3\overline{)9}$ Divide 9 by 3.

$3\overline{)3}$ Divide 3 by 3.

Quotient is 1.

$54 = 2 \cdot 3 \cdot 3 \cdot 3 = 2 \cdot 3^3$

(c)

$2\overline{)60}$ Divide 60 by 2.

$2\overline{)30}$ Divide 30 by 2.

$3\overline{)15}$ Divide 15 by 3.

$5\overline{)5}$ Divide 5 by 5.

Quotient is 1.

$60 = 2 \cdot 2 \cdot 3 \cdot 5 = 2^2 \cdot 3 \cdot 5$

(d)

$3\overline{)81}$ Divide 81 by 3.

$3\overline{)27}$ Divide 27 by 3.

$3\overline{)9}$ Divide 9 by 3.

$3\overline{)3}$ Divide 3 by 3.

Quotient is 1.

$81 = 3 \cdot 3 \cdot 3 \cdot 3 = 3^4$

6.   (a)

$2\overline{)48}$ Divide 48 by 2.

$2\overline{)24}$ Divide 24 by 2.

$2\overline{)12}$ Divide 12 by 2.

$2\overline{)6}$ Divide 6 by 2.

$3\overline{)3}$ Divide 3 by 3.

Quotient is 1.

$48 = 2 \cdot 2 \cdot 2 \cdot 2 \cdot 3 = 2^4 \cdot 3$

(b)

$2\overline{)44}$ Divide 44 by 2.

$2\overline{)22}$ Divide 22 by 2.

$11\overline{)11}$ Divide 11 by 11.

Quotient is 1.

$44 = 2 \cdot 2 \cdot 11 = 2^2 \cdot 11$

(c)

$2\overline{)90}$ Divide 90 by 2.

$3\overline{)45}$ Divide 45 by 3.

$3\overline{)15}$ Divide 15 by 3.

$5\overline{)5}$ Divide 5 by 5.

Quotient is 1.

$90 = 2 \cdot 3 \cdot 3 \cdot 5 = 2 \cdot 3^2 \cdot 5$

(d)

$2\overline{)120}$ Divide 120 by 2.

$2\overline{)60}$ Divide 60 by 2.

$2\overline{)30}$ Divide 30 by 2.

$3\overline{)15}$ Divide 15 by 3.

$5\overline{)5}$ Divide 5 by 5.

Quotient is 1.

$120 = 2 \cdot 2 \cdot 2 \cdot 3 \cdot 5 = 2^3 \cdot 3 \cdot 5$

(e)

$2\overline{)180}$ Divide 180 by 2.

$2\overline{)90}$ Divide 90 by 2.

$3\overline{)45}$ Divide 45 by 3.

$3\overline{)15}$ Divide 15 by 3.

$5\overline{)5}$ Divide 5 by 5.

Quotient is 1.

$180 = 2 \cdot 2 \cdot 3 \cdot 3 \cdot 5 = 2^2 \cdot 3^2 \cdot 5$

Copyright © 2018 Pearson Education, Inc.

7. (a)

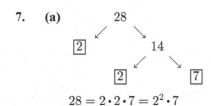

$$28 = 2 \cdot 2 \cdot 7 = 2^2 \cdot 7$$

(b)

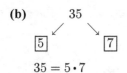

$$35 = 5 \cdot 7$$

(c)

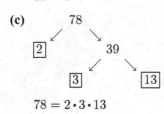

$$78 = 2 \cdot 3 \cdot 13$$

2.3 Section Exercises

1. Factorizations of 8:

    $$1 \cdot 8 = 8 \quad 2 \cdot 4 = 8$$

    The factors of 8 are 1, 2, 4, and 8. The statement is *false* (missing 1 and includes 6).

3. Factorizations of 15:

    $$1 \cdot 15 = 15 \quad 3 \cdot 5 = 15$$

    The factors of 15 are 1, 3, 5, and 15. The statement is *true*.

5. Factorizations of 48:

    $$1 \cdot 48 = 48 \quad 2 \cdot 24 = 48 \quad 3 \cdot 16 = 48$$
    $$4 \cdot 12 = 48 \quad 6 \cdot 8 = 48$$

    The factors of 48 are 1, 2, 3, 4, 6, 8, 12, 16, 24, and 48.

7. Factorizations of 56:

    $$1 \cdot 56 = 56 \quad 2 \cdot 28 = 56 \quad 4 \cdot 14 = 56 \quad 7 \cdot 8 = 56$$

    The factors of 56 are 1, 2, 4, 7, 8, 14, 28, and 56.

9. Factorizations of 36:

    $$1 \cdot 36 = 36 \quad 2 \cdot 18 = 36 \quad 3 \cdot 12 = 36 \quad 4 \cdot 9 = 36$$
    $$6 \cdot 6 = 36$$

    The factors of 36 are 1, 2, 3, 4, 6, 9, 12, 18, and 36.

11. Factorizations of 40:

    $$1 \cdot 40 = 40 \quad 2 \cdot 20 = 40 \quad 4 \cdot 10 = 40 \quad 5 \cdot 8 = 40$$

    The factors of 40 are 1, 2, 4, 5, 8, 10, 20, and 40.

13. Factorizations of 64:

    $$1 \cdot 64 = 64 \quad 2 \cdot 32 = 64 \quad 4 \cdot 16 = 64 \quad 8 \cdot 8 = 64$$

    The factors of 64 are 1, 2, 4, 8, 16, 32, and 64.

15. Factorizations of 82:

    $$1 \cdot 82 = 82 \quad 2 \cdot 41 = 82$$

    The factors of 82 are 1, 2, 41, and 82.

17. 6 is divisible by 2 and 3, so 6 is composite.

19. 5 is only divisible by itself and 1, so it is prime.

21. 10 is divisible by 2 and 5, so 10 is composite.

23. 19 is only divisible by itself and 1, so it is prime.

25. 25 is divisible by 5, so 25 is composite.

27. 47 is only divisible by itself and 1, so it is prime.

29. 40

    $$\begin{array}{r} 20 \\ 2\overline{)40} \end{array}$$ Divide 40 by 2.

    $$\begin{array}{r} 10 \\ 2\overline{)20} \end{array}$$ Divide 20 by 2.

    $$\begin{array}{r} 5 \\ 2\overline{)10} \end{array}$$ Divide 10 by 2.

    $$\begin{array}{r} 1 \\ 5\overline{)5} \end{array}$$ Divide 5 by 5.

    Quotient is 1.

    $$40 = 2 \cdot 2 \cdot 2 \cdot 5 = 2^3 \cdot 5$$

    The correct choice is **(b)**.

31. 100

    $$\begin{array}{r} 50 \\ 2\overline{)100} \end{array}$$ Divide 100 by 2.

    $$\begin{array}{r} 25 \\ 2\overline{)50} \end{array}$$ Divide 50 by 2.

    $$\begin{array}{r} 5 \\ 5\overline{)25} \end{array}$$ Divide 25 by 5.

    $$\begin{array}{r} 1 \\ 5\overline{)5} \end{array}$$ Divide 5 by 5.

    Quotient is 1.

    $$100 = 2 \cdot 2 \cdot 5 \cdot 5 = 2^2 \cdot 5^2$$

    The correct choice is **(a)**.

33. 6

    $$\begin{array}{r} 3 \\ 2\overline{)6} \end{array}$$ Divide 6 by 2.

    $$\begin{array}{r} 1 \\ 3\overline{)3} \end{array}$$ Divide 3 by 3.

    Quotient is 1.

    $$6 = 2 \cdot 3$$

Copyright © 2018 Pearson Education, Inc.

35.

$$25$$

$$5 \qquad 5$$

$$25 = 5 \cdot \underline{5} = 5^2$$

37.

$$68$$

$$2 \qquad 34$$

$$2 \qquad 17$$

$$68 = 2 \cdot 2 \cdot 17 = 2^2 \cdot 17$$

39. 72

$$\begin{array}{r} 36 \\ 2\overline{)72} \end{array} \qquad \text{Divide 72 by 2.}$$

$$\begin{array}{r} 18 \\ 2\overline{)36} \end{array} \qquad \text{Divide 36 by 2.}$$

$$\begin{array}{r} 9 \\ 2\overline{)18} \end{array} \qquad \text{Divide 18 by 2.}$$

$$\begin{array}{r} 3 \\ 3\overline{)9} \end{array} \qquad \text{Divide 9 by 3.}$$

$$\begin{array}{r} 1 \\ 3\overline{)3} \end{array} \qquad \text{Divide 3 by 3.}$$

Quotient is 1.

$$72 = 2 \cdot 2 \cdot 2 \cdot 3 \cdot 3 = 2^3 \cdot 3^2$$

41. 44

$$\begin{array}{r} 1 \\ 11\overline{)11} \end{array} \qquad \text{Divide 11 by 11.}$$

$$\begin{array}{r} 11 \\ 2\overline{)22} \end{array} \qquad \text{Divide 22 by 2.}$$

$$\begin{array}{r} 22 \\ 2\overline{)44} \end{array} \qquad \text{Divide 44 by 2.}$$

Quotient is 1.

Because all factors (divisors) are prime, the prime factorization of 44 is $2 \cdot 2 \cdot 11 = 2^2 \cdot 11$.

43. 100

$$\begin{array}{r} 50 \\ 2\overline{)100} \end{array} \qquad \text{Divide 100 by 2.}$$

$$\begin{array}{r} 25 \\ 2\overline{)50} \end{array} \qquad \text{Divide 50 by 2.}$$

$$\begin{array}{r} 5 \\ 5\overline{)25} \end{array} \qquad \text{Divide 25 by 5.}$$

$$\begin{array}{r} 1 \\ 5\overline{)5} \end{array} \qquad \text{Divide 5 by 5.}$$

Quotient is 1.

$$100 = 2 \cdot 2 \cdot 5 \cdot 5 = 2^2 \cdot 5^2$$

45. 125

$$\begin{array}{r} 25 \\ 5\overline{)125} \end{array} \qquad \text{Divide 125 by 5.}$$

$$\begin{array}{r} 5 \\ 5\overline{)25} \end{array} \qquad \text{Divide 25 by 5.}$$

$$\begin{array}{r} 1 \\ 5\overline{)5} \end{array} \qquad \text{Divide 5 by 5.}$$

Quotient is 1.

$$125 = 5 \cdot 5 \cdot 5 = 5^3$$

47. 180

$$\begin{array}{r} 90 \\ 2\overline{)180} \end{array} \qquad \text{Divide 180 by 2.}$$

$$\begin{array}{r} 45 \\ 2\overline{)90} \end{array} \qquad \text{Divide 90 by 2.}$$

$$\begin{array}{r} 15 \\ 3\overline{)45} \end{array} \qquad \text{Divide 45 by 3.}$$

$$\begin{array}{r} 5 \\ 3\overline{)15} \end{array} \qquad \text{Divide 15 by 3.}$$

$$\begin{array}{r} 1 \\ 5\overline{)5} \end{array} \qquad \text{Divide 5 by 5.}$$

Quotient is 1.

$$180 = 2 \cdot 2 \cdot 3 \cdot 3 \cdot 5 = 2^2 \cdot 3^2 \cdot 5$$

49. 320

$$\begin{array}{r} 160 \\ 2\overline{)320} \end{array} \qquad \text{Divide 320 by 2.}$$

$$\begin{array}{r} 80 \\ 2\overline{)160} \end{array} \qquad \text{Divide 160 by 2.}$$

$$\begin{array}{r} 40 \\ 2\overline{)80} \end{array} \qquad \text{Divide 80 by 2.}$$

$$\begin{array}{r} 20 \\ 2\overline{)40} \end{array} \qquad \text{Divide 40 by 2.}$$

$$\begin{array}{r} 10 \\ 2\overline{)20} \end{array} \qquad \text{Divide 20 by 2.}$$

$$\begin{array}{r} 5 \\ 2\overline{)10} \end{array} \qquad \text{Divide 10 by 2.}$$

$$\begin{array}{r} 1 \\ 5\overline{)5} \end{array} \qquad \text{Divide 5 by 5.}$$

Quotient is 1.

$$320 = 2 \cdot 2 \cdot 2 \cdot 2 \cdot 2 \cdot 2 \cdot 5 = 2^6 \cdot 5$$

Copyright © 2018 Pearson Education, Inc.

51. 360

$$\begin{array}{r} 180 \\ 2\overline{)360} \end{array}$$ Divide 360 by 2.

$$\begin{array}{r} 90 \\ 2\overline{)180} \end{array}$$ Divide 180 by 2.

$$\begin{array}{r} 45 \\ 2\overline{)90} \end{array}$$ Divide 90 by 2.

$$\begin{array}{r} 15 \\ 3\overline{)45} \end{array}$$ Divide 45 by 3.

$$\begin{array}{r} 5 \\ 3\overline{)15} \end{array}$$ Divide 15 by 3.

$$\begin{array}{r} 1 \\ 5\overline{)5} \end{array}$$ Divide 5 by 5.

Quotient is 1.

$360 = 2 \cdot 2 \cdot 2 \cdot 3 \cdot 3 \cdot 5 = 2^3 \cdot 3^2 \cdot 5$

53. Answers will vary. A sample answer follows. A prime number is a whole number that has exactly two *different* factors, itself and 1. Examples include 2, 3, 5, 7, and 11. A composite number has a factor(s) other than itself or 1. Examples include 4, 6, 8, 9, and 10. The numbers 0 and 1 are neither prime nor composite.

55. All the possible factors of 24 are 1, 2, 3, 4, 6, 8, 12, and 24. This list includes both prime numbers and composite numbers. The prime factors of 24 include only prime numbers. The prime factorization of 24 is

$$24 = 2 \cdot 2 \cdot 2 \cdot 3 = 2^3 \cdot 3.$$

57. 350

$$\begin{array}{r} 175 \\ 2\overline{)350} \end{array}$$ Divide 350 by 2.

$$\begin{array}{r} 35 \\ 5\overline{)175} \end{array}$$ Divide 175 by 5.

$$\begin{array}{r} 7 \\ 5\overline{)35} \end{array}$$ Divide 35 by 5.

$$\begin{array}{r} 1 \\ 7\overline{)7} \end{array}$$ Divide 7 by 7.

Quotient is 1.

$350 = 2 \cdot 5 \cdot 5 \cdot 7 = 2 \cdot 5^2 \cdot 7$

59. 960

$$\begin{array}{r} 480 \\ 2\overline{)960} \end{array}$$ Divide 960 by 2.

$$\begin{array}{r} 240 \\ 2\overline{)480} \end{array}$$ Divide 480 by 2.

$$\begin{array}{r} 120 \\ 2\overline{)240} \end{array}$$ Divide 240 by 2.

$$\begin{array}{r} 60 \\ 2\overline{)120} \end{array}$$ Divide 120 by 2.

$$\begin{array}{r} 30 \\ 2\overline{)60} \end{array}$$ Divide 60 by 2.

$$\begin{array}{r} 15 \\ 2\overline{)30} \end{array}$$ Divide 30 by 2.

$$\begin{array}{r} 5 \\ 3\overline{)15} \end{array}$$ Divide 15 by 3.

$$\begin{array}{r} 1 \\ 5\overline{)5} \end{array}$$ Divide 5 by 5. *Quotient is 1.*

$960 = 2 \cdot 2 \cdot 2 \cdot 2 \cdot 2 \cdot 2 \cdot 3 \cdot 5 = 2^6 \cdot 3 \cdot 5$

61. 1560

$$\begin{array}{r} 780 \\ 2\overline{)1560} \end{array}$$ Divide 1560 by 2.

$$\begin{array}{r} 390 \\ 2\overline{)780} \end{array}$$ Divide 780 by 2.

$$\begin{array}{r} 195 \\ 2\overline{)390} \end{array}$$ Divide 390 by 2.

$$\begin{array}{r} 65 \\ 3\overline{)195} \end{array}$$ Divide 195 by 3.

$$\begin{array}{r} 13 \\ 5\overline{)65} \end{array}$$ Divide 65 by 5.

$$\begin{array}{r} 1 \\ 13\overline{)13} \end{array}$$ Divide 13 by 13.

Quotient is 1.

$1560 = 2 \cdot 2 \cdot 2 \cdot 3 \cdot 5 \cdot 13 = 2^3 \cdot 3 \cdot 5 \cdot 13$

63.

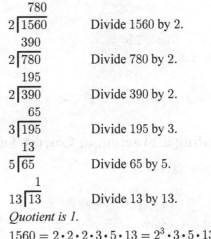

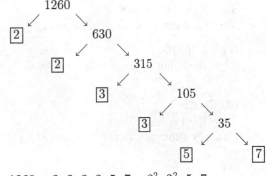

$1260 = 2 \cdot 2 \cdot 3 \cdot 3 \cdot 5 \cdot 7 = 2^2 \cdot 3^2 \cdot 5 \cdot 7$

Copyright © 2018 Pearson Education, Inc.

65. The prime numbers less than 50 are 2, 3, 5, 7, 11, 13, 17, 19, 23, 29, 31, 37, 41, 43, and 47.

66. A prime number is a whole number that is evenly divisible by itself and 1 only.

67. No. Every other even number is divisible by 2 in addition to being divisible by itself and 1.

68. No. A multiple of a prime number can never be prime because it will always be divisible by the prime number.

69. 2100

$$
\begin{array}{r}
1050 \\
2\overline{)2100}
\end{array}
\quad \text{Divide 2100 by 2.}
$$

$$
\begin{array}{r}
525 \\
2\overline{)1050}
\end{array}
\quad \text{Divide 1050 by 2.}
$$

$$
\begin{array}{r}
175 \\
3\overline{)525}
\end{array}
\quad \text{Divide 525 by 3.}
$$

$$
\begin{array}{r}
35 \\
5\overline{)175}
\end{array}
\quad \text{Divide 175 by 5.}
$$

$$
\begin{array}{r}
7 \\
5\overline{)35}
\end{array}
\quad \text{Divide 35 by 5.}
$$

$$
\begin{array}{r}
1 \\
7\overline{)7}
\end{array}
\quad \text{Divide 7 by 7.}
$$

Quotient is 1.

$2100 = 2 \cdot 2 \cdot 3 \cdot 5 \cdot 5 \cdot 7$

70. $2100 = 2 \cdot 2 \cdot 3 \cdot 5 \cdot 5 \cdot 7 = 2^2 \cdot 3 \cdot 5^2 \cdot 7$

2.4 Writing a Fraction in Lowest Terms

2.4 Margin Exercises

1. **(a)** 6, 12; 2
$6 = 2 \cdot 3 \quad 12 = 2 \cdot 6$
Yes, 2 is a common factor of 6 and 12.

(b) 32, 64; 8
$32 = 8 \cdot 4 \quad 64 = 8 \cdot 8$
Yes, 8 is a common factor of 32 and 64.

(c) 32, 56; 16
$32 = 16 \cdot 2$, but 16 is not a factor of 56.
No.

(d) 75, 81; 1
$75 = 75 \cdot 1 \quad 81 = 81 \cdot 1$
Yes, 1 is a common factor of 75 and 81.

2. **(a)** $\dfrac{4}{5}$

4 and 5 have no common factor other than 1.
Yes, it is in lowest terms.

(b) $\dfrac{6}{18}$

6 and 18 have a common factor of 6.
No, it is not in lowest terms.

(c) $\dfrac{9}{15}$

9 and 15 have a common factor of 3.
No, it is not in lowest terms.

(d) $\dfrac{17}{46}$

17 and 46 have no common factor other than 1.
Yes, it is in lowest terms.

3. **(a)** $\dfrac{8}{16} = \dfrac{8 \div 8}{16 \div 8} = \dfrac{1}{2}$

(b) $\dfrac{9}{12} = \dfrac{9 \div 3}{12 \div 3} = \dfrac{3}{4}$

(c) $\dfrac{28}{42} = \dfrac{28 \div 14}{42 \div 14} = \dfrac{2}{3}$

(d) $\dfrac{30}{80} = \dfrac{30 \div 10}{80 \div 10} = \dfrac{3}{8}$

(e) $\dfrac{16}{40} = \dfrac{16 \div 8}{40 \div 8} = \dfrac{2}{5}$

4. **(a)** $\dfrac{12}{36} = \dfrac{2 \cdot 2 \cdot 3}{2 \cdot 2 \cdot 3 \cdot 3}$

$= \dfrac{\overset{1}{\cancel{2}} \cdot \overset{1}{\cancel{2}} \cdot \overset{1}{\cancel{3}}}{\underset{1}{\cancel{2}} \cdot \underset{1}{\cancel{2}} \cdot \underset{1}{\cancel{3}} \cdot 3} = \dfrac{1 \cdot 1 \cdot 1}{1 \cdot 1 \cdot 1 \cdot 3} = \dfrac{1}{3}$

(b) $\dfrac{32}{56} = \dfrac{2 \cdot 2 \cdot 2 \cdot 2 \cdot 2}{2 \cdot 2 \cdot 2 \cdot 7}$

$= \dfrac{\overset{1}{\cancel{2}} \cdot \overset{1}{\cancel{2}} \cdot \overset{1}{\cancel{2}} \cdot 2 \cdot 2}{\underset{1}{\cancel{2}} \cdot \underset{1}{\cancel{2}} \cdot \underset{1}{\cancel{2}} \cdot 7} = \dfrac{1 \cdot 1 \cdot 1 \cdot 2 \cdot 2}{1 \cdot 1 \cdot 1 \cdot 7} = \dfrac{4}{7}$

(c) $\dfrac{74}{111} = \dfrac{2 \cdot \overset{1}{\cancel{37}}}{3 \cdot \underset{1}{\cancel{37}}} = \dfrac{2 \cdot 1}{3 \cdot 1} = \dfrac{2}{3}$

(d) $\dfrac{124}{340} = \dfrac{\overset{1}{\cancel{2}} \cdot \overset{1}{\cancel{2}} \cdot 31}{\underset{1}{\cancel{2}} \cdot \underset{1}{\cancel{2}} \cdot 5 \cdot 17} = \dfrac{1 \cdot 1 \cdot 31}{1 \cdot 1 \cdot 5 \cdot 17} = \dfrac{31}{85}$

5. **(a)** $\dfrac{24}{48}$ and $\dfrac{36}{72}$

$\dfrac{24}{48} = \dfrac{\overset{1}{\cancel{2}} \cdot \overset{1}{\cancel{2}} \cdot \overset{1}{\cancel{2}} \cdot \overset{1}{\cancel{3}}}{\underset{1}{\cancel{2}} \cdot \underset{1}{\cancel{2}} \cdot \underset{1}{\cancel{2}} \cdot 2 \cdot \underset{1}{\cancel{3}}} = \dfrac{1 \cdot 1 \cdot 1 \cdot 1}{1 \cdot 1 \cdot 1 \cdot 2 \cdot 1} = \dfrac{1}{2}$

(continued)

Copyright © 2018 Pearson Education, Inc.

$$\frac{36}{72} = \frac{\overset{1}{\cancel{2}} \cdot \overset{1}{\cancel{2}} \cdot \overset{1}{\cancel{3}} \cdot \overset{1}{\cancel{3}}}{\cancel{2} \cdot \cancel{2} \cdot 2 \cdot \cancel{3} \cdot \cancel{3}} = \frac{1 \cdot 1 \cdot 1 \cdot 1}{1 \cdot 1 \cdot 2 \cdot 1 \cdot 1} = \frac{1}{2}$$

The fractions are *equivalent* $(\frac{1}{2} = \frac{1}{2})$.

(b) $\frac{45}{60}$ and $\frac{50}{75}$

$$\frac{45}{60} = \frac{3 \cdot \overset{1}{\cancel{3}} \cdot \overset{1}{\cancel{5}}}{2 \cdot 2 \cdot \cancel{3} \cdot \cancel{5}} = \frac{3 \cdot 1 \cdot 1}{2 \cdot 2 \cdot 1 \cdot 1} = \frac{3}{4}$$

$$\frac{50}{75} = \frac{2 \cdot \overset{1}{\cancel{5}} \cdot \overset{1}{\cancel{5}}}{3 \cdot \cancel{5} \cdot \cancel{5}} = \frac{2 \cdot 1 \cdot 1}{3 \cdot 1 \cdot 1} = \frac{2}{3}$$

The fractions are *not equivalent* $(\frac{3}{4} \neq \frac{2}{3})$.

(c) $\frac{20}{4}$ and $\frac{110}{22}$

$$\frac{20}{4} = \frac{\overset{1}{\cancel{2}} \cdot \overset{1}{\cancel{2}} \cdot 5}{\cancel{2} \cdot \cancel{2}} = \frac{1 \cdot 1 \cdot 5}{1 \cdot 1} = 5$$

$$\frac{110}{22} = \frac{\overset{1}{\cancel{2}} \cdot 5 \cdot \overset{1}{\cancel{11}}}{\cancel{2} \cdot \cancel{11}} = \frac{1 \cdot 5 \cdot 1}{1 \cdot 1} = 5$$

The fractions are *equivalent* $(5 = 5)$.

(d) $\frac{120}{220}$ and $\frac{180}{320}$

$$\frac{120}{220} = \frac{\overset{1}{\cancel{2}} \cdot \overset{1}{\cancel{2}} \cdot 2 \cdot 3 \cdot \overset{1}{\cancel{5}}}{\cancel{2} \cdot \cancel{2} \cdot \cancel{5} \cdot 11} = \frac{1 \cdot 1 \cdot 2 \cdot 3 \cdot 1}{1 \cdot 1 \cdot 1 \cdot 11} = \frac{6}{11}$$

$$\frac{180}{320} = \frac{\overset{1}{\cancel{2}} \cdot \overset{1}{\cancel{2}} \cdot 3 \cdot 3 \cdot \overset{1}{\cancel{5}}}{\cancel{2} \cdot \cancel{2} \cdot 2 \cdot 2 \cdot 2 \cdot 2 \cdot \cancel{5}}$$
$$= \frac{1 \cdot 1 \cdot 3 \cdot 3 \cdot 1}{1 \cdot 1 \cdot 2 \cdot 2 \cdot 2 \cdot 2 \cdot 1} = \frac{9}{16}$$

The fractions are *not equivalent* $(\frac{6}{11} \neq \frac{9}{16})$.

2.4 Section Exercises

1. A number can be divided by 2 if the number is an <u>even</u> number.

3. Any number can be divided by 10 if the number ends in <u>0</u>.

		2	3	5	10
5.	60	✓	✓	✓	✓
7.	48	✓	✓	X	X
9.	160	✓	X	✓	✓
11.	138	✓	✓	X	X

13. $\frac{6}{8} = \frac{6 \div 2}{8 \div 2} = \frac{3}{4}$, *true*

15. $\frac{5}{16}$ is in lowest terms, so the fractions $\frac{5}{16}$ and $\frac{3}{8}$ are not equivalent. *false*

17. $\frac{15}{25} = \frac{15 \div 5}{25 \div 5} = \frac{3}{5}$

19. $\frac{36}{42} = \frac{36 \div 6}{42 \div 6} = \frac{6}{7}$

21. Because the greatest common factor of 56 and 64 is 8, divide both numerator and denominator by 8.

$$\frac{56}{64} = \frac{56 \div 8}{64 \div 8} = \frac{7}{8}$$

23. $\frac{180}{210} = \frac{180 \div 30}{210 \div 30} = \frac{6}{7}$

25. $\frac{72}{126} = \frac{72 \div 18}{126 \div 18} = \frac{4}{7}$

27. $\frac{12}{600} = \frac{12 \div 12}{600 \div 12} = \frac{1}{50}$

29. $\frac{96}{132} = \frac{96 \div 12}{132 \div 12} = \frac{8}{11}$

31. $\frac{60}{108} = \frac{60 \div 12}{108 \div 12} = \frac{5}{9}$

33. Write the prime factorization of both numerator and denominator. Then divide both numerator and denominator by any common factors, and write a 1 by each factor that has been divided. Finally, multiply the remaining factors in both numerator and denominator.

$$\frac{18}{24} = \frac{\overset{1}{\cancel{2}} \cdot \overset{1}{\cancel{3}} \cdot 3}{\cancel{2} \cdot 2 \cdot 2 \cdot \cancel{3}} = \frac{1 \cdot 1 \cdot 3}{1 \cdot 2 \cdot 2 \cdot 1} = \frac{3}{4}$$

35. $\frac{35}{40} = \frac{\overset{1}{\cancel{5}} \cdot 7}{2 \cdot 2 \cdot 2 \cdot \cancel{5}} = \frac{1 \cdot 7}{2 \cdot 2 \cdot 2 \cdot 1} = \frac{7}{8}$

37. $\frac{90}{180} = \frac{\overset{1}{\cancel{2}} \cdot \overset{1}{\cancel{3}} \cdot \overset{1}{\cancel{3}} \cdot \overset{1}{\cancel{5}}}{\cancel{2} \cdot 2 \cdot \cancel{3} \cdot \cancel{3} \cdot \cancel{5}} = \frac{1 \cdot 1 \cdot 1 \cdot 1}{1 \cdot 2 \cdot 1 \cdot 1 \cdot 1} = \frac{1}{2}$

39. $\frac{36}{12} = \frac{\overset{1}{\cancel{2}} \cdot \overset{1}{\cancel{2}} \cdot \overset{1}{\cancel{3}} \cdot 3}{\cancel{2} \cdot \cancel{2} \cdot \cancel{3}} = \frac{1 \cdot 1 \cdot 1 \cdot 3}{1 \cdot 1 \cdot 1} = 3$

41. $\frac{72}{225} = \frac{2 \cdot 2 \cdot 2 \cdot \overset{1}{\cancel{3}} \cdot \overset{1}{\cancel{3}}}{\cancel{3} \cdot \cancel{3} \cdot 5 \cdot 5} = \frac{2 \cdot 2 \cdot 2 \cdot 1 \cdot 1}{1 \cdot 1 \cdot 5 \cdot 5} = \frac{8}{25}$

Copyright © 2018 Pearson Education, Inc.

43. $\dfrac{3}{6}$ and $\dfrac{18}{36}$ Write each fraction in lowest terms.

$$\frac{3}{6} = \frac{1 \cdot \overset{1}{\cancel{3}}}{2 \cdot \underset{1}{\cancel{3}}} = \frac{1}{2}$$

$$\frac{18}{36} = \frac{\overset{1}{\cancel{2}} \cdot \overset{1}{\cancel{3}} \cdot \overset{1}{\cancel{3}}}{\underset{1}{\cancel{2}} \cdot 2 \cdot \underset{1}{\cancel{3}} \cdot \underset{1}{\cancel{3}}} = \frac{1}{2}$$

The fractions are *equivalent* $\left(\frac{1}{2} = \frac{1}{2}\right)$.

45. $\dfrac{15}{35}$ and $\dfrac{18}{40}$

$$\frac{15}{35} = \frac{3 \cdot \overset{1}{\cancel{5}}}{\underset{1}{\cancel{5}} \cdot 7} = \frac{3}{7}$$

$$\frac{18}{40} = \frac{\overset{1}{\cancel{2}} \cdot 3 \cdot 3}{\underset{1}{\cancel{2}} \cdot 2 \cdot 2 \cdot 5} = \frac{9}{20}$$

The fractions are *not equivalent* $\left(\frac{3}{7} \neq \frac{9}{20}\right)$.

47. $\dfrac{14}{16}$ and $\dfrac{35}{40}$

$$\frac{14}{16} = \frac{\overset{1}{\cancel{2}} \cdot 7}{\underset{1}{\cancel{2}} \cdot 2 \cdot 2 \cdot 2} = \frac{7}{8}$$

$$\frac{35}{40} = \frac{\overset{1}{\cancel{5}} \cdot 7}{2 \cdot 2 \cdot 2 \cdot \underset{1}{\cancel{5}}} = \frac{7}{8}$$

The fractions are *equivalent* $\left(\frac{7}{8} = \frac{7}{8}\right)$.

49. $\dfrac{48}{6}$ and $\dfrac{72}{8}$

$$\frac{48}{6} = \frac{2 \cdot 2 \cdot 2 \cdot \overset{1}{\cancel{2}} \cdot \overset{1}{\cancel{3}}}{\underset{1}{\cancel{2}} \cdot \underset{1}{\cancel{3}}} = 8$$

$$\frac{72}{8} = \frac{\overset{1}{\cancel{2}} \cdot \overset{1}{\cancel{2}} \cdot \overset{1}{\cancel{2}} \cdot 3 \cdot 3}{\underset{1}{\cancel{2}} \cdot \underset{1}{\cancel{2}} \cdot \underset{1}{\cancel{2}}} = 9$$

The fractions are *not equivalent* $(8 \neq 9)$.

51. A fraction is in lowest terms when the numerator and the denominator have no common factors other than 1. Some examples are $\frac{1}{2}$, $\frac{3}{8}$, and $\frac{2}{3}$.

52. Two fractions are equivalent when they represent the same portion of a whole. For example, the fractions $\frac{10}{15}$ and $\frac{8}{12}$ are equivalent.

$$\frac{10}{15} = \frac{2 \cdot \overset{1}{\cancel{5}}}{3 \cdot \underset{1}{\cancel{5}}} = \frac{2 \cdot 1}{3 \cdot 1} = \frac{2}{3}$$

$$\frac{8}{12} = \frac{\overset{1}{\cancel{2}} \cdot \overset{1}{\cancel{2}} \cdot 2}{\underset{1}{\cancel{2}} \cdot \underset{1}{\cancel{2}} \cdot 3} = \frac{1 \cdot 1 \cdot 2}{1 \cdot 1 \cdot 3} = \frac{2}{3}$$

The fractions are *equivalent* $\left(\frac{2}{3} = \frac{2}{3}\right)$.

53. $\dfrac{160}{256} = \dfrac{\overset{1}{\cancel{2}} \cdot \overset{1}{\cancel{2}} \cdot \overset{1}{\cancel{2}} \cdot \overset{1}{\cancel{2}} \cdot \overset{1}{\cancel{2}} \cdot 5}{\underset{1}{\cancel{2}} \cdot \underset{1}{\cancel{2}} \cdot \underset{1}{\cancel{2}} \cdot \underset{1}{\cancel{2}} \cdot \underset{1}{\cancel{2}} \cdot 2 \cdot 2 \cdot 2} = \dfrac{5}{8}$

54. $\dfrac{363}{528} = \dfrac{\overset{1}{\cancel{3}} \cdot \overset{1}{\cancel{11}} \cdot 11}{2 \cdot 2 \cdot 2 \cdot 2 \cdot \underset{1}{\cancel{3}} \cdot \underset{1}{\cancel{11}}} = \dfrac{11}{16}$

55. $\dfrac{238}{119} = \dfrac{2 \cdot \overset{1}{\cancel{7}} \cdot \overset{1}{\cancel{17}}}{\underset{1}{\cancel{7}} \cdot \underset{1}{\cancel{17}}} = \dfrac{2}{1} = 2$

56. $\dfrac{570}{95} = \dfrac{2 \cdot 3 \cdot \overset{1}{\cancel{5}} \cdot \overset{1}{\cancel{19}}}{\underset{1}{\cancel{5}} \cdot \underset{1}{\cancel{19}}} = \dfrac{6}{1} = 6$

57. $\dfrac{65}{234} = \dfrac{5 \cdot \overset{1}{\cancel{13}}}{2 \cdot 3 \cdot 3 \cdot \underset{1}{\cancel{13}}} = \dfrac{5}{18}$

58. $\dfrac{105}{180} = \dfrac{\overset{1}{\cancel{3}} \cdot 7 \cdot \overset{1}{\cancel{5}}}{2 \cdot 2 \cdot 3 \cdot \underset{1}{\cancel{3}} \cdot \underset{1}{\cancel{5}}} = \dfrac{7}{12}$

59. $\dfrac{19}{171} = \dfrac{\overset{1}{\cancel{19}}}{3 \cdot 3 \cdot \underset{1}{\cancel{19}}} = \dfrac{1}{9}$

60. $\dfrac{36}{468} = \dfrac{\overset{1}{\cancel{2}} \cdot \overset{1}{\cancel{2}} \cdot \overset{1}{\cancel{3}} \cdot \overset{1}{\cancel{3}}}{\underset{1}{\cancel{2}} \cdot \underset{1}{\cancel{2}} \cdot \underset{1}{\cancel{3}} \cdot \underset{1}{\cancel{3}} \cdot 13} = \dfrac{1}{13}$

Summary Exercises *Fraction Basics*

1. The figure has 6 equal parts.
Five parts are shaded: $\frac{5}{6}$

One part is unshaded: $\frac{1}{6}$

3. The figure has 8 equal parts.
Five parts are shaded: $\frac{5}{8}$

Three parts are unshaded: $\frac{3}{8}$

5. $\dfrac{8}{5}$ $\leftarrow$ Numerator
        $\leftarrow$ Denominator

Copyright © 2018 Pearson Education, Inc.

7. There are 40 winners in total
$(12 + 10 + 8 + 4 + 3 + 3)$. Eight of the 40
winners were from Switzerland. $\frac{8}{40} = \frac{1}{5}$

9. $3 + 12 = 15$ of the winners were from either
Japan or the United States. $\frac{15}{40} = \frac{3}{8}$

11. $\frac{5}{2}$

$$\begin{array}{r} 2 \leftarrow \text{Whole number part} \\ 2\overline{)5} \\ \underline{4} \\ 1 \leftarrow \text{Remainder} \end{array}$$

$\frac{5}{2} = 2\frac{1}{2}$

13. $\frac{9}{7}$

$$\begin{array}{r} 1 \leftarrow \text{Whole number part} \\ 7\overline{)9} \\ \underline{7} \\ 2 \leftarrow \text{Remainder} \end{array}$$

$\frac{9}{7} = 1\frac{2}{7}$

15. $\frac{64}{8}$

$$\begin{array}{r} 8 \leftarrow \text{Whole number part} \\ 8\overline{)6\ 4} \\ \underline{6\ 4} \\ 0 \leftarrow \text{Remainder} \end{array}$$

$\frac{64}{8} = 8$

17. $\frac{36}{5}$

$$\begin{array}{r} 7 \leftarrow \text{Whole number part} \\ 5\overline{)3\ 6} \\ \underline{3\ 5} \\ 1 \leftarrow \text{Remainder} \end{array}$$

$\frac{36}{5} = 7\frac{1}{5}$

19. $3\frac{1}{3}$ $3 \cdot 3 = 9$ Multiply 3 and 3.

$9 + 1 = 10$ Add 1.

$3\frac{1}{3} = \frac{10}{3}$

21. $6\frac{4}{5}$ $6 \cdot 5 = 30$ Multiply 6 and 5.

$30 + 4 = 34$ Add 4.

$6\frac{4}{5} = \frac{34}{5}$

23. $12\frac{3}{4}$ $12 \cdot 4 = 48$ Multiply 12 and 4.

$48 + 3 = 51$ Add 3.

$12\frac{3}{4} = \frac{51}{4}$

25. $11\frac{5}{6}$ $11 \cdot 6 = 66$ Multiply 11 and 6.

$66 + 5 = 71$ Add 5.

$11\frac{5}{6} = \frac{71}{6}$

27.

$10 = 2 \cdot 5$

29.

$36 = 2^2 \cdot 3^2$

31.

$280 = 2^3 \cdot 5 \cdot 7$

33. $\frac{4}{12} = \frac{4 \div 4}{12 \div 4} = \frac{1}{3}$

35. $\frac{4}{16} = \frac{4 \div 4}{16 \div 4} = \frac{1}{4}$

37. $\frac{18}{24} = \frac{18 \div 6}{24 \div 6} = \frac{3}{4}$

39. $\frac{56}{64} = \frac{56 \div 8}{64 \div 8} = \frac{7}{8}$

41. $\frac{25}{200} = \frac{25 \div 25}{200 \div 25} = \frac{1}{8}$

43. $\frac{88}{154} = \frac{88 \div 22}{154 \div 22} = \frac{4}{7}$

45. $\frac{24}{36} = \frac{\overset{1}{\cancel{2}} \cdot \overset{1}{\cancel{2}} \cdot 2 \cdot \overset{1}{\cancel{3}}}{\underset{1}{\cancel{2}} \cdot \underset{1}{\cancel{2}} \cdot \underset{1}{\cancel{3}} \cdot 3} = \frac{2}{3}$

Copyright © 2018 Pearson Education, Inc.

47. $\dfrac{126}{42} = \dfrac{\overset{1}{\cancel{2}} \cdot \overset{1}{\cancel{3}} \cdot 3 \cdot \overset{1}{\cancel{7}}}{\underset{1}{\cancel{2}} \cdot \underset{1}{\cancel{3}} \cdot \underset{1}{\cancel{7}}} = \dfrac{3}{1} = 3$

2.5 Multiplying Fractions

2.5 Margin Exercises

1. $\frac{1}{4}$ of $\frac{1}{2}$ as read from the figure is the darker shaded part of the second figure. One of eight equal parts is shaded, or $\frac{1}{8}$.

$$\frac{1}{4} \cdot \frac{1}{2} = \frac{1}{\underline{8}}$$

2. **(a)** $\dfrac{1}{2} \cdot \dfrac{3}{4} = \dfrac{1 \cdot 3}{2 \cdot 4} = \dfrac{3}{\underline{8}}$

(b) $\dfrac{3}{5} \cdot \dfrac{1}{3} = \dfrac{\overset{1}{\cancel{3}} \cdot 1}{5 \cdot \underset{1}{\cancel{3}}} = \dfrac{1}{5}$

(c) $\dfrac{5}{6} \cdot \dfrac{1}{2} \cdot \dfrac{1}{8} = \dfrac{5 \cdot 1 \cdot 1}{6 \cdot 2 \cdot 8} = \dfrac{5}{96}$

(d) $\dfrac{1}{2} \cdot \dfrac{3}{4} \cdot \dfrac{3}{8} = \dfrac{1 \cdot 3 \cdot 3}{2 \cdot 4 \cdot 8} = \dfrac{9}{64}$

3. **(a)** $\dfrac{3}{\underset{2}{\cancel{4}}} \cdot \dfrac{\overset{1}{\cancel{2}}}{5} = \dfrac{3 \cdot 1}{2 \cdot 5} = \dfrac{3}{10}$

(b) $\dfrac{\overset{2}{\cancel{6}}}{\underset{1}{\cancel{11}}} \cdot \dfrac{\overset{3}{\cancel{33}}}{\underset{7}{\cancel{21}}} = \dfrac{2 \cdot 3}{1 \cdot 7} = \dfrac{6}{7}$

(c) $\dfrac{\overset{1}{\cancel{20}}}{4} \cdot \dfrac{\overset{1}{\cancel{3}}}{\underset{2}{\cancel{40}}} \cdot \dfrac{1}{\underset{1}{\cancel{3}}} = \dfrac{1 \cdot 1 \cdot 1}{4 \cdot 2 \cdot 1} = \dfrac{1}{8}$

(d) $\dfrac{\overset{1}{\cancel{18}}}{17} \cdot \dfrac{1}{\underset{2}{\cancel{36}}} \cdot \dfrac{\overset{1}{\cancel{2}}}{3} = \dfrac{1 \cdot 1 \cdot 1}{17 \cdot 1 \cdot 3} = \dfrac{1}{51}$

4. **(a)** $8 \cdot \dfrac{1}{8} = \dfrac{\overset{1}{\cancel{8}}}{1} \cdot \dfrac{1}{\underset{1}{\cancel{8}}} = \dfrac{1 \cdot 1}{1 \cdot 1} = \dfrac{1}{1} = 1$

(b) $\dfrac{\overset{1}{\cancel{3}}}{4} \cdot 5 \cdot \dfrac{5}{\underset{1}{\cancel{3}}} = \dfrac{1 \cdot 5 \cdot 5}{4 \cdot 1} = \dfrac{25}{4}$ or $6\dfrac{1}{4}$

(c) $\dfrac{3}{5} \cdot 40 = \dfrac{3}{\underset{1}{\cancel{5}}} \cdot \dfrac{\overset{8}{\cancel{40}}}{1} = \dfrac{3 \cdot 8}{1 \cdot 1} = \dfrac{24}{1} = 24$

(d) $\dfrac{3}{25} \cdot \dfrac{5}{11} \cdot 99 = \dfrac{3}{\underset{5}{\cancel{25}}} \cdot \dfrac{\overset{1}{\cancel{5}}}{\underset{1}{\cancel{11}}} \cdot \dfrac{\overset{9}{\cancel{99}}}{1}$

$$= \dfrac{3 \cdot 1 \cdot 9}{5 \cdot 1 \cdot 1}$$

$$= \dfrac{27}{5} = 5\dfrac{2}{5}$$

5. **(a)** Area = length · width

$$= \dfrac{\overset{1}{\cancel{3}}}{4} \cdot \dfrac{1}{\underset{1}{\cancel{3}}}$$

$$= \dfrac{1}{4}\text{yd}^2$$

(b) Area = length · width

$$= \dfrac{\overset{1}{\cancel{3}}}{8} \cdot \dfrac{1}{\underset{1}{\cancel{3}}}$$

$$= \dfrac{1}{8}\text{mi}^2$$

(c) Area = length · width

$$= \dfrac{\overset{3}{\cancel{9}}}{\underset{1}{\cancel{7}}} \cdot \dfrac{\overset{1}{\cancel{7}}}{\underset{4}{\cancel{12}}} = \dfrac{3 \cdot 1}{1 \cdot 4}$$

$$= \dfrac{3}{4}\text{mi}^2$$

2.5 Section Exercises

1. To multiply two or more fractions, you <u>multiply</u> the numerators and you multiply the <u>denominators</u>.

3. A shortcut when multiplying fractions is to <u>divide</u> both a numerator and a <u>denominator</u> by the same number.

5. $\dfrac{1}{3} \cdot \dfrac{3}{4} = \dfrac{1}{\underset{1}{\cancel{3}}} \cdot \dfrac{\overset{1}{\cancel{3}}}{4} = \dfrac{1 \cdot 1}{1 \cdot 4} = \dfrac{1}{4}$

7. $\dfrac{2}{7} \cdot \dfrac{1}{5} = \dfrac{2 \cdot 1}{7 \cdot 5} = \dfrac{2}{35}$

9. $\dfrac{8}{5} \cdot \dfrac{15}{32} = \dfrac{\overset{1}{\cancel{8}}}{\underset{1}{\cancel{5}}} \cdot \dfrac{\overset{3}{\cancel{15}}}{\underset{4}{\cancel{32}}} = \dfrac{1 \cdot 3}{1 \cdot 4} = \dfrac{3}{4}$

11. Divide a numerator and denominator by the same number to use the multiplication shortcut.

$\dfrac{2}{3} \cdot \dfrac{7}{12} \cdot \dfrac{9}{14} = \dfrac{\overset{1}{\cancel{2}}}{\underset{1}{\cancel{3}}} \cdot \dfrac{\overset{1}{\cancel{7}}}{\underset{\underset{2}{\cancel{6}}}{\cancel{12}}} \cdot \dfrac{\overset{\overset{1}{\cancel{3}}}{\cancel{9}}}{\underset{2}{\cancel{14}}} = \dfrac{1 \cdot 1 \cdot 1}{1 \cdot 2 \cdot 2} = \dfrac{1}{4}$

Copyright © 2018 Pearson Education, Inc.

13. $\frac{3}{4}\cdot\frac{5}{6}\cdot\frac{2}{3}=\frac{\overset{1}{\cancel{3}}}{\underset{2}{\cancel{4}}}\cdot\frac{5}{6}\cdot\frac{\overset{1}{\cancel{2}}}{\underset{1}{\cancel{3}}}=\frac{1\cdot5\cdot1}{2\cdot6\cdot1}=\frac{5}{12}$

15. $\frac{6}{11}\cdot\frac{22}{15}=\frac{\overset{2}{\cancel{6}}}{\underset{1}{\cancel{11}}}\cdot\frac{\overset{2}{\cancel{22}}}{\underset{5}{\cancel{15}}}=\frac{2\cdot2}{1\cdot5}=\frac{4}{5}$

17. $\frac{35}{64}\cdot\frac{32}{15}\cdot\frac{27}{72}=\frac{\overset{7}{\cancel{35}}}{\underset{2}{\cancel{64}}}\cdot\frac{\overset{1}{\cancel{32}}}{\underset{\underset{1}{\cancel{3}}}{\cancel{15}}}\cdot\frac{\overset{\overset{1}{\cancel{3}}}{\cancel{27}}}{\underset{8}{\cancel{72}}}=\frac{7\cdot1\cdot1}{2\cdot1\cdot8}=\frac{7}{16}$

19. $\frac{39}{42}\cdot\frac{7}{13}\cdot\frac{7}{24}=\frac{\overset{\overset{1}{\cancel{3}}}{\cancel{39}}}{\underset{\underset{2}{\cancel{6}}}{\cancel{42}}}\cdot\frac{\overset{1}{\cancel{7}}}{\underset{1}{\cancel{13}}}\cdot\frac{7}{24}=\frac{1\cdot1\cdot7}{2\cdot1\cdot24}=\frac{7}{48}$

21. $\frac{4}{5}\cdot8=\frac{\overset{1}{\cancel{4}}}{5}\cdot\frac{1}{\underset{2}{\cancel{8}}}=\frac{1}{10}$ is *false*.

The correct method is as follows:

$$\frac{4}{5}\cdot8=\frac{4}{5}\cdot\frac{8}{1}=\frac{32}{5}=6\frac{2}{5}$$

23. $20\cdot\frac{3}{4}=\frac{\overset{5}{\cancel{20}}}{1}\cdot\frac{3}{\underset{1}{\cancel{4}}}=\frac{15}{1}=15$

25. Write 36 as $\frac{36}{1}$ and multiply.

$$36\cdot\frac{5}{8}\cdot\frac{9}{15}=\frac{\overset{9}{\cancel{36}}}{1}\cdot\frac{\overset{1}{\cancel{5}}}{\underset{2}{\cancel{8}}}\cdot\frac{\overset{3}{\cancel{9}}}{\underset{\underset{1}{\cancel{3}}}{\cancel{15}}}=\frac{27}{2}=13\frac{1}{2}$$

27. $100\cdot\frac{21}{50}\cdot\frac{3}{4}=\frac{\overset{\overset{1}{\cancel{2}}}{\cancel{100}}}{1}\cdot\frac{21}{\underset{1}{\cancel{50}}}\cdot\frac{3}{\underset{2}{\cancel{4}}}=\frac{63}{2}=31\frac{1}{2}$

29. $\frac{2}{5}\cdot200=\frac{2}{\underset{1}{\cancel{5}}}\cdot\frac{\overset{40}{\cancel{200}}}{1}=\frac{80}{1}=80$

31. $142\cdot\frac{2}{3}=\frac{142}{1}\cdot\frac{2}{3}=\frac{284}{3}=94\frac{2}{3}$

33. $\frac{28}{21}\cdot640\cdot\frac{15}{32}=\frac{\overset{4}{\cancel{28}}}{\underset{\underset{1}{\cancel{3}}}{\cancel{21}}}\cdot\frac{\overset{20}{\cancel{640}}}{1}\cdot\frac{\overset{5}{\cancel{15}}}{\underset{1}{\cancel{32}}}$

$$=\frac{4\cdot20\cdot5}{1\cdot1\cdot1}$$

$$=\frac{400}{1}=400$$

35. $\frac{54}{38}\cdot684\cdot\frac{5}{6}=\frac{\overset{9}{\cancel{54}}}{\underset{1}{\cancel{38}}}\cdot\frac{\overset{18}{\cancel{684}}}{1}\cdot\frac{5}{\underset{1}{\cancel{6}}}$

$$=\frac{9\cdot18\cdot5}{1\cdot1\cdot1}$$

$$=\frac{810}{1}=810$$

37. Area = length • width

$$=\frac{3}{4}\cdot\frac{1}{3}=\frac{\overset{1}{\cancel{3}}}{4}\cdot\frac{1}{\underset{1}{\cancel{3}}}=\frac{1}{4}\text{ mi}^2$$

39. Area = length • width

$$=12\cdot\frac{3}{4}=\frac{\overset{3}{\cancel{12}}}{1}\cdot\frac{3}{\underset{1}{\cancel{4}}}=\frac{9}{1}=9\text{ m}^2$$

41. Area = length • width

$$\text{Area}=\frac{5}{6}\cdot\frac{3}{10}$$

$$=\frac{\overset{1}{\cancel{5}}}{\underset{2}{\cancel{6}}}\cdot\frac{\overset{1}{\cancel{3}}}{\underset{2}{\cancel{10}}}$$ Divide 5 and 10 by 5. Divide 3 and 6 by 3.

$$=\frac{1}{4}\text{ mi}^2$$

43. Multiply the numerators and multiply the denominators. An example is

$$\frac{3}{4}\cdot\frac{1}{2}=\frac{3\cdot1}{4\cdot2}=\frac{3}{8}.$$

45. Area = length • width

$$\text{Area}=2\cdot\frac{3}{4}$$

$$=\frac{\overset{1}{\cancel{2}}}{1}\cdot\frac{3}{\underset{2}{\cancel{4}}}$$ Divide numerator and denominator by 2.

$$=\frac{3}{2}$$ Write as a mixed number.

$$=1\frac{1}{2}\text{ yd}^2$$ Square yards

47. Area = length • width

$$=4\cdot\frac{7}{8}=\frac{\overset{1}{\cancel{4}}}{1}\cdot\frac{7}{\underset{2}{\cancel{8}}}=\frac{7}{2}=3\frac{1}{2}\text{ mi}^2$$

49.

Sunny Side Soccer Park	**Creekside Soc. Park**
Area = length • width	Area = length • width
$=\frac{1}{4}\cdot\frac{3}{16}$	$=\frac{3}{8}\cdot\frac{1}{8}$
$=\frac{3}{64}\text{ mi}^2$	$=\frac{3}{64}\text{ mi}^2$

They are both the same size.

Copyright © 2018 Pearson Education, Inc.

51. $4000 + 2000 + 1000 + 800 + 400 + 200$
$+ 100 + 80 = \underline{8580 \text{ supermarkets}}$
The estimate of the total number of supermarkets in these states is 8580.

52. $3753 + 2202 + 1349 + 826 + 441 + 163$
$+ 98 + 80 = \underline{8912 \text{ supermarkets}}$, which is the exact total number of supermarkets in these states.

53. An estimate of the number of supermarkets in medium to large population areas in New York is
$$\frac{4}{5} \cdot 2000 = \frac{4}{\cancel{5}} \cdot \frac{\cancel{2000}^{400}}{1} = 1600.$$

The exact value is
$$\frac{4}{5} \cdot 2202 = \frac{4}{5} \cdot \frac{2202}{1} = \frac{8808}{5} = 1761\frac{3}{5}.$$

Rounding gives us 1762 supermarkets.

54. An estimate of the number of supermarkets in New Hampshire which are in shopping centers, is
$$\frac{3}{8} \cdot 200 = \frac{3}{\cancel{8}} \cdot \frac{\cancel{200}^{25}}{1} = 75.$$

The exact value is
$$\frac{3}{8} \cdot 163 = \frac{3}{8} \cdot \frac{163}{1} = \frac{489}{8} = 61\frac{1}{8}.$$

Rounding gives us 61 supermarkets.

55. We need a multiple of 5 with *two* nonzero digits that is close to 2202. A reasonable choice is 2200 and an estimate is
$$\frac{4}{5} \cdot 2200 = \frac{4}{\cancel{5}} \cdot \frac{\cancel{2200}^{440}}{1} = 1760 \text{ supermarkets.}$$

This value is closer to the exact value because using 2200 as a rounded guess is closer to 2202 than using 2000 as a rounded guess.

56. We need a multiple of 8 with *two* nonzero digits that is close to 163. A reasonable choice is 160 and an estimate is
$$\frac{3}{8} \cdot 160 = \frac{3}{\cancel{8}} \cdot \frac{\cancel{160}^{20}}{1} = 60 \text{ supermarkets.}$$

This value is closer to the exact value because using 160 as a rounded guess is closer to 163 than using 200 as a rounded guess.

2.6   Applications of Multiplication

2.6   Margin Exercises

1.   (a) *Step 1*
The problem asks us to find the amount of money they can save in a year.

Step 2
Find the amount they can save by multiplying $\frac{3}{8}$ and 81,576.

Step 3
We can estimate this amount using $\frac{1}{2}$ and 82,000.
$$\frac{1}{2} \cdot 82,000 = \frac{1}{\cancel{2}} \cdot \frac{\cancel{82,000}^{41,000}}{1} = \underline{41,000}$$

Step 4
Now solve the problem using the original values.
$$\frac{3}{8} \cdot 81,576 = \frac{3}{\cancel{8}} \cdot \frac{\cancel{81,576}^{10,197}}{1} = \underline{30,591}$$

Step 5
They can save $30,591 in a year.

Step 6
The answer is reasonably close to our estimate.

(b) *Step 1*
The problem asks us to find the amount of money she will receive as retirement income.

Step 2
To find her retirement income, multiply $\frac{5}{8}$ and $62,504.

Step 3
We can estimate this amount using $\frac{1}{2}$ and 63,000.
$$\frac{1}{2} \cdot 63,000 = \frac{1}{\cancel{2}} \cdot \frac{\cancel{63,000}^{31,500}}{1} = 31,500$$

Step 4
Now solve the problem using the original values.
$$\frac{5}{8} \cdot 62,504 = \frac{5}{\cancel{8}} \cdot \frac{\cancel{62,504}^{7813}}{1} = 39,065$$

Step 5
She will receive $39,065 as retirement income.

Step 6
The answer is reasonably close to our estimate.

Copyright © 2018 Pearson Education, Inc.

2. *Step 1*
The problem asks for the number of prescriptions paid for by a third party.

Step 2
A third party pays for $\frac{5}{16}$ of the total number of prescriptions, 3696.

Step 3
An estimate is $\frac{1}{4} \cdot 4000 = \frac{1}{\cancel{4}} \cdot \frac{\cancel{4000}^{1000}}{1} = 1000.$

Step 4
The exact value is
$$\frac{5}{16} \cdot 3696 = \frac{5}{\cancel{16}} \cdot \frac{\cancel{3696}^{231}}{1} = 1155.$$

Step 5
A third party pays for 1155 prescriptions.

Step 6
The answer is reasonably close to our estimate.

3. *Step 1*
The problem asks for the fraction of students who speak Spanish.

Step 2
$\frac{3}{4}$ of the $\frac{1}{3}$ of the students who speak a foreign language, speak Spanish.

Step 3
An estimate is $\frac{3}{4} \cdot \frac{1}{3} = \frac{\cancel{3}^{1}}{4} \cdot \frac{1}{\cancel{3}_{1}} = \frac{1}{4}.$

Step 4
The exact value is $\frac{1}{4}$, which is the same as the estimate since we didn't round.

Step 5
The fraction of students who speak Spanish is $\frac{1}{4}$.

Step 6
The answer, $\frac{1}{4}$, matches our estimate.

4. **(a)** From the circle graph, the fraction is $\frac{1}{5}$.

(b) Multiply $\frac{1}{5}$ by the number of people in the survey, 2500. Since we can estimate the answer using the exact values, our estimated answer will be the same as the exact answer.
$$\frac{1}{5} \cdot 2500 = \frac{1}{\cancel{5}_{1}} \cdot \frac{\cancel{2500}^{500}}{1} = \frac{500}{1} = \underline{500}$$

500 children buy food from vending machines.

(c) From the circle graph, the fraction is $\frac{1}{10}$.

(d) Multiply $\frac{1}{10}$ by 2500. Since we can estimate the answer using the exact values, our estimated answer will be the same as the exact answer.
$$\frac{1}{10} \cdot 2500 = \frac{1}{\cancel{10}_{1}} \cdot \frac{\cancel{2500}^{250}}{1} = \frac{250}{1} = 250$$

250 children buy food from a convenience store or street vendor.

2.6 Section Exercises

1. The words that are indicator words for multiplication are *of*, *times*, *twice*, *triple*, *product*, and *twice as much*.

3. When you multiply length by width you are finding the <u>area</u> of a rectangular surface.

5. *Step 1*: The problem asks for the area of the oil painting.

Step 2: To find area, multiply the length of $\frac{3}{4}$ foot by the width of $\frac{2}{3}$ foot.

Step 3: An estimate is $\frac{6}{12}$ square foot. (ft^2)
$$\frac{3}{4} \cdot \frac{2}{3} = \frac{\cancel{3}^{1} \cdot \cancel{2}^{1}}{\cancel{4}_{2} \cdot \cancel{3}_{1}} = \frac{1}{2}$$

Step 4: The exact value is $\frac{1}{2}$, which is the same as the estimate since we didn't round.

Step 5: The area of the oil painting is $\frac{1}{2}$ square foot. (ft^2)

Step 6: The answer, $\frac{1}{2}$ square foot (ft^2), matches our estimate.

7. Multiply the length and the width.
$$\frac{4}{3} \cdot \frac{2}{3} = \frac{8}{9}$$

The area of the cookie sheet is $\frac{8}{9}$ ft^2.

9. Multiply the length and the width.
$$\frac{4}{5} \cdot \frac{3}{8} = \frac{\cancel{4}^{1} \cdot 3}{5 \cdot \cancel{8}_{2}} = \frac{3}{10}$$

The area of the top of the table is $\frac{3}{10}$ yd^2.

11. Multiply $\frac{3}{8}$ by 6848.
$$\frac{3}{8} \cdot 6848 = \frac{3}{\cancel{8}_{1}} \cdot \frac{\cancel{6848}^{856}}{1} = 2568$$

He earned $2568 on his job.

Copyright © 2018 Pearson Education, Inc.

13. Multiply the daily parking fee by the fraction.

$$40 \cdot \frac{7}{8} = \frac{\overset{5}{\cancel{40}}}{1} \cdot \frac{7}{\cancel{8}} = 35$$

The daily parking fee in Boston is $35.

15. (a) *Step 1:* The problem asks for the number of runners who are women.

Step 2: $\frac{7}{12}$ of the 1560 runners are women, so multiply $\frac{7}{12}$ by 1560 to find the number of runners who are women.

Step 3: Round the number of runners to 1600; then, $\frac{1}{2}$ of 1600 is 800. Since $\frac{7}{12}$ is more than $\frac{1}{2}$, our estimate is that "more than 800 runners" are women.

Step 4: $\frac{7}{12} \cdot 1560 = \frac{7 \cdot \overset{130}{\cancel{1560}}}{\underset{1}{\cancel{12}} \cdot 1} = 910$

Step 5: 910 runners are women.

Step 6: The exact answer, 910, fits our estimate of "more than 800."

(b) The number of runners that are men is

1560 total runners $-$ 910 women $=$ 650 men.

17. The smallest sector of the circle graph is the *4 hours* group, so this response was given by the least number of people. To find how many people gave this response, multiply $\frac{3}{20}$ by the total number of people, 1020.

$$\frac{3}{20} \cdot 1020 = \frac{3 \cdot \overset{51}{\cancel{1020}}}{\underset{1}{\cancel{20}} \cdot 1} = 153$$

153 people gave this response.

19. The only group that is *not* willing to wait 4 hours or less is the *8 hours* group, and the fraction corresponding to that group is $\frac{1}{4}$. Thus, the fraction willing to wait 4 hours or less is

$$1 - \frac{1}{4} = \frac{4}{4} - \frac{1}{4} = \frac{3}{4}.$$

The total number of people willing to wait 4 hours or less is

$$\frac{3}{4} \cdot 1020 = \frac{3 \cdot \overset{255}{\cancel{1020}}}{\underset{1}{4 \cdot 1}} = 765.$$

21. Because everyone is included and fractions are given for *all* groups, the sum of the fractions must be *1*, or *all* of the people.

23. Add the income for all twelve months to find the income for the year.

$6075 + 5812 + 6488 + 6030 + 5820 + 6398$
$+ 7040 + 5232 + 5670 + 7012 + 6465 + 7958$
$= 76{,}000$

The Owens family had income of $76,000 for the year.

25. From Exercise 23, the total income is $76,000. The circle graph shows that $\frac{1}{5}$ of the income is for rent.

$$\frac{1}{5} \cdot 76{,}000 = \frac{1}{\underset{1}{\cancel{5}}} \cdot \frac{\overset{15{,}200}{\cancel{76{,}000}}}{1} = 15{,}200$$

The amount of their rent is $15,200.

27. Multiply the total income by the fraction saved.

$$\frac{1}{16} \cdot 76{,}000 = \frac{1}{\underset{1}{\cancel{16}}} \cdot \frac{\overset{4750}{\cancel{76{,}000}}}{1} = 4750$$

The Owens family saved $4750 for the year.

29. The error was made when dividing 21 by 3 and writing 3 instead of 7. The correct solution is

$$\frac{9}{10} \times \frac{20}{21} = \frac{\overset{3}{\cancel{9}}}{\underset{1}{\cancel{10}}} \times \frac{\overset{2}{\cancel{20}}}{\underset{7}{\cancel{21}}} = \frac{6}{7}.$$

31. We want $\frac{1}{128}$ of the actual length.

$$\frac{1}{128} \cdot 256 = \frac{1}{\underset{1}{\cancel{128}}} \cdot \frac{\overset{2}{\cancel{256}}}{1} = 2$$

The length of the scale model is 2 feet.

33. Multiply the number of views by $\frac{3}{8}$.

$$2000 \cdot \frac{3}{8} = \frac{\overset{250}{\cancel{2000}}}{1} \cdot \frac{3}{\cancel{8}} = 750$$

750 dancers viewed the You Tube video.

35. First multiply $\frac{2}{3}$ and 27,000 to find the number of votes she needs from senior citizens.

$$\frac{2}{3} \cdot 27{,}000 = \frac{2}{\underset{1}{\cancel{3}}} \cdot \frac{\overset{9000}{\cancel{27{,}000}}}{1} = 18{,}000$$

To find the votes needed from voters other than the senior citizens, subtract:

$$27{,}000 - 18{,}000 = 9000 \text{ votes.}$$

She needs 9000 votes from voters other than the senior citizens.

Copyright © 2018 Pearson Education, Inc.

37. To find the remaining amount of the estate, subtract $\frac{7}{8}$ from 1.

$$1 - \frac{7}{8} = \frac{8}{8} - \frac{7}{8} = \frac{1}{8}$$

Multiply the remaining $\frac{1}{8}$ of the estate by the fraction going to the American Cancer Society.

$$\frac{1}{4} \cdot \frac{1}{8} = \frac{1}{32}$$

$\frac{1}{32}$ of the estate goes to the American Cancer Society.

2.7 Dividing Fractions

2.7 Margin Exercises

1. **(a)** $\frac{4}{5} \diagup\!\!\!\!\diagdown \frac{5}{4}$; The reciprocal of $\frac{4}{5}$ is $\frac{5}{4}$ because

$$\frac{4}{5} \cdot \frac{5}{4} = \frac{20}{20} = 1.$$

(b) $\frac{3}{8} \diagup\!\!\!\!\diagdown \frac{8}{3}$; The reciprocal of $\frac{3}{8}$ is $\frac{8}{3}$ because

$$\frac{3}{8} \cdot \frac{8}{3} = \frac{24}{24} = 1.$$

(c) The reciprocal of $\frac{9}{4}$ is $\frac{4}{9}$ because

$$\frac{9}{4} \cdot \frac{4}{9} = \frac{36}{36} = 1.$$

(d) The reciprocal of 16 is $\frac{1}{16}$ because

$$\frac{16}{1} \cdot \frac{1}{16} = \frac{16}{16} = 1.$$

2. **(a)** $\frac{1}{4} \div \frac{2}{3} = \frac{1}{4} \cdot \frac{3}{2} = \frac{1 \cdot 3}{4 \cdot 2} = \frac{3}{8}$

(b) $\frac{3}{8} \div \frac{5}{8} = \frac{3}{8} \cdot \frac{8}{5} = \frac{3 \cdot \overset{1}{\cancel{8}}}{\underset{1}{\cancel{8}} \cdot 5} = \frac{3}{5}$

(c) $\dfrac{\frac{2}{3}}{\frac{4}{5}} = \frac{2}{3} \div \frac{4}{5} = \frac{\overset{1}{\cancel{2}}}{3} \cdot \frac{5}{\underset{2}{\cancel{4}}} = \frac{1 \cdot 5}{3 \cdot 2} = \frac{5}{6}$

(d) $\dfrac{\frac{5}{6}}{\frac{7}{12}} = \frac{5}{6} \div \frac{7}{12}$

$$= \frac{5}{\underset{1}{\cancel{6}}} \cdot \frac{\overset{2}{\cancel{12}}}{7} = \frac{5 \cdot 2}{1 \cdot 7} = \frac{10}{7} = 1\frac{3}{7}$$

3. **(a)** $10 \div \frac{1}{2} = 10 \cdot \frac{2}{1} = \frac{10}{1} \cdot \frac{2}{1}$

$$= \frac{10 \cdot 2}{1 \cdot 1} = \frac{20}{1} = 20$$

(b) $6 \div \frac{6}{7} = \frac{6}{1} \div \frac{6}{7} = \frac{\overset{1}{\cancel{6}}}{1} \cdot \frac{7}{\underset{1}{\cancel{6}}} = \frac{7}{1} = 7$

(c) $\frac{4}{5} \div 6 = \frac{4}{5} \div \frac{6}{1} = \frac{\overset{2}{\cancel{4}}}{5} \cdot \frac{1}{\underset{3}{\cancel{6}}} = \frac{2 \cdot 1}{5 \cdot 3} = \frac{2}{15}$

(d) $\frac{3}{8} \div 4 = \frac{3}{8} \div \frac{4}{1} = \frac{3}{8} \cdot \frac{1}{4} = \frac{3 \cdot 1}{8 \cdot 4} = \frac{3}{32}$

4. **(a)** *Step 1*
The problem asks for the number of $\frac{5}{6}$-ounce dispensers that can be filled using 40 ounces of eye drops.

Step 2
Divide the total number of ounces of eye drops by the fraction of an ounce each dispenser holds.

Step 3
An estimate is $40 \div 1 = 40$.

Step 4
Solving gives us

$$40 \div \frac{5}{6} = \frac{40}{1} \div \frac{5}{6}$$

$$= \frac{\overset{8}{\cancel{40}}}{1} \cdot \frac{6}{\underset{1}{\cancel{5}}} = \frac{8 \cdot 6}{1 \cdot 1} = \frac{48}{1} = 48$$

Step 5
48 $\frac{5}{6}$-ounce dispensers can be filled.

Step 6
The answer is reasonably close to our estimate.

(b) *Step 1*
The problem asks for the number of $\frac{4}{5}$-quart bottles that can be filled from a 120-quart cask.

Step 2
Divide the total number of quarts in the cask by the size of the bottles.

Step 3
An estimate is $120 \div 1 = 120$.

Step 4
Solving gives us

Copyright © 2018 Pearson Education, Inc.

$$120 \div \frac{4}{5} = \frac{120}{1} \div \frac{4}{5}$$

$$= \frac{\overset{30}{\cancel{120}}}{1} \cdot \frac{5}{\underset{1}{\cancel{4}}} = \frac{30 \cdot 5}{1 \cdot 1} = \frac{150}{1} = 150$$

Step 5
150 $\frac{4}{5}$-quart bottles can be filled.

Step 6
The answer is reasonably close to our estimate.

5.   **(a)** *Step 1*
The problem asks for the fraction of the bonus money that each employee will receive.

Step 2
Divide the fraction of the bonus money, $\frac{3}{4}$, by the number of employees, 12.

Step 3
An estimate is $1 \div 12 = \frac{1}{12}$.

Step 4
Solving gives us

$$\frac{3}{4} \div 12 = \frac{3}{4} \div \frac{12}{1} = \frac{3}{4} \cdot \frac{1}{\underset{4}{\cancel{12}}} = \frac{1 \cdot 1}{4 \cdot 4} = \frac{1}{16}$$

Step 5
Each employee will receive $\frac{1}{16}$ of the bonus money.

Step 6
The answer is reasonably close to our estimate.

(b) *Step 1*
The problem asks for the fraction of the prize money that each employee will receive.

Step 2
Since they donate $\frac{1}{5}$ of the winnings, they have

$$1 - \frac{1}{5} = \frac{5}{5} - \frac{1}{5} = \frac{4}{5}$$

of the winnings left to divide. Divide the fraction of the winnings that remain, $\frac{4}{5}$, by the number of employees, 8.

Step 3
An estimate is $1 \div 8 = \frac{1}{8}$.

Step 4
Solving gives us

$$\frac{4}{5} \div 8 = \frac{4}{5} \div \frac{8}{1} = \frac{4}{5} \cdot \frac{1}{\underset{2}{\cancel{8}}} = \frac{1}{10}$$

Step 5
Each employee will receive $\frac{1}{10}$ of the prize money.

Step 6
The answer is reasonably close to our estimate.

2.7   Section Exercises

1.   When you invert or flip a fraction, you have the <u>reciprocal</u> of the fraction.

3.   To divide by a fraction, you must first <u>invert</u> the divisor and then change division to <u>multiplication</u>.

5.   The reciprocal of $\frac{3}{8}$ is $\frac{8}{3}$ because $\frac{3}{8} \cdot \frac{8}{3} = \frac{24}{24} = 1$.

7.   The reciprocal of $\frac{5}{6}$ is $\frac{6}{5}$ because $\frac{5}{6} \cdot \frac{6}{5} = \frac{30}{30} = 1$.

9.   The reciprocal of $\frac{8}{5}$ is $\frac{5}{8}$ because $\frac{8}{5} \cdot \frac{5}{8} = \frac{40}{40} = 1$.

11.   The reciprocal of 4 is $\frac{1}{4}$ because $\frac{4}{1} \cdot \frac{1}{4} = \frac{4}{4} = 1$.

13. $\frac{1}{2} \div \frac{3}{4} = \frac{1}{\underset{1}{\cancel{2}}} \cdot \frac{\overset{2}{\cancel{4}}}{3} = \frac{2}{3}$

15. $\frac{7}{8} \div \frac{1}{3} = \frac{7}{8} \cdot \frac{3}{1} = \frac{21}{8} = 2\frac{5}{8}$

17. $\frac{3}{4} \div \frac{5}{3} = \frac{3}{4} \cdot \frac{3}{5} = \frac{9}{20}$

19. $\frac{7}{9} \div \frac{7}{36} = \frac{\overset{1}{\cancel{7}}}{\underset{1}{\cancel{9}}} \cdot \frac{\overset{4}{\cancel{36}}}{\underset{1}{\cancel{7}}} = \frac{4}{1} = 4$

21. $\frac{15}{32} \div \frac{5}{64} = \frac{\overset{3}{\cancel{15}}}{\underset{1}{\cancel{32}}} \cdot \frac{\overset{2}{\cancel{64}}}{\underset{1}{\cancel{5}}} = \frac{6}{1} = 6$

23. $\dfrac{\frac{13}{20}}{\frac{4}{5}} = \frac{13}{20} \div \frac{4}{5} = \frac{13}{\underset{4}{\cancel{20}}} \cdot \frac{\overset{1}{\cancel{5}}}{4} = \frac{13}{16}$

25. $\dfrac{\frac{5}{6}}{\frac{25}{24}} = \frac{5}{6} \div \frac{25}{24} = \frac{\overset{1}{\cancel{5}}}{\underset{1}{\cancel{6}}} \cdot \frac{\overset{4}{\cancel{24}}}{\underset{5}{\cancel{25}}} = \frac{4}{5}$

Copyright © 2018 Pearson Education, Inc.

27. $12 \div \dfrac{2}{3} = \dfrac{12}{1} \div \dfrac{2}{3} = \dfrac{\overset{6}{\cancel{12}}}{1} \cdot \dfrac{3}{\underset{1}{\cancel{2}}} = \dfrac{18}{1} = 18$

29. $\dfrac{\frac{18}{3}}{4} = 18 \div \dfrac{3}{4}$ Rewrite using the $\div$ symbol for division.

$= \dfrac{18}{1} \div \dfrac{3}{4}$ Write 18 as $\dfrac{18}{1}$.

$= \dfrac{\overset{6}{\cancel{18}}}{1} \cdot \dfrac{4}{\underset{1}{\cancel{3}}}$ Change "$\div$" to "$\cdot$". Divide the numerator and denominator by 3.

$= \dfrac{6 \cdot 4}{1 \cdot 1}$ Multiply.

$= \dfrac{24}{1} = 24$

31. $\dfrac{\frac{4}{7}}{8} = \dfrac{4}{7} \div 8 = \dfrac{4}{7} \div \dfrac{8}{1} = \dfrac{\overset{1}{\cancel{4}}}{7} \cdot \dfrac{1}{\underset{2}{\cancel{8}}} = \dfrac{1}{14}$

33. $\frac{8}{9}$ of a quart divided into 4 parts:

$\dfrac{8}{9} \div 4 = \dfrac{8}{9} \div \dfrac{4}{1} = \dfrac{\overset{2}{\cancel{8}}}{9} \cdot \dfrac{1}{\underset{1}{\cancel{4}}} = \dfrac{2}{9}$

Each horse will get $\frac{2}{9}$ of a quart.

35. *Step 1*: The problem asks for the number of times a measuring cup needs to be filled.

Step 2: Solve the problem by dividing the total number of cups (5) by the size of the measuring cup $\left(\frac{1}{3}\right)$.

Step 3: Because there are three $\frac{1}{3}$-cups in one cup, multiply 3 by 5 to get 15. So, our estimate is 15.

Step 4: $5 \div \dfrac{1}{3} = \dfrac{5}{1} \div \dfrac{1}{3} = \dfrac{5}{1} \cdot \dfrac{3}{1} = 15$

Step 5: They need to fill the measuring cup 15 times.

Step 6: The exact answer, 15, is the same as our estimate.

37. Divide the total number of cups of sugar by the amount of sugar in each loaf.

$11 \div \dfrac{1}{4} = \dfrac{11}{1} \div \dfrac{1}{4} = \dfrac{11}{1} \cdot \dfrac{4}{1} = 44$

You can make 44 loaves.

39. Divide the total weight of a carton by the weight per fastener.

$25 \div \dfrac{5}{32} = \dfrac{\overset{5}{\cancel{25}}}{1} \cdot \dfrac{32}{\underset{1}{\cancel{5}}} = \dfrac{160}{1} = 160$

There are 160 $\frac{5}{32}$-pound fasteners in each carton.

41. Answers will vary. A sample answer follows: You can divide two fractions by multiplying the first fraction by the reciprocal of the second fraction (divisor).

43. Each loafcake requires $\frac{3}{4}$ pound of jellybeans, so to make 16 loafcakes, use multiplication.

$\dfrac{3}{4} \cdot 16 = \dfrac{3}{\cancel{4}} \cdot \dfrac{\overset{4}{\cancel{16}}}{1} = 12$

12 pounds will be needed.

45. Divide the 156 cans of compound by the fraction of a can needed for each new home.

$156 \div \dfrac{3}{4} = \dfrac{156}{1} \cdot \dfrac{4}{3} = \dfrac{\overset{52}{\cancel{156}}}{1} \cdot \dfrac{4}{\underset{1}{\cancel{3}}} = 208$

208 homes can be plumbed.

47. **(a)** In $\frac{2}{3}$ of the 186 visits, doctors failed to discuss the issues—use multiplication.

$\dfrac{2}{3} \cdot 186 = \dfrac{2}{\cancel{3}} \cdot \dfrac{\overset{62}{\cancel{186}}}{1} = 124$

The doctors failed to discuss the issues in 124 visits.

(b) The doctors *did* discuss the issues in $186 - 124 = 62$ visits.

49. *Step 1*: The problem asks for the number of towels that can be made.

Step 2: Solve the problem by dividing the 912 yards of fabric by the fraction of a yard $\left(\frac{3}{8}\right)$ needed for each dish towel.

Step 3: Round 912 yards to 900 yards. Round $\frac{3}{8}$ to $\frac{1}{2}$.

Step 4: $912 \div \dfrac{3}{8} = \dfrac{912}{1} \cdot \dfrac{8}{3} = \dfrac{\overset{304}{\cancel{912}}}{1} \cdot \dfrac{8}{\underset{1}{\cancel{3}}} = 2432$

Step 5: 2432 towels can be made.

Step 6: The exact answer, 2432, is close to our estimate, 1800.

Copyright © 2018 Pearson Education, Inc.

51. The indicator words for multiplication are underlined below.

more than	per
<u>double</u>	<u>twice</u>
<u>times</u>	<u>product</u>
less than	difference
equals	<u>twice as much</u>

52. The indicator words for division are underlined below.

fewer	sum of
<u>goes into</u>	<u>divide</u>
<u>per</u>	<u>quotient</u>
equals	double
loss of	<u>divided by</u>

53. Two numbers are reciprocals of each other if their product is <u>1</u>.

54. The number 0 has no reciprocal because there is no number that can be multiplied by 0 to get 1.

55. To divide two fractions, multiply the first fraction by the <u>reciprocal</u> of the second fraction.

56. The reciprocal of $\frac{3}{4}$ is $\frac{4}{3}$ because
$$\frac{3}{4} \cdot \frac{4}{3} = \frac{12}{12} = 1.$$
The reciprocal of $\frac{7}{8}$ is $\frac{8}{7}$ because
$$\frac{7}{8} \cdot \frac{8}{7} = \frac{56}{56} = 1.$$
The reciprocal of 5 is $\frac{1}{5}$ because $\frac{5}{1} \cdot \frac{1}{5} = \frac{5}{5} = 1.$
The reciprocal of $\frac{12}{19}$ is $\frac{19}{12}$ because
$$\frac{12}{19} \cdot \frac{19}{12} = \frac{228}{228} = 1.$$

57. (a) To find the perimeter of any flat equal-sided 3-, 4-, 5-, or 6-sided figure, multiply the length of one side by 3, 4, 5, or 6, respectively.

(b) The stamp has four sides, so multiply $\frac{15}{16}$ by 4.
$$\frac{15}{16} \times 4 = \frac{15}{\cancel{16}} \times \frac{\cancel{4}^{1}}{1} = \frac{15}{4} = 3\frac{3}{4} \text{ in.}$$
The perimeter of the stamp is $3\frac{3}{4}$ inches.

58. Area $=$ length $\cdot$ width
$$= \frac{15}{16} \cdot \frac{15}{16} = \frac{225}{256}$$
The area is $\frac{225}{256}$ in.^{2}. Multiply the length by the width to find the area of any rectangle.

2.8 Multiplying and Dividing Mixed Numbers

2.8 Margin Exercises

1. (a) $4\frac{2}{3}$ ← Half of 3 is $1\frac{1}{2}$.

2 is <u>more than</u> half of 3.

$4\frac{2}{3}$ rounds up to <u>5</u>.

(b) $3\frac{2}{5}$ ← Half of 5 is $2\frac{1}{2}$.

2 is <u>less than</u> of 5.

$3\frac{2}{5}$ rounds down to <u>3</u>.

(c) $5\frac{3}{4}$ ← Half of 4 is 2.

3 is more than half of 4.

$5\frac{3}{4}$ rounds up to 6.

(d) $4\frac{7}{12}$ ← Half of 12 is 6.

7 is more than half of 12.

$4\frac{7}{12}$ rounds up to 5.

(e) $1\frac{1}{2}$ ← Half of 2 is 1.

1 is the same as half of 2.

$1\frac{1}{2}$ rounds up to 2.

(f) $8\frac{4}{9}$ ← Half of 9 is $4\frac{1}{2}$.

4 is less than half of 9.

$8\frac{4}{9}$ rounds down to 8.

2. (a) $3\frac{1}{4} \cdot 6\frac{2}{3}$

Estimate: $3\frac{1}{4}$ rounds to 3. $6\frac{2}{3}$ rounds to 7.
$\underline{3} \cdot \underline{7} = \underline{21}$

Exact: $3\frac{1}{4} \cdot 6\frac{2}{3} = \frac{13}{\cancel{4}_1} \cdot \frac{\cancel{20}^5}{3} = \frac{65}{3} = 21\frac{2}{3}$

Copyright © 2018 Pearson Education, Inc.

(b) $4\frac{2}{3} \cdot 2\frac{3}{4}$

Estimate: $4\frac{2}{3}$ rounds to 5. $2\frac{3}{4}$ rounds to 3.
$\underline{5} \cdot \underline{3} = \underline{15}$

Exact:

$$4\frac{2}{3} \cdot 2\frac{3}{4} = \frac{14}{3} \cdot \frac{11}{4} = \frac{\overset{7}{\cancel{14}}}{3} \cdot \frac{11}{\underset{2}{\cancel{4}}} = \frac{77}{6} = 12\frac{5}{6}$$

(c) $3\frac{3}{5} \cdot 4\frac{4}{9}$

Estimate: $3\frac{3}{5}$ rounds to 4. $4\frac{4}{9}$ rounds to 4.
$\underline{4} \cdot \underline{4} = \underline{16}$

Exact:

$$3\frac{3}{5} \cdot 4\frac{4}{9} = \frac{18}{5} \cdot \frac{40}{9} = \frac{\overset{2}{\cancel{18}}}{\underset{1}{\cancel{5}}} \cdot \frac{\overset{8}{\cancel{40}}}{\underset{1}{\cancel{9}}} = 16$$

(d) $5\frac{1}{4} \cdot 3\frac{3}{5}$

Estimate: $5\frac{1}{4}$ rounds to 5. $3\frac{3}{5}$ rounds to 4.
$\underline{5} \cdot \underline{4} = \underline{20}$

Exact:

$$5\frac{1}{4} \cdot 3\frac{3}{5} = \frac{21}{4} \cdot \frac{18}{5} = \frac{21}{\underset{2}{\cancel{4}}} \cdot \frac{\overset{9}{\cancel{18}}}{5} = \frac{189}{10} = 18\frac{9}{10}$$

3. **(a)** $3\frac{1}{8} \div 6\frac{1}{4}$

Estimate: $3\frac{1}{8}$ rounds to 3. $6\frac{1}{4}$ rounds to 6.
$\underline{3} \div \underline{6} = \underline{\frac{1}{2}}$

Exact:

$$3\frac{1}{8} \div 6\frac{1}{4} = \frac{25}{8} \div \frac{25}{4} = \frac{25}{8} \cdot \frac{4}{25} = \frac{\overset{1}{\cancel{25}}}{\underset{2}{\cancel{8}}} \cdot \frac{\overset{1}{\cancel{4}}}{\underset{1}{\cancel{25}}} = \frac{1}{2}$$

(b) $10\frac{1}{3} \div 2\frac{1}{2}$

Estimate: $10\frac{1}{3}$ rounds to 10. $2\frac{1}{2}$ rounds to 3.
$\underline{10} \div \underline{3} = \underline{3\frac{1}{3}}$

Exact:
$$10\frac{1}{3} \div 2\frac{1}{2} = \frac{31}{3} \div \frac{5}{2} = \frac{31}{3} \cdot \frac{2}{5} = \frac{62}{15} = 4\frac{2}{15}$$

(c) $8 \div 5\frac{1}{3}$

Estimate: 8 rounds to 8. $5\frac{1}{3}$ rounds to 5.
$\underline{8} \div \underline{5} = \underline{1\frac{3}{5}}$

Exact:

$$8 \div 5\frac{1}{3} = \frac{8}{1} \div \frac{16}{3} = \frac{\overset{1}{\cancel{8}}}{1} \cdot \frac{3}{\underset{2}{\cancel{16}}} = \frac{3}{2} = 1\frac{1}{2}$$

(d) $13\frac{1}{2} \div 18$

Estimate: $13\frac{1}{2}$ rounds to 14. 18 rounds to 18.
$\underline{14} \div \underline{18} = \frac{14}{18} = \frac{7}{9}$

Exact:

$$13\frac{1}{2} \div 18 = \frac{27}{2} \div \frac{18}{1} = \frac{\overset{3}{\cancel{27}}}{2} \cdot \frac{1}{\underset{2}{\cancel{18}}} = \frac{3}{4}$$

4. The problem asks how many quarts of paint are needed to paint 16 cars.

Multiply the amount of paint needed for each car by the number of cars.

Estimate: $2\frac{5}{8}$ rounds to 3. 16 rounds to 16.
$\underline{3} \cdot \underline{16} = \underline{48}$

Exact: $2\frac{5}{8} \cdot 16 = \frac{21}{\underset{1}{\cancel{8}}} \cdot \frac{\overset{2}{\cancel{16}}}{1} = \frac{42}{1} = 42$

42 quarts are needed for 16 cars. The answer is reasonably close to the estimate.

5. **(a)** The problem asks how many propellers can be manufactured from 57 pounds of brass.

Divide the total pounds of brass by the number of pounds needed for one engine.

Estimate: 57 rounds to 57. $4\frac{3}{4}$ rounds to 5.
$57 \div 5 \approx 11$

Exact:

$$57 \div 4\frac{3}{4} = \frac{57}{1} \div \frac{19}{4} = \frac{\overset{3}{\cancel{57}}}{1} \cdot \frac{4}{\underset{1}{\cancel{19}}} = \frac{12}{1} = 12$$

12 propellers can be manufactured from 57 pounds of brass. The answer is reasonably close to the estimate.

(b) The problem asks the number of oil changes that can be made with 609 quarts of oil.

Divide the total number of quarts by the number of quarts needed for each oil change.

Estimate: 609 rounds to 600. $21\frac{3}{4}$ rounds to 22.
$600 \div 22 \approx 27$

(continued)

Copyright © 2018 Pearson Education, Inc.

Exact:

$$609 \div 21\frac{3}{4} = 609 \div \frac{87}{4} = \frac{\overset{7}{\cancel{609}}}{1} \cdot \frac{4}{\cancel{87}_{1}} = 28$$

28 oil changes can be made with 609 quarts of oil. The answer is reasonably close to the estimate.

2.8   Section Exercises

1. The statement "When multiplying two mixed numbers, the reciprocal of the second mixed number must be used." is *false*. A reciprocal is used when *dividing* fractions, not *multiplying* fractions.

3. The statement "When rounding a mixed number before estimating the answer, round up to the next whole number if the numerator of the fraction part is half of the denominator or more" is *true*.

5. $4\frac{1}{2} \cdot 1\frac{3}{4}$

Estimate: $5 \cdot 2 = 10$

To find the exact answer, change each mixed number to an improper fraction and then multiply.

Exact: $4\frac{1}{2} \cdot 1\frac{3}{4} = \frac{9}{2} \cdot \frac{7}{4} = \frac{63}{8} = 7\frac{7}{8}$

7. $1\frac{2}{3} \cdot 2\frac{7}{10}$

Estimate: $2 \cdot 3 = 6$

Exact: $1\frac{2}{3} \cdot 2\frac{7}{10} = \frac{5}{3} \cdot \frac{27}{10} = \frac{\overset{1}{\cancel{5}}}{\cancel{3}_{1}} \cdot \frac{\overset{9}{\cancel{27}}}{\cancel{10}_{2}} = \frac{9}{2} = 4\frac{1}{2}$

9. $3\frac{1}{9} \cdot 1\frac{2}{7}$

Estimate: $3 \cdot 1 = 3$

Exact: $3\frac{1}{9} \cdot 1\frac{2}{7} = \frac{28}{9} \cdot \frac{9}{7} = \frac{\overset{4}{\cancel{28}}}{\cancel{9}_{1}} \cdot \frac{\overset{1}{\cancel{9}}}{7} = \frac{4}{1} = 4$

11. $8 \cdot 6\frac{1}{4}$

Estimate: $8 \cdot 6 = 48$

Exact: $8 \cdot 6\frac{1}{4} = \frac{8}{1} \cdot \frac{25}{4} = \frac{\overset{2}{\cancel{8}}}{1} \cdot \frac{25}{\cancel{4}_{1}} = \frac{50}{1} = 50$

13. $4\frac{1}{2} \cdot 2\frac{1}{5} \cdot 5$

Estimate: $5 \cdot 2 \cdot 5 = 10 \cdot 5 = 50$

Exact: $4\frac{1}{2} \cdot 2\frac{1}{5} \cdot 5 = \frac{9}{2} \cdot \frac{11}{\cancel{5}_{1}} \cdot \frac{\overset{1}{\cancel{5}}}{1} = \frac{99}{2} = 49\frac{1}{2}$

15. $3 \cdot 1\frac{1}{2} \cdot 2\frac{2}{3}$

Estimate: $3 \cdot 2 \cdot 3 = 6 \cdot 3 = 18$

Exact: $3 \cdot 1\frac{1}{2} \cdot 2\frac{2}{3} = \frac{\overset{1}{\cancel{3}}}{1} \cdot \frac{3}{\cancel{2}_{1}} \cdot \frac{\overset{4}{\cancel{8}}}{\cancel{3}_{1}} = \frac{12}{1} = 12$

17. $3\frac{1}{4} \cdot 7\frac{5}{8}$ *Estimate*: $3 \cdot 8 = 24$

The best estimate is choice **(d)**.

19. $2\frac{1}{8} \div 1\frac{3}{4}$ *Estimate*: $2 \div 2 = 1$

The best estimate is choice **(b)**.

21. $1\frac{1}{4} \div 3\frac{3}{4}$

Estimate: $1 \div 4 = \frac{1}{4}$

To find the exact answer, first change each mixed number to an improper fraction. Then, use the reciprocal of the divisor (the second fraction) and multiply.

Exact: $1\frac{1}{4} \div 3\frac{3}{4} = \frac{5}{4} \div \frac{15}{4} = \frac{\overset{1}{\cancel{5}}}{\cancel{4}_{1}} \cdot \frac{\overset{1}{\cancel{4}}}{\cancel{15}_{3}} = \frac{1}{3}$

23. $2\frac{1}{2} \div 3$

Estimate: $3 \div 3 = 1$

Exact: $2\frac{1}{2} \div 3 = \frac{5}{2} \div \frac{3}{1} = \frac{5}{2} \cdot \frac{1}{3} = \frac{5}{6}$

25. $9 \div 2\frac{1}{2}$

Estimate: $9 \div 3 = 3$

Exact: $9 \div 2\frac{1}{2} = \frac{9}{1} \div \frac{5}{2} = \frac{9}{1} \cdot \frac{2}{5} = \frac{18}{5} = 3\frac{3}{5}$

27. $\frac{5}{8} \div 1\frac{1}{2}$

Estimate: $1 \div 2 = \frac{1}{2}$

Exact: $\frac{5}{8} \div 1\frac{1}{2} = \frac{5}{8} \div \frac{3}{2} = \frac{5}{\cancel{8}_{4}} \cdot \frac{\overset{1}{\cancel{2}}}{3} = \frac{5}{12}$

Copyright © 2018 Pearson Education, Inc.

29. $1\frac{7}{8} \div 6\frac{1}{4}$

Estimate: $2 \div 6 = \frac{2}{6} = \frac{1}{3}$

Exact: $1\frac{7}{8} \div 6\frac{1}{4} = \frac{15}{8} \div \frac{25}{4} = \frac{\overset{3}{\cancel{15}}}{\underset{2}{\cancel{8}}} \cdot \frac{\overset{1}{\cancel{4}}}{\underset{5}{\cancel{25}}} = \frac{3}{10}$

31. $5\frac{2}{3} \div 6$

Estimate: $6 \div 6 = 1$

Exact: $5\frac{2}{3} \div 6 = \frac{17}{3} \div \frac{6}{1} = \frac{17}{3} \cdot \frac{1}{6} = \frac{17}{18}$

33. Multiply each amount by $2\frac{1}{2}$.

(a) Blueberries: $\frac{2}{3}$ cup

Estimate: $1 \cdot 3 = 3$ cups

Exact: $\frac{2}{3} \cdot 2\frac{1}{2} = \frac{\overset{1}{\cancel{2}}}{3} \cdot \frac{5}{\underset{1}{\cancel{2}}} = \frac{5}{3} = 1\frac{2}{3}$ cups

(b) Salt: $\frac{1}{2}$ tsp.

Estimate: $1 \cdot 3 = 3$ tsp.

Exact: $\frac{1}{2} \cdot 2\frac{1}{2} = \frac{1}{2} \cdot \frac{5}{2} = \frac{5}{4} = 1\frac{1}{4}$ tsp.

(c) Flour: $1\frac{1}{4}$ cups

Estimate: $1 \cdot 3 = 3$ cups

Exact: $1\frac{1}{4} \cdot 2\frac{1}{2} = \frac{5}{4} \cdot \frac{5}{2} = \frac{25}{8} = 3\frac{1}{8}$ cups

35. Divide each amount by 2.

(a) Vanilla extract: $\frac{1}{2}$ tsp.

Estimate: $1 \div 2 = \frac{1}{2}$ tsp.

Exact: $\frac{1}{2} \div 2 = \frac{1}{2} \div \frac{2}{1} = \frac{1}{2} \cdot \frac{1}{2} = \frac{1}{4}$ tsp.

(b) Blueberries: $\frac{2}{3}$ cup

Estimate: $1 \div 2 = \frac{1}{2}$ cup

Exact: $\frac{2}{3} \div 2 = \frac{2}{3} \div \frac{2}{1} = \frac{\overset{1}{\cancel{2}}}{3} \cdot \frac{1}{\underset{1}{\cancel{2}}} = \frac{1}{3}$ cup

(c) Flour: $1\frac{1}{4}$ cups

Estimate: $1 \div 2 = \frac{1}{2}$ cup

Exact: $1\frac{1}{4} \div 2 = \frac{5}{4} \div \frac{2}{1} = \frac{5}{4} \cdot \frac{1}{2} = \frac{5}{8}$ cup

37. Divide the number of gallons available by the number of gallons needed for each unit.

Estimate: $329 \div 12 \approx 27$ units

Exact:

$329 \div 11\frac{3}{4} = \frac{329}{1} \div \frac{47}{4} = \frac{\overset{7}{\cancel{329}}}{1} \cdot \frac{4}{\underset{1}{\cancel{47}}} = 28$

28 units can be painted with 329 gallons of paint.

39. Each handle requires $19\frac{1}{2}$ inches of steel tubing. Use multiplication.

Estimate: $45 \cdot 20 = 900$ in.

Exact: $45 \cdot 19\frac{1}{2} = \frac{45}{1} \cdot \frac{39}{2} = \frac{1755}{2} = 877\frac{1}{2}$ in.

$877\frac{1}{2}$ inches of steel tubing is needed to make 45 jacks.

41. The answer should include:

Step 1
Change mixed numbers to improper fractions.

Step 2
Multiply the fractions.

Step 3
Write the answer in lowest terms, changing to mixed or whole numbers where possible.

43. Multiply the amount of money for each cell phone times the number of cell phones to get the total amount of money from the sale of gold.

Estimate: $\$1 \cdot 130$ million $= \$130$ million

Exact:
$1\frac{2}{5} \cdot 130$ million $= \frac{7}{5} \cdot \frac{130}{1} = \frac{910}{5} = 182$
$= 182$ million

You would have $182 million from the sale of the gold.

45. Divide the total amount of firewood to be delivered by the amount of firewood that can be delivered per trip.

Estimate: $140 \div 1 = 140$ trips

Exact: $140 \div 1\frac{1}{4} = \frac{140}{1} \div \frac{5}{4} = \frac{\overset{28}{\cancel{140}}}{1} \cdot \frac{4}{\underset{1}{\cancel{5}}} = 112$

112 trips will be needed to deliver 140 cords of firewood.

Copyright © 2018 Pearson Education, Inc.

47. **(a)** The maximum height of the standard jack is $17\frac{3}{4}$ inches. Use multiplication.

Estimate: $18 \cdot 4 = 72$ in.

Exact: $17\frac{3}{4} \cdot 4 = \frac{71}{\overset{1}{\cancel{4}}} \cdot \frac{\overset{1}{\cancel{4}}}{1} = \frac{71}{1} = 71$ in.

The hydraulic lift must raise the car 71 inches.

 (b) There are 12 inches in a foot, so the 6-foot-tall mechanic is $6 \times 12 = 72$ inches tall. So no, the mechanic can not stand under the car without bending.

49. Multiply the swimming speed of the person times the number of times faster that a shark can swim than a person.

Estimate: $6 \cdot 6 = 36$ miles per hour

Exact: $6\frac{1}{8} \cdot 5\frac{1}{2} = \frac{49}{8} \cdot \frac{11}{2} = \frac{539}{16} = 33\frac{11}{16}$

The shark can swim $33\frac{11}{16}$ miles per hour.

51. **(a)** The recipe makes four Alaska burgers, so divide the $\frac{1}{3}$-pound by 4.

$$\frac{1}{3} \div 4 = \frac{1}{3} \cdot \frac{1}{4} = \frac{1}{12}$$

Each $\frac{1}{2}$-inch slice of cheese weighs $\frac{1}{12}$ pound.

(b) There are four $\frac{1}{2}$-inch slices in the $\frac{1}{3}$-pound brick, so multiply 4 times $\frac{1}{2}$.

$$4 \cdot \frac{1}{2} = \frac{\overset{2}{\cancel{4}}}{1} \cdot \frac{1}{\underset{1}{\cancel{2}}} = \frac{2}{1} = 2$$

A $\frac{1}{3}$-pound brick of smoked cheddar cheese is 2 inches thick.

52. **(a)** One good pinch of allspice is equal to $\frac{1}{4}$ teaspoon, so multiply 8 times $\frac{1}{4}$.

$$8 \cdot \frac{1}{4} = \frac{\overset{2}{\cancel{8}}}{1} \cdot \frac{1}{\underset{1}{\cancel{4}}} = \frac{2}{1} = 2$$

There are 2 teaspoons of allspice in 8 good pinches.

(b) Two good pinches of ground cumin is equal to $\frac{1}{2}$ teaspoon, so first divide 7 by 2 to find the number of "two-pinches" there are.

$$7 \div 2 = \frac{7}{2} \text{ "two-pinches"}$$

Now multiply $\frac{7}{2}$ by $\frac{1}{2}$ to find the number of teaspoons.

$$\frac{7}{2} \cdot \frac{1}{2} = \frac{7}{4} = 1\frac{3}{4} \text{ teaspoons}$$

There are $1\frac{3}{4}$ teaspoons of ground cumin in 7 good pinches.

53. The recipe makes four Alaska burgers, so to prepare 18 burgers, multiply the ingredient amounts by $\frac{18}{4} = \frac{9}{2} = 4\frac{1}{2}$.

54. **(a)** The five large eaters will eat

$$5 \cdot 1\frac{1}{2} = \frac{5}{1} \cdot \frac{3}{2} = \frac{15}{2} = 7\frac{1}{2} \text{ burgers.}$$

The five children will eat

$$5 \cdot \frac{1}{2} = \frac{5}{1} \cdot \frac{1}{2} = \frac{5}{2} = 2\frac{1}{2} \text{ burgers.}$$

There are $15 - (5 + 5) = 5$ guests who will eat 1 burger each, for a total of 5 burgers. So, you will need

$$7\frac{1}{2} + 2\frac{1}{2} + 5 = 15 \text{ burgers.}$$

(b) The recipe makes four Alaska burgers, so to prepare 15 burgers, multiply the ingredient amounts by $\frac{15}{4} = 3\frac{3}{4}$.

55. To find the ingredient amounts for 9 servings, multiply each amount in the recipe by $\frac{9}{4}$.

Lean ground beef: $1 \cdot \frac{9}{4} = \frac{9}{4} = 2\frac{1}{4}$ pounds

Spanish onions: $\frac{1}{2} \cdot \frac{9}{4} = \frac{9}{8} = 1\frac{1}{8}$ onions

Worcestershire sauce:

$$4 \cdot \frac{9}{4} = \frac{\overset{1}{\cancel{4}}}{1} \cdot \frac{9}{\underset{1}{\cancel{4}}} = \frac{9}{1} = 9 \text{ shakes}$$

Allspice: $\frac{1}{4} \cdot \frac{9}{4} = \frac{9}{16}$ tsp

Ground cumin: $\frac{1}{2} \cdot \frac{9}{4} = \frac{9}{8} = 1\frac{1}{8}$ tsp

Cheddar cheese: $\frac{1}{\underset{1}{\cancel{3}}} \cdot \frac{\overset{3}{\cancel{9}}}{4} = \frac{3}{4}$ pound

56. One pound of beef makes 4 Alaska burgers, so $5\frac{3}{4}$ pounds make

$$4 \cdot 5\frac{3}{4} = \frac{\overset{1}{\cancel{4}}}{1} \cdot \frac{23}{\underset{1}{\cancel{4}}} = 23 \text{ burgers}$$

Copyright © 2018 Pearson Education, Inc.

Chapter 2 Review Exercises

1. $\frac{1}{3}$ There are 3 parts, and 1 is shaded.

2. $\frac{5}{8}$ There are 8 parts, and 5 are shaded.

3. $\frac{2}{4}$ There are 4 parts, and 2 are shaded.

4. Proper fractions have numerator (top) smaller than denominator (bottom). They are $\frac{1}{8}, \frac{3}{4}, \frac{2}{3}$.

Improper fractions have numerator (top) larger than or equal to the denominator (bottom). They are $\frac{4}{3}, \frac{5}{5}$.

5. Proper fractions have numerator (top) smaller than denominator (bottom). They are $\frac{15}{16}, \frac{1}{8}$.

Improper fractions have numerator (top) larger than or equal to the denominator (bottom). They are $\frac{6}{5}, \frac{16}{13}, \frac{5}{3}$.

6. $4\frac{3}{4}$ $\quad 4 \cdot 4 = 16$
$\qquad 16 + 3 = 19$
$4\frac{3}{4} = \frac{19}{4}$

7. $9\frac{5}{6}$ $\quad 9 \cdot 6 = 54$
$\qquad 54 + 5 = 59$
$9\frac{5}{6} = \frac{59}{6}$

8. $\frac{27}{8}$

$\begin{array}{r} 3 \leftarrow \text{Whole number part} \\ 8\overline{)2\ 7} \\ \underline{2\ 4} \\ 3 \leftarrow \text{Remainder} \end{array}$

$\frac{27}{8} = 3\frac{3}{8}$

9. $\frac{63}{5}$

$\begin{array}{r} 1\ 2 \leftarrow \text{Whole number part} \\ 5\overline{)6\ 3} \\ \underline{5} \\ 1\ 3 \\ \underline{1\ 0} \\ 3 \leftarrow \text{Remainder} \end{array}$

$\frac{63}{5} = 12\frac{3}{5}$

10. Factorizations of 6:
$$1 \cdot 6 = 6 \quad 2 \cdot 3 = 6$$
The factors of 6 are 1, 2, 3, and 6.

11. Factorizations of 24:
$$1 \cdot 24 = 24 \quad 2 \cdot 12 = 24 \quad 3 \cdot 8 = 24 \quad 4 \cdot 6 = 24$$
The factors of 24 are 1, 2, 3, 4, 6, 8, 12, and 24.

12. Factorizations of 55:
$$1 \cdot 55 = 55 \quad 5 \cdot 11 = 55$$
The factors of 55 are 1, 5, 11, and 55.

13. Factorizations of 90:
$$1 \cdot 90 = 90 \quad 2 \cdot 45 = 90 \quad 3 \cdot 30 = 90$$
$$5 \cdot 18 = 90 \quad 6 \cdot 15 = 90 \quad 9 \cdot 10 = 90$$
The factors of 90 are 1, 2, 3, 5, 6, 9, 10, 15, 18, 30, 45, and 90.

14.

$27 = 3 \cdot 3 \cdot 3 = 3^3$

15.

$150 = 2 \cdot 3 \cdot 5 \cdot 5 = 2 \cdot 3 \cdot 5^2$

16.

$420 = 2 \cdot 2 \cdot 3 \cdot 5 \cdot 7 = 2^2 \cdot 3 \cdot 5 \cdot 7$

17. $5^2 = 5 \cdot 5 = 25$

18. $6^2 \cdot 2^3 = 36 \cdot 8 = 288$

19. $8^2 \cdot 3^3 = 64 \cdot 27 = 1728$

20. $4^3 \cdot 2^5 = 64 \cdot 32 = 2048$

21. All 24 parts out of a possible 24 parts are gold.
$$\frac{24}{24} = 1$$

Copyright © 2018 Pearson Education, Inc.

22. 18 of the possible $18 + 6 = 24$ parts are gold.

$$\frac{18}{24} = \frac{18 \div 6}{24 \div 6} = \frac{3}{4}$$

23. 14 of the possible $14 + 10 = 24$ parts are gold.

$$\frac{14}{24} = \frac{14 \div 2}{24 \div 2} = \frac{7}{12}$$

24. 10 of the possible $10 + 14 = 24$ parts are gold.

$$\frac{10}{24} = \frac{10 \div 2}{24 \div 2} = \frac{5}{12}$$

25. $\dfrac{25}{60} = \dfrac{\overset{1}{\cancel{5}} \cdot 5}{2 \cdot 2 \cdot 3 \cdot \underset{1}{\cancel{5}}} = \dfrac{5}{12}$

26. $\dfrac{384}{96} = \dfrac{\overset{1}{\cancel{2}} \cdot \overset{1}{\cancel{2}} \cdot \overset{1}{\cancel{2}} \cdot \overset{1}{\cancel{2}} \cdot \overset{1}{\cancel{2}} \cdot 2 \cdot 2 \cdot \overset{1}{\cancel{3}}}{\underset{1}{\cancel{2}} \cdot \underset{1}{\cancel{2}} \cdot \underset{1}{\cancel{2}} \cdot \underset{1}{\cancel{2}} \cdot \underset{1}{\cancel{2}} \cdot \underset{1}{\cancel{3}}} = \dfrac{4}{1} = 4$

27. $\dfrac{3}{4}$ and $\dfrac{48}{64}$

$$\frac{48}{64} = \frac{48 \div 16}{64 \div 16} = \frac{3}{4}$$

The fractions are equivalent $(\frac{3}{4} = \frac{3}{4})$.

28. $\dfrac{5}{8}$ and $\dfrac{70}{120}$

$$\frac{70}{120} = \frac{70 \div 10}{120 \div 10} = \frac{7}{12}$$

The fractions are not equivalent $(\frac{5}{8} \neq \frac{7}{12})$.

29. $\dfrac{2}{3}$ and $\dfrac{360}{540}$

$$\frac{360}{540} = \frac{360 \div 180}{540 \div 180} = \frac{2}{3}$$

The fractions are equivalent $(\frac{2}{3} = \frac{2}{3})$.

30. $\dfrac{4}{5} \cdot \dfrac{3}{4} = \dfrac{4}{5} \cdot \dfrac{3}{\underset{1}{\cancel{4}}} = \dfrac{1 \cdot 3}{5 \cdot 1} = \dfrac{3}{5}$

31. $\dfrac{3}{10} \cdot \dfrac{5}{8} = \dfrac{3}{\underset{2}{\cancel{10}}} \cdot \dfrac{\overset{1}{\cancel{5}}}{8} = \dfrac{3 \cdot 1}{2 \cdot 8} = \dfrac{3}{16}$

32. $\dfrac{70}{175} \cdot \dfrac{5}{14} = \dfrac{\overset{\overset{1}{\cancel{5}}}{\cancel{70}}}{\underset{\underset{7}{\cancel{35}}}{\cancel{175}}} \cdot \dfrac{\overset{1}{\cancel{5}}}{\underset{1}{\cancel{14}}} = \dfrac{1 \cdot 1}{7 \cdot 1} = \dfrac{1}{7}$

33. $\dfrac{44}{63} \cdot \dfrac{3}{11} = \dfrac{\overset{4}{\cancel{44}}}{\underset{21}{\cancel{63}}} \cdot \dfrac{\overset{1}{\cancel{3}}}{\underset{1}{\cancel{11}}} = \dfrac{4 \cdot 1}{21 \cdot 1} = \dfrac{4}{21}$

34. $\dfrac{5}{16} \cdot 48 = \dfrac{5}{\underset{1}{\cancel{16}}} \cdot \dfrac{\overset{3}{\cancel{48}}}{1} = \dfrac{5 \cdot 3}{1 \cdot 1} = \dfrac{15}{1} = 15$

35. $\dfrac{5}{8} \cdot 1000 = \dfrac{5}{\underset{1}{\cancel{8}}} \cdot \dfrac{\overset{125}{\cancel{1000}}}{1} = \dfrac{5 \cdot 125}{1 \cdot 1} = \dfrac{625}{1} = 625$

36. $\dfrac{2}{3} \div \dfrac{1}{2} = \dfrac{2}{3} \cdot \dfrac{2}{1} = \dfrac{2 \cdot 2}{3 \cdot 1} = \dfrac{4}{3} = 1\dfrac{1}{3}$

37. $\dfrac{5}{6} \div \dfrac{1}{2} = \dfrac{5}{\underset{3}{\cancel{6}}} \cdot \dfrac{\overset{1}{\cancel{2}}}{1} = \dfrac{5}{3} = 1\dfrac{2}{3}$

38. $\dfrac{\frac{15}{18}}{\frac{10}{30}} = \dfrac{15}{18} \div \dfrac{10}{30}$

$= \dfrac{\overset{5}{\cancel{15}}}{\underset{6}{\cancel{18}}} \cdot \dfrac{\overset{3}{\cancel{30}}}{\underset{1}{\cancel{10}}} = \dfrac{5 \cdot 3}{\underset{2}{\cancel{6}} \cdot 1} = \dfrac{5 \cdot 1}{2 \cdot 1} = \dfrac{5}{2} = 2\dfrac{1}{2}$

39. $\dfrac{\frac{3}{4}}{\frac{3}{8}} = \dfrac{3}{4} \div \dfrac{3}{8} = \dfrac{\overset{1}{\cancel{3}}}{\underset{1}{\cancel{4}}} \cdot \dfrac{\overset{2}{\cancel{8}}}{\underset{1}{\cancel{3}}} = \dfrac{1 \cdot 2}{1 \cdot 1} = 2$

40. $7 \div \dfrac{7}{8} = \dfrac{7}{1} \div \dfrac{7}{8} = \dfrac{\overset{1}{\cancel{7}}}{1} \cdot \dfrac{8}{\underset{1}{\cancel{7}}} = \dfrac{1 \cdot 8}{1 \cdot 1} = 8$

41. $18 \div \dfrac{3}{4} = \dfrac{\overset{6}{\cancel{18}}}{1} \cdot \dfrac{4}{\underset{1}{\cancel{3}}} = \dfrac{6 \cdot 4}{1 \cdot 1} = 24$

42. $\dfrac{5}{8} \div 3 = \dfrac{5}{8} \div \dfrac{3}{1} = \dfrac{5}{8} \cdot \dfrac{1}{3} = \dfrac{5 \cdot 1}{8 \cdot 3} = \dfrac{5}{24}$

43. $\dfrac{2}{3} \div 5 = \dfrac{2}{3} \div \dfrac{5}{1} = \dfrac{2}{3} \cdot \dfrac{1}{5} = \dfrac{2 \cdot 1}{3 \cdot 5} = \dfrac{2}{15}$

44. $\dfrac{\frac{12}{13}}{3} = \dfrac{12}{13} \div 3 = \dfrac{12}{13} \div \dfrac{3}{1} = \dfrac{\overset{4}{\cancel{12}}}{13} \cdot \dfrac{1}{\underset{1}{\cancel{3}}} = \dfrac{4 \cdot 1}{13 \cdot 1} = \dfrac{4}{13}$

45. To find the area, multiply the length and the width.

$$2\frac{3}{4} \cdot \frac{1}{2} = \frac{11}{4} \cdot \frac{1}{2} = \frac{11}{8} = 1\frac{3}{8}$$

The area is $1\frac{3}{8}$ ft^2.

Copyright © 2018 Pearson Education, Inc.

46. To find the area, multiply the length and the width.

$$4\frac{1}{2}\cdot\frac{7}{8}=\frac{9}{2}\cdot\frac{7}{8}=\frac{63}{16}=3\frac{15}{16}$$

The area is $3\frac{15}{16}$ yd^2.

47. Multiply the length and width.

$$108\cdot72\frac{3}{4}=\frac{\overset{27}{\cancel{108}}}{1}\cdot\frac{291}{\underset{1}{\cancel{4}}}=7857$$

The area is 7857 ft^2.

48. Multiply the length and width.

$$6\cdot\frac{11}{12}=\frac{\overset{1}{\cancel{6}}}{1}\cdot\frac{11}{\underset{2}{\cancel{12}}}=\frac{11}{2}$$

The area is $5\frac{1}{2}$ ft^2.

49. $5\frac{1}{2}\cdot1\frac{1}{4}$

Estimate: $6\cdot1=6$

Exact: $5\frac{1}{2}\cdot1\frac{1}{4}=\frac{11}{2}\cdot\frac{5}{4}=\frac{55}{8}=6\frac{7}{8}$

50. $2\frac{1}{4}\cdot7\frac{1}{8}\cdot1\frac{1}{3}$

Estimate: $2\cdot7\cdot1=14$

Exact: $2\frac{1}{4}\cdot7\frac{1}{8}\cdot1\frac{1}{3}=\frac{9}{4}\cdot\frac{57}{\underset{2}{\cancel{8}}}^{3}\cdot\frac{\overset{1}{\cancel{4}}}{\underset{1}{\cancel{3}}}=\frac{171}{8}=21\frac{3}{8}$

51. $15\frac{1}{2}\div3$

Estimate: $16\div3=\frac{16}{3}=5\frac{1}{3}$

Exact: $15\frac{1}{2}\div3=\frac{31}{2}\cdot\frac{1}{3}=\frac{31}{6}=5\frac{1}{6}$

52. $4\frac{3}{4}\div6\frac{1}{3}$

Estimate: $5\div6=\frac{5}{6}$

Exact: $4\frac{3}{4}\div6\frac{1}{3}=\frac{19}{4}\div\frac{19}{3}=\frac{\overset{1}{\cancel{19}}}{4}\cdot\frac{3}{\underset{1}{\cancel{19}}}=\frac{3}{4}$

53. Divide the total tons of almonds by the size of the bins.

Estimate: $300\div\frac{1}{2}=300\cdot2=600$ bins

Exact: $320\div\frac{5}{8}=\frac{\overset{64}{\cancel{320}}}{1}\cdot\frac{8}{\underset{1}{\cancel{5}}}=512$

512 bins will be needed to store the almonds.

54. The 4 other equal partners own

$$1-\tfrac{2}{5}=\tfrac{3}{5}$$

of the business. Divide that amount by 4.

$$\frac{3}{5}\div4=\frac{3}{5}\div\frac{4}{1}=\frac{3}{5}\cdot\frac{1}{4}=\frac{3}{20}$$

Each of the other partners owns $\frac{3}{20}$ of the business.

55. Divide the total yardage by the amount needed for each pull cord.

Estimate: $158\div4\approx40$ pull cords

Exact:

$$157\frac{1}{2}\div4\frac{3}{8}=\frac{315}{2}\div\frac{35}{8}=\frac{\overset{9}{\cancel{315}}}{\underset{1}{\cancel{2}}}\cdot\frac{\overset{4}{\cancel{8}}}{\underset{1}{\cancel{35}}}=36$$

36 pull cords can be made.

56. Multiply the weight per gallon times the number of aquariums times the gallons per aquarium.

Estimate: $8\cdot2\cdot50=800$

Exact: $8\frac{1}{3}\cdot2\cdot50=\frac{25}{3}\cdot\frac{2}{1}\cdot\frac{50}{1}=\frac{2500}{3}$

The weight of the water is $\frac{2500}{3}$, or $833\frac{1}{3}$ pounds.

57. Ebony gave $\frac{1}{4}$ of 100 cookies to her neighbor.

$$\frac{1}{4}\cdot100=\frac{1}{\underset{1}{\cancel{4}}}\cdot\frac{\overset{25}{\cancel{100}}}{1}=\frac{1\cdot25}{1\cdot1}=\frac{25}{1}=25\text{ cookies}$$

Thus, $100-25=75$ cookies remain. She used $\frac{2}{3}$ of 75 cookies for the bake sale.

$$\frac{2}{3}\cdot75=\frac{2}{\underset{1}{\cancel{3}}}\cdot\frac{\overset{25}{\cancel{75}}}{1}=\frac{2\cdot25}{1\cdot1}=\frac{50}{1}=50\text{ cookies}$$

Ebony used 50 cookies for the bake sale. She has $75-50=25$ cookies left for her family.

Copyright © 2018 Pearson Education, Inc.

58. Sheila paid $\frac{3}{8}$ of $2976 for taxes, social security, and a retirement plan.

$$\frac{3}{8} \cdot 2976 = \frac{3}{\overset{1}{\cancel{8}}} \cdot \frac{\overset{372}{\cancel{2976}}}{1} = 1116$$

She paid $1116 for taxes, social security, and a retirement plan.

She paid $\frac{9}{10}$ of the remainder, $2976 - $1116 = $1860, for basic living expenses.

$$\frac{9}{10} \cdot 1860 = \frac{9}{\overset{1}{\cancel{10}}} \cdot \frac{\overset{186}{\cancel{1860}}}{1} = \frac{9 \cdot 186}{1 \cdot 1} = 1674$$

She has $1860 - $1674 = $186 left.

59. $\frac{7}{8}$ must be divided by 6.

$$\frac{7}{8} \div 6 = \frac{7}{8} \div \frac{6}{1} = \frac{7}{8} \cdot \frac{1}{6} = \frac{7 \cdot 1}{8 \cdot 6} = \frac{7}{48}$$

Each school will receive $\frac{7}{48}$ of the amount raised.

60. $\frac{4}{5}$ of the catch must be divided evenly among 5 fishermen.

$$\frac{4}{5} \div 5 = \frac{4}{5} \div \frac{5}{1} = \frac{4}{5} \cdot \frac{1}{5} = \frac{4}{25}$$

Each fisherman receives $\frac{4}{25}$ ton.

Chapter 2 Mixed Review Exercises

1. $\dfrac{1}{2} \cdot \dfrac{3}{4} = \dfrac{1 \cdot 3}{2 \cdot 4} = \dfrac{3}{8}$

2. $\dfrac{2}{3} \cdot \dfrac{3}{5} = \dfrac{2}{\underset{1}{\cancel{3}}} \cdot \dfrac{\overset{1}{\cancel{3}}}{5} = \dfrac{2 \cdot 1}{1 \cdot 5} = \dfrac{2}{5}$

3. $12\dfrac{1}{2} \cdot 2\dfrac{1}{2} = \dfrac{25}{2} \cdot \dfrac{5}{2} = \dfrac{125}{4} = 31\dfrac{1}{4}$

4. $8\dfrac{1}{3} \cdot 3\dfrac{2}{5} = \dfrac{25}{3} \cdot \dfrac{17}{5} = \dfrac{425}{15} = \dfrac{85}{3} = 28\dfrac{1}{3}$

5. $\dfrac{\frac{4}{5}}{8} = \dfrac{4}{5} \div 8 = \dfrac{4}{5} \div \dfrac{8}{1} = \dfrac{\overset{1}{\cancel{4}}}{5} \cdot \dfrac{1}{\underset{2}{\cancel{8}}} = \dfrac{1 \cdot 1}{5 \cdot 2} = \dfrac{1}{10}$

6. $\dfrac{\frac{5}{8}}{4} = \dfrac{5}{8} \div \dfrac{4}{1} = \dfrac{5}{8} \cdot \dfrac{1}{4} = \dfrac{5}{32}$

7. $\dfrac{15}{31} \cdot 62 = \dfrac{15}{\underset{1}{\cancel{31}}} \cdot \dfrac{\overset{2}{\cancel{62}}}{1} = \dfrac{15 \cdot 2}{1 \cdot 1} = \dfrac{30}{1} = 30$

8. $3\dfrac{1}{4} \div 1\dfrac{1}{4} = \dfrac{13}{4} \div \dfrac{5}{4} = \dfrac{13}{\underset{1}{\cancel{4}}} \cdot \dfrac{\overset{1}{\cancel{4}}}{5} = \dfrac{13 \cdot 1}{1 \cdot 5} = \dfrac{13}{5} = 2\dfrac{3}{5}$

9. $\dfrac{8}{5} = 1\dfrac{3}{5}$

$$\begin{array}{r} 1 \\ 5\overline{)8} \\ \underline{5} \\ 3 \end{array}$$ $\begin{array}{l} \leftarrow \text{Whole number part} \\ \\ \leftarrow \text{Remainder} \end{array}$

10. $\dfrac{153}{4} = 38\dfrac{1}{4}$

$$\begin{array}{r} 3\ 8 \\ 4\overline{)1\ 5\ 3} \\ \underline{1\ 2} \\ 3\ 3 \\ \underline{3\ 2} \\ 1 \end{array}$$ $\begin{array}{l} \leftarrow \text{Whole number part} \\ \\ \\ \\ \leftarrow \text{Remainder} \end{array}$

11. $5\dfrac{2}{3}$ $\quad 5 \cdot 3 = 15$

$\qquad\quad 15 + 2 = 17$

$5\dfrac{2}{3} = \dfrac{17}{3}$

12. $38\dfrac{3}{8}$ $\quad 38 \cdot 8 = 304$

$\qquad\quad 304 + 3 = 307$

$38\dfrac{3}{8} = \dfrac{307}{8}$

13. $\dfrac{8}{12} = \dfrac{\overset{1}{\cancel{2}} \cdot \overset{1}{\cancel{2}} \cdot 2}{\underset{1}{\cancel{2}} \cdot \underset{1}{\cancel{2}} \cdot 3} = \dfrac{1 \cdot 1 \cdot 2}{1 \cdot 1 \cdot 3} = \dfrac{2}{3}$

14. $\dfrac{108}{210} = \dfrac{\overset{1}{\cancel{2}} \cdot 2 \cdot \overset{1}{\cancel{3}} \cdot 3 \cdot 3}{\underset{1}{\cancel{2}} \cdot \underset{1}{\cancel{3}} \cdot 5 \cdot 7} = \dfrac{1 \cdot 2 \cdot 1 \cdot 3 \cdot 3}{1 \cdot 1 \cdot 5 \cdot 7} = \dfrac{18}{35}$

15. $\dfrac{75}{90} = \dfrac{75 \div 15}{90 \div 15} = \dfrac{5}{6}$

16. $\dfrac{48}{72} = \dfrac{48 \div 24}{72 \div 24} = \dfrac{2}{3}$

17. $\dfrac{44}{110} = \dfrac{44 \div 22}{110 \div 22} = \dfrac{2}{5}$

18. $\dfrac{87}{261} = \dfrac{87 \div 87}{261 \div 87} = \dfrac{1}{3}$

Copyright © 2018 Pearson Education, Inc.

19. Multiply $2\frac{1}{2}$ ounces per gallon by the number of gallons.

Estimate: $3 \cdot 50 = 150$ ounces

Exact: $2\frac{1}{2} \cdot 50 = \frac{5}{\cancel{2}} \cdot \frac{\overset{25}{\cancel{50}}}{1} = \frac{5 \cdot 25}{1 \cdot 1} = \frac{125}{1} = 125$

125 ounces of the product are needed.

20. Multiply the number of tanks by the number of quarts needed for each tank.

Estimate: $7 \cdot 9 = 63$ qt

Exact:
$7\frac{1}{4} \cdot 9\frac{1}{3} = \frac{29}{4} \cdot \frac{28}{3} = \frac{812}{12} = \frac{203}{3} = 67\frac{2}{3}$

$67\frac{2}{3}$ quarts are needed.

21. To find the area, multiply the length and the width.

$$1\frac{3}{4} \cdot \frac{7}{8} = \frac{7}{4} \cdot \frac{7}{8} = \frac{49}{32} = 1\frac{17}{32}$$

The area of the stamp is $1\frac{17}{32}$ in.^{2}.

22. To find the area, multiply the length and the width.

$$\frac{7}{8} \cdot 2\frac{1}{4} = \frac{7}{8} \cdot \frac{9}{4} = \frac{63}{32} = 1\frac{31}{32}$$

The area of the patio table top is $1\frac{31}{32}$ yd^2.

Chapter 2 Test

1. $\dfrac{5}{6}$ There are 6 parts, and 5 are shaded.

2. $\dfrac{3}{8}$ There are 8 parts, and 3 are shaded.

3. Proper fractions have the numerator (top) smaller than the denominator (bottom).

$$\frac{2}{3}, \frac{6}{7}, \frac{1}{4}, \frac{5}{8}$$

4. $3\dfrac{3}{8}$ $3 \cdot 8 = 24$

$24 + 3 = 27$

$3\dfrac{3}{8} = \dfrac{27}{8}$

5. $\dfrac{123}{4} = 30\dfrac{3}{4}$

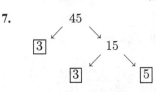

$$\begin{array}{r} 3\ 0 \leftarrow \text{Whole number part} \\ 4\overline{\smash{)}1\ 2\ 3} \\ \underline{1\ 2} \\ 3 \leftarrow \text{Remainder} \end{array}$$

6. Factorizations of 18:

$1 \cdot 18 = 18 \quad 2 \cdot 9 = 18 \quad 3 \cdot 6 = 18$

The factors of 18 are 1, 2, 3, 6, 9, and 18.

7.

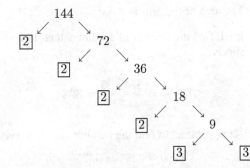

$45 = 3 \cdot 3 \cdot 5 = 3^2 \cdot 5$

8.

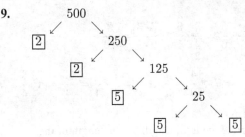

$144 = 2 \cdot 2 \cdot 2 \cdot 2 \cdot 3 \cdot 3 = 2^4 \cdot 3^2$

9.

$500 = 2 \cdot 2 \cdot 5 \cdot 5 \cdot 5 = 2^2 \cdot 5^3$

10. $\dfrac{36}{48} = \dfrac{36 \div 12}{48 \div 12} = \dfrac{3}{4}$

11. $\dfrac{60}{72} = \dfrac{60 \div 12}{72 \div 12} = \dfrac{5}{6}$

12. Write the prime factorization of both numerator and denominator. Divide the numerator and denominator by any common factors. Multiply the remaining factors in the numerator and denominator.

$$\frac{56}{84} = \frac{\overset{1}{\cancel{2}} \cdot \overset{1}{\cancel{2}} \cdot 2 \cdot \overset{1}{\cancel{7}}}{\underset{1}{\cancel{2}} \cdot \underset{1}{\cancel{2}} \cdot 3 \cdot \underset{1}{\cancel{7}}} = \frac{2}{3}$$

Copyright © 2018 Pearson Education, Inc.

13. Multiply fractions by multiplying the numerators and multiplying the denominators. Divide two fractions by using the reciprocal of the divisor (the second fraction) and then changing division to multiplication.

14. $\dfrac{3}{4} \cdot \dfrac{4}{9} = \dfrac{\cancel{3}^{1}}{\cancel{4}_{1}} \cdot \dfrac{\cancel{4}^{1}}{\cancel{9}_{3}} = \dfrac{1}{3}$

15. $54 \cdot \dfrac{2}{3} = \dfrac{\cancel{54}^{18}}{1} \cdot \dfrac{2}{\cancel{3}_{1}} = 36$

16. Multiply the length and the width.

$$\dfrac{15}{16} \cdot \dfrac{4}{9} = \dfrac{\cancel{15}^{5}}{\cancel{16}_{4}} \cdot \dfrac{\cancel{4}^{1}}{\cancel{9}_{3}} = \dfrac{5 \cdot 1}{4 \cdot 3} = \dfrac{5}{12}$$

The area of the grill is $\frac{5}{12}$ yd^2.

17. First, find the number of seedlings that don't survive.

$$8760 \cdot \dfrac{3}{8} = \dfrac{\cancel{8760}^{1095}}{1} \cdot \dfrac{3}{\cancel{8}_{1}} = 3285$$

Next, subtract to find the number that do survive.

$$8760 - 3285 = 5475$$

5475 seedlings do survive.

18. $\dfrac{3}{4} \div \dfrac{5}{6} = \dfrac{3}{\cancel{4}_{2}} \cdot \dfrac{\cancel{6}^{3}}{5} = \dfrac{3 \cdot 3}{2 \cdot 5} = \dfrac{9}{10}$

19. $\dfrac{7}{\dfrac{4}{9}} = 7 \div \dfrac{4}{9} = \dfrac{7}{1} \div \dfrac{4}{9} = \dfrac{7}{1} \cdot \dfrac{9}{4} = \dfrac{63}{4} = 15\dfrac{3}{4}$

20. Divide the total length by the length of the pieces.

$$54 \div 2\dfrac{1}{4} = \dfrac{54}{1} \div \dfrac{9}{4} = \dfrac{\cancel{54}^{6}}{1} \cdot \dfrac{4}{\cancel{9}_{1}} = 24$$

24 pieces can be cut.

21. $4\dfrac{1}{8} \cdot 3\dfrac{1}{2}$

Estimate: $4 \cdot 4 = 16$

Exact: $4\dfrac{1}{8} \cdot 3\dfrac{1}{2} = \dfrac{33}{8} \cdot \dfrac{7}{2} = \dfrac{231}{16} = 14\dfrac{7}{16}$

22. $1\dfrac{5}{6} \cdot 4\dfrac{1}{3}$

Estimate: $2 \cdot 4 = 8$

Exact: $1\dfrac{5}{6} \cdot 4\dfrac{1}{3} = \dfrac{11}{6} \cdot \dfrac{13}{3} = \dfrac{143}{18} = 7\dfrac{17}{18}$

23. $9\dfrac{3}{5} \div 2\dfrac{1}{4}$

Estimate: $10 \div 2 = 5$

Exact:

$$9\dfrac{3}{5} \div 2\dfrac{1}{4} = \dfrac{48}{5} \div \dfrac{9}{4} = \dfrac{\cancel{48}^{16}}{5} \cdot \dfrac{4}{\cancel{9}_{3}} = \dfrac{64}{15} = 4\dfrac{4}{15}$$

24. $\dfrac{8\dfrac{1}{2}}{1\dfrac{3}{4}}$

Estimate: $9 \div 2 = \dfrac{9}{2} = 4\dfrac{1}{2}$

Exact:

$$\dfrac{8\dfrac{1}{2}}{1\dfrac{3}{4}} = 8\dfrac{1}{2} \div 1\dfrac{3}{4} = \dfrac{17}{2} \div \dfrac{7}{4}$$

$$= \dfrac{17}{2} \cdot \dfrac{4}{7} = \dfrac{17 \cdot 4}{2 \cdot 7} = \dfrac{68}{14} = \dfrac{34}{7} = 4\dfrac{6}{7}$$

25. If $2\dfrac{1}{2}$ grams can be synthesized per day, multiply to find the amount synthesized in $12\dfrac{1}{4}$ days.

Estimate: $3 \cdot 12 = 36$ grams

Exact:
$$2\dfrac{1}{2} \cdot 12\dfrac{1}{4} = \dfrac{5}{2} \cdot \dfrac{49}{4} = \dfrac{5 \cdot 49}{2 \cdot 4} = \dfrac{245}{8} = 30\dfrac{5}{8}$$

$30\dfrac{5}{8}$ grams can be synthesized.

Cumulative Review Exercises (Chapters 1–2)

1. 7̲8̲3

hundreds: 7
tens: 8

2. 8̲,6̲2̲1,785

millions: 8
ten-thousands: 2

Copyright © 2018 Pearson Education, Inc.

3.

$$
\begin{array}{r}
\overset{1}{7}1 \\
23 \\
47 \\
+\,36 \\
\hline
177
\end{array}
$$

4.

$$
\begin{array}{r}
\overset{1}{8}\overset{11}{2},121 \\
5\,468 \\
316 \\
+\,61,294 \\
\hline
149,199
\end{array}
$$

5.

$$
\begin{array}{r}
6\overset{4}{\cancel{5}}\overset{13}{\cancel{3}}7 \\
-\,2\,085 \\
\hline
4\,4\,5\,2
\end{array}
$$

6.

$$
\begin{array}{r}
4,\overset{7}{\cancel{8}}\overset{18}{\cancel{1}}\overset{15}{\cancel{9}},\overset{10}{\cancel{6}}\,0\,4 \\
-\,1,597,783 \\
\hline
3,2\,2\,1,8\,2\,1
\end{array}
$$

7.

$$
\begin{array}{r}
\overset{2}{8}3 \\
\times\ 9 \\
\hline
747
\end{array}
$$

8. $9\cdot4\cdot2=(9\cdot4)\cdot2=36\cdot2=72$

9.

```
3784*573
         2168232
```

10.

$$
\begin{array}{r}
563 \\
\times\ 800 \\
\hline
\end{array}
\qquad
\begin{array}{r}
\overset{5}{5}\overset{2}{6}3 \\
\times\ 8 \\
\hline
45\,04
\end{array}
\qquad
\begin{array}{r}
563 \\
\times\ 800 \\
\hline
450,400 \ \textit{Attach } 00.
\end{array}
$$

11. $\dfrac{63}{7}$ $7\overline{)63}$ $\dfrac{63}{7}=9$

12.

```
136458/18
         7581
```

13. $33,886\div4$

$$
\begin{array}{r}
8\ 4\ 7\ 1\ \ \mathbf{R2} \\
4\overline{)3\ 3,\!8\ 8\ 6} \\
\underline{3\ 2} \\
1\ 8 \\
\underline{1\ 6} \\
2\ 8 \\
\underline{2\ 8} \\
0\ 6 \\
\underline{4} \\
2
\end{array}
$$

Check:

$$
\begin{array}{r}
8471 \\
\times\ \ \ 4 \\
\hline
33,884 \\
+\ \ \ \ 2 \\
\hline
33,886
\end{array}
$$

14.

$$
\begin{array}{r}
2\ 2\ \ \mathbf{R26} \\
492\overline{)1\ 0,\!8\ 5\ 0} \\
\underline{9\ 8\ 4} \\
1\ 0\ 1\ 0 \\
\underline{9\ 8\ 4} \\
2\ 6
\end{array}
$$

Check:

$$
\begin{array}{r}
492 \\
\times\ \ 22 \\
\hline
984 \\
984 \\
\hline
10,824 \\
+\ \ 26 \\
\hline
10,850
\end{array}
$$

15. **To the nearest ten:** $65\underline{8}3$
Next digit is 4 or less.
Tens place does not change.
All digits to the right of the underlined place change to zero. **6580**

To the nearest hundred: $6\underline{5}83$
Next digit is 5 or more.
Hundreds place changes ($5+1=6$).
All digits to the right of the underlined place change to zero. **6600**

To the nearest thousand: $\underline{6}583$
Next digit is 5 or more.
Thousands place changes ($6+1=7$).
All digits to the right of the underlined place change to zero. **7000**

Copyright © 2018 Pearson Education, Inc.

16. **To the nearest ten:** 76,2$\underline{7}$1
Next digit is 4 or less.
Tens place does not change.
All digits to the right of the underlined place change to zero. **76,270**

To the nearest hundred: 76,$\underline{2}$71
Next digit is 5 or more.
Hundreds place changes ($2 + 1 = 3$).
All digits to the right of the underlined place change to zero. **76,300**

To the nearest thousand: 7$\underline{6}$,271
Next digit is 4 or less.
Thousands place does not change.
All digits to the right of the underlined place change to zero. **76,000**

17. $2^5 - 6(4)$ *Exponent*
$= 32 - 6(4)$ *Multiply.*
$= 32 - 24 = 8$ *Subtract.*

18. $\sqrt{36} - 2 \cdot 3 + 5$ *Square root*
$= 6 - 2 \cdot 3 + 5$ *Multiply.*
$= 6 - 6 + 5$ *Subtract.*
$= 0 + 5 = 5$ *Add.*

19. Multiply to find the amount used for the half-day and full-day tours; then add to find the total.

$$
\begin{array}{r} 26 \\ \times\ 9 \\ \hline 234 \end{array}
\qquad
\begin{array}{r} 18 \\ \times\ 17 \\ \hline 126 \\ 18 \\ \hline 306 \end{array}
\qquad
\begin{array}{r} 234 \\ +306 \\ \hline 540 \end{array}
$$

540 gallons of fuel are needed.

20. Subtract to find the difference in cases.

$$
\begin{array}{r} {}^{1}2\,{}^{10}1,\,{}^{12}2{}^{8}9\,{}^{11}1 \\ -\ \ 4,\,5\,1\,9 \\ \hline 1\,6,\,7\,7\,2 \end{array}
$$

There were 16,772 more cases of pertussis than mumps.

21. Find the number of hairs lost in 2 years and subtract to find the hairs remaining.

$$
\begin{array}{r} 365 \\ \times\ 100 \\ \hline 36,500 \end{array}
\qquad
\begin{array}{r} 36,500 \\ \times\ 2 \\ \hline 73,000 \end{array}
\qquad
\begin{array}{r} 120,000 \\ -\ 73,000 \\ \hline 47,000 \end{array}
$$

47,000 hairs remain.

22. Divide the total number of hours by the number of workers.

$$
\begin{array}{r}
1\,8\,8 \\
22\overline{)4\,1\,3\,6} \\
\underline{2\,2} \\
1\,9\,3 \\
\underline{1\,7\,6} \\
1\,7\,6 \\
\underline{1\,7\,6} \\
0
\end{array}
$$

Each health care worker will work 188 hours.

23. Multiply the number of flushes and the amount of water used per flush to find the number of gallons of water used.

$$
160 \cdot 1\frac{3}{5} = \frac{160}{1} \cdot \frac{8}{5} = \frac{\overset{32}{\cancel{160}}}{1} \cdot \frac{8}{\underset{1}{\cancel{5}}} = 256
$$

256 gallons of water are used in 160 flushes.

24. Divide the total length by the length of the pieces.

$$
70 \div 3\frac{1}{3} = \frac{70}{1} \div \frac{10}{3} = \frac{\overset{7}{\cancel{70}}}{1} \cdot \frac{3}{\underset{1}{\cancel{10}}} = 21
$$

21 pieces can be cut.

25. $\frac{2}{3}$ is *proper* because the numerator (2) is smaller than the denominator (3).

26. $\frac{6}{6}$ is *improper* because the numerator (6) is larger than or the same as the denominator (6).

27. $\frac{9}{18}$ is *proper* because the numerator (9) is smaller than the denominator (18).

28. $3\frac{3}{8}$ $3 \cdot 8 = 24$
$24 + 3 = 27$
$3\frac{3}{8} = \frac{27}{8}$

29. $6\frac{2}{5}$ $6 \cdot 5 = 30$
$30 + 2 = 32$
$6\frac{2}{5} = \frac{32}{5}$

Copyright © 2018 Pearson Education, Inc.

30. $\dfrac{14}{7}$

$$\begin{array}{r} 2 \\ 7\overline{\smash{)}1\ 4} \\ 1\ 4 \\ \hline 0 \end{array}$$ ← Whole number part

← Remainder

$\dfrac{14}{7} = 2$

31. $\dfrac{103}{8}$

$$\begin{array}{r} 1\ 2 \\ 8\overline{\smash{)}1\ 0\ 3} \\ 8 \\ \hline 2\ 3 \\ 1\ 6 \\ \hline 7 \end{array}$$ ← Whole number part

← Remainder

$\dfrac{103}{8} = 12\dfrac{7}{8}$

32. 72

$$\begin{array}{r} 36 \\ 2\overline{\smash{)}72} \end{array}$$
$$\begin{array}{r} 18 \\ 2\overline{\smash{)}36} \end{array}$$
$$\begin{array}{r} 9 \\ 2\overline{\smash{)}18} \end{array}$$
$$\begin{array}{r} 3 \\ 3\overline{\smash{)}9} \end{array}$$
$$\begin{array}{r} 1 \\ 3\overline{\smash{)}3} \end{array}$$

$72 = 2 \cdot 2 \cdot 2 \cdot 3 \cdot 3 = 2^3 \cdot 3^2$

33.

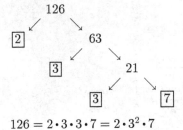

$126 = 2 \cdot 3 \cdot 3 \cdot 7 = 2 \cdot 3^2 \cdot 7$

34.

$$350 = 2 \cdot 5 \cdot 5 \cdot 7 = 2 \cdot 5^2 \cdot 7$$

35. $4^2 \cdot 2^2 = 16 \cdot 4 = 64$

36. $2^3 \cdot 6^2 = 8 \cdot 36 = 288$

37. $2^3 \cdot 4^2 \cdot 5 = 8 \cdot 16 \cdot 5$
$$= 128 \cdot 5$$
$$= 640$$

38. $\dfrac{42}{48} = \dfrac{42 \div 6}{48 \div 6} = \dfrac{7}{8}$

39. $\dfrac{24}{36} = \dfrac{24 \div 12}{36 \div 12} = \dfrac{2}{3}$

40. $\dfrac{30}{54} = \dfrac{30 \div 6}{54 \div 6} = \dfrac{5}{9}$

41. $\dfrac{1}{2} \cdot \dfrac{3}{4} = \dfrac{1 \cdot 3}{2 \cdot 4} = \dfrac{3}{8}$

42. $30 \cdot \dfrac{2}{3} \cdot \dfrac{3}{5} = \dfrac{30}{1} \cdot \dfrac{2}{3} \cdot \dfrac{3}{5} = \dfrac{\overset{6}{\cancel{30}} \cdot 2 \cdot \overset{1}{\cancel{3}}}{\underset{1}{\cancel{3}} \cdot \underset{1}{\cancel{5}}} = 12$

43. $7\dfrac{1}{2} \cdot 3\dfrac{1}{3} = \dfrac{\overset{5}{\cancel{15}}}{\underset{1}{\cancel{2}}} \cdot \dfrac{\overset{5}{\cancel{10}}}{\underset{1}{\cancel{3}}} = \dfrac{25}{1} = 25$

44. $\dfrac{3}{5} \div \dfrac{5}{8} = \dfrac{3}{5} \cdot \dfrac{8}{5} = \dfrac{24}{25}$

45. $\dfrac{7}{8} \div 1\dfrac{1}{2} = \dfrac{7}{8} \div \dfrac{3}{2} = \dfrac{7}{\underset{4}{\cancel{8}}} \cdot \dfrac{\overset{1}{\cancel{2}}}{3} = \dfrac{7 \cdot 1}{4 \cdot 3} = \dfrac{7}{12}$

46. $3 \div 1\dfrac{1}{4} = \dfrac{3}{1} \div \dfrac{5}{4} = \dfrac{3}{1} \cdot \dfrac{4}{5} = \dfrac{12}{5} = 2\dfrac{2}{5}$

Copyright © 2018 Pearson Education, Inc.

CHAPTER 3   ADDING AND SUBTRACTING FRACTIONS

3.1   Adding and Subtracting Like Fractions

3.1   Margin Exercises

1. **(a)** $\dfrac{2}{5} \quad \dfrac{3}{5}$

The denominators are the same, so $\frac{2}{5}$ and $\frac{3}{5}$ are *like* fractions.

(b) $\dfrac{2}{3} \quad \dfrac{3}{4}$

The denominators are different, so $\frac{2}{3}$ and $\frac{3}{4}$ are *unlike* fractions.

(c) $\dfrac{7}{12} \quad \dfrac{11}{12}$

The denominators are the same, so $\frac{7}{12}$ and $\frac{11}{12}$ are *like* fractions.

(d) $\dfrac{3}{8} \quad \dfrac{3}{16}$

The denominators are different, so $\frac{3}{8}$ and $\frac{3}{16}$ are *unlike* fractions.

2. **(a)** $\dfrac{3}{8} + \dfrac{1}{8} = \dfrac{3+1}{8}$
$$= \dfrac{4}{8} = \dfrac{4 \div 4}{8 \div 4} = \dfrac{1}{2} \quad \textit{Lowest terms}$$

(b) $\quad \dfrac{2}{9}$
$$+ \dfrac{5}{9}$$

Add numerators. The denominator stays the same.
$$\dfrac{2+5}{9} = \dfrac{7}{9}$$

(c) $\dfrac{3}{16} + \dfrac{1}{16} = \dfrac{3+1}{16}$
$$= \dfrac{4}{16} = \dfrac{4 \div 4}{16 \div 4} = \dfrac{1}{4} \quad \textit{Lowest terms}$$

(d) $\dfrac{3}{10} + \dfrac{1}{10} + \dfrac{4}{10}$
$$= \dfrac{3+1+4}{10}$$
$$= \dfrac{8}{10} = \dfrac{8 \div 2}{10 \div 2} = \dfrac{4}{5} \quad \textit{Lowest terms}$$

3. **(a)** $\dfrac{5}{6} - \dfrac{1}{6} = \dfrac{5-1}{6} = \dfrac{4}{6} = \dfrac{4 \div 2}{6 \div 2} = \dfrac{2}{3}$

(b) $\quad \dfrac{16}{10}$
$$- \dfrac{7}{10}$$
$$\dfrac{16 - 7}{10} = \dfrac{9}{10}$$

(c) $\dfrac{15}{3} - \dfrac{5}{3} = \dfrac{15-5}{3} = \dfrac{10}{3} = 3\dfrac{1}{3}$

(d) $\quad \dfrac{25}{32}$
$$- \dfrac{6}{32}$$
$$\dfrac{25 - 6}{32} = \dfrac{19}{32}$$

3.1   Section Exercises

1. $\dfrac{3}{8} \quad \dfrac{5}{8}$

The denominators are the same, so $\frac{3}{8}$ and $\frac{5}{8}$ are *like* fractions.

3. $\dfrac{3}{5} \quad \dfrac{3}{4}$

The denominators are different, so $\frac{3}{5}$ and $\frac{3}{4}$ are *unlike* fractions.

5. In order to add or subtract fractions, they must be <u>like</u> fractions.

7. $\dfrac{3}{8} + \dfrac{2}{8} = \dfrac{3+2}{8} = \dfrac{5}{8}$

9. $\dfrac{1}{4} + \dfrac{1}{4} = \dfrac{1+1}{4} = \dfrac{2}{4}$
$$= \dfrac{2 \div 2}{4 \div 2} = \dfrac{1}{2} \quad \textit{Lowest terms}$$

11. $\quad \dfrac{13}{12}$
$$+ \dfrac{5}{12}$$
$$\dfrac{13}{12} + \dfrac{5}{12} = \dfrac{13+5}{12} = \dfrac{18}{12} = \dfrac{18 \div 6}{12 \div 6} = \dfrac{3}{2} = 1\dfrac{1}{2}$$

13. $\dfrac{7}{12} + \dfrac{3}{12} = \dfrac{7+3}{12} = \dfrac{10}{12} = \dfrac{10 \div 2}{12 \div 2} = \dfrac{5}{6}$

Copyright © 2018 Pearson Education, Inc.

15. Add the numerators. Keep the denominator.

$$\frac{3}{8} + \frac{7}{8} + \frac{2}{8} = \frac{3+7+2}{8} = \frac{12}{8}$$
$$= \frac{12 \div 4}{8 \div 4} = \frac{3}{2} = 1\frac{1}{2}$$

17. $\dfrac{2}{54} + \dfrac{8}{54} + \dfrac{12}{54} = \dfrac{2+8+12}{54} = \dfrac{22}{54}$
$$= \frac{22 \div 2}{54 \div 2} = \frac{11}{27}$$

19. $\dfrac{7}{8} - \dfrac{4}{8} = \dfrac{7-4}{8} = \dfrac{3}{8}$

21. $\dfrac{10}{11} - \dfrac{4}{11} = \dfrac{10-4}{11} = \dfrac{6}{11}$

23. $\dfrac{9}{10} - \dfrac{3}{10} = \dfrac{9-3}{10} = \dfrac{6}{10} = \dfrac{6 \div 2}{10 \div 2} = \dfrac{3}{5}$

25.
$$\frac{31}{21}$$
$$-\frac{7}{21}$$
$$\frac{31}{21} - \frac{7}{21} = \frac{31-7}{21} = \frac{24}{21} = \frac{24 \div 3}{21 \div 3} = \frac{8}{7} = 1\frac{1}{7}$$

27.
$$\frac{27}{40}$$
$$-\frac{19}{40}$$
$$\frac{27}{40} - \frac{19}{40} = \frac{27-19}{40} = \frac{8}{40} = \frac{8 \div 8}{40 \div 8} = \frac{1}{5}$$

29. First, subtract the numerators and keep the denominator. Then write the fraction in lowest terms.

$$\frac{47}{36} - \frac{5}{36} = \frac{47-5}{36} = \frac{42}{36} = \frac{42 \div 6}{36 \div 6} = \frac{7}{6} = 1\frac{1}{6}$$

31. $\dfrac{73}{60} - \dfrac{7}{60} = \dfrac{73-7}{60}$
$$= \frac{66}{60} = \frac{66 \div 6}{60 \div 6} = \frac{11}{10} = 1\frac{1}{10}$$

33. Three steps to add like fractions are:
1. Add the numerators of the fractions to find the numerator of the sum (the answer).
2. Use the denominator of the fractions as the denominator of the sum.
3. Write the answer in lowest terms.

35. Add the two fractions to find the total fraction raised.

$$\frac{2}{9} + \frac{5}{9} = \frac{2+5}{9} = \frac{7}{9}$$

They raised $\frac{7}{9}$ of their target goal.

37. Add the two amounts to find the total fraction completed.

$$\frac{1}{8} + \frac{5}{8} = \frac{1+5}{8} = \frac{6}{8} = \frac{6 \div 2}{8 \div 2} = \frac{3}{4}$$

She has completed $\frac{3}{4}$ of the irrigation system.

39. First, add the two fractions of land that were purchased.

$$\frac{9}{10} + \frac{3}{10} = \frac{9+3}{10} = \frac{12}{10}$$

Next, subtract the fraction of land that was planted in carrots.

$$\frac{12}{10} - \frac{7}{10} = \frac{12-7}{10} = \frac{5}{10} = \frac{5 \div 5}{10 \div 5} = \frac{1}{2}$$

$\frac{1}{2}$ acre is planted with squash.

3.2 Least Common Multiples

3.2 Margin Exercises

1. (a) The multiples of 5 are
$$5, 10, 15, \underline{20}, \underline{25}, \underline{30}, 35, \underline{40}, \ldots.$$

(b) The multiples of 8 are
$$8, \underline{16}, \underline{24}, \underline{32}, \underline{40}, \underline{48}, \underline{56}, \ldots.$$

(c) Look at the answers for (a) and (b). 40 is the only number found in both lists, so it is the least common multiple.

2. (a) 2 and 5

The multiples of 5 are
$$5, 10, 15, 20, \ldots.$$

The first multiple of 5 that is divisible by 2 is $\underline{10}$, so the least common multiple of the numbers 2 and 5 is $\underline{10}$.

(b) 3 and 9

The multiples of 9 are
$$9, 18, 27, 36, 45, \ldots.$$

The first multiple of 9 that is divisible by 3 is 9, so the least common multiple of the numbers 3 and 9 is 9.

(c) 6 and 8

The multiples of 8 are
$$8, 16, 24, 32, 40, 48, \ldots.$$

The first multiple of 8 that is divisible by 6 is 24, so the least common multiple of the numbers 6 and 8 is 24.

Copyright © 2018 Pearson Education, Inc.

(d) 4 and 7

The multiples of 7 are

$$7, 14, 21, 28, 35, 42, \ldots.$$

The first multiple of 7 that is divisible by 4 is 28, so the least common multiple of the numbers 4 and 7 is 28.

3. **(a)** 15 and 18

$$15 = 3 \cdot 5$$
$$18 = 2 \cdot 3 \cdot 3$$

To find the LCM of 15 and 18, we'll start with the factors of 18, $2 \cdot 3 \cdot 3$. Now 15 only has 3 and 5 as factors and we already have a 3 in the LCM, so we just need to include a 5.

$$LCM = 2 \cdot 3 \cdot 3 \cdot 5 = 90$$

The least common multiple of 15 and 18 is 90.

(b) 12 and 20

$$12 = 2 \cdot 2 \cdot 3$$
$$20 = 2 \cdot 2 \cdot 5$$

$$LCM = 2 \cdot 2 \cdot 3 \cdot 5 = 60$$

The least common multiple of 12 and 20 is 60.

4. **(a)** $\dfrac{3}{8}$ and $\dfrac{6}{5}$

$$8 = 2 \cdot 2 \cdot 2$$
$$5 = 1 \cdot 5$$

$$LCM = 2 \cdot 2 \cdot 2 \cdot 5 = 40$$

The least common multiple of 8 and 5 is 40.

(b) $\dfrac{5}{6}$ and $\dfrac{1}{14}$

$$6 = 2 \cdot 3$$
$$14 = 2 \cdot 7$$

$$LCM = 2 \cdot 3 \cdot 7 = 42$$

The least common multiple of 6 and 14 is 42.

(c) $\dfrac{4}{9}, \dfrac{5}{18},$ and $\dfrac{7}{24}$

$$9 = 3 \cdot 3$$
$$18 = 2 \cdot 3 \cdot 3$$
$$24 = 2 \cdot 2 \cdot 2 \cdot 3$$

$$LCM = 2 \cdot 2 \cdot 2 \cdot 3 \cdot 3 = 72$$

The least common multiple of 9, 18, and 24 is 72.

5. **(a)** $4, 8, 9$

$$4 = 2 \cdot 2$$
$$8 = 2 \cdot 2 \cdot 2$$
$$9 = 3 \cdot 3$$

$$LCM = \underline{2} \cdot \underline{2} \cdot \underline{2} \cdot \underline{3} \cdot \underline{3} = \underline{72}$$

The least common multiple of 4, 8, and 9 is 72.

(b) $3, 6, 8$

$$3 = 1 \cdot 3$$
$$6 = 2 \cdot 3$$
$$8 = 2 \cdot 2 \cdot 2$$

$$LCM = 2 \cdot 2 \cdot 2 \cdot 3 = 24$$

The least common multiple of 3, 6, and 8 is 24.

(c) $15, 20, 30, 40$

$$15 = 3 \cdot 5$$
$$20 = 2 \cdot 2 \cdot 5$$
$$30 = 2 \cdot 3 \cdot 5$$
$$40 = 2 \cdot 2 \cdot 2 \cdot 5$$

$$LCM = 2 \cdot 2 \cdot 2 \cdot 3 \cdot 5 = 120$$

The least common multiple of 15, 20, 30, and 40 is 120.

6. **(a)** The least common multiple of 6 and 15 is the product of the numbers on the left side.

$$LCM = 2 \cdot \underline{3} \cdot \underline{5} = \underline{30}$$

(b) The least common multiple of 20 and 36 is the product of the numbers on the left side.

$$LCM = 2 \cdot 2 \cdot 3 \cdot 3 \cdot 5 = 180$$

7. **(a)** $3, 6,$ and 10

$$
\begin{array}{c|ccc}
2 & \not{3} & 6 & 10 \\
3 & 3 & 3 & \not{5} \\
5 & \not{1} & \not{1} & 5 \\
\hline
 & 1 & 1 & 1
\end{array}
$$

The LCM of 3, 6, and 10 is

$$2 \cdot \underline{3} \cdot 5 = \underline{30}.$$

(b) 15 and 40

$$
\begin{array}{c|cc}
2 & \not{15} & 40 \\
2 & \not{15} & 20 \\
2 & \not{15} & 10 \\
3 & 15 & \not{5} \\
5 & 5 & 5 \\
\hline
 & 1 & 1
\end{array}
$$

The LCM of 15 and 40 is

$$2 \cdot 2 \cdot 2 \cdot 3 \cdot 5 = 120.$$

Copyright © 2018 Pearson Education, Inc.

(c) 9 and 24

$$
\begin{array}{c|cc}
2 & \cancel{9} & 24 \\
2 & \cancel{9} & 12 \\
2 & \cancel{9} & 6 \\
3 & 9 & 3 \\
3 & 3 & \cancel{1} \\
\hline
& 1 & 1
\end{array}
$$

The LCM of 9 and 24 is

$$2 \cdot 2 \cdot 2 \cdot 3 \cdot 3 = 72.$$

(d) 8, 21, and 24

$$
\begin{array}{c|ccc}
2 & 8 & \cancel{21} & 24 \\
2 & 4 & \cancel{21} & 12 \\
2 & 2 & \cancel{21} & 6 \\
3 & \cancel{1} & 21 & 3 \\
7 & \cancel{1} & 7 & \cancel{1} \\
\hline
& 1 & 1 & 1
\end{array}
$$

The LCM of 8, 21, and 24 is

$$2 \cdot 2 \cdot 2 \cdot 3 \cdot 7 = 168.$$

8.   **(a)** $\dfrac{1}{4} = \dfrac{?}{16}$

Divide 16 by 4, getting 4. Now multiply both the numerator and the denominator of $\frac{1}{4}$ by 4.

$$\frac{1}{4} = \frac{1 \cdot 4}{4 \cdot 4} = \frac{4}{16}$$

(b) $\dfrac{5}{3} = \dfrac{?}{15}$

Divide 15 by 3, getting 5. Now multiply both the numerator and the denominator of $\frac{5}{3}$ by 5.

$$\frac{5}{3} = \frac{5 \cdot 5}{3 \cdot 5} = \frac{25}{15}$$

(c) $\dfrac{7}{16} = \dfrac{?}{32}$

Divide 32 by 16, getting 2. Now multiply both the numerator and the denominator of $\frac{7}{16}$ by 2.

$$\frac{7}{16} = \frac{7 \cdot 2}{16 \cdot 2} = \frac{14}{32}$$

(d) $\dfrac{6}{11} = \dfrac{?}{33}$

Divide 33 by 11, getting 3. Now multiply both the numerator and the denominator of $\frac{6}{11}$ by 3.

$$\frac{6}{11} = \frac{6 \cdot 3}{11 \cdot 3} = \frac{18}{33}$$

3.2 Section Exercises

1. The least common multiple of 4 and 8 is 8, so the statement is *true*.

3. The least common multiple of 3 and 7 is 21, not 28, so the statement is *false*.

5. 3 and 6

Multiples of 6:

$$\underline{6}, 12, 18, 24, 30, \ldots$$

6 is the first number divisible by 3. $(6 \div 3 = 2)$

The least common multiple of 3 and 6 is 6.

7. 3 and 5

Multiples of 5:

$$5, 10, \underline{15}, 20, 25, \ldots$$

15 is the first number divisible by 3.
$(15 \div 3 = 5)$

The least common multiple of 3 and 5 is 15.

9. 4 and 9

Multiples of 9:

$$9, 18, 27, \underline{36}, 45, \ldots$$

36 is the first number divisible by 4.
$(36 \div 4 = 9)$

The least common multiple of 4 and 9 is 36.

11. 12 and 16

Multiples of 16:

$$16, 32, \underline{48}, 64, 80, \ldots$$

48 is the first number divisible by 12.
$(48 \div 12 = 4)$

The least common multiple of 12 and 16 is 48.

13. 20 and 50

Multiples of 50:

$$50, \underline{100}, 150, 200, 250, \ldots$$

100 is the first number divisible by 20.
$(100 \div 20 = 5)$

The least common multiple of 20 and 50 is 100.

15. 8, 10

$$
\begin{array}{c|cc}
2 & 8 & 10 \\
2 & 4 & \cancel{5} \\
2 & 2 & 5 \\
5 & \cancel{1} & 5 \\
\hline
& 1 & 1
\end{array}
$$

The LCM of 8 and 10 is

$$2 \cdot 2 \cdot 2 \cdot 5 = 40.$$

Copyright © 2018 Pearson Education, Inc.

17. 9 and 15

$$\begin{array}{c|cc} 3 & 9 & 15 \\ 3 & 3 & \cancel{5} \\ 5 & \cancel{1} & 5 \\ \hline & 1 & 1 \end{array}$$

The LCM of 9 and 15 is

$$3 \cdot 3 \cdot 5 = 45.$$

19. 20, 24, 30

$$\begin{array}{c|ccc} 2 & 20 & 24 & 30 \\ 2 & 10 & 12 & \cancel{15} \\ 2 & \cancel{5} & 6 & \cancel{15} \\ 3 & \cancel{5} & 3 & 15 \\ 5 & 5 & \cancel{1} & 5 \\ \hline & 1 & 1 & 1 \end{array}$$

The LCM of 20, 24, and 30 is

$$2 \cdot 2 \cdot 2 \cdot 3 \cdot 5 = 120.$$

21. 4, 6, 8, 10

$$\begin{aligned} 4 &= 2 \cdot 2 \\ 6 &= 2 \cdot 3 \\ 8 &= 2 \cdot 2 \cdot 2 \\ 10 &= 2 \cdot 5 \end{aligned}$$

$$\text{LCM} = 2 \cdot 2 \cdot 2 \cdot 3 \cdot 5 = 120$$

The LCM of 4, 6, 8, and 10 is 120.

23. 6, 8, 9, 27, 36

$$\begin{aligned} 6 &= 2 \cdot 3 \\ 8 &= 2 \cdot 2 \cdot 2 \\ 9 &= 3 \cdot 3 \\ 27 &= 3 \cdot 3 \cdot 3 \\ 36 &= 2 \cdot 2 \cdot 3 \cdot 3 \end{aligned}$$

$$\text{LCM} = 2 \cdot 2 \cdot 2 \cdot 3 \cdot 3 \cdot 3 = 216$$

The LCM of 6, 8, 9, 27, and 36 is 216.

25. 5, 6, 8, 25, 30

$$\begin{aligned} 5 &= 1 \cdot 5 \\ 6 &= 2 \cdot 3 \\ 8 &= 2 \cdot 2 \cdot 2 \\ 25 &= 5 \cdot 5 \\ 30 &= 2 \cdot 3 \cdot 5 \end{aligned}$$

$$\text{LCM} = 2 \cdot 2 \cdot 2 \cdot 3 \cdot 5 \cdot 5 = 600$$

The LCM of 5, 6, 8, 25, and 30 is 600.

27. $\dfrac{3}{8} = \dfrac{?}{24}$ $24 \div 8 = 3$

$$\frac{3}{8} = \frac{3 \cdot 3}{8 \cdot 3} = \frac{9}{24}$$

29. $\dfrac{5}{12} = \dfrac{?}{24}$ $24 \div 12 = 2$

$$\frac{5}{12} = \frac{5 \cdot 2}{12 \cdot 2} = \frac{10}{24}$$

31. $\dfrac{7}{8} = \dfrac{?}{24}$ $24 \div 8 = 3$

$$\frac{7}{8} = \frac{7 \cdot 3}{8 \cdot 3} = \frac{21}{24}$$

33. Write the given fractions (that are not already in lowest terms) in lowest terms.

$$\frac{18}{22} = \frac{18 \div 2}{22 \div 2} = \frac{9}{11}$$

$$\frac{21}{28} = \frac{21 \div 7}{28 \div 7} = \frac{3}{4} \quad match$$

$\frac{21}{28}$ is equivalent to $\frac{3}{4}$.

35. Write the given fractions (that are not already in lowest terms) in lowest terms.

$$\frac{21}{24} = \frac{21 \div 3}{24 \div 3} = \frac{7}{8} \quad match$$

$\frac{21}{24}$ is equivalent to $\frac{7}{8}$.

37. $\dfrac{2}{3} = \dfrac{?}{9}$ $9 \div 3 = 3$

Divide 9 by 3, to get 3. Now multiply both numerator and denominator of $\frac{2}{3}$ by 3.

$$\frac{2}{3} = \frac{2 \cdot 3}{3 \cdot 3} = \frac{6}{9}$$

39. $\dfrac{7}{8} = \dfrac{?}{32}$ $32 \div 8 = 4$

$$\frac{7}{8} = \frac{7 \cdot 4}{8 \cdot 4} = \frac{28}{32}$$

41. $\dfrac{3}{16} = \dfrac{?}{64}$ $64 \div 16 = 4$

Divide 64 by 16, to get 4. Now multiply both numerator and denominator of $\frac{3}{16}$ by 4.

$$\frac{3}{16} = \frac{3 \cdot 4}{16 \cdot 4} = \frac{12}{64}$$

43. $\dfrac{9}{7} = \dfrac{?}{56}$ $56 \div 7 = 8$

$$\frac{9}{7} = \frac{9 \cdot 8}{7 \cdot 8} = \frac{72}{56}$$

45. $\dfrac{7}{4} = \dfrac{?}{48}$ $48 \div 4 = 12$

$$\frac{7}{4} = \frac{7 \cdot 12}{4 \cdot 12} = \frac{84}{48}$$

Divide 48 by 4 to get <u>12</u>.
Now multiply 7 by <u>12</u> to get the new numerator, which is <u>84</u>.

Copyright © 2018 Pearson Education, Inc.

47. $\dfrac{8}{11} = \dfrac{?}{132}$ $132 \div 11 = 12$

$\dfrac{8}{11} = \dfrac{8 \cdot 12}{11 \cdot 12} = \dfrac{96}{132}$

49. $\dfrac{3}{16} = \dfrac{?}{144}$ $144 \div 16 = 9$

$\dfrac{3}{16} = \dfrac{3 \cdot 9}{16 \cdot 9} = \dfrac{27}{144}$

51. Answers will vary. A sample answer follows:
It probably depends on how large the numbers
are. If the numbers are small, the method using
multiples of the largest number seems easiest. If
the numbers are larger, or if there are more than
two numbers, then the factorization method or
the alternative method will be better.

53. Find the prime factorization for each number.

$5, 7, 14, 10$

$5 = 1 \cdot 5$

$7 = 1 \cdot 7$

$14 = 2 \cdot 7$

$10 = 2 \cdot 5$

$\text{LCM} = 2 \cdot 5 \cdot 7 = 70$

The LCM of 5, 7, 14, and 10 is 70.

55. $8 = 2 \cdot 2 \cdot 2$

$3 = 1 \cdot 3$

$5 = 1 \cdot 5$

$4 = 2 \cdot 2$

$10 = 2 \cdot 5$

$\text{LCM} = 2 \cdot 2 \cdot 2 \cdot 3 \cdot 5 = 120$

The LCM of 8, 3, 5, 4, and 10 is 120. 240 is a
common multiple, but it is not the *least* common
multiple.

57. Fractions with the same denominators are <u>like</u>
fractions and fractions with different
denominators are <u>unlike</u> fractions.

58. To subtract like fractions, first subtract the
<u>numerators</u> to find the numerator of the
difference. Write the denominator of the like
fractions as the <u>denominator</u> of the difference.
Finally, write the answer in <u>lowest</u> terms.

59. The <u>least</u> common multiple (LCM) of two
numbers is the <u>smallest</u> whole number divisible
by both of those numbers.

60. The smallest number in both lists is 40, so 40 is
the least common multiple of 8 and 10.

61. $\dfrac{25}{400}, \dfrac{38}{1800}$

Multiples of 1800:

$1800, \underline{3600}, 5400, \ldots$

3600 is the first number divisible by 400.
$(3600 \div 400 = 9)$

The LCM of the denominators is 3600.

62. $\dfrac{53}{600}, \dfrac{115}{4000}$

Multiples of 4000:

$4000, 8000, \underline{12{,}000}, \ldots$

12,000 is the first number divisible by 600.
$(12{,}000 \div 600 = 20)$

The LCM of the denominators is 12,000.

63. $\dfrac{109}{1512}, \dfrac{23}{392}$

2	1512	392
2	756	196
2	378	98
3	189	4̶9̶
3	63	4̶9̶
3	21	4̶9̶
7	7	49
7	1̶	7
	1	1

The LCM of the denominators is

$2 \cdot 2 \cdot 2 \cdot 3 \cdot 3 \cdot 3 \cdot 7 \cdot 7 = 10{,}584.$

64. $\dfrac{61}{810}, \dfrac{37}{1170}$

2	810	1170
3	405	585
3	135	195
3	45	6̶5̶
3	15	6̶5̶
5	5	65
13	1̶	13
	1	1

The LCM of the denominators is

$2 \cdot 3 \cdot 3 \cdot 3 \cdot 3 \cdot 5 \cdot 13 = 10{,}530.$

65. $\dfrac{46}{125}, \dfrac{18}{2525}$

5	125	2525
5	25	505
5	5	1̶0̶1̶
101	1̶	101
	1	1

The LCM of the denominators is

$5 \cdot 5 \cdot 5 \cdot 101 = 12{,}625.$

Copyright © 2018 Pearson Education, Inc.

66. $\dfrac{80}{445}, \dfrac{75}{110}$

$$
\begin{array}{c|cc}
2 & 445 & 110 \\
5 & 445 & 55 \\
11 & 89 & 11 \\
89 & 89 & 1 \\
\hline
 & 1 & 1
\end{array}
$$

The LCM of the denominators is

$$2 \cdot 5 \cdot 11 \cdot 89 = 9790.$$

3.3   Adding and Subtracting Unlike Fractions

3.3   Margin Exercises

1.   (a) *Step 1* $\dfrac{1}{2} + \dfrac{3}{8}$ (LCD = 8)

$$\dfrac{1}{2} = \dfrac{1 \cdot 4}{2 \cdot 4} = \dfrac{4}{8}$$

Step 2 $\dfrac{1}{2} + \dfrac{3}{8} = \dfrac{4}{8} + \dfrac{3}{8} = \dfrac{4+3}{8} = \dfrac{7}{8}$

(b) *Step 1* $\dfrac{3}{4} + \dfrac{1}{8}$ (LCD = 8)

$$\dfrac{3}{4} = \dfrac{3 \cdot 2}{4 \cdot 2} = \dfrac{6}{8}$$

Step 2 $\dfrac{3}{4} + \dfrac{1}{8} = \dfrac{6}{8} + \dfrac{1}{8} = \dfrac{6+1}{8} = \dfrac{7}{8}$

(c) *Step 1* $\dfrac{3}{5} + \dfrac{3}{10}$ (LCD = 10)

$$\dfrac{3}{5} = \dfrac{3 \cdot 2}{5 \cdot 2} = \dfrac{6}{10}$$

Step 2 $\dfrac{3}{5} + \dfrac{3}{10} = \dfrac{6}{10} + \dfrac{3}{10} = \dfrac{6+3}{10} = \dfrac{9}{10}$

(d) *Step 1* $\dfrac{1}{12} + \dfrac{5}{6}$ (LCD = 12)

$$\dfrac{5}{6} = \dfrac{5 \cdot 2}{6 \cdot 2} = \dfrac{10}{12}$$

Step 2 $\dfrac{1}{12} + \dfrac{5}{6} = \dfrac{1}{12} + \dfrac{10}{12} = \dfrac{1+10}{12} = \dfrac{11}{12}$

2.   (a) *Step 1* $\dfrac{3}{10} + \dfrac{1}{5} = \dfrac{3}{10} + \dfrac{2}{10}$

Step 2 $\dfrac{3}{10} + \dfrac{2}{10} = \dfrac{3+2}{10} = \dfrac{5}{10}$

Step 3 $\dfrac{5}{10} = \dfrac{1}{2}$

(b) *Step 1* $\dfrac{5}{8} + \dfrac{1}{3} = \dfrac{15}{24} + \dfrac{8}{24}$

Step 2 $\dfrac{15}{24} + \dfrac{8}{24} = \dfrac{15+8}{24} = \dfrac{23}{24}$

(c) *Step 1* $\dfrac{1}{10} + \dfrac{1}{3} + \dfrac{1}{6} = \dfrac{3}{30} + \dfrac{10}{30} + \dfrac{5}{30}$

Step 2 $\dfrac{3}{30} + \dfrac{10}{30} + \dfrac{5}{30} = \dfrac{3+10+5}{30} = \dfrac{18}{30}$

Step 3 $\dfrac{18}{30} = \dfrac{3}{5}$

3.   (a)

$$
\begin{aligned}
\dfrac{5}{8} &= \dfrac{5 \cdot 3}{8 \cdot 3} = \dfrac{15}{24} \\
+ \dfrac{1}{12} &= \dfrac{1 \cdot 2}{12 \cdot 2} = \dfrac{2}{24} \\
\hline
& \qquad\qquad\;\; \dfrac{17}{24}
\end{aligned}
$$

(b)

$$
\begin{aligned}
\dfrac{7}{16} &= \qquad\quad \dfrac{7}{16} \\
+ \dfrac{1}{4} &= \dfrac{1 \cdot 4}{4 \cdot 4} = \dfrac{4}{16} \\
\hline
& \qquad\qquad\;\; \dfrac{11}{16}
\end{aligned}
$$

(c) The LCM of 8, 24, and 16 is 48.

$$
\begin{aligned}
\dfrac{1}{8} &= \dfrac{1 \cdot 6}{8 \cdot 6} = \dfrac{6}{48} \\
\dfrac{5}{24} &= \dfrac{5 \cdot 2}{24 \cdot 2} = \dfrac{10}{48} \\
+ \dfrac{7}{16} &= \dfrac{7 \cdot 3}{16 \cdot 3} = \dfrac{21}{48} \\
\hline
& \qquad\qquad\;\; \dfrac{37P}{48}
\end{aligned}
$$

4.   (a) *Step 1* $\dfrac{5}{8} - \dfrac{1}{4} = \dfrac{5}{8} - \dfrac{2}{8}$

Step 2 $\dfrac{5}{8} - \dfrac{2}{8} = \dfrac{5-2}{8} = \dfrac{3}{8}$

(b) *Step 1* $\dfrac{4}{5} - \dfrac{3}{4} = \dfrac{16}{20} - \dfrac{15}{20}$

Step 2 $\dfrac{16}{20} - \dfrac{15}{20} = \dfrac{16-15}{20} = \dfrac{1}{20}$

5.   (a)

$$
\begin{aligned}
\dfrac{7}{8} &= \dfrac{7 \cdot 3}{8 \cdot 3} = \dfrac{21}{24} \\
- \dfrac{2}{3} &= \dfrac{2 \cdot 8}{3 \cdot 8} = \dfrac{16}{24} \\
\hline
& \qquad\qquad\;\; \dfrac{5}{24}
\end{aligned}
$$

(b)

$$
\begin{aligned}
\dfrac{5}{6} &= \dfrac{5 \cdot 2}{6 \cdot 2} = \dfrac{10}{12} \\
- \dfrac{1}{12} &= \qquad\quad \dfrac{1}{12} \\
\hline
& \qquad\qquad\;\; \dfrac{9}{12} = \dfrac{3}{4}
\end{aligned}
$$

Copyright © 2018 Pearson Education, Inc.

3.3 Section Exercises

1. To add or subtract unlike fractions, the first step is to rewrite the fractions as <u>like</u> fractions.

3. The least common multiple of 4 and 8 is 8. Rewrite both fractions as fractions with a least common denominator of 8. Then add the numerators.

$$\frac{3}{4} + \frac{1}{8} = \frac{6}{8} + \frac{1}{8} \quad \textit{LCD is 8}$$
$$= \frac{6+1}{8}$$
$$= \frac{7}{8}$$

5. $\frac{2}{3} + \frac{2}{9} = \frac{6}{9} + \frac{2}{9} \quad \textit{LCD is 9}$
$$= \frac{6+2}{9}$$
$$= \frac{8}{9}$$

7. $\frac{9}{20} + \frac{3}{10} = \frac{9}{20} + \frac{6}{20} \quad \textit{LCD is 20}$
$$= \frac{9+6}{20}$$
$$= \frac{15}{20} = \frac{3}{4} \quad \textit{lowest terms}$$

9. $\frac{3}{5} + \frac{3}{8} = \frac{24}{40} + \frac{15}{40} \quad \textit{LCD is 40}$
$$= \frac{24+15}{40}$$
$$= \frac{39}{40}$$

11. $\frac{2}{9} + \frac{5}{12} = \frac{8}{36} + \frac{15}{36} \quad \textit{LCD is 36}$
$$= \frac{8+15}{36}$$
$$= \frac{23}{36}$$

13. $\frac{3}{7} + \frac{2}{5} + \frac{1}{10} = \frac{30}{70} + \frac{28}{70} + \frac{7}{70} \quad \textit{LCD is 70}$
$$= \frac{30+28+7}{70}$$
$$= \frac{65}{70} = \frac{13}{14} \quad \textit{lowest terms}$$

15. $\frac{1}{3} + \frac{3}{8} + \frac{1}{4} = \frac{8}{24} + \frac{9}{24} + \frac{6}{24} \quad \textit{LCD is 24}$
$$= \frac{8+9+6}{24} = \frac{23}{24}$$

17. $\frac{5}{12} + \frac{2}{9} + \frac{1}{6} = \frac{15}{36} + \frac{8}{36} + \frac{6}{36} \quad \textit{LCD is 36}$
$$= \frac{15+8+6}{36} = \frac{29}{36}$$

19.
$$\frac{1}{4} = \frac{1\cdot 2}{4\cdot 2} = \frac{2}{8} \quad \textit{LCD is 8}$$
$$+\frac{1}{8} = \qquad\qquad \frac{1}{8}$$
$$\overline{\qquad\qquad\qquad \frac{3}{8}}$$

21.
$$\frac{5}{12} = \frac{5\cdot 4}{12\cdot 4} = \frac{20}{48} \quad \textit{LCD is 48}$$
$$+\frac{1}{16} = \frac{1\cdot 3}{16\cdot 3} = \frac{3}{48}$$
$$\overline{\qquad\qquad\qquad\quad \frac{23}{48}}$$

23. $\frac{5}{6} - \frac{1}{3} = \frac{5}{6} - \frac{2}{6} \quad \textit{LCD is 6}$
$$= \frac{5-2}{6}$$
$$= \frac{3}{6} = \frac{1}{2} \quad \textit{lowest terms}$$

25. $\frac{2}{3} - \frac{1}{6} = \frac{4}{6} - \frac{1}{6} \quad \textit{LCD is 6}$
$$= \frac{4-1}{6}$$
$$= \frac{3}{6} = \frac{1}{2} \quad \textit{lowest terms}$$

27. $\frac{2}{3} - \frac{1}{5} = \frac{10}{15} - \frac{3}{15} \quad \textit{LCD is 15}$
$$= \frac{10-3}{15}$$
$$= \frac{7}{15}$$

29. The least common multiple of 12 and 4 is 12. Rewrite both fractions as fractions with a least common denominator of 12. Then subtract the numerators and write the resulting fraction in lowest terms.

$$\frac{5}{12} - \frac{1}{4} = \frac{5}{12} - \frac{3}{12} \quad \textit{LCD is 12}$$
$$= \frac{5-3}{12}$$
$$= \frac{2}{12} = \frac{1}{6} \quad \textit{lowest terms}$$

31. $\frac{8}{9} - \frac{7}{15} = \frac{40}{45} - \frac{21}{45} \quad \textit{LCD is 45}$
$$= \frac{40-21}{45}$$
$$= \frac{19}{45}$$

Copyright © 2018 Pearson Education, Inc.

33.
$$\frac{7}{8} = \frac{7 \cdot 5}{8 \cdot 5} = \frac{35}{40} \quad LCD \text{ is } 40$$
$$-\frac{4}{5} = \frac{4 \cdot 8}{5 \cdot 8} = \frac{32}{40}$$
$$\frac{3}{40}$$

35.
$$\frac{5}{12} = \frac{5 \cdot 4}{12 \cdot 4} = \frac{20}{48} \quad LCD \text{ is } 48$$
$$-\frac{1}{16} = \frac{1 \cdot 3}{16 \cdot 3} = \frac{3}{48}$$
$$\frac{17}{48}$$

37. The widest blades have cutting-edge widths 1" and $\frac{3}{4}$". The difference is
$$1 - \frac{3}{4} = \frac{4}{4} - \frac{3}{4} = \frac{4-3}{4} = \frac{1}{4} \text{ inch.}$$

39. Subtract the fraction of area used for general admission from the fraction of area devoted to reserved seating.
$$\frac{4}{5} = \frac{4 \cdot 8}{5 \cdot 8} = \frac{32}{40} \quad LCD \text{ is } 40$$
$$-\frac{3}{8} = \frac{3 \cdot 5}{8 \cdot 5} = \frac{15}{40}$$
$$\frac{17}{40}$$
The fraction of the total area used for reserved seating is $\frac{17}{40}$.

41. Add the fractions to find the total length. The least common denominator of the fractions is 40. Rewrite all three fractions with a denominator of 40. Then add the numerators.
$$\frac{1}{8} + \frac{1}{4} + \frac{2}{5} = \frac{5}{40} + \frac{10}{40} + \frac{16}{40} \quad LCD \text{ is } 40$$
$$= \frac{5 + 10 + 16}{40}$$
$$= \frac{31}{40}$$
The total length of the screw is $\frac{31}{40}$ inch.

43. First add to find the amount used.
$$\frac{1}{3} + \frac{3}{8} = \frac{8}{24} + \frac{9}{24} = \frac{8+9}{24} = \frac{17}{24}$$
Then subtract to find the amount remaining.
$$\frac{3}{4} - \frac{17}{24} = \frac{18}{24} - \frac{17}{24} = \frac{18-17}{24} = \frac{1}{24}$$
The fraction of the tank of fuel that remains is $\frac{1}{24}$.

45. Answers will vary. A sample answer follows: You cannot add or subtract until all the fractional pieces are the same size. For example, halves are larger than fourths, so you cannot add $\frac{1}{2} + \frac{1}{4}$ until you rewrite $\frac{1}{2}$ as $\frac{2}{4}$.

47. The total of the two lengths on either side of the hole must be subtracted from the total length $\left(\frac{15}{16} \text{ in.}\right)$ The least common denominator of the fractions is 16.
$$\frac{3}{8} + \frac{3}{8} = \frac{6+6}{16} = \frac{12}{16}$$
$$\frac{15}{16} - \frac{12}{16} = \frac{3}{16}$$
The diameter at the hole is $\frac{3}{16}$ inch.

49. From the circle graph, the fraction of those surveyed that are totally honest is $\frac{1}{3}$.

50. From the circle graph, the fraction of those surveyed that fib a little is $\frac{1}{4}$.

51. **(a)** The response that was given most often is represented by the largest piece of the circle graph, that is, "totally honest."

(b) To find the number of people that gave this response, multiply.
$$\frac{1}{3} \cdot 1200 = \frac{1}{\overset{}{\underset{1}{3}}} \cdot \frac{\overset{400}{\cancel{1200}}}{1} = \frac{400}{1} = 400$$
400 people gave the "totally honest" response.

(c) Add the fractions for "totally honest" and "fib a little."
$$\frac{1}{3} + \frac{1}{4} = \frac{4}{12} + \frac{3}{12} = \frac{4+3}{12} = \frac{7}{12}$$
The fraction of users that gave these responses is $\frac{7}{12}$.

52. **(a)** The response that was given least often is represented by the smallest piece of the circle graph, that is, "flat-out lie."

(b) To find the number of people that gave this response, multiply.
$$\frac{1}{5} \cdot 1200 = \frac{1}{\overset{}{\underset{1}{5}}} \cdot \frac{\overset{240}{\cancel{1200}}}{1} = \frac{240}{1} = 240$$
240 people gave the "flat-out lie" response.

(c) Add the fractions for "flat-out lie" and "total fabrication."
$$\frac{1}{5} + \frac{13}{60} = \frac{12}{60} + \frac{13}{60} = \frac{12+13}{60} = \frac{25}{60} = \frac{5}{12}$$
The fraction of users that gave these responses is $\frac{5}{12}$.

Copyright © 2018 Pearson Education, Inc.

3.4 Adding and Subtracting Mixed Numbers

3.4 Margin Exercises

1. (a) $\dfrac{9}{2}$ $\begin{array}{r} 4 \leftarrow \text{Whole number part} \\ 2\overline{)9} \\ \underline{8} \\ 1 \leftarrow \text{Remainder} \end{array}$

$$\dfrac{9}{2} = 4\dfrac{1}{2}$$

   (b) $3\dfrac{7}{8}$ $3 \cdot 8 = 24$

   $24 + 7 = 31$

$$3\dfrac{7}{8} = \dfrac{31}{8}$$

2. (a) *Estimate:*   *Exact:*

   $\begin{array}{r} 7 \\ +2 \\ \hline 9 \end{array}$ $\begin{array}{r} 6\dfrac{7}{8} = 6\dfrac{7}{8} \\ +2\dfrac{1}{4} = 2\dfrac{2}{8} \\ \hline 8\dfrac{9}{8} \end{array}$

$$8 + \dfrac{9}{8} = 8 + 1\dfrac{1}{8} = 9\dfrac{1}{8}$$

   (b) *Estimate:*   *Exact:*

   $\begin{array}{r} 5 \\ -3 \\ \hline 2 \end{array}$ $\begin{array}{r} 4\dfrac{7}{9} = 4\dfrac{7}{9} \\ -2\dfrac{2}{3} = 2\dfrac{6}{9} \\ \hline 2\dfrac{1}{9} \end{array}$

3. (a) *Estimate:*   *Exact:*

   $\begin{array}{r} 10 \\ +8 \\ \hline 18 \end{array}$ $\begin{array}{r} 9\dfrac{3}{4} = 9\dfrac{3}{4} \\ +7\dfrac{1}{2} = 7\dfrac{2}{4} \\ \hline 16\dfrac{5}{4} \end{array}$

$$16\dfrac{5}{4} = 16 + \dfrac{5}{4} = 16 + 1\dfrac{1}{4} = 17\dfrac{1}{4}$$

   (b) *Estimate:*   *Exact:*

   $\begin{array}{r} 16 \\ +13 \\ \hline 29 \end{array}$ $\begin{array}{r} 15\dfrac{4}{5} = 15\dfrac{12}{15} \\ +12\dfrac{2}{3} = 12\dfrac{10}{15} \\ \hline 27\dfrac{22}{15} \end{array}$

$$27\dfrac{22}{15} = 27 + \dfrac{22}{15} = 27 + 1\dfrac{7}{15} = 28\dfrac{7}{15}$$

4. (a) *Estimate:*   *Exact:*

   $\begin{array}{r} 7 \\ -5 \\ \hline 2 \end{array}$ $\begin{array}{r} 7\dfrac{1}{3} = 7\dfrac{2}{6} \\ -4\dfrac{5}{6} = 4\dfrac{5}{6} \\ \hline \end{array}$

Regroup:

$$7\dfrac{2}{6} = 7 + \dfrac{2}{6} = 6 + 1 + \dfrac{2}{6} = 6 + \dfrac{6}{6} + \dfrac{2}{6} = 6\dfrac{8}{6}$$

$$\begin{array}{r} 6\dfrac{8}{6} \\ -4\dfrac{5}{6} \\ \hline 2\dfrac{3}{6} = 2\dfrac{1}{2} \end{array}$$

   (b) *Estimate:*   *Exact:*

   $\begin{array}{r} 5 \\ -3 \\ \hline 2 \end{array}$ $\begin{array}{r} 4\dfrac{5}{8} = 4\dfrac{10}{16} \\ -2\dfrac{15}{16} = 2\dfrac{15}{16} \\ \hline \end{array}$

Regroup:

$$4\dfrac{10}{16} = 4 + \dfrac{10}{16} = 3 + 1 + \dfrac{10}{16} = 3 + \dfrac{16}{16} + \dfrac{10}{16} = 3\dfrac{26}{16}$$

$$\begin{array}{r} 3\dfrac{26}{16} \\ -2\dfrac{15}{16} \\ \hline 1\dfrac{11}{16} \end{array}$$

   (c) *Estimate:*   *Exact:*

   $\begin{array}{r} 15 \\ -6 \\ \hline 9 \end{array}$ $\begin{array}{r} 15 \\ -6\dfrac{4}{9} \\ \hline \end{array}$

Regroup:

$$15 = 14 + 1 = 14 + \dfrac{9}{9} = 14\dfrac{9}{9}$$

$$\begin{array}{r} 14\dfrac{9}{9} \\ -6\dfrac{4}{9} \\ \hline 8\dfrac{5}{9} \end{array}$$

5. (a) $\begin{array}{r} 3\dfrac{3}{8} = \dfrac{27}{8} = \dfrac{27}{8} \\ +2\dfrac{1}{2} = \dfrac{5}{2} = \dfrac{20}{8} \\ \hline \dfrac{47}{8} = 5\dfrac{7}{8} \end{array}$

Copyright © 2018 Pearson Education, Inc.

(b)
$$6\frac{3}{4} = \frac{27}{4} = \frac{81}{12}$$
$$-4\frac{2}{3} = \frac{14}{3} = \frac{56}{12}$$
$$\frac{25}{12} = 2\frac{1}{12}$$

3.4 Section Exercises

1. Note: This rounding concept was covered in Section 2.8.

$5\frac{1}{3}$ ← 1 is less than half of 3.
$\quad$ ← Half of 3 is $1\frac{1}{2}$.

$5\frac{1}{3}$ rounds down to 5.

3. $8\frac{4}{5}$ ← 4 is more than half of 5.
$\quad$ ← Half of 5 is $2\frac{1}{2}$.

$8\frac{4}{5}$ rounds up to 9.

5. $15\frac{7}{15}$ ← 7 is less than half of 15.
$\quad$ ← Half of 15 is $7\frac{1}{2}$.

$15\frac{7}{15}$ rounds down to 15.

7. $16\frac{2}{3}$ ← 2 is more than half of 3.
$\quad$ ← Half of 3 is $1\frac{1}{2}$.

$16\frac{2}{3}$ rounds up to 17.

9. *Estimate:*   *Exact:*

$$\begin{array}{r} 6 \\ +3 \\ \hline 9 \end{array} \qquad \begin{array}{r} 5\frac{1}{2} = 5\frac{3}{6} \\ +3\frac{1}{3} = 3\frac{2}{6} \\ \hline 8\frac{5}{6} \end{array}$$

11. *Estimate:*   *Exact:*

$$\begin{array}{r} 7 \\ +4 \\ \hline 11 \end{array} \qquad \begin{array}{r} 7\frac{1}{3} = 7\frac{2}{6} \\ +4\frac{1}{6} = 4\frac{1}{6} \\ \hline 11\frac{3}{6} = 11\frac{1}{2} \end{array}$$

13. *Estimate:*   *Exact:*

$$\begin{array}{r} 1 \\ +4 \\ \hline 5 \end{array} \qquad \begin{array}{r} \frac{5}{8} = \frac{15}{24} \\ +3\frac{7}{12} = 3\frac{14}{24} \\ \hline 3\frac{29}{24} \end{array}$$

$$3\frac{29}{24} = 3 + \frac{29}{24} = 3 + 1\frac{5}{24} = 4\frac{5}{24}$$

15. *Estimate:*   *Exact:*

$$\begin{array}{r} 25 \\ +19 \\ \hline 44 \end{array} \qquad \begin{array}{r} 24\frac{5}{6} \\ +18\frac{5}{6} \\ \hline 42\frac{10}{6} \end{array}$$

$$42\frac{10}{6} = 42 + 1\frac{4}{6} = 43\frac{2}{3}$$

17. *Estimate:*   *Exact:*

$$\begin{array}{r} 34 \\ +19 \\ \hline 53 \end{array} \qquad \begin{array}{r} 33\frac{3}{5} = 33\frac{6}{10} \\ +18\frac{1}{2} = 18\frac{5}{10} \\ \hline 51\frac{11}{10} \end{array}$$

$$51\frac{11}{10} = 51 + 1\frac{1}{10} = 52\frac{1}{10}$$

19. *Estimate:*   *Exact:*

$$\begin{array}{r} 23 \underline{\text{ Rounds to}} \\ +15 \underline{\text{ Rounds to}} \\ \hline 38 \end{array} \qquad \begin{array}{r} 22\frac{3}{4} = 22\frac{21}{28} \\ +15\frac{3}{7} = 15\frac{12}{28} \\ \hline 37\frac{33}{28} \end{array}$$

$$37\frac{33}{28} = 37 + 1\frac{5}{28} = 38\frac{5}{28}$$

21. *Estimate:*   *Exact:*

$$\begin{array}{r} 13 \\ 19 \\ +15 \\ \hline 47 \end{array} \qquad \begin{array}{r} 12\frac{8}{15} = 12\frac{16}{30} \\ 18\frac{3}{5} = 18\frac{18}{30} \\ +14\frac{7}{10} = 14\frac{21}{30} \\ \hline 44\frac{55}{30} \end{array}$$

$$44\frac{55}{30} = 44 + 1\frac{25}{30} = 45\frac{5}{6}$$

23. *Estimate:*   *Exact:*

$$\begin{array}{r} 15 \\ -12 \\ \hline 3 \end{array} \qquad \begin{array}{r} 14\frac{7}{8} = 14\frac{7}{8} \\ -12\frac{1}{4} = 12\frac{2}{8} \\ \hline 2\frac{5}{8} \end{array}$$

Copyright © 2018 Pearson Education, Inc.

25. *Estimate:*   *Exact:*

$$13$$
$$\underline{-\ 1\ \ }$$
$$12$$

$$12\frac{2}{3} = 12\frac{10}{15}$$
$$\underline{-1\frac{1}{5}\ = \ 1\frac{3}{15}}$$
$$11\frac{7}{15}$$

27. *Estimate:*   *Exact:*

$$28$$
$$\underline{-\ 6\ \ }$$
$$22$$

$$28\frac{3}{10} = 28\frac{9}{30}$$
$$\underline{-6\frac{1}{15}\ = \ 6\frac{2}{30}}$$
$$22\frac{7}{30}$$

29. *Estimate:*   *Exact:*

$$17 \underset{\text{Rounds to}}{\longleftarrow} 17$$
$$\underline{-\ 7} \underset{\text{Rounds to}}{\longleftarrow} \underline{-\ 6\frac{5}{8}}$$
$$10$$

To subtract $\frac{5}{8}$, first regroup the whole number 17 into $16 + 1$.

$$17 = 16 + 1 = 16 + \frac{8}{8}$$

Now you can subtract.

$$16\frac{8}{8}$$
$$\underline{-\ 6\frac{5}{8}}$$
$$10\frac{3}{8}$$

31. *Estimate:*   *Exact:*

$$19$$
$$\underline{-\ 6\ \ }$$
$$13$$

$$18\frac{3}{4} = 18\frac{15}{20}$$
$$\underline{-5\frac{4}{5}\ = \ 5\frac{16}{20}}$$

Regroup:

$$18\frac{15}{20} = 17 + 1 + \frac{15}{20} = 17 + \frac{20}{20} + \frac{15}{20} = 17\frac{35}{20}$$

$$17\frac{35}{20}$$
$$\underline{-\ 5\frac{16}{20}}$$
$$12\frac{19}{20}$$

33. *Estimate:*   *Exact:*

$$20$$
$$\underline{-\ 12\ \ }$$
$$8$$

$$19\frac{2}{3} = 19\frac{8}{12}$$
$$\underline{-11\frac{3}{4}\ = \ 11\frac{9}{12}}$$

Regroup:

$$19\frac{8}{12} = 18 + 1 + \frac{8}{12} = 18 + \frac{12}{12} + \frac{8}{12} = 18\frac{20}{12}$$

$$18\frac{20}{12}$$
$$\underline{-\ 11\frac{9}{12}}$$
$$7\frac{11}{12}$$

35. $3\frac{3}{4}$ $3 \cdot 4 = 12$
         $12 + 3 = 15$

$$3\frac{3}{4} = \frac{15}{4}$$

37. $\frac{12}{5}$

$$\begin{array}{r} 2 \leftarrow \text{Whole number part} \\ 5\overline{)12} \\ \underline{10} \\ 2 \leftarrow \text{Remainder} \end{array}$$

$$\frac{12}{5} = 2\frac{2}{5}$$

39. $5\frac{3}{8}$ $5 \cdot 8 = 40$
         $40 + 3 = 43$

$$5\frac{3}{8} = \frac{43}{8}$$

41. $\frac{56}{3}$

$$\begin{array}{r} 18 \leftarrow \text{Whole number part} \\ 3\overline{)56} \\ \underline{54} \\ 2 \leftarrow \text{Remainder} \end{array}$$

$$\frac{56}{3} = 18\frac{2}{3}$$

43.

$$7\frac{5}{8} = \frac{61}{8} = \frac{61}{8}$$
$$\underline{+1\frac{1}{2}\ = \ \frac{3}{2}\ = \ \frac{12}{8}}$$
$$\frac{73}{8} = 9\frac{1}{8}$$

45.

$$4\frac{2}{3} = \frac{14}{3} = \frac{28}{6}$$
$$\underline{+6\frac{5}{6}\ = \ \frac{41}{6}\ = \ \frac{41}{6}}$$
$$\frac{69}{6} = 11\frac{3}{6} = 11\frac{1}{2}$$

Copyright © 2018 Pearson Education, Inc.

47.
$$2\frac{2}{3} = \frac{8}{3} = \frac{16}{6}$$
$$+1\frac{1}{6} = \frac{7}{6} = \frac{7}{6}$$
$$\frac{23}{6} = 3\frac{5}{6}$$

49.
$$3\frac{1}{4} = \frac{13}{4} = \frac{39}{12}$$
$$+3\frac{2}{3} = \frac{11}{3} = \frac{44}{12}$$
$$\frac{83}{12} = 6\frac{11}{12}$$

51.
$$1\frac{3}{8} = \frac{11}{8} = \frac{11}{8}$$
$$+6\frac{3}{4} = \frac{27}{4} = \frac{54}{8}$$
$$\frac{65}{8} = 8\frac{1}{8}$$

53.
$$3\frac{1}{2} = \frac{7}{2} = \frac{21}{6}$$
$$-2\frac{2}{3} = \frac{8}{3} = \frac{16}{6}$$
$$\frac{5}{6}$$

55.
$$8\frac{3}{4} = \frac{35}{4} = \frac{70}{8}$$
$$-5\frac{7}{8} = \frac{47}{8} = \frac{47}{8}$$
$$\frac{23}{8} = 2\frac{7}{8}$$

57.
$$8 \quad = 7\frac{8}{8}$$
$$-4\frac{3}{8} = 4\frac{3}{8}$$
$$3\frac{5}{8}$$

59.
$$9\frac{1}{5} = \frac{46}{5} = \frac{184}{20}$$
$$-3\frac{3}{4} = \frac{15}{4} = \frac{75}{20}$$
$$\frac{109}{20} = 5\frac{9}{20}$$

61.
$$6\frac{3}{7} = \frac{45}{7} = \frac{135}{21}$$
$$-2\frac{2}{3} = \frac{8}{3} = \frac{56}{21}$$
$$\frac{79}{21} = 3\frac{16}{21}$$

63. The sum in the units column was not regrouped to the tens column, and the fraction part was not simplified to $1\frac{11}{20}$. The correct answer is $105\frac{11}{20}$.

$$78\frac{3}{4} = \quad 78\frac{15}{20}$$
$$+26\frac{4}{5} = \quad +26\frac{16}{20}$$
$$104\frac{31}{20} = 104\frac{20+11}{20} = 105\frac{11}{20}$$

65. Subtract $15\frac{3}{4}$ from $24\frac{1}{2}$ to determine the difference.

Estimate: $25 - 16 = 9$ feet

Exact:
$$24\frac{1}{2} = \frac{49}{2} = \frac{98}{4}$$
$$-15\frac{3}{4} = \frac{63}{4} = \frac{63}{4}$$
$$\frac{35}{4} = 8\frac{3}{4}$$

The garage is $8\frac{3}{4}$ feet longer than the truck.

67. The 8" slip joint pliers has a maximum jaw capacity of $3\frac{3}{8}$" and the 6" slip joint pliers has a maximum jaw capacity of $2\frac{15}{16}$". Subtract to find the difference.

Estimate: $3 - 3 = 0$ inches

Exact:
$$3\frac{3}{8} = 3\frac{6}{16} = 2\frac{22}{16}$$
$$-2\frac{15}{16} = 2\frac{15}{16} = 2\frac{15}{16}$$
$$\frac{7}{16}$$

The maximum jaw capacity of the 8" slip joint pliers is $\frac{7}{16}$ inch larger than that of the 6" slip joint pliers.

69. The 8" slip joint pliers has the largest maximum jaw capacity, $3\frac{3}{8}$", and the 7" diagonal pliers has the smallest maximum jaw capacity of $\frac{7}{8}$". Subtract to find the difference.

Estimate: $3 - 1 = 2$ inches

Exact:
$$3\frac{3}{8} = 2\frac{11}{8}$$
$$-\frac{7}{8} = \frac{7}{8}$$
$$2\frac{4}{8} = 2\frac{1}{2}$$

The difference in the largest maximum jaw capacity and the smallest maximum jaw capacity is $2\frac{1}{2}$ inches.

Copyright © 2018 Pearson Education, Inc.

71. The largest hose clamp is $2\frac{3}{4}$" and the smallest hose clamp is $\frac{9}{16}$". Subtract the size of the smallest hose clamp from the largest hose clamp.

Estimate: $3 - 1 = 2$ inches

Exact:
$$2\frac{3}{4} = 2\frac{12}{16}$$
$$-\ \frac{9}{16} = \frac{9}{16}$$
$$\overline{\qquad\quad 2\frac{3}{16}}$$

The difference in size is $2\frac{3}{16}$ inches.

73. Add the four measurements.

Estimate: $16 + 19 + 24 + 31 = 90$ ft

Exact:
$$15\frac{1}{2} = 15\frac{2}{4}$$
$$18\frac{3}{4} = 18\frac{3}{4}$$
$$24\frac{1}{4} = 24\frac{1}{4}$$
$$+\ 30\frac{1}{2} = 30\frac{2}{4}$$
$$\overline{\qquad 87\frac{8}{4}} = 87 + 2 = 89$$

Pam needs 89 feet of fencing to go around the garden.

75. Add the lengths of the four sides.

Estimate: $24 + 35 + 24 + 35 = 118$ in.

Exact:
$$23\frac{3}{4} = 23\frac{3}{4}$$
$$34\frac{1}{2} = 34\frac{2}{4}$$
$$23\frac{3}{4} = 23\frac{3}{4}$$
$$+\ 34\frac{1}{2} = 34\frac{2}{4}$$
$$\overline{\qquad 114\frac{10}{4}} = 114 + 2\frac{2}{4} = 116\frac{1}{2}$$

The craftsperson needs $116\frac{1}{2}$ inches of lead stripping.

77. Subtract the amounts used from the total amount.

Estimate: $100 - 10 - 14 - 9 - 19 - 12 - 10 - 14 = 12$ gallons

Exact: Add up the amounts used.

$$10\frac{1}{4} = 10\frac{2}{8}$$
$$13\frac{1}{2} = 13\frac{4}{8}$$
$$8\frac{7}{8} = 8\frac{7}{8}$$
$$18\frac{3}{4} = 18\frac{6}{8}$$
$$12\frac{3}{8} = 12\frac{3}{8}$$
$$9\frac{1}{2} = 9\frac{4}{8}$$
$$+\ 14\frac{1}{8} = 14\frac{1}{8}$$
$$\overline{\qquad 84\frac{27}{8}} = 84 + 3\frac{3}{8} = 87\frac{3}{8}$$

Now subtract from 100.

$$100\ = 99\frac{8}{8}$$
$$-\ 87\frac{3}{8} = 87\frac{3}{8}$$
$$\overline{\qquad\quad 12\frac{5}{8}}$$

There are $12\frac{5}{8}$ gallons of milk remaining.

79. Subtract the known lengths from the total length.

Estimate: $527 - 108 - 151 - 139 = 129$ ft

Exact: Add up the lengths of the known sides.

$$107\frac{2}{3} = 107\frac{16}{24}$$
$$150\frac{3}{4} = 150\frac{18}{24}$$
$$+\ 138\frac{5}{8} = 138\frac{15}{24}$$
$$\overline{\qquad 395\frac{49}{24}} = 395 + 2\frac{1}{24} = 397\frac{1}{24}$$

Now subtract from $527\frac{1}{24}$.

$$527\frac{1}{24}$$
$$-\ 397\frac{1}{24}$$
$$\overline{\qquad\ 130}$$

The length of the fourth side is 130 feet.

Copyright © 2018 Pearson Education, Inc.

81. Add the weights.

Estimate: $59 + 24 + 17 + 29 + 58 = 187$ tons

Exact:

$$58\frac{1}{2} = 58\frac{12}{24}$$
$$23\frac{5}{8} = 23\frac{15}{24}$$
$$16\frac{5}{6} = 16\frac{20}{24}$$
$$29\frac{1}{4} = 29\frac{6}{24}$$
$$+58\frac{1}{3} = 58\frac{8}{24}$$

$$184\frac{61}{24} = 184 + 2\frac{13}{24} = 186\frac{13}{24}$$

The total weight is $186\frac{13}{24}$ tons.

83. First add the two given portions of the line.

$$2\frac{3}{8}$$
$$+2\frac{3}{8}$$
$$4\frac{6}{8}$$

Then subtract to find the unknown length.

$$9\frac{7}{16} = 9\frac{7}{16} = 8\frac{23}{16}$$
$$-4\frac{6}{8} = 4\frac{12}{16} = 4\frac{12}{16}$$
$$4\frac{11}{16}$$

The unknown length is $4\frac{11}{16}$ inches.

85. Add the length of the sections at each end.

$$6\frac{1}{4} = 6\frac{2}{8}$$
$$+1\frac{7}{8} = 1\frac{7}{8}$$
$$7\frac{9}{8} = 7 + 1\frac{1}{8} = 8\frac{1}{8}$$

Then subtract this total from the total length of the arrow.

$$29\frac{1}{2} = 29\frac{4}{8}$$
$$-8\frac{1}{8} = 8\frac{1}{8}$$
$$21\frac{3}{8}$$

The unknown length is $21\frac{3}{8}$ inches.

87. (a) $\dfrac{5}{9} = \dfrac{?}{54} \quad 54 \div 9 = 6$

$$\frac{5}{9} = \frac{5 \cdot 6}{9 \cdot 6} = \frac{30}{54}$$

(b) $\dfrac{7}{12} = \dfrac{?}{48} \quad 48 \div 12 = 4$

$$\frac{7}{12} = \frac{7 \cdot 4}{12 \cdot 4} = \frac{28}{48}$$

(c) $\dfrac{5}{8} = \dfrac{?}{40} \quad 40 \div 8 = 5$

$$\frac{5}{8} = \frac{5 \cdot 5}{8 \cdot 5} = \frac{25}{40}$$

(d) $\dfrac{11}{5} = \dfrac{?}{120} \quad 120 \div 5 = 24$

$$\frac{11}{5} = \frac{11 \cdot 24}{5 \cdot 24} = \frac{264}{120}$$

88. When rewriting unlike fractions as like fractions with the least common multiple as a denominator, the new denominator is called the <u>least</u> <u>common</u> <u>denominator</u>, or LCD.

89. (a) $\dfrac{5}{8} + \dfrac{1}{3} = \dfrac{15}{24} + \dfrac{8}{24} = \dfrac{15 + 8}{24} = \dfrac{23}{24}$

(b)
$$\frac{19}{20} - \frac{5}{12} = \frac{57}{60} - \frac{25}{60} = \frac{57 - 25}{60} = \frac{32}{60} = \frac{8}{15}$$

(c)
$$\frac{7}{12} = \frac{28}{48}$$
$$\frac{3}{16} = \frac{9}{48}$$
$$+\frac{3}{24} = \frac{6}{48}$$
$$\frac{43}{48}$$

(d)
$$\frac{6}{7} = \frac{18}{21}$$
$$-\frac{2}{3} = \frac{14}{21}$$
$$\frac{4}{21}$$

90. A common method for adding or subtracting mixed numbers is to add or subtract the <u>fraction parts</u> and then add or subtract the whole number parts.

91. Another method for adding or subtracting mixed numbers is to first change the mixed numbers to <u>improper</u> fractions. After adding or subtracting, write the answer in lowest terms and as a mixed number when possible. This method is difficult to use if the mixed numbers are <u>large</u>.

Copyright © 2018 Pearson Education, Inc.

92. **(a)** *First Method:*

$$
\begin{aligned}
4\frac{5}{8} &= 4\frac{5}{8}\\
+3\frac{3}{4} &= 3\frac{6}{8}\\
\hline
7\frac{11}{8} &= 7+1\frac{3}{8} = 8\frac{3}{8}
\end{aligned}
$$

Second Method:

$$
\begin{aligned}
4\frac{5}{8} &= \frac{37}{8} = \frac{37}{8}\\
+3\frac{3}{4} &= \frac{15}{4} = \frac{30}{8}\\
\hline
&\qquad\quad \frac{67}{8} = 8\frac{3}{8}
\end{aligned}
$$

(b) *First Method:*

$$
\begin{aligned}
12\frac{2}{5} &= 12\frac{16}{40} = 11\frac{56}{40}\\
-8\frac{7}{8} &= 8\frac{35}{40} = 8\frac{35}{40}\\
\hline
& \qquad\qquad\quad 3\frac{21}{40}
\end{aligned}
$$

Second Method:

$$
\begin{aligned}
12\frac{2}{5} &= \frac{62}{5} = \frac{496}{40}\\
-8\frac{7}{8} &= \frac{71}{8} = \frac{355}{40}\\
\hline
& \qquad\qquad \frac{141}{40} = 3\frac{21}{40}
\end{aligned}
$$

Both methods give the same answer.
Preferences will vary. A sample answer follows:
When a problem requires regrouping, it is easier
to change all the numbers to improper fractions.
Otherwise, adding the whole numbers and then
adding the fractions seems easier.

Summary Exercises
Operations with Fractions

1. $\frac{3}{4}$ is a *proper* fraction since the numerator is less than the denominator.

3. $\frac{10}{10}$ is an *improper* fraction since the numerator is greater than or equal to the denominator.

5. $\dfrac{30}{36} = \dfrac{30 \div 6}{36 \div 6} = \dfrac{5}{6}$

7. $\dfrac{15}{35} = \dfrac{15 \div 5}{35 \div 5} = \dfrac{3}{7}$

9. $\dfrac{3}{4} \cdot \dfrac{2}{3} = \dfrac{\overset{1}{\cancel{3}} \cdot \overset{1}{\cancel{2}}}{\underset{2}{\cancel{4}} \cdot \underset{1}{\cancel{3}}} = \dfrac{1 \cdot 1}{2 \cdot 1} = \dfrac{1}{2}$

11. $56 \cdot \dfrac{5}{8} = \dfrac{\overset{7}{\cancel{56}}}{1} \cdot \dfrac{5}{\cancel{8}} = \dfrac{7 \cdot 5}{1 \cdot 1} = \dfrac{35}{1} = 35$

13. $\dfrac{35}{45} \div \dfrac{10}{15} = \dfrac{35}{45} \cdot \dfrac{15}{10}$

$$
= \dfrac{\overset{7}{\cancel{35}}}{\underset{3}{\cancel{45}}} \cdot \dfrac{\overset{1}{\cancel{15}}}{\underset{2}{\cancel{10}}}
$$

$$
= \dfrac{7 \cdot 1}{3 \cdot 2} = \dfrac{7}{6} = 1\dfrac{1}{6}
$$

15. $\dfrac{7}{8} + \dfrac{2}{3} = \dfrac{7 \cdot 3}{8 \cdot 3} + \dfrac{2 \cdot 8}{3 \cdot 8} = \dfrac{21}{24} + \dfrac{16}{24}$

$$
= \dfrac{21+16}{24} = \dfrac{37}{24} = 1\dfrac{13}{24}
$$

17. $\dfrac{7}{12} + \dfrac{5}{6} + \dfrac{2}{3} = \dfrac{7}{12} + \dfrac{10}{12} + \dfrac{8}{12} = \dfrac{7+10+8}{12}$

$$
= \dfrac{25}{12} = 2\dfrac{1}{12}
$$

19. $\dfrac{7}{8} - \dfrac{5}{12} = \dfrac{21}{24} - \dfrac{10}{24} = \dfrac{21-10}{24} = \dfrac{11}{24}$

21. $3\frac{1}{2} \cdot 2\frac{1}{4}$

Estimate: $4 \cdot 2 = 8$

Exact: $3\dfrac{1}{2} \cdot 2\dfrac{1}{4} = \dfrac{7}{2} \cdot \dfrac{9}{4} = \dfrac{63}{8} = 7\dfrac{7}{8}$

23. $8 \cdot 5\frac{2}{3} \cdot 2\frac{3}{8}$

Estimate: $8 \cdot 6 \cdot 2 = 48 \cdot 2 = 96$

Exact: $8 \cdot 5\dfrac{2}{3} \cdot 2\dfrac{3}{8} = \dfrac{\overset{1}{\cancel{8}}}{1} \cdot \dfrac{17}{3} \cdot \dfrac{19}{\underset{1}{\cancel{8}}} = \dfrac{323}{3} = 107\dfrac{2}{3}$

25. $6\frac{7}{8} \div 2$

Estimate: $7 \div 2 = 3\dfrac{1}{2}$

Exact: $6\dfrac{7}{8} \div 2 = \dfrac{55}{8} \div \dfrac{2}{1}$

$$
= \dfrac{55}{8} \cdot \dfrac{1}{2}
$$

$$
= \dfrac{55}{16} = 3\dfrac{7}{16}
$$

27. *Estimate:*   *Exact:*

$$
\begin{aligned}
6 \qquad\qquad & 5\frac{2}{3} = 5\frac{8}{12}\\
+4 \qquad\qquad & +4\frac{1}{4} = 4\frac{3}{12}\\
\hline
10 \qquad\qquad & \qquad\quad 9\frac{11}{12}
\end{aligned}
$$

Copyright © 2018 Pearson Education, Inc.

29. *Estimate:* *Exact:*

$$15$$
$$\underline{+\ 11}$$
$$26$$

$$14\frac{3}{5} = 14\frac{9}{15}$$
$$+\ 10\frac{2}{3} = 10\frac{10}{15}$$
$$\overline{\qquad\qquad 24\frac{19}{15}} = 24 + 1\frac{4}{15} = 25\frac{4}{15}$$

31. *Estimate:* *Exact:*

$$14 \qquad 14 = 13\frac{8}{8}$$
$$\underline{-\ 7} \qquad -7\frac{3}{8} = 7\frac{3}{8}$$
$$7 \qquad\qquad\qquad 6\frac{5}{8}$$

33. $8, 10$

$$8 = 2 \cdot 2 \cdot 2$$
$$10 = 2 \cdot 5$$

$$\text{LCM} = 2 \cdot 2 \cdot 2 \cdot 5 = 40$$

The LCM of 8 and 10 is 40.

35. $3, 5, 10$

$$3 = 1 \cdot 3$$
$$5 = 1 \cdot 5$$
$$10 = 2 \cdot 5$$

$$\text{LCM} = 2 \cdot 3 \cdot 5 = 30$$

The LCM of 3, 5, and 10 is 30.

37. $9, 18, 24$

$$9 = 3 \cdot 3$$
$$18 = 2 \cdot 3 \cdot 3$$
$$24 = 2 \cdot 2 \cdot 2 \cdot 3$$

$$\text{LCM} = 2 \cdot 2 \cdot 2 \cdot 3 \cdot 3 = 72$$

The LCM of 9, 18, and 24 is 72.

39. $\dfrac{5}{6} = \dfrac{?}{42}$ $\quad 42 \div 6 = 7$

$$\frac{5}{6} = \frac{5 \cdot 7}{6 \cdot 7} = \frac{35}{42}$$

41. $\dfrac{3}{7} = \dfrac{?}{28}$ $\quad 28 \div 7 = 4$

$$\frac{3}{7} = \frac{3 \cdot 4}{7 \cdot 4} = \frac{12}{28}$$

43. $\dfrac{3}{9} = \dfrac{?}{45}$ $\quad 45 \div 9 = 5$

$$\frac{3}{9} = \frac{3 \cdot 5}{9 \cdot 5} = \frac{15}{45}$$

3.5 Order Relations and the Order of Operations

3.5 Margin Exercises

1. **(a)–(c)** Put dots on the number line between 0 and 1 for $\frac{2}{3}$, between 1 and 2 for $1\frac{1}{2}$, and between 2 and 3 for $2\frac{3}{4}$, as shown below.

2. **(a)** 1 is to the left of $\frac{5}{4} = 1\frac{1}{4}$ on the number line, so 1 is less than $\frac{5}{4}$.

$$1 \boxed{<} \frac{5}{4}$$

(b) $\frac{8}{3} = 2\frac{2}{3}$ is to the right of $\frac{3}{2} = 1\frac{1}{2}$ on the number line, so $\frac{8}{3}$ is greater than $\frac{3}{2}$.

$$\frac{8}{3} \boxed{>} \frac{3}{2}$$

(c) 0 is to the left of 1 on the number line, so 0 is less than 1.

$$0 \boxed{<} 1$$

(d) $\frac{17}{8} = 2\frac{1}{8}$ is to the right of $\frac{8}{4} = 2$ on the number line, so $\frac{17}{8}$ is greater than $\frac{8}{4}$.

$$\frac{17}{8} \boxed{>} \frac{8}{4}$$

3. **(a)** $\dfrac{7}{8} - \dfrac{3}{4}$ $\qquad$ *LCD is 8*

$$\frac{3}{4} = \frac{6}{8}$$
$$\frac{7}{8} > \frac{6}{8}, \text{ so } \frac{7}{8} \boxed{>} \frac{3}{4}.$$

(b) $\dfrac{13}{8} - \dfrac{15}{9}$ $\qquad$ *LCD is 72*

$$\frac{13}{8} = \frac{117}{72} \text{ and } \frac{15}{9} = \frac{120}{72}$$
$$\frac{117}{72} < \frac{120}{72}, \text{ so } \frac{13}{8} \boxed{<} \frac{15}{9}.$$

Copyright © 2018 Pearson Education, Inc.

(c) $\dfrac{9}{4} - \dfrac{7}{3}$ *LCD is 12*

$\dfrac{9}{4} = \dfrac{27}{12}$ and $\dfrac{7}{3} = \dfrac{28}{12}$

$\dfrac{27}{12} < \dfrac{28}{12}$, so $\dfrac{9}{4}$ $\boxed{<}$ $\dfrac{7}{3}$.

(d) $\dfrac{9}{10} - \dfrac{14}{15}$ *LCD is 30*

$\dfrac{9}{10} = \dfrac{27}{30}$ and $\dfrac{14}{15} = \dfrac{28}{30}$

$\dfrac{27}{30} < \dfrac{28}{30}$, so $\dfrac{9}{10}$ $\boxed{<}$ $\dfrac{14}{15}$.

Four factors of $\frac{1}{2}$

4. (a) $\left(\dfrac{1}{2}\right)^4 = \overbrace{\dfrac{1}{2} \cdot \dfrac{1}{2} \cdot \dfrac{1}{2} \cdot \dfrac{1}{2}} = \dfrac{1}{16}$

(b) $\left(\dfrac{3}{4}\right)^2 = \dfrac{3}{4} \cdot \dfrac{3}{4} = \dfrac{9}{16}$

(c) $\left(\dfrac{1}{2}\right)^3 \cdot \left(\dfrac{2}{3}\right)^2 = \left(\dfrac{1}{2} \cdot \dfrac{1}{2} \cdot \dfrac{1}{2}\right) \cdot \left(\dfrac{2}{3} \cdot \dfrac{2}{3}\right)$

$= \dfrac{1 \cdot 1 \cdot 1 \cdot \overset{1}{\cancel{2}} \cdot \overset{1}{\cancel{2}}}{2 \cdot \cancel{2} \cdot \cancel{2} \cdot 3 \cdot 3}$

$= \dfrac{1}{18}$

(d) $\left(\dfrac{1}{5}\right)^2 \cdot \left(\dfrac{5}{3}\right)^2 = \left(\dfrac{1}{5} \cdot \dfrac{1}{5}\right) \cdot \left(\dfrac{5}{3} \cdot \dfrac{5}{3}\right)$

$= \dfrac{1 \cdot 1 \cdot \overset{1}{\cancel{5}} \cdot \overset{1}{\cancel{5}}}{\cancel{5} \cdot \cancel{5} \cdot 3 \cdot 3}$

$= \dfrac{1}{9}$

5. (a) $\dfrac{5}{9} - \dfrac{3}{4}\left(\dfrac{2}{3}\right) = \dfrac{5}{9} - \dfrac{\overset{1}{\cancel{3}}}{\underset{2}{\cancel{4}}}\left(\dfrac{\overset{1}{\cancel{2}}}{3}\right)$

$= \dfrac{5}{9} - \dfrac{1}{2}$ *LCD is 18*

$= \dfrac{10}{18} - \dfrac{9}{18}$

$= \dfrac{1}{18}$

(b) $\dfrac{3}{4}\left(\dfrac{2}{3} \cdot \dfrac{3}{5}\right) = \dfrac{3}{4}\left(\dfrac{2}{\cancel{3}} \cdot \dfrac{\overset{1}{\cancel{3}}}{5}\right)$

$= \dfrac{3}{\underset{2}{\cancel{4}}} \cdot \dfrac{\overset{1}{\cancel{2}}}{5}$

$= \dfrac{3}{10}$

(c) $\dfrac{7}{8}\left(\dfrac{2}{3}\right) - \left(\dfrac{1}{2}\right)^2 = \dfrac{7}{\underset{4}{\cancel{8}}}\left(\dfrac{\overset{1}{\cancel{2}}}{3}\right) - \dfrac{1}{2} \cdot \dfrac{1}{2}$

$= \dfrac{7}{12} - \dfrac{1}{4}$ *LCD is 12*

$= \dfrac{7}{12} - \dfrac{3}{12}$

$= \dfrac{4}{12} = \dfrac{1}{3}$

(d) $\dfrac{\left(\frac{5}{6}\right)^2}{\frac{4}{3}} = \left(\dfrac{5}{6} \cdot \dfrac{5}{6}\right) \div \dfrac{4}{3}$

$= \dfrac{5 \cdot 5}{6 \cdot 6} \cdot \dfrac{3}{4}$

$= \dfrac{5 \cdot 5 \cdot \overset{1}{\cancel{3}}}{6 \cdot \underset{2}{\cancel{6}} \cdot 4}$

$= \dfrac{25}{48}$

3.5 Section Exercises

1.–12.

$\begin{array}{ccccccccccc} \frac{1}{4} & \frac{1}{2} & \frac{7}{8} & \frac{5}{4} & \frac{3}{2} & 1\frac{7}{8} & 2\frac{1}{6} & \frac{7}{3} & \frac{11}{4} & 3\frac{1}{4} & \frac{7}{2} & 3\frac{4}{5} \end{array}$

2. 1. 10. 4. 3. 12. 7. 5. 6. 11. 9. 8.

13. $\dfrac{1}{2} - \dfrac{3}{8}$ *LCD is 8*

$\dfrac{1}{2} = \dfrac{4}{8}$

$\dfrac{4}{8} > \dfrac{3}{8}$, so $\dfrac{1}{2}$ $\boxed{>}$ $\dfrac{3}{8}$.

15. $\dfrac{5}{6} - \dfrac{11}{12}$ *LCD is 12*

$\dfrac{5}{6} = \dfrac{10}{12}$

$\dfrac{10}{12} < \dfrac{11}{12}$, so $\dfrac{5}{6}$ $\boxed{<}$ $\dfrac{11}{12}$.

17. $\dfrac{5}{12} - \dfrac{3}{8}$ *LCD is 24*

$\dfrac{5}{12} = \dfrac{10}{24}$ and $\dfrac{3}{8} = \dfrac{9}{24}$

$\dfrac{10}{24} > \dfrac{9}{24}$, so $\dfrac{5}{12}$ $\boxed{>}$ $\dfrac{3}{8}$.

19. $\dfrac{11}{18} - \dfrac{5}{9}$ *LCD is 18*

$\dfrac{5}{9} = \dfrac{10}{18}$

$\dfrac{11}{18} > \dfrac{10}{18}$, so $\dfrac{11}{18}$ $\boxed{>}$ $\dfrac{5}{9}$.

21. $\left(\dfrac{1}{2}\right)^2 = \dfrac{1}{2}\cdot\dfrac{1}{2} = \dfrac{1\cdot 1}{2\cdot 2} = \dfrac{1}{4}$

The statement, $\left(\frac{1}{2}\right)^2 = \frac{1}{4}$, is *true*.

23. $\left(\dfrac{2}{5}\right)^3 = \dfrac{2}{5}\cdot\dfrac{2}{5}\cdot\dfrac{2}{5} = \dfrac{8}{125}$

The statement, $\left(\frac{2}{5}\right)^3 = \frac{6}{15}$, is *false*.

25. $\left(\dfrac{1}{3}\right)^2 = \dfrac{1}{3}\cdot\dfrac{1}{3} = \dfrac{1\cdot 1}{3\cdot 3} = \dfrac{1}{9}$

27. $\left(\dfrac{5}{8}\right)^2 = \dfrac{5}{8}\cdot\dfrac{5}{8} = \dfrac{5\cdot 5}{8\cdot 8} = \dfrac{25}{64}$

29. $\left(\dfrac{3}{4}\right)^2 = \dfrac{3}{4}\cdot\dfrac{3}{4} = \dfrac{9}{16}$

31. $\left(\dfrac{4}{5}\right)^3 = \dfrac{4}{5}\cdot\dfrac{4}{5}\cdot\dfrac{4}{5} = \dfrac{64}{125}$

33. $\left(\dfrac{3}{2}\right)^4 = \dfrac{3}{2}\cdot\dfrac{3}{2}\cdot\dfrac{3}{2}\cdot\dfrac{3}{2} = \dfrac{81}{16} = 5\dfrac{1}{16}$

35. $\left(\dfrac{3}{4}\right)^4 = \dfrac{3}{4}\cdot\dfrac{3}{4}\cdot\dfrac{3}{4}\cdot\dfrac{3}{4} = \dfrac{81}{256}$

37. Answers will vary. A sample answer follows: A number line is a horizontal line with a range of numbers placed on it. The lowest number is on the left and the greatest number is on the right. It can be used to compare the size or value of numbers.

39. $2^4 - 4(3) = 16 - 4(3)$
$= 16 - 12$
$= 4$

41. $3\cdot 2^2 - \dfrac{6}{3} = 3\cdot 4 - \dfrac{6}{3}$
$= 12 - \dfrac{6}{3}$
$= 12 - 2$
$= 10$

43. $\left(\dfrac{1}{2}\right)^2 \cdot 4 = \dfrac{1}{2}\cdot\dfrac{1}{2}\cdot 4$
$= \dfrac{1}{4}\cdot 4$
$= \dfrac{1}{4}\cdot\dfrac{4}{1} = 1$

45. $\left(\dfrac{3}{4}\right)^2 \cdot \left(\dfrac{1}{3}\right) = \left(\dfrac{3}{4}\cdot\dfrac{3}{4}\right)\cdot\dfrac{1}{3}$
$= \dfrac{3\cdot\overset{1}{\cancel{3}}\cdot 1}{4\cdot 4\cdot\underset{1}{\cancel{3}}}$
$= \dfrac{3}{16}$

47. $\left(\dfrac{4}{5}\right)^2 \cdot \left(\dfrac{5}{6}\right)^2 = \left(\dfrac{4}{5}\cdot\dfrac{4}{5}\right)\cdot\left(\dfrac{5}{6}\cdot\dfrac{5}{6}\right)$
$= \dfrac{\overset{2}{\cancel{4}}\cdot\overset{2}{\cancel{4}}\cdot\overset{1}{\cancel{5}}\cdot\overset{1}{\cancel{5}}}{\underset{1}{\cancel{5}}\cdot\underset{1}{\cancel{5}}\cdot\underset{3}{\cancel{6}}\cdot\underset{3}{\cancel{6}}}$
$= \dfrac{4}{9}$

49. $6\left(\dfrac{2}{3}\right)^2\left(\dfrac{1}{2}\right)^3 = 6\left(\dfrac{2}{3}\cdot\dfrac{2}{3}\right)\left(\dfrac{1}{2}\cdot\dfrac{1}{2}\cdot\dfrac{1}{2}\right)$
$= \dfrac{\overset{\overset{1}{\cancel{2}}}{\cancel{6}}\cdot\overset{1}{\cancel{2}}\cdot\overset{1}{\cancel{2}}\cdot 1\cdot 1\cdot 1}{3\cdot 3\cdot\underset{1}{\cancel{2}}\cdot\underset{1}{\cancel{2}}\cdot\underset{1}{\cancel{2}}}$ *Simplify.*
$= \dfrac{1}{3}$

51. $\dfrac{3}{5}\left(\dfrac{1}{3}\right) + \dfrac{2}{5}\left(\dfrac{3}{4}\right) = \dfrac{\overset{1}{\cancel{3}}}{5}\left(\dfrac{1}{\underset{1}{\cancel{3}}}\right) + \dfrac{\overset{1}{\cancel{2}}}{5}\left(\dfrac{3}{\underset{2}{\cancel{4}}}\right)$
$= \dfrac{1}{5} + \dfrac{3}{10}$ *LCD is 10*
$= \dfrac{2}{10} + \dfrac{3}{10}$
$= \dfrac{5}{10} = \dfrac{1}{2}$

53. $\dfrac{1}{2} + \left(\dfrac{1}{2}\right)^2 - \dfrac{3}{8} = \dfrac{1}{2} + \dfrac{1}{2}\cdot\dfrac{1}{2} - \dfrac{3}{8}$
$= \dfrac{2}{4} + \dfrac{1}{4} - \dfrac{3}{8}$
$= \dfrac{3}{4} - \dfrac{3}{8}$ *LCD is 8*
$= \dfrac{6}{8} - \dfrac{3}{8} = \dfrac{3}{8}$

55. $\left(\dfrac{1}{3} + \dfrac{1}{6}\right)\cdot\dfrac{1}{2} = \left(\dfrac{2}{6} + \dfrac{1}{6}\right)\cdot\dfrac{1}{2}$
$= \dfrac{3}{6}\cdot\dfrac{1}{2}$
$= \dfrac{\overset{1}{\cancel{3}}\cdot 1}{\underset{2}{\cancel{6}}\cdot 2}$
$= \dfrac{1}{4}$

57.
$$\frac{9}{8} \div \left(\frac{2}{3} + \frac{1}{12} \right) = \frac{9}{8} \div \left(\frac{8}{12} + \frac{1}{12} \right)$$
$$= \frac{9}{8} \div \frac{9}{12}$$
$$= \frac{\overset{1}{\cancel{9}}}{\cancel{8}} \cdot \frac{\overset{3}{\cancel{12}}}{\cancel{9}}$$
$$ \quad {}_2 \quad {}_1$$
$$= \frac{3}{2} = 1\frac{1}{2}$$

59.
$$\left(\frac{7}{8} - \frac{3}{4} \right) \div \frac{3}{2} = \left(\frac{7}{8} - \frac{6}{8} \right) \div \frac{3}{2}$$
$$= \frac{1}{8} \div \frac{3}{2}$$
$$= \frac{1}{\cancel{8}} \cdot \frac{\overset{1}{\cancel{2}}}{3}$$
$$ \quad {}_4$$
$$= \frac{1}{12}$$

61.
$$\frac{3}{8} \left(\frac{1}{4} + \frac{1}{2} \right) \cdot \frac{32}{3} = \frac{3}{8} \left(\frac{1}{4} + \frac{2}{4} \right) \cdot \frac{32}{3}$$
$$= \frac{\overset{1}{\cancel{3}}}{\cancel{8}} \cdot \frac{\overset{1}{\cancel{3}}}{\cancel{4}} \cdot \frac{\overset{\overset{1}{\cancel{8}}}{\cancel{32}}}{\cancel{3}}$$
$$ \quad {}_1 \quad {}_1 \quad {}_1$$
$$= 3$$

63.
$$\left(\frac{3}{4} \right)^2 - \left(\frac{1}{2} - \frac{1}{6} \right) \div \frac{4}{3}$$
$$= \left(\frac{3}{4} \right)^2 - \left(\frac{3}{6} - \frac{1}{6} \right) \div \frac{4}{3}$$

Work inside parentheses first.

$$= \left(\frac{3}{4} \right)^2 - \left(\frac{2}{6} \right) \cdot \frac{3}{4}$$

Simplify the expression with the exponent.

$$= \frac{3}{4} \cdot \frac{3}{4} - \left(\frac{2}{6} \right) \cdot \frac{3}{4} \quad \textit{Multiply.}$$
$$= \frac{9}{16} - \frac{\overset{1}{\cancel{2}}}{\cancel{6}} \cdot \frac{\overset{1}{\cancel{3}}}{\cancel{4}}$$
$$ \quad\quad\quad {}_2 \quad {}_2$$
$$= \frac{9}{16} - \frac{1}{4} \quad \textit{LCD is 16}$$
$$= \frac{9}{16} - \frac{4}{16} = \frac{5}{16}$$

65.
$$\left(\frac{7}{8} - \frac{1}{4} \right) - \frac{2}{3} \left(\frac{3}{4} \right)^2 = \left(\frac{7}{8} - \frac{2}{8} \right) - \frac{2}{3} \left(\frac{3}{4} \right)^2$$
$$= \frac{5}{8} - \frac{\overset{1}{\cancel{2}}}{3} \cdot \frac{\overset{3}{\cancel{9}}}{\overset{\cancel{16}}{}}$$
$$ \quad\quad\quad {}_1 \quad {}_8$$
$$= \frac{5}{8} - \frac{3}{8}$$
$$= \frac{2}{8} = \frac{1}{4}$$

67.
$$\left(\frac{3}{4} \right)^2 \left(\frac{2}{3} - \frac{5}{9} \right) - \frac{1}{4} \left(\frac{1}{8} \right)$$
$$= \left(\frac{3}{4} \right)^2 \left(\frac{6}{9} - \frac{5}{9} \right) - \frac{1}{4} \left(\frac{1}{8} \right)$$
$$= \left(\frac{3}{4} \right)^2 \cdot \frac{1}{9} - \frac{1}{4} \cdot \frac{1}{8}$$
$$= \frac{\overset{1}{\cancel{3}}}{4} \cdot \frac{\overset{1}{\cancel{3}}}{4} \cdot \frac{1}{\cancel{9}} - \frac{1}{4} \cdot \frac{1}{8}$$
$$ \quad\quad\quad\quad {}_{\overset{\cancel{3}}{1}}$$
$$= \frac{1}{16} - \frac{1}{32}$$
$$= \frac{2}{32} - \frac{1}{32} = \frac{1}{32}$$

69.
$$\frac{11}{50} - \frac{5}{30} \quad \textit{LCD is 150}$$
$$\frac{11}{50} = \frac{33}{150} \quad \text{and} \quad \frac{5}{30} = \frac{25}{150}$$
$$\frac{33}{150} > \frac{25}{150}, \text{ so } \frac{11}{50} > \frac{5}{30}.$$

$\frac{11}{50}$ in Las Vegas is greater.

71. When comparing the size of two numbers, the symbol <u>$<$</u> means **is less than** and the symbol <u>$>$</u> means **is greater than**.

72. **(a)** To identify the greater of two fractions, we must first write the fractions as <u>like</u> fractions and then compare the <u>numerators</u>.
The fraction with the greater <u>numerator</u> is the greater fraction.

(b) Answers will vary. A sample answer follows:

$$\frac{1}{4} < \frac{1}{2}, \ \ \frac{1}{10} > \frac{1}{16}, \ \ \frac{3}{8} < \frac{3}{2}, \ \ \frac{5}{6} > \frac{5}{7}$$

Copyright © 2018 Pearson Education, Inc.

73. $\left(\dfrac{2}{3}\right)^2 - \left(\dfrac{4}{5} - \dfrac{3}{10}\right) \div \dfrac{5}{4}$

$= \left(\dfrac{2}{3}\right)^2 - \left(\dfrac{8}{10} - \dfrac{3}{10}\right) \div \dfrac{5}{4}$

$= \left(\dfrac{2}{3}\right)^2 - \dfrac{5}{10} \div \dfrac{5}{4}$

$= \dfrac{4}{9} - \dfrac{5}{10} \cdot \dfrac{4}{5}$

$= \dfrac{4}{9} - \dfrac{\overset{1}{\cancel{5}}}{\underset{5}{\cancel{10}}} \cdot \dfrac{\overset{2}{\cancel{4}}}{\underset{1}{\cancel{5}}}$

$= \dfrac{4}{9} - \dfrac{2}{5} \qquad LCD\ is\ 45$

$= \dfrac{20}{45} - \dfrac{18}{45} = \dfrac{2}{45}$

74. $\left(\dfrac{1}{2}\right)^2 \left(\dfrac{3}{8} - \dfrac{1}{4}\right) + \dfrac{2}{3} \div \dfrac{1}{3}$

$= \left(\dfrac{1}{2}\right)^2 \left(\dfrac{3}{8} - \dfrac{2}{8}\right) + \dfrac{2}{3} \div \dfrac{1}{3}$

$= \left(\dfrac{1}{4}\right) \left(\dfrac{1}{8}\right) + \dfrac{2}{3} \cdot \dfrac{3}{1}$

$= \dfrac{1}{32} + \dfrac{2}{1} = 2\dfrac{1}{32}$

For Exercises 75–80, see the number line following Exercise 80.

75. $\left(\dfrac{2}{3}\right)^2 = \dfrac{2}{3} \cdot \dfrac{2}{3} = \dfrac{4}{9}$

76. $\left(\dfrac{3}{2}\right)^2 = \dfrac{3}{2} \cdot \dfrac{3}{2} = \dfrac{9}{4} = 2\dfrac{1}{4}$

77. $\left(\dfrac{1}{2}\right)^3 = \dfrac{1}{2} \cdot \dfrac{1}{2} \cdot \dfrac{1}{2} = \dfrac{1}{8}$

78. $\left(\dfrac{5}{3}\right)^2 = \dfrac{5}{3} \cdot \dfrac{5}{3} = \dfrac{25}{9} = 2\dfrac{7}{9}$

79. $4 + 2 - 2^2 = 4 + 2 - 4 = 6 - 4 = 2$

80. $\left(\dfrac{3}{4}\right)^2 + \left(\dfrac{5}{8} - \dfrac{1}{4}\right) \div \dfrac{2}{3}$

$= \left(\dfrac{3}{4}\right)^2 + \left(\dfrac{5}{8} - \dfrac{2}{8}\right) \div \dfrac{2}{3}$

$= \left(\dfrac{3}{4}\right)^2 + \dfrac{3}{8} \div \dfrac{2}{3}$

$= \dfrac{9}{16} + \dfrac{3}{8} \cdot \dfrac{3}{2}$

$= \dfrac{9}{16} + \dfrac{9}{16}$

$= \dfrac{18}{16} = \dfrac{9}{8} = 1\dfrac{1}{8}$

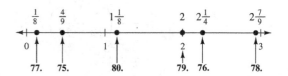

Chapter 3 Review Exercises

1. $\dfrac{5}{7} + \dfrac{1}{7} = \dfrac{5+1}{7} = \dfrac{6}{7}$

2. $\dfrac{4}{9} + \dfrac{3}{9} = \dfrac{4+3}{9} = \dfrac{7}{9}$

3. $\dfrac{1}{8} + \dfrac{3}{8} + \dfrac{2}{8} = \dfrac{1+3+2}{8} = \dfrac{6}{8} = \dfrac{3}{4}$

4. $\dfrac{5}{16} - \dfrac{3}{16} = \dfrac{5-3}{16} = \dfrac{2}{16} = \dfrac{1}{8}$

5. $\dfrac{5}{10} + \dfrac{3}{10} = \dfrac{5+3}{10} = \dfrac{8}{10} = \dfrac{4}{5}$

6. $\dfrac{5}{12} - \dfrac{3}{12} = \dfrac{5-3}{12} = \dfrac{2}{12} = \dfrac{1}{6}$

7. $\dfrac{36}{62} - \dfrac{10}{62} = \dfrac{36-10}{62} = \dfrac{26}{62} = \dfrac{13}{31}$

8. $\dfrac{68}{75} - \dfrac{43}{75} = \dfrac{68-43}{75} = \dfrac{25}{75} = \dfrac{1}{3}$

9. Add to find what fraction of his total income comes from the two jobs.

$$\dfrac{7}{12} + \dfrac{4}{12} = \dfrac{7+4}{12} = \dfrac{11}{12}$$

$\dfrac{11}{12}$ of his total income comes from the two jobs.

10. Subtract to find the answer.

$$\dfrac{5}{8} - \dfrac{3}{8} = \dfrac{5-3}{8} = \dfrac{2}{8} = \dfrac{1}{4}$$

$\dfrac{1}{4}$ more of the events were completed after lunch.

11. 5, 2

Multiples of 5:

$5, \underline{10}, 15, 20, 25, 30, \ldots$

10 is the first number divisible by 2. ($10 \div 2 = 5$) The least common multiple of 5 and 2 is 10.

12. 3, 4

Multiples of 4:

$4, 8, \underline{12}, 16, 20, 24, \ldots$

12 is the first number divisible by 3. ($12 \div 3 = 4$) The least common multiple of 3 and 4 is 12.

Copyright © 2018 Pearson Education, Inc.

13. $10, 12, 20$

$$
\begin{array}{r|ccc}
2 & 10 & 12 & 20 \\
\hline
2 & \cancel{5} & 6 & 10 \\
\hline
3 & \cancel{5} & 3 & \cancel{5} \\
\hline
5 & 5 & \cancel{1} & 5 \\
\hline
& 1 & 1 & 1
\end{array}
$$

$\text{LCM} = 2 \cdot 2 \cdot 3 \cdot 5 = 60$

14. $3, 8, 4$

$$
\begin{array}{r|ccc}
2 & \cancel{3} & 8 & 4 \\
\hline
2 & \cancel{3} & 4 & 2 \\
\hline
2 & \cancel{3} & 2 & \cancel{1} \\
\hline
3 & 3 & \cancel{1} & \cancel{1} \\
\hline
& 1 & 1 & 1
\end{array}
$$

$\text{LCM} = 2 \cdot 2 \cdot 2 \cdot 3 = 24$

15. $6, 8, 5, 15$

$$6 = 2 \cdot 3$$
$$8 = 2 \cdot 2 \cdot 2$$
$$5 = 1 \cdot 5$$
$$15 = 3 \cdot 5$$

$\text{LCM} = 2 \cdot 2 \cdot 2 \cdot 3 \cdot 5 = 120$

The LCM of 6, 8, 5, and 15 is 120.

16. $15, 9, 20$

$$
\begin{array}{r|ccc}
2 & \cancel{15} & \cancel{9} & 20 \\
\hline
2 & \cancel{15} & \cancel{9} & 10 \\
\hline
3 & 15 & 9 & \cancel{5} \\
\hline
3 & \cancel{5} & 3 & \cancel{5} \\
\hline
5 & 5 & \cancel{1} & 5 \\
\hline
& 1 & 1 & 1
\end{array}
$$

$\text{LCM} = 2 \cdot 2 \cdot 3 \cdot 3 \cdot 5 = 180$

17. $\dfrac{2}{3} = \dfrac{?}{12}$ $12 \div 3 = 4$

$$\frac{2}{3} = \frac{2 \cdot 4}{3 \cdot 4} = \frac{8}{12}$$

18. $\dfrac{3}{8} = \dfrac{?}{56}$ $56 \div 8 = 7$

$$\frac{3}{8} = \frac{3 \cdot 7}{8 \cdot 7} = \frac{21}{56}$$

19. $\dfrac{2}{5} = \dfrac{?}{25}$ $25 \div 5 = 5$

$$\frac{2}{5} = \frac{2 \cdot 5}{5 \cdot 5} = \frac{10}{25}$$

20. $\dfrac{5}{9} = \dfrac{?}{81}$ $81 \div 9 = 9$

$$\frac{5}{9} = \frac{5 \cdot 9}{9 \cdot 9} = \frac{45}{81}$$

21. $\dfrac{4}{5} = \dfrac{?}{40}$ $40 \div 5 = 8$

$$\frac{4}{5} = \frac{4 \cdot 8}{5 \cdot 8} = \frac{32}{40}$$

22. $\dfrac{5}{16} = \dfrac{?}{64}$ $64 \div 16 = 4$

$$\frac{5}{16} = \frac{5 \cdot 4}{16 \cdot 4} = \frac{20}{64}$$

23.
$$\frac{1}{2} + \frac{1}{3} = \frac{3}{6} + \frac{2}{6}$$
$$= \frac{3 + 2}{6}$$
$$= \frac{5}{6}$$

24.
$$\frac{1}{5} + \frac{3}{10} + \frac{3}{8} = \frac{8}{40} + \frac{12}{40} + \frac{15}{40}$$
$$= \frac{8 + 12 + 15}{40}$$
$$= \frac{35}{40} = \frac{7}{8}$$

25.
$$
\begin{array}{r}
\dfrac{5}{12} = \dfrac{10}{24} \\[2mm]
+\ \dfrac{5}{24} = \dfrac{5}{24} \\[1mm]
\hline
\dfrac{15}{24} = \dfrac{5}{8}
\end{array}
$$

26. $\dfrac{2}{3} - \dfrac{1}{4} = \dfrac{8}{12} - \dfrac{3}{12} = \dfrac{5}{12}$

27.
$$
\begin{array}{r}
\dfrac{7}{8} = \dfrac{21}{24} \\[2mm]
-\ \dfrac{1}{3} = \dfrac{8}{24} \\[1mm]
\hline
\dfrac{13}{24}
\end{array}
$$

28.
$$
\begin{array}{r}
\dfrac{11}{12} = \dfrac{33}{36} \\[2mm]
-\ \dfrac{4}{9} = \dfrac{16}{36} \\[1mm]
\hline
\dfrac{17}{36}
\end{array}
$$

29. Add the fractions.

$$
\begin{array}{r}
\dfrac{2}{5} = \dfrac{12}{30} \\[2mm]
\dfrac{1}{6} = \dfrac{5}{30} \\[2mm]
+\ \dfrac{1}{3} = \dfrac{10}{30} \\[1mm]
\hline
\dfrac{27}{30} = \dfrac{9}{10}
\end{array}
$$

$\dfrac{9}{10}$ of the total students participated in these activities.

Copyright © 2018 Pearson Education, Inc.

30. Add the fractions.

$$\frac{1}{3} = \frac{16}{48}$$
$$\frac{3}{8} = \frac{18}{48}$$
$$\frac{3}{16} = \frac{9}{48}$$
$$+\frac{1}{16} = \frac{3}{48}$$
$$\frac{46}{48} = \frac{23}{24}$$

$\frac{23}{24}$ of her budget will be spent on these four categories.

31. *Estimate:*   *Exact:*

$$\begin{array}{r} 19 \\ +\,14 \\ \hline 33 \end{array}$$

$$18\frac{5}{8} = 18\frac{5}{8}$$
$$+13\frac{3}{4} = 13\frac{6}{8}$$
$$\overline{31\frac{11}{8} = 31 + 1\frac{3}{8} = 32\frac{3}{8}}$$

32. *Estimate:*   *Exact:*

$$\begin{array}{r} 23 \\ +\,15 \\ \hline 38 \end{array}$$

$$22\frac{2}{3} = 22\frac{6}{9}$$
$$+15\frac{4}{9} = 15\frac{4}{9}$$
$$\overline{37\frac{10}{9} = 37 + 1\frac{1}{9} = 38\frac{1}{9}}$$

33. *Estimate:*   *Exact:*

$$\begin{array}{r} 13 \\ 9 \\ +\,10 \\ \hline 32 \end{array}$$

$$12\frac{3}{5} = 12\frac{48}{80}$$
$$8\frac{5}{8} = 8\frac{50}{80}$$
$$+10\frac{5}{16} = 10\frac{25}{80}$$
$$\overline{30\frac{123}{80} = 30 + 1\frac{43}{80} = 31\frac{43}{80}}$$

34. *Estimate:*   *Exact:*

$$\begin{array}{r} 32 \\ -\,15 \\ \hline 17 \end{array}$$

$$31\frac{3}{4} = 31\frac{9}{12}$$
$$-14\frac{2}{3} = 14\frac{8}{12}$$
$$\overline{17\frac{1}{12}}$$

35. *Estimate:*   *Exact:*

$$\begin{array}{r} 34 \\ -\,16 \\ \hline 18 \end{array}$$

$$34 = 33\frac{3}{3}$$
$$-15\frac{2}{3} = 15\frac{2}{3}$$
$$\overline{18\frac{1}{3}}$$

36. *Estimate:*   *Exact:*

$$\begin{array}{r} 215 \\ -\,136 \\ \hline 79 \end{array}$$

$$215\frac{7}{16}$$
$$-136$$
$$\overline{79\frac{7}{16}}$$

37.
$$5\frac{2}{5} = \frac{27}{5} = \frac{54}{10}$$
$$+3\frac{7}{10} = \frac{37}{10} = \frac{37}{10}$$
$$\overline{\frac{91}{10} = 9\frac{1}{10}}$$

38.
$$4\frac{3}{4} = \frac{19}{4} = \frac{57}{12}$$
$$+5\frac{2}{3} = \frac{17}{3} = \frac{68}{12}$$
$$\overline{\frac{125}{12} = 10\frac{5}{12}}$$

39.
$$5 = \frac{5}{1} = \frac{20}{4}$$
$$-1\frac{3}{4} = \frac{7}{4} = \frac{7}{4}$$
$$\overline{\frac{13}{4} = 3\frac{1}{4}}$$

40.
$$6\frac{1}{2} = \frac{13}{2} = \frac{39}{6}$$
$$-4\frac{5}{6} = \frac{29}{6} = \frac{29}{6}$$
$$\overline{\frac{10}{6} = 1\frac{4}{6} = 1\frac{2}{3}}$$

41.
$$8\frac{1}{3} = \frac{25}{3} = \frac{50}{6}$$
$$-2\frac{5}{6} = \frac{17}{6} = \frac{17}{6}$$
$$\overline{\frac{33}{6} = 5\frac{3}{6} = 5\frac{1}{2}}$$

42.
$$5\frac{5}{12} = \frac{65}{12} = \frac{130}{24}$$
$$-2\frac{5}{8} = \frac{21}{8} = \frac{63}{24}$$
$$\overline{\frac{67}{24} = 2\frac{19}{24}}$$

Copyright © 2018 Pearson Education, Inc.

43. Subtract the uphill and downhill distances from the total distance.

Estimate: $19 - 6 - 7 = 6$ miles

Exact: Add the uphill and downhill distances.

$$5\frac{5}{8} = 5\frac{15}{24}$$
$$+7\frac{1}{3} = 7\frac{8}{24}$$
$$\overline{\phantom{+7\frac{1}{3}} \quad 12\frac{23}{24}}$$

Now subtract from $18\frac{3}{4}$.

$$18\frac{3}{4} = 18\frac{18}{24} = 17\frac{42}{24}$$
$$-12\frac{23}{24} = 12\frac{23}{24} = 12\frac{23}{24}$$
$$\overline{\phantom{-12\frac{23}{24}} \quad 5\frac{19}{24}}$$

The level portion of the course was $5\frac{19}{24}$ miles.

44. Add to find the total weight.

Estimate: $29 + 25 = 54$ tons

Exact: $28\frac{2}{3} = 28\frac{8}{12}$
$$+24\frac{3}{4} = 24\frac{9}{12}$$
$$\overline{\phantom{+24\frac{3}{4}} \quad 52\frac{17}{12}} = 52 + 1\frac{5}{12} = 53\frac{5}{12}$$

The total weight of the newspapers was $53\frac{5}{12}$ tons.

45. Add to find the total weight.

Estimate: $8 + 3 + 5 + 3 = 19$ pounds

Exact: $7\frac{1}{2} = 7\frac{4}{8}$
$$2\frac{3}{4} = 2\frac{6}{8}$$
$$4\frac{7}{8} = 4\frac{7}{8}$$
$$+3\frac{3}{8} = 3\frac{3}{8}$$
$$\overline{\phantom{+3\frac{3}{8}} \quad 16\frac{20}{8}} = 16 + 2\frac{4}{8} = 18\frac{1}{2}$$

The total weight of the four fish was $18\frac{1}{2}$ pounds.

46. Subtract the weight of the third-place pumpkin from the weight of the first-place pumpkin to find the difference in weights.

Estimate: $2058 - 1393 = 665$ pounds

Exact:
$$2058\frac{3}{8} = 2058\frac{3}{8} = 2057\frac{11}{8}$$
$$-1392\frac{1}{2} = 1392\frac{4}{8} = 1392\frac{4}{8}$$
$$\overline{\phantom{-1392\frac{1}{2}} \quad 665\frac{7}{8}}$$

The first place pumpkin weighed $665\frac{7}{8}$ pounds more than the third place pumpkin.

47.–50.

$$\begin{array}{c} \frac{3}{8} \quad \frac{7}{4} \quad \frac{8}{3} \; 3\frac{1}{5} \\ \overset{\bullet\bullet\bullet\bullet}{\longmapsto\!\!\!\!\!\!-\!\!\!-\!\!\!-\!\!\!\!\!\longmapsto} \\ 0 \;\; 1 \;\; 2 \;\; 3 \;\; 4 \\ \uparrow \;\;\; \uparrow \;\;\; \uparrow\uparrow \\ 47. \quad 48. \; 49.\,50. \end{array}$$

51. $\frac{2}{3} - \frac{3}{4}$ *LCD is 12*

$\frac{2}{3} = \frac{8}{12}$ and $\frac{3}{4} = \frac{9}{12}$

$\frac{8}{12} < \frac{9}{12}$, so $\frac{2}{3}$ $\boxed{<}$ $\frac{3}{4}$.

52. $\frac{3}{4} - \frac{7}{8}$ *LCD is 8*

$\frac{3}{4} = \frac{6}{8}$

$\frac{6}{8} < \frac{7}{8}$, so $\frac{3}{4}$ $\boxed{<}$ $\frac{7}{8}$.

53. $\frac{1}{2} - \frac{7}{15}$ *LCD is 30*

$\frac{1}{2} = \frac{15}{30}$ and $\frac{7}{15} = \frac{14}{30}$

$\frac{15}{30} > \frac{14}{30}$, so $\frac{1}{2}$ $\boxed{>}$ $\frac{7}{15}$.

54. $\frac{7}{10} - \frac{8}{15}$ *LCD is 30*

$\frac{7}{10} = \frac{21}{30}$ and $\frac{8}{15} = \frac{16}{30}$

$\frac{21}{30} > \frac{16}{30}$, so $\frac{7}{10}$ $\boxed{>}$ $\frac{8}{15}$.

55. $\frac{9}{16} - \frac{5}{8}$ *LCD is 16*

$\frac{5}{8} = \frac{10}{16}$

$\frac{9}{16} < \frac{10}{16}$, so $\frac{9}{16}$ $\boxed{<}$ $\frac{5}{8}$.

56. $\frac{7}{20} - \frac{8}{25}$ *LCD is 100*

$\frac{7}{20} = \frac{35}{100}$ and $\frac{8}{25} = \frac{32}{100}$

$\frac{35}{100} > \frac{32}{100}$, so $\frac{7}{20}$ $\boxed{>}$ $\frac{8}{25}$.

Copyright © 2018 Pearson Education, Inc.

57. $\dfrac{19}{36} - \dfrac{29}{54}$ *LCD is 108*

$\dfrac{19}{36} = \dfrac{57}{108}$ and $\dfrac{29}{54} = \dfrac{58}{108}$

$\dfrac{57}{108} < \dfrac{58}{108}$, so $\dfrac{19}{36}$ $\boxed{<}$ $\dfrac{29}{54}$.

58. $\dfrac{19}{132} - \dfrac{7}{55}$ *LCD is 660*

$\dfrac{19}{132} = \dfrac{95}{660}$ and $\dfrac{7}{55} = \dfrac{84}{660}$

$\dfrac{95}{660} > \dfrac{84}{660}$, so $\dfrac{19}{132}$ $\boxed{>}$ $\dfrac{7}{55}$.

59. $\left(\dfrac{1}{2}\right)^2 = \dfrac{1}{2} \cdot \dfrac{1}{2} = \dfrac{1}{4}$

60. $\left(\dfrac{2}{3}\right)^2 = \dfrac{2}{3} \cdot \dfrac{2}{3} = \dfrac{4}{9}$

61. $\left(\dfrac{3}{10}\right)^3 = \dfrac{3}{10} \cdot \dfrac{3}{10} \cdot \dfrac{3}{10} = \dfrac{27}{1000}$

62. $\left(\dfrac{3}{8}\right)^4 = \dfrac{3}{8} \cdot \dfrac{3}{8} \cdot \dfrac{3}{8} \cdot \dfrac{3}{8} = \dfrac{81}{4096}$

63. $12\left(\dfrac{3}{4}\right)^2 = \dfrac{12}{1} \cdot \dfrac{3}{4} \cdot \dfrac{3}{4}$

$= \dfrac{\overset{3}{\cancel{12}} \cdot 3 \cdot 3}{1 \cdot \underset{1}{\cancel{4}} \cdot 4}$

$= \dfrac{27}{4} = 6\dfrac{3}{4}$

64. $\dfrac{7}{8} \div \left(\dfrac{1}{8} + \dfrac{3}{4}\right) = \dfrac{7}{8} \div \left(\dfrac{1}{8} + \dfrac{6}{8}\right)$

$= \dfrac{7}{8} \div \dfrac{7}{8}$

$= \dfrac{\overset{1}{\cancel{7}}}{\underset{1}{\cancel{8}}} \cdot \dfrac{\overset{1}{\cancel{8}}}{\underset{1}{\cancel{7}}} = 1$

65. $\left(\dfrac{1}{2}\right)^2 \cdot \left(\dfrac{1}{4} + \dfrac{1}{2}\right) = \dfrac{1}{2} \cdot \dfrac{1}{2} \cdot \left(\dfrac{1}{4} + \dfrac{2}{4}\right)$

$= \dfrac{1}{2} \cdot \dfrac{1}{2} \cdot \dfrac{3}{4}$

$= \dfrac{3}{16}$

66. $\left(\dfrac{1}{4}\right)^3 + \left(\dfrac{5}{8} + \dfrac{3}{4}\right) = \dfrac{1}{4} \cdot \dfrac{1}{4} \cdot \dfrac{1}{4} + \left(\dfrac{5}{8} + \dfrac{6}{8}\right)$

$= \dfrac{1}{64} + \dfrac{11}{8}$ *LCD is 64*

$= \dfrac{1}{64} + \dfrac{88}{64}$

$= \dfrac{89}{64} = 1\dfrac{25}{64}$

Chapter 3 Mixed Review Exercises

1. $\dfrac{7}{8} - \dfrac{1}{8} = \dfrac{7-1}{8} = \dfrac{6}{8} = \dfrac{3}{4}$

2. $\dfrac{7}{10} - \dfrac{3}{10} = \dfrac{7-3}{10} = \dfrac{4}{10} = \dfrac{2}{5}$

3. $\dfrac{29}{32} - \dfrac{5}{16} = \dfrac{29}{32} - \dfrac{10}{32} = \dfrac{29-10}{32} = \dfrac{19}{32}$

4. $\dfrac{1}{4} + \dfrac{1}{8} + \dfrac{5}{16} = \dfrac{4}{16} + \dfrac{2}{16} + \dfrac{5}{16} = \dfrac{4+2+5}{16} = \dfrac{11}{16}$

5. $\begin{aligned} 6\dfrac{2}{3} &= 6\dfrac{4}{6} \\ -4\dfrac{1}{2} &= 4\dfrac{3}{6} \\ \hline &\ \ 2\dfrac{1}{6} \end{aligned}$

6. $\begin{aligned} 9\dfrac{1}{2} &= 9\dfrac{2}{4} \\ +16\dfrac{3}{4} &= 16\dfrac{3}{4} \\ \hline 25\dfrac{5}{4} &= 25 + 1\dfrac{1}{4} = 26\dfrac{1}{4} \end{aligned}$

7. $\begin{aligned} 7 &= 6\dfrac{8}{8} \\ -1\dfrac{5}{8} &= 1\dfrac{5}{8} \\ \hline &\ \ 5\dfrac{3}{8} \end{aligned}$

8. $\begin{aligned} 2\dfrac{3}{5} &= 2\dfrac{48}{80} \\ 8\dfrac{5}{8} &= 8\dfrac{50}{80} \\ +\dfrac{5}{16} &= \dfrac{25}{80} \\ \hline 10\dfrac{123}{80} &= 10 + 1\dfrac{43}{80} = 11\dfrac{43}{80} \end{aligned}$

9. $\begin{aligned} 32\dfrac{5}{12} \\ -17 \\ \hline 15\dfrac{5}{12} \end{aligned}$

Copyright © 2018 Pearson Education, Inc.

10. $\dfrac{7}{22} + \dfrac{3}{22} + \dfrac{3}{11} = \dfrac{7}{22} + \dfrac{3}{22} + \dfrac{6}{22}$

$= \dfrac{7+3+6}{22}$

$= \dfrac{16}{22} = \dfrac{8}{11}$

11. $\left(\dfrac{1}{4}\right)^2 \cdot \left(\dfrac{2}{5}\right)^3 = \dfrac{1}{\overset{}{\underset{2}{\underset{1}{4}}}} \cdot \dfrac{1}{\overset{}{\underset{2}{4}}} \cdot \dfrac{\overset{1}{2}}{5} \cdot \dfrac{\overset{1}{2}}{5} \cdot \dfrac{\overset{1}{2}}{5}$

$= \dfrac{1}{250}$

12. $\dfrac{3}{8} \div \left(\dfrac{1}{2} + \dfrac{1}{4}\right) = \dfrac{3}{8} \div \left(\dfrac{2}{4} + \dfrac{1}{4}\right)$

$= \dfrac{3}{8} \div \dfrac{3}{4}$

$= \dfrac{\overset{1}{3}}{\underset{2}{8}} \cdot \dfrac{\overset{1}{4}}{\underset{1}{3}}$

$= \dfrac{1}{2}$

13. $\left(\dfrac{2}{3}\right)^2 \cdot \left(\dfrac{1}{3} + \dfrac{1}{6}\right) = \left(\dfrac{2}{3}\right)^2 \cdot \left(\dfrac{2}{6} + \dfrac{1}{6}\right)$

$= \dfrac{2}{3} \cdot \dfrac{2}{3} \cdot \dfrac{1}{\overset{}{\underset{1}{2}}} \qquad \dfrac{3}{6} = \dfrac{1}{2}$

$= \dfrac{2}{9}$

14. $\left(\dfrac{2}{3}\right)^3 + \left(\dfrac{2}{3} - \dfrac{5}{9}\right) = \left(\dfrac{2}{3}\right)^3 + \left(\dfrac{6}{9} - \dfrac{5}{9}\right)$

$= \dfrac{2}{3} \cdot \dfrac{2}{3} \cdot \dfrac{2}{3} + \dfrac{1}{9}$

$= \dfrac{8}{27} + \dfrac{1}{9}$

$= \dfrac{8}{27} + \dfrac{3}{27} = \dfrac{11}{27}$

15. $\dfrac{2}{3} - \dfrac{7}{12}$ *LCD is 12*

$\dfrac{2}{3} = \dfrac{8}{12}$

$\dfrac{8}{12} > \dfrac{7}{12}$, so $\dfrac{2}{3}\ \boxed{>}\ \dfrac{7}{12}$.

16. $\dfrac{8}{9} - \dfrac{15}{8}$ *LCD is 72*

$\dfrac{8}{9} = \dfrac{64}{72}$ and $\dfrac{15}{8} = \dfrac{135}{72}$

$\dfrac{64}{72} < \dfrac{135}{72}$, so $\dfrac{8}{9}\ \boxed{<}\ \dfrac{15}{8}$.

17. $\dfrac{17}{30} - \dfrac{36}{60}$ *LCD is 60*

$\dfrac{17}{30} = \dfrac{34}{60}$

$\dfrac{34}{60} < \dfrac{36}{60}$, so $\dfrac{17}{30}\ \boxed{<}\ \dfrac{36}{60}$.

18. $\dfrac{5}{8} - \dfrac{17}{30}$ *LCD is 120*

$\dfrac{5}{8} = \dfrac{75}{120}$ and $\dfrac{17}{30} = \dfrac{68}{120}$

$\dfrac{75}{120} > \dfrac{68}{120}$, so $\dfrac{5}{8}\ \boxed{>}\ \dfrac{17}{30}$.

19. 12, 18

$$
\begin{array}{r|cc}
2 & 12 & 18 \\
2 & 6 & \cancel{9} \\
3 & 3 & 9 \\
3 & \cancel{1} & 3 \\
\hline
 & 1 & 1
\end{array}
$$

LCM $= 2 \cdot 2 \cdot 3 \cdot 3 = 36$

20. 6, 8, 10, 12

$6 = 2 \cdot 3$

$8 = 2 \cdot 2 \cdot 2$

$10 = 2 \cdot 5$

$12 = 2 \cdot 2 \cdot 3$

LCM $= 2 \cdot 2 \cdot 2 \cdot 3 \cdot 5 = 120$

21. 9, 14, 21

$$
\begin{array}{r|ccc}
2 & \cancel{9} & 14 & \cancel{21} \\
3 & 9 & 7 & 21 \\
3 & 3 & \cancel{7} & \cancel{7} \\
7 & \cancel{1} & 7 & 7 \\
\hline
 & 1 & 1 & 1
\end{array}
$$

LCM $= 2 \cdot 3 \cdot 3 \cdot 7 = 126$

22. $\dfrac{2}{3} = \dfrac{?}{27}$ $27 \div 3 = 9$

$\dfrac{2}{3} = \dfrac{2 \cdot 9}{3 \cdot 9} = \dfrac{18}{27}$

23. $\dfrac{9}{12} = \dfrac{?}{144}$ $144 \div 12 = 12$

$\dfrac{9}{12} = \dfrac{9 \cdot 12}{12 \cdot 12} = \dfrac{108}{144}$

24. $\dfrac{4}{5} = \dfrac{?}{75}$ $75 \div 5 = 15$

$\dfrac{4}{5} = \dfrac{4 \cdot 15}{5 \cdot 15} = \dfrac{60}{75}$

Copyright © 2018 Pearson Education, Inc.

25. Subtract the amounts of wire mesh from the total length of the roll.

Estimate: $93 - 14 - 22 = 57$ feet

Exact: Add the amounts of wire mesh needed.

$$13\frac{1}{2} = 13\frac{4}{8}$$
$$+22\frac{3}{8} = 22\frac{3}{8}$$
$$35\frac{7}{8}$$

Now subtract from $92\frac{3}{4}$.

$$92\frac{3}{4} = 92\frac{6}{8} = 91\frac{14}{8}$$
$$-35\frac{7}{8} = 35\frac{7}{8} = 35\frac{7}{8}$$
$$56\frac{7}{8}$$

After completing the two jobs, $56\frac{7}{8}$ feet of wire mesh remain.

26. Multiply to determine the total amount of sugar available. Then subtract the amounts of sugar used.

Estimate: $4 \cdot 50 = 200$ pounds of sugar
$200 - 69 - 77 - 33 = 21$ pounds of sugar

Exact: Add the amounts of sugar used.

$$68\frac{1}{2} = 68\frac{4}{8}$$
$$76\frac{5}{8} = 76\frac{5}{8}$$
$$+33\frac{1}{4} = 33\frac{2}{8}$$
$$177\frac{11}{8} = 177 + 1\frac{3}{8} = 178\frac{3}{8}$$

Now subtract from 200.

$$200 = 199\frac{8}{8}$$
$$-178\frac{3}{8} = 178\frac{3}{8}$$
$$21\frac{5}{8}$$

$21\frac{5}{8}$ pounds of sugar remain.

27. Subtract the today's population from the population at the end of the century.

Estimate: $4 - 1 = 3$ billion people

Exact:
$$3\frac{3}{5} = 3\frac{6}{10}$$
$$-1\frac{1}{10} = -1\frac{1}{10}$$
$$2\frac{5}{10} = 2\frac{1}{2} \text{ billion people}$$

The population will increase by $2\frac{1}{2}$ billion people.

28. Subtract the today's population from the population in 2100.

Estimate: $10 - 7 = 3$ billion people

Exact:
$$10\frac{1}{4} = 10\frac{5}{20}$$
$$-7\frac{1}{5} = -7\frac{4}{20}$$
$$3\frac{1}{20} \text{ billion people}$$

The expected increase in world population is $3\frac{1}{20}$ billion people.

Chapter 3 Test

1. $\dfrac{5}{8} + \dfrac{1}{8} = \dfrac{5+1}{8} = \dfrac{6}{8} = \dfrac{3}{4}$

2. $\dfrac{1}{16} + \dfrac{7}{16} = \dfrac{1+7}{16} = \dfrac{8}{16} = \dfrac{1}{2}$

3. $\dfrac{7}{10} - \dfrac{3}{10} = \dfrac{7-3}{10} = \dfrac{4}{10} = \dfrac{2}{5}$

4. $\dfrac{7}{12} - \dfrac{5}{12} = \dfrac{7-5}{12} = \dfrac{2}{12} = \dfrac{1}{6}$

5. $2, 3, 4$

$$
\begin{array}{c|ccc}
2 & 2 & 3 & 4 \\
2 & 1 & 3 & 2 \\
3 & 1 & 3 & 1 \\
\hline
  & 1 & 1 & 1
\end{array}
$$

$\text{LCM} = 2 \cdot 2 \cdot 3 = 12$

The LCM of 2, 3, and 4 is 12.

6. $6, 3, 5, 15$

$$6 = 2 \cdot 3$$
$$3 = 1 \cdot 3$$
$$5 = 1 \cdot 5$$
$$15 = 3 \cdot 5$$

$\text{LCM} = 2 \cdot 3 \cdot 5 = 30$

The LCM of 6, 3, 5, and 15 is 30.

7. $6, 9, 27, 36$

$$6 = 2 \cdot 3$$
$$9 = 3 \cdot 3$$
$$27 = 3 \cdot 3 \cdot 3$$
$$36 = 2 \cdot 2 \cdot 3 \cdot 3$$

$\text{LCM} = 2 \cdot 2 \cdot 3 \cdot 3 \cdot 3 = 108$

The LCM of 6, 9, 27, and 36 is 108.

Copyright © 2018 Pearson Education, Inc.

8. $\dfrac{3}{8} + \dfrac{1}{4} = \dfrac{3}{8} + \dfrac{2}{8}$ *LCD is 8*

$\phantom{\dfrac{3}{8} + \dfrac{1}{4}} = \dfrac{3+2}{8} = \dfrac{5}{8}$

9. $\dfrac{2}{9} + \dfrac{5}{12} = \dfrac{8}{36} + \dfrac{15}{36}$ *LCD is 36*

$\phantom{\dfrac{2}{9} + \dfrac{5}{12}} = \dfrac{8+15}{36} = \dfrac{23}{36}$

10. $\dfrac{7}{8} - \dfrac{2}{3} = \dfrac{21}{24} - \dfrac{16}{24}$ *LCD is 24*

$\phantom{\dfrac{7}{8} - \dfrac{2}{3}} = \dfrac{21-16}{24} = \dfrac{5}{24}$

11. $\dfrac{2}{5} - \dfrac{3}{8} = \dfrac{16}{40} - \dfrac{15}{40}$ *LCD is 40*

$\phantom{\dfrac{2}{5} - \dfrac{3}{8}} = \dfrac{16-15}{40} = \dfrac{1}{40}$

12. $7\dfrac{2}{3} + 4\dfrac{5}{6}$

Estimate: $8 + 5 = 13$

Exact: $7\dfrac{2}{3} + 4\dfrac{5}{6} = 7\dfrac{4}{6} + 4\dfrac{5}{6}$

$\phantom{Exact: 7\dfrac{2}{3} + 4\dfrac{5}{6}} = 11\dfrac{9}{6} = 11 + 1\dfrac{3}{6} = 12\dfrac{1}{2}$

13. $16\dfrac{2}{5} - 11\dfrac{2}{3}$

Estimate: $16 - 12 = 4$

Exact:
$$16\dfrac{2}{5} = 16\dfrac{6}{15} = 15\dfrac{21}{15}$$
$$-11\dfrac{2}{3} = 11\dfrac{10}{15} = 11\dfrac{10}{15}$$
$$\overline{\phantom{-11\dfrac{2}{3} = 11\dfrac{10}{15} = {}}4\dfrac{11}{15}}$$

14. $18\dfrac{3}{4} + 9\dfrac{2}{5} + 12\dfrac{1}{3}$

Estimate: $19 + 9 + 12 = 40$

Exact:
$$18\dfrac{3}{4} = 18\dfrac{45}{60}$$
$$9\dfrac{2}{5} = 9\dfrac{24}{60}$$
$$+12\dfrac{1}{3} = 12\dfrac{20}{60}$$
$$\overline{\phantom{+12\dfrac{1}{3} = {}}39\dfrac{89}{60}} = 39 + 1\dfrac{29}{60} = 40\dfrac{29}{60}$$

15. $24 - 18\dfrac{3}{8}$

Estimate: $24 - 18 = 6$

Exact:
$$24 = 23\dfrac{8}{8}$$
$$-18\dfrac{3}{8} = 18\dfrac{3}{8}$$
$$\overline{\phantom{-18\dfrac{3}{8} = {}}5\dfrac{5}{8}}$$

16. Answers will vary. One possibility is: Probably addition and subtraction of fractions are more difficult because you have to find the least common denominator and then change the fractions to the same denominator.

17. Answers will vary. One possibility is: Round mixed numbers to the nearest whole number. Then add or subtract to estimate the answer. The estimate may vary from the exact answer but it lets you know if your answer is reasonable.

18. Add the number of pounds.

Estimate: $10 + 85 + 37 + 8 = 140$ pounds

Exact:
$$10\dfrac{3}{8} = 10\dfrac{9}{24}$$
$$84\dfrac{1}{2} = 84\dfrac{12}{24}$$
$$36\dfrac{5}{6} = 36\dfrac{20}{24}$$
$$+ 8\dfrac{1}{3} = 8\dfrac{8}{24}$$
$$\overline{\phantom{+ 8\dfrac{1}{3} = {}}138\dfrac{49}{24}} = 138 + 2\dfrac{1}{24} = 140\dfrac{1}{24}$$

The total number of pounds used was $140\dfrac{1}{24}$.

19. Subtract the number of gallons that were used from the amount the contractor had when he arrived to find the number of gallons remaining.

Estimate: $148 - 69 - 37 = 42$ gallons

Exact: Add the number of gallons that were used.
$$68\dfrac{1}{2} = 68\dfrac{4}{8}$$
$$+37\dfrac{3}{8} = 37\dfrac{3}{8}$$
$$\overline{\phantom{+37\dfrac{3}{8} = {}}105\dfrac{7}{8}}$$

Now subtract from $147\dfrac{1}{2}$.
$$147\dfrac{1}{2} = 147\dfrac{4}{8} = 146\dfrac{12}{8}$$
$$-105\dfrac{7}{8} = 105\dfrac{7}{8} = 105\dfrac{7}{8}$$
$$\overline{\phantom{-105\dfrac{7}{8} = 105\dfrac{7}{8} = {}}41\dfrac{5}{8}}$$

The number of gallons remaining is $41\dfrac{5}{8}$.

Copyright © 2018 Pearson Education, Inc.

20. $\dfrac{3}{4} - \dfrac{17}{24}$ *LCD is 24*

$\dfrac{3}{4} = \dfrac{18}{24}$

$\dfrac{18}{24} > \dfrac{17}{24}$, so $\dfrac{3}{4} \boxed{>} \dfrac{17}{24}$.

21. $\dfrac{19}{24} - \dfrac{17}{36}$ *LCD is 72*

$\dfrac{19}{24} = \dfrac{57}{72}$ and $\dfrac{17}{36} = \dfrac{34}{72}$

$\dfrac{57}{72} > \dfrac{34}{72}$, so $\dfrac{19}{24} \boxed{>} \dfrac{17}{36}$.

22. $\left(\dfrac{1}{3}\right)^3 \cdot 54 = \left(\dfrac{1}{3} \cdot \dfrac{1}{3} \cdot \dfrac{1}{3}\right) \cdot 54$

$= \dfrac{1}{\overset{}{\underset{1}{\cancel{27}}}} \cdot \dfrac{\overset{2}{\cancel{54}}}{1}$

$= \dfrac{2}{1} = 2$

23. $\left(\dfrac{3}{4}\right)^2 - \left(\dfrac{7}{8} \cdot \dfrac{1}{3}\right) = \dfrac{3}{4} \cdot \dfrac{3}{4} - \dfrac{7}{24}$

$= \dfrac{9}{16} - \dfrac{7}{24}$ *LCD is 48*

$= \dfrac{27}{48} - \dfrac{14}{48} = \dfrac{13}{48}$

24. $4\left(\dfrac{7}{8} - \dfrac{7}{16}\right) = 4\left(\dfrac{14}{16} - \dfrac{7}{16}\right)$

$= 4 \cdot \dfrac{7}{16}$

$= \dfrac{\overset{1}{\cancel{4}}}{1} \cdot \dfrac{7}{\underset{4}{\cancel{16}}}$

$= \dfrac{7}{4} = 1\dfrac{3}{4}$

25. $\dfrac{5}{6} + \dfrac{4}{3}\left(\dfrac{3}{8}\right) = \dfrac{5}{6} + \dfrac{\overset{1}{\cancel{4}}}{\underset{1}{\cancel{3}}}\left(\dfrac{\overset{1}{\cancel{3}}}{\underset{2}{\cancel{8}}}\right)$

$= \dfrac{5}{6} + \dfrac{1}{2}$ *LCD is 6*

$= \dfrac{5}{6} + \dfrac{3}{6}$

$= \dfrac{5+3}{6}$

$= \dfrac{8}{6} = 1\dfrac{2}{6} = 1\dfrac{1}{3}$

Cumulative Review Exercises (Chapters 1–3)

1. $\underline{5},6\underline{3}9,\underline{4}2\,8$

5 is in the millions place.
3 is in the ten-thousands place.
9 is in the thousands place.
4 is in the hundreds place.

2. **To the nearest ten:** 59,8$\underline{0}$3

Next digit is 4 or less. Tens place does not change. All digits to the right of the underlined place are changed to zero. **59,800**

To the nearest hundred: 59,$\underline{8}$03

Next digit is 4 or less. Hundreds place does not change. All digits to the right of the underlined place are changed to zero. **59,800**

To the nearest thousand: 5$\underline{9}$,803

Next digit is 5 or more. All digits to the right of the underlined place are changed to zero. Add 1 to 9. Write 0 and carry 1. **60,000**

3.   *Estimate:*   *Exact:*

20,000$$

$\underline{-\ 10,000}$

10,000$$

$\overset{\overset{13\ \ 1116}{1\ \cancel{3}\ \ \cancel{1}\ \cancel{6}\ 16}}{2\ 4\,,2\ \cancel{7}\ \cancel{6}}$

$\underline{-\ \ \ \ 9\ 8\ 8\ 7}$

$1\ 4\,,3\ 8\ 9$

4.   *Estimate:*   *Exact:*

$2\ 5\ 0\ 0$

$40\overline{)1\ 0\ 0\,,0\ 0\ 0}$

$3\ 2\ 1\ 1$

$35\overline{)1\ 1\ 2\,,3\ 8\ 5}$

$\underline{1\ 0\ 5}$

$7\ 3$

$\underline{7\ 0}$

$3\ 8$

$\underline{3\ 5}$

$3\ 5$

$\underline{3\ 5}$

0

5. $\overset{1\,11\,2\ \ 21}{375,899}$

$521,742$

$\underline{+\ 357,968}$

$1,255,609$

6. $\overset{\overset{\ \ \ \ 14}{5\ \ \cancel{4}\ \ 9\,12}}{3,89\cancel{6},\cancel{5}\ \cancel{0}\ \cancel{2}}$

$\underline{-\ 1,094,8\ 0\ 7}$

$2,801,6\ 9\ 5$

Copyright © 2018 Pearson Education, Inc.

7. $5(8)(4) = 40(4) = 160$

8.
$$
\begin{array}{r}
962 \\
\times\ 384 \\
\hline
3848 \leftarrow 4 \times 962 \\
7696 \leftarrow 8 \times 962 \\
2886 \leftarrow 3 \times 962 \\
\hline
369{,}408
\end{array}
$$

9.
$$
\begin{array}{r}
135 \\
8\overline{)1080} \\
\underline{8} \\
28 \\
\underline{24} \\
40 \\
\underline{40} \\
0
\end{array}
$$

10. $13{,}467 \div 5$

$$
5\overline{)13,\,{}^34\,{}^46\,{}^17} \quad 2\ 6\ 9\ 3\ \mathbf{R2}
$$

11. To find the perimeter, add the six measurements of the four sides.

Estimate: $20 + 9 + 5 + 20 + 9 + 5 = 68$ ft

Exact: $18 + 9 + 5 + 18 + 9 + 5 = 64$ ft

The perimeter of this parking space is 64 feet.

12. To find the area of the parking space, multiply the length (18) and the width (14).

Estimate: $20 \cdot 10 = 200$ ft^2

Exact:
$$
\begin{array}{r}
18 \\
\times\ 14 \\
\hline
72 \\
18 \\
\hline
252
\end{array}
$$

The area of the parking space is 252 ft^2.

13. Multiply the miles per gallon and the gallons to find the total miles.

Estimate: $47 \cdot 9 = 423$ miles

Exact: $47\frac{3}{10} \cdot 8\frac{3}{4} = \frac{473}{10} \cdot \frac{35}{4}$

$$
= \frac{473}{\underset{2}{\cancel{10}}} \cdot \frac{\overset{7}{\cancel{35}}}{4}
$$

$$
= \frac{3311}{8} = 413\frac{7}{8}
$$

The car could go $413\frac{7}{8}$ miles on a full tank of gasoline.

14. Multiply the height of the person by 130.

Estimate: $70 \cdot 130 = 9100$ in.

Exact:
$$
70\frac{1}{4} \cdot 130 = \frac{281}{4} \cdot \frac{130}{1} = \frac{36{,}530}{4} = 9132\frac{1}{2}
$$

A human could jump $9132\frac{1}{2}$ inches high.

15. $2^4 \cdot 3^2 = (2 \cdot 2 \cdot 2 \cdot 2) \cdot (3 \cdot 3) = 16 \cdot 9 = 144$

16. $5^2 - 2(8) = 25 - 2(8) = 25 - 16 = 9$

17. $\sqrt{25} + 5 \cdot 9 - 6 = 5 + 5 \cdot 9 - 6$
$$
= 5 + 45 - 6
$$
$$
= 50 - 6 = 44
$$

18. $\frac{2}{3}\left(\frac{4}{5} - \frac{2}{3}\right) = \frac{2}{3}\left(\frac{12}{15} - \frac{10}{15}\right)$
$$
= \frac{2}{3} \cdot \frac{2}{15} = \frac{4}{45}
$$

19. $\frac{3}{4} \div \left(\frac{1}{3} + \frac{1}{2}\right) = \frac{3}{4} \div \left(\frac{2}{6} + \frac{3}{6}\right)$
$$
= \frac{3}{4} \div \frac{5}{6}
$$
$$
= \frac{3}{\underset{2}{\cancel{4}}} \cdot \frac{\overset{3}{\cancel{6}}}{5} = \frac{9}{10}
$$

20. $\frac{7}{8} + \left(\frac{3}{4}\right)^2 - \frac{3}{8} = \frac{7}{8} + \left(\frac{3}{4} \cdot \frac{3}{4}\right) - \frac{3}{8}$
$$
= \frac{7}{8} + \frac{9}{16} - \frac{3}{8}
$$
$$
= \frac{14}{16} + \frac{9}{16} - \frac{6}{16}
$$
$$
= \frac{14 + 9}{16} - \frac{6}{16}
$$
$$
= \frac{23}{16} - \frac{6}{16}
$$
$$
= \frac{23 - 6}{16}
$$
$$
= \frac{17}{16} = 1\frac{1}{16}
$$

21. $\frac{3}{4} \cdot \frac{2}{3} = \frac{\overset{1}{\cancel{3}}}{\underset{2}{\cancel{4}}} \cdot \frac{\overset{1}{\cancel{2}}}{\underset{1}{\cancel{3}}} = \frac{1 \cdot 1}{2 \cdot 1} = \frac{1}{2}$

22. $42 \cdot \frac{7}{8} = \frac{\overset{21}{\cancel{42}}}{1} \cdot \frac{7}{\underset{4}{\cancel{8}}} = \frac{21 \cdot 7}{1 \cdot 4} = \frac{147}{4} = 36\frac{3}{4}$

23. $\frac{25}{40} \div \frac{10}{35} = \frac{\overset{5}{\cancel{25}}}{\underset{8}{\cancel{40}}} \cdot \frac{35}{\underset{2}{\cancel{10}}} = \frac{35}{16} = 2\frac{3}{16}$

Copyright © 2018 Pearson Education, Inc.

24. $9 \div \dfrac{2}{3} = \dfrac{9}{1} \cdot \dfrac{3}{2} = \dfrac{27}{2} = 13\dfrac{1}{2}$

25. *Estimate:*   *Exact:*

$$\begin{array}{r} 3 \\ +5 \\ \hline 8 \end{array} \qquad \begin{array}{r} 3\dfrac{3}{8} = 3\dfrac{3}{8} \\ +4\dfrac{1}{2} = 4\dfrac{4}{8} \\ \hline 7\dfrac{7}{8} \end{array}$$

26. *Estimate:*   *Exact:*

$$\begin{array}{r} 22 \\ +4 \\ \hline 26 \end{array} \qquad \begin{array}{r} 21\dfrac{7}{8} = 21\dfrac{21}{24} \\ +4\dfrac{5}{12} = 4\dfrac{10}{24} \\ \hline 25\dfrac{31}{24} = 25 + 1\dfrac{7}{24} = 26\dfrac{7}{24} \end{array}$$

27. *Estimate:*   *Exact:*

$$\begin{array}{r} 5 \\ -2 \\ \hline 3 \end{array} \qquad \begin{array}{r} 5 = 4\dfrac{8}{8} \\ -2\dfrac{3}{8} = 2\dfrac{3}{8} \\ \hline 2\dfrac{5}{8} \end{array}$$

28–31.

32. $\dfrac{3}{5} - \dfrac{5}{8}$ *LCD is 40*

$\dfrac{3}{5} = \dfrac{24}{40}$ and $\dfrac{5}{8} = \dfrac{25}{40}$

$\dfrac{24}{40} < \dfrac{25}{40}$, so $\dfrac{3}{5} \boxed{<} \dfrac{5}{8}$.

33. $\dfrac{17}{20} - \dfrac{3}{4}$ *LCD is 20*

$\dfrac{3}{4} = \dfrac{15}{20}$

$\dfrac{17}{20} > \dfrac{15}{20}$, so $\dfrac{17}{20} \boxed{>} \dfrac{3}{4}$.

34. $\dfrac{7}{12} - \dfrac{11}{18}$ *LCD is 36*

$\dfrac{7}{12} = \dfrac{21}{36}$ and $\dfrac{11}{18} = \dfrac{22}{36}$

$\dfrac{21}{36} < \dfrac{22}{36}$, so $\dfrac{7}{12} \boxed{<} \dfrac{11}{18}$.

Copyright © 2018 Pearson Education, Inc.

CHAPTER 4   DECIMALS

4.1   Reading and Writing Decimal Numbers

4.1 Margin Exercises

1. (a) The figure has 10 equal parts; 1 part is shaded.

$\frac{1}{10}$; 0.1; one tenth

(b) The figure has 10 equal parts; 3 parts are shaded.

$\frac{3}{10}$; 0.3; three tenths

(c) The figure has 10 equal parts; 9 parts are shaded.

$\frac{9}{10}$; 0.9; nine tenths

2. (a) The figure has 10 equal parts; 3 parts are shaded.

$\frac{3}{10} = 0.\underline{3} = $ <u>three</u> tenths

(b) The figure has 100 equal parts; 41 parts are shaded.

$\frac{41}{100}$; 0.41; forty-one hundredths

3. (a) $0.7 = \frac{7}{10}$

(b) $0.2 = \frac{2}{10}$

(c) $0.03 = \frac{3}{100}$

(d) $0.69 = \frac{69}{100}$

(e) $0.047 = \frac{47}{1000}$

(f) $0.351 = \frac{351}{1000}$

4. (a)

hundreds	tens	ones	.	tenths	hundredths
9	7	1	.	5	4

(b)

ones	.	tenths
0	.	4

(c)

ones	.	tenths	hundredths
5	.	6	0

(d)

ones	.	tenths	hundredths	thousandths	ten-thousandths
0	.	0	8	3	5

5. (a) 0.6 is six <u>tenths</u>.

(b) 0.46 is forty-six <u>hundredths</u>.

(c) 0.05 is five hundredths.

(d) 0.409 is four hundred nine thousandths.

(e) 0.0003 is three ten-thousandths.

(f) 0.2703 is two thousand seven hundred three ten-thousandths.

6. (a) 3.8 is three and eight <u>tenths</u>.

(b) 15.001 is fifteen and one thousandth.

(c) 0.0073 is seventy-three ten-thousandths.

(d) 64.309 is sixty-four and three hundred nine thousandths.

7. (a) $0.7 = \frac{7}{10}$

(b) $12.21 = 12\frac{21}{100}$

(c) $0.101 = \frac{101}{1000}$

(d) $0.007 = \frac{7}{1000}$

(e) $1.3717 = 1\frac{3717}{10,000}$

8. (a) $0.5 = \frac{5}{10} = \frac{5 \div 5}{10 \div 5} = \frac{1}{2}$

(b) $12.6 = 12\frac{6}{10} = 12\frac{6 \div 2}{10 \div 2} = 12\frac{3}{5}$

(c) $0.85 = \frac{85}{100} = \frac{85 \div 5}{100 \div 5} = \frac{17}{20}$

(d) $3.05 = 3\frac{5}{100} = 3\frac{5 \div 5}{100 \div 5} = 3\frac{1}{20}$

(e) $0.225 = \frac{225}{1000} = \frac{225 \div 25}{1000 \div 25} = \frac{9}{40}$

(f) $420.0802 = 420\frac{802}{10,000} = 420\frac{802 \div 2}{10,000 \div 2}$
$= 420\frac{401}{5000}$

Copyright © 2018 Pearson Education, Inc.

4.1   Section Exercises

1. There are three numbers after the decimal point in 11.084, so the number has 3 decimal places.

3. In 22.85, the 8 means 8 tenths.

5.

	tens	ones		tenths		
	7	0	.	4	8	9

7.

		tenths		thousandths	ten-thousandths
0	.	8	3	4	7	2

9.

	hundreds		ones		hundredths	
1	4	9	.	0	8	3	2

11.

	thousands					hundredths	thousandths
6	2	8	5	.	7	1	2	5

13. 0 ones, 5 hundredths, 1 ten, 4 hundreds, 2 tenths: **410.25**

15. 3 thousandths, 4 hundredths, 6 ones, 2 ten-thousandths, 5 tenths: **6.5432**

17. 4 hundredths, 4 hundreds, 0 tens, 0 tenths, 5 thousandths, 5 thousands, 6 ones

Reorder as 5 thousands, 4 hundreds, 0 tens, 6 ones, 0 tenths, 4 hundredths, 5 thousandths: **5406.045**

19. $0.7 = \dfrac{7}{10}$

21. $13.4 = 13\dfrac{4}{10} = 13\dfrac{4 \div 2}{10 \div 2} = 13\dfrac{2}{5}$

23. $0.25 = \dfrac{25}{100} = \dfrac{25 \div 25}{100 \div 25} = \dfrac{1}{4}$

25. $0.66 = \dfrac{66}{100} = \dfrac{66 \div 2}{100 \div 2} = \dfrac{33}{50}$

27. $10.17 = 10\dfrac{17}{100}$

29. $0.06 = \dfrac{6}{100} = \dfrac{6 \div 2}{100 \div 2} = \dfrac{3}{50}$

31. $0.205 = \dfrac{205}{1000} = \dfrac{205 \div 5}{1000 \div 5} = \dfrac{41}{200}$

33. $5.002 = 5\dfrac{2}{1000} = 5\dfrac{2 \div 2}{1000 \div 2} = 5\dfrac{1}{500}$

35. $0.686 = \dfrac{686}{1000} = \dfrac{686 \div 2}{1000 \div 2} = \dfrac{343}{500}$

37. 0.5 is five tenths.

39. 0.78 is seventy-eight hundredths.

41. 0.105 is one hundred five thousandths.

43. 12.04 is twelve and four hundredths.

45. Three decimal places is *thousandths*, so 1.075 is one and seventy-five thousandths.

47. six and seven tenths: $6\dfrac{7}{10} = \underline{6.7}$

49. thirty-two hundredths: $\dfrac{32}{100} = 0.32$

51. **(a)** four hundred twenty and eight thousandths: $420\dfrac{8}{1000} = 420.008$

(b) four hundred twenty eight thousandths: 0.428

53. **(a)** seven hundred three ten-thousandths: $\dfrac{703}{10,000} = 0.0703$

(b) seven hundred and three ten-thousandths: $700\dfrac{3}{10,000} = 700.0003$

55. seventy-five and thirty thousandths: $75\dfrac{30}{1000} = 75.030$

57. Anne should not say "and" because that denotes a decimal point. She should have read it as "four thousand three hundred two."

59. 8-pound test line has a diameter of 0.010 inch. 0.010 inch is read ten thousandths of an inch.

$0.010 = \dfrac{10}{1000} = \dfrac{10 \div 10}{1000 \div 10} = \dfrac{1}{100}$ inch

61. **(a)** $\dfrac{13}{1000}$ inch $= 0.013$ inch

A diameter of 0.013 inch corresponds to a test strength of 12 pounds.

(b) eight thousandths inch: 0.008 inch
The test strength of a fishing line with a diameter of eight thousandths inch is 4 pounds.

63. Six tenths is 0.6, so the correct part number is 3-C.

65. One and six thousandths is 1.006, which is part number 4-A.

Copyright © 2018 Pearson Education, Inc.

67. The size of part number 4-E is 1.602 centimeters, which in words is one and six hundred two thousandths centimeters.

69. **(a)** Use the whole number place value chart on text page 4 to find the names after hundred thousands. Then change "s" to "*ths*" in those names. So millions becomes million*ths* when used on the right side of the decimal point, and so on for ten-millionths, hundred-millionths, and billionths.

(b) These match the words on the left side of the chart with "*ths*" attached.

71. $20,500,000,000$

72. $14,600$

73. **(a)** 7.82 million is $7,820,000$.

(b) 9.6 billion is $9,600,000,000$.

(c) 42.7 thousand is $42,700$

74. **(a)** 123.2 thousand is $123,200$.

(b) 503.42 million is $503,420,000$.

(c) 36.51 billion is $36,510,000,000$.

75. **(a)** $3,079,000 = 3.079$ million

(b) $25,600 = 25.6$ thousand

(c) $590,800,000,000 = 590.8$ billion

76. **(a)** $608,900,000,000 = 608.9$ billion

(b) $42,387,000 = 42.387$ million

(c) $85,200 = 85.2$ thousand

4.2   Rounding Decimal Numbers

4.2   Margin Exercises

1. Round to the nearest thousandth.

(a) Draw a cut-off line: 0.334|92
The first digit cut is 9, which is 5 or more, so round up.

$$\begin{array}{r} 0.334 \\ + 0.001 \\ \hline 0.335 \end{array}$$

Answer: ≈ 0.335

(b) Draw a cut-off line: 8.008|51
The first digit cut is 5, which is 5 or more, so round up.

$$\begin{array}{r} 8.008 \\ + 0.001 \\ \hline 8.009 \end{array}$$

Answer: ≈ 8.009

(c) Draw a cut-off line: 265.420|38
The first digit cut is 3, which is 4 or less. The part you keep stays the same.

Answer: ≈ 265.420

(d) Draw a cut-off line: 10.701|80
The first digit cut is 8, which is 5 or more, so round up.

$$\begin{array}{r} 10.701 \\ + 0.001 \\ \hline 10.702 \end{array}$$

Answer: ≈ 10.702

2. **(a)** 0.8988 to the nearest hundredth
Draw a cut-off line: 0.89|88
The first digit cut is 8, which is 5 or more, so round up.

$$\begin{array}{r} 0.89 \\ + 0.01 \\ \hline 0.\underline{90} \end{array}$$

Answer: ≈ 0.90

(b) 5.8903 to the nearest hundredth
Draw a cut-off line: 5.89|03
The first digit cut is 0, which is 4 or less. The part you keep stays the same.

Answer: ≈ 5.89

(c) 11.0299 to the nearest thousandth
Draw a cut-off line: 11.029|9
The first digit cut is 9, which is 5 or more, so round up.

$$\begin{array}{r} 11.029 \\ + 0.001 \\ \hline 11.030 \end{array}$$

Answer: ≈ 11.030

(d) 0.545 to the nearest tenth
Draw a cut-off line: 0.5|45
The first digit cut is 4, which is 4 or less. The part you keep stays the same.

Answer: ≈ 0.5

Copyright © 2018 Pearson Education, Inc.

3.   Round to the nearest cent (hundredths).

(a) Draw a cut-off line: $14.59|5
The first digit cut is 5, which is 5 or more, so round up.

$14.59
$\underline{+\ 0.01}$
$14.<u>60</u>

Answer: ≈ $14.60

(b) Draw a cut-off line: $578.06|63
The first digit cut is 6, which is 5 or more, so round up.

$578.06
$\underline{+\ 0.01}$
$578.07

Answer: ≈ $578.07

(c) Draw a cut-off line: $0.84|9
The first digit cut is 9, which is 5 or more, so round up.

$0.84
$\underline{+\ 0.01}$
$0.85

Answer: ≈ $0.85

(d) Draw a cut-off line: $0.05|48
The first digit cut is 4, which is 4 or less. The part you keep stays the same.

Answer: ≈ $0.05

4.   Round to the nearest dollar.

(a) Draw a cut-off line: $29|.10
The first digit cut is 1, which is 4 or less. The part you keep stays the same.

Answer: ≈ $29

(b) Draw a cut-off line: $136|.49
The first digit cut is 4, which is 4 or less. The part you keep stays the same.

Answer: ≈ $136

(c) Draw a cut-off line: $990|.91
The first digit cut is 9, which is 5 or more, so round up by adding $1.

$990
$\underline{+\ \ 1}$
$991

Answer: ≈ $991

(d) Draw a cut-off line: $5949|.88
The first digit cut is 8, which is 5 or more, so round up by adding $1.

$5949
$\underline{+\ \ 1}$
$5950

Answer: ≈ $5950

(e) Draw a cut-off line: $49|.60
The first digit cut is 6, which is 5 or more, so round up by adding $1.

$49
$\underline{+\ 1}$
$50

Answer: ≈ $50

(f) Draw a cut-off line: $0|.55
The first digit cut is 5, which is 5 or more, so round up by adding $1.

$0
$\underline{+\ 1}$
$1

Answer: ≈ $1

(g) Draw a cut-off line: $1|.08
The first digit cut is 0, which is 4 or less. The part you keep stays the same.

Answer: ≈ $1

4.2   Section Exercises

1.   In 4.8073, 8 is in the tenths place, so you would look at the next digit, 0.

3.   Look only at the 6, which is *5 or more*. So round up by adding 1 thousandth to 5.709 to get 5.710.

5.   16.8974 to the nearest tenth
Draw a cut-off line after the tenths place:
16.8|974
The first digit cut is 9, which is 5 or more, so round up the tenths place.

16.8
$\underline{+\ 0.1}$
16.9

Answer: ≈ 16.9

7.   0.95647 to the nearest thousandth
Draw a cut-off line after the thousandths place:
0.956|47
The first digit cut is 4, which is 4 or less. The part you keep stays the same.

Answer: ≈ 0.956

Copyright © 2018 Pearson Education, Inc.

9. 0.799 to the nearest hundredth
Draw a cut-off line after the hundredths place:
0.79|9
The first digit cut is 9, which is 5 or more, so round up the hundredths place.

    0.79
  + 0.01
  ─────
    0.80

Answer: ≈ 0.80

11. 3.66062 to the nearest thousandth
Draw a cut-off line after the thousandths place:
3.660|62
The first digit cut is 6, which is 5 or more, so round up the thousandths place.

    3.660
  + 0.001
  ─────
    3.661

Answer: ≈ 3.661

13. 793.988 to the nearest tenth
Draw a cut-off line after the tenths place:
793.9|88
The first digit cut is 8, which is 5 or more, so round up the tenths place.

    793.9
  +   0.1
  ─────
    794.0

Answer: ≈ 794.0

15. 0.09804 to the nearest ten-thousandth
Draw a cut-off line after the ten-thousandths place: 0.0980|4
The first digit cut is 4, which is 4 or less. The part you keep stays the same.

Answer: ≈ 0.0980

17. 9.0906 to the nearest hundredth
Draw a cut-off line after the hundredths place:
9.09|06
The first digit cut is 0, which is 4 or less. The part you keep stays the same.

Answer: ≈ 9.09

19. 82.000151 to the nearest ten-thousandth
Draw a cut-off line after the ten-thousandths place: 82.0001|51
The first digit cut is 5, which is 5 or more, so round up the ten-thousandths place.

    82.0001
  + 0.0001
  ─────
    82.0002

Answer: ≈ 82.0002

21. Round $0.81666 to the nearest cent.
Draw a cut-off line: $0.81|666
The first digit cut is 6, which is 5 or more, so round up.

    $0.81
  + 0.01
  ─────
    $0.82

Nardos pays $0.82.

23. Round $1.2225 to the nearest cent.
Draw a cut-off line: $1.22|25
The first digit cut is 2, which is 4 or less. The part you keep stays the same.

Nardos pays $1.22.

25. Round $0.6983 to the nearest cent.
Draw a cut-off line: $0.69|83
The first digit cut is 8, which is 5 or more, so round up.

    $0.69
  + 0.01
  ─────
    $0.70

Nardos pays $0.70.

27. Round $48,649.60 to the nearest dollar.
Draw a cut-off line: $48,649|.60
The first digit cut is 6, which is 5 or more, so round up.

    $48,649
  +   1
  ─────
    $48,650

Income from job: ≈ $48,650

29. Round $840.08 to the nearest dollar.
Draw a cut-off line: $840|.08
The first digit cut is 0, which is 4 or less. The part you keep stays the same.

Donations to charity: ≈ $840

31. $499.98 to the nearest dollar
Draw a cut-off line: $499|.98
The first digit cut is 9, which is 5 or more, so round up.

    $499
  +   1
  ─────
    $500

Answer: ≈ $500

Copyright © 2018 Pearson Education, Inc.

33. $0.996 to the nearest cent
Draw a cut-off line: $0.99|6
The first digit cut is 6, which is 5 or more, so round up.

$0.99
$+ 0.01$
_____
$1.00

Answer: ≈ $1.00

35. $999.73 to the nearest dollar
Draw a cut-off line: $999|.73
The first digit cut is 7, which is 5 or more, so round up.

$999
$+ 1$
_____
$1000

Answer: $999.73 ≈ $1000

37. (a) Round 311.945 to the nearest whole number.
Draw a cut-off line: 311|.945
The first digit cut is 9, which is 5 or more, so round up.

311
$+1$
_____
312

The record speed for a conventional motorcycle is about 312 miles per hour.

(b) Round 149.1 to the nearest whole number.
Draw a cut-off line: 149|.1
The first digit cut is 1, which is 4 or less. The part you keep stays the same.

The fastest roller coaster speed record is about 149 miles per hour.

39. (a) Round 185.981 to the nearest tenth.
Draw a cut-off line: 185.9|81
The first digit cut is 8, which is 5 or more, so round up.

185.9
$+ 0.1$
_____
186.0

The Indianapolis 500 fastest average winning speed is about 186.0 miles per hour.

(b) Round 763.04 to the nearest tenth.
Draw a cut-off line: 763.0|4
The first digit cut is 4, which is 4 or less. The part you keep stays the same.

The land speed record is about 763.0 miles per hour.

41. If you round $0.499 to the nearest dollar, it will round to $0 (zero dollars) because $0.499 is closer to $0 than to $1.

42. Rounding $0.499 to $0.50 would be more helpful than rounding to $0.

Round amounts less than $1.00 to the nearest cent instead of the nearest dollar.

43. If you round $0.0015 to the nearest cent, it will round to $0.00 (zero cents) because $0.0015 is closer to $0.00 than to $0.01.

44. Both $0.5968 and $0.6014 round to $0.60. Rounding to nearest thousandth (tenth of a cent) would help you to identify $0.597 as less than $0.601.

4.3 Adding and Subtracting Decimal Numbers

4.3 Margin Exercises

1. **(a)** $2.86 + 7.09$

$$\begin{array}{r} \overset{1}{2.86} \\ + 7.09 \quad \textit{Line up decimal points.} \\ \hline 9.95 \end{array}$$

(b) $13.761 + 8.325$

$$\begin{array}{r} \overset{11}{13.761} \\ + 8.325 \quad \textit{Line up decimal points.} \\ \hline 22.086 \end{array}$$

(c) $0.319 + 56.007 + 8.252$

$$\begin{array}{r} \overset{1}{}\overset{1}{0.319} \\ 56.007 \\ + 8.252 \quad \textit{Line up decimal points.} \\ \hline 64.578 \end{array}$$

(d) $39.4 + 0.4 + 177.2$

$$\begin{array}{r} \overset{1\,11}{39.4} \\ 0.4 \\ + 177.2 \quad \textit{Line up decimal points.} \\ \hline 217.0 \end{array}$$

2. **(a)** $6.54 + 9.8$

$$\begin{array}{r} \overset{1}{6.54} \quad \textit{Line up decimal points.} \\ + 9.80 \quad \textit{Write in zeros.} \\ \hline \underline{16.34} \end{array}$$

Copyright © 2018 Pearson Education, Inc.

(b) $0.831 + 222.2 + 10$

$$
\begin{array}{r}
\overset{1}{0.831} \quad \textit{Line up decimal points.} \\
222.200 \quad \textit{Write in zeros.} \\
+\,10.000 \\
\hline
233.031
\end{array}
$$

(c) $8.64 + 39.115 + 3.0076$

$$
\begin{array}{r}
\overset{2}{}\overset{1}{8}.6400 \quad \textit{Line up decimal points.} \\
39.1150 \quad \textit{Write in zeros.} \\
+\,3.0076 \\
\hline
50.7626
\end{array}
$$

(d) $5 + 429.823 + 0.76$

$$
\begin{array}{r}
\overset{11}{}5.000 \quad \textit{Line up decimal points.} \\
429.823 \quad \textit{Write in zeros.} \\
+\,0.760 \\
\hline
435.583
\end{array}
$$

3. (a) Subtract 22.7 from 72.9.

$$
\begin{array}{rr}
72.9 & \text{Check:} \quad 22.7 \\
-\,22.7 & +\,50.2 \\
\hline
50.2 & 72.9
\end{array}
$$

(b) Subtract 6.425 from 11.813.

$$
\begin{array}{rr}
\overset{71013}{11.8\cancel{1}\cancel{3}} & \text{Check:} \quad 6.425 \\
-\,6.425 & +\,5.388 \\
\hline
5.388 & 11.813
\end{array}
$$

(c) $20.15 - 19.67$

$$
\begin{array}{rr}
\overset{1\,9\,1015}{2\cancel{0}.\cancel{1}\cancel{5}} & \text{Check:} \quad 19.67 \\
-\,19.67 & +\,0.48 \\
\hline
0.48 & 20.15
\end{array}
$$

4. (a) Subtract 18.651 from 25.3.

$$
\begin{array}{r}
\overset{1\,14\,12\,9\,10}{\cancel{2}\cancel{5}.\cancel{3}\cancel{0}\cancel{0}} \quad \textit{Write two zeros.} \\
-\,18.651 \\
\hline
6.649
\end{array}
$$

$$
\begin{array}{r}
\text{Check:} \quad 18.651 \\
+\,6.649 \\
\hline
25.300
\end{array}
$$

(b) $5.816 - 4.98$

$$
\begin{array}{r}
\overset{4\,1711}{\cancel{5}.\cancel{8}\cancel{1}6} \\
-\,4.980 \quad \textit{Write one zero.} \\
\hline
0.836
\end{array}
$$

Check:

$$
\begin{array}{r}
4.980 \\
+\,0.836 \\
\hline
5.816
\end{array}
$$

(c) 40 less 3.66

$$
\begin{array}{rr}
\overset{3\,9\,910}{\cancel{4}\cancel{0}.\cancel{0}\cancel{0}} & \text{Check:} \quad 36.34 \\
-\,3.66 & +\,3.66 \\
\hline
36.34 & 40.00
\end{array}
$$

(d) $1 - 0.325$

$$
\begin{array}{rr}
\overset{0\,9\,910}{\cancel{1}.\cancel{0}\cancel{0}\cancel{0}} & \text{Check:} \quad 0.325 \\
-\,0.325 & +\,0.675 \\
\hline
0.675 & 1.000
\end{array}
$$

5. (a) $2.83 + 5.009 + 76.1$

Estimate:	*Exact:*
3	2.830
5	5.009
+ 80	+ 76.100
88	83.939

(b) 11.365 from 58

Estimate:	*Exact:*
60	58.000
− 10	− 11.365
50	46.635

(c) $398.81 + 47.658 + 4158.7$

Estimate:	*Exact:*
400	398.810
50	47.658
+ 4000	+ 4158.700
4450	4605.168

(d) Find the difference between 12.837 meters and 46.091 meters.

Estimate:	*Exact:*
50	46.091
− 10	− 12.837
40	33.254 meters

(e) $19.28 plus $1.53

Estimate:	*Exact:*
$20	$19.28
+ 2	+ 1.53
$22	$20.81

Copyright © 2018 Pearson Education, Inc.

4.3   Section Exercises

1. $6.42 + 10.163$

$$
\begin{array}{r}
6.420 \quad \textit{Line up decimal points.} \\
+\,10.163 \\
\hline
\end{array}
$$

3. $20 - 9.1263$

$$
\begin{array}{r}
20.0000 \quad \textit{Line up decimal points.} \\
-\,9.1263 \\
\hline
\end{array}
$$

5. $5.69 + 11.79$

$$
\begin{array}{r}
5.69 \quad \textit{Line up decimal points.} \\
+\,11.79 \\
\hline
17.48
\end{array}
$$

7. $8.263 - 0.5$

$$
\begin{array}{r}
8.263 \quad \textit{Line up decimal points.} \\
-\,0.500 \quad \textit{Write in zeros.} \\
\hline
7.763
\end{array}
$$

9. $76.5 + 0.506$

$$
\begin{array}{r}
76.500 \quad \textit{Line up decimal points.} \\
+\,0.506 \quad \textit{Write in zeros.} \\
\hline
77.006
\end{array}
$$

11. $21 - 0.896$

$$
\begin{array}{r}
21.000 \quad \textit{Line up decimal points.} \\
-\,0.896 \quad \textit{Write in zeros.} \\
\hline
20.104
\end{array}
$$

13.   Subtract 0.291 from 0.4

$$
\begin{array}{r}
0.400 \quad \textit{Line up decimal points.} \\
-\,0.291 \quad \textit{Write in zeros.} \\
\hline
0.109
\end{array}
$$

15. $39.76005 + 182 + 4.799 + 98.31 + 5.9999$

$$
\begin{array}{r}
39.76005 \quad \textit{Line up decimal points.} \\
182.00000 \quad \textit{Write in zeros.} \\
4.79900 \\
98.31000 \\
+\,5.99990 \\
\hline
330.86895
\end{array}
$$

17.   (a) Humerus 14.35 in., radius 10.4 in.

$$
\begin{array}{r}
14.35 \quad \textit{Line up decimal points.} \\
+\,10.40 \quad \textit{Write in zeros.} \\
\hline
24.75
\end{array}
$$

The combined length is 24.75 inches.

(b)
$$
\begin{array}{r}
14.35 \\
-\,10.40 \\
\hline
3.95
\end{array}
$$

The difference is 3.95 inches.

19.   (a) Humerus 14.35 in., ulna 11.1 in., femur 19.88 in., tibia 16.94 in.

$$
\begin{array}{r}
14.35 \quad \textit{Line up decimal points.} \\
11.10 \quad \textit{Write in zeros.} \\
19.88 \\
+\,16.94 \\
\hline
62.27
\end{array}
$$

The sum of the lengths is 62.27 inches.

(b) 8th rib 9.06 in., 7th rib 9.45 in.

$$
\begin{array}{r}
9.45 \\
-\,9.06 \\
\hline
0.39
\end{array}
$$

The 8th rib is 0.39 in. shorter than the 7th rib.

21.   6 should be written 6.00. The sum should be

$$
\begin{array}{r}
0.72 \\
6.00 \\
+\,39.50 \\
\hline
46.22
\end{array}
$$

23.   *Estimate:*   *Exact:*

$$
\begin{array}{r}
\$20 \\
-\,7 \\
\hline
\$13
\end{array}
\qquad
\begin{array}{r}
\$19.74 \\
-\,6.58 \\
\hline
\$13.16
\end{array}
$$

25.   *Estimate:*   *Exact:*

$$
\begin{array}{r}
400 \\
1 \\
+\,20 \\
\hline
421
\end{array}
\qquad
\begin{array}{r}
392.700 \\
0.865 \\
+\,21.080 \\
\hline
414.645
\end{array}
$$

27.   What is 8.6 less 3.751?

Estimate: *Exact:*

$$
\begin{array}{r}
9 \\
-\,4 \\
\hline
5
\end{array}
\qquad
\begin{array}{r}
8.600 \\
-\,3.751 \\
\hline
4.849
\end{array}
$$

29.   *Estimate:*   *Exact:*

$$
\begin{array}{r}
60 \\
500 \\
+\,6 \\
\hline
566
\end{array}
\qquad
\begin{array}{r}
62.8173 \\
539.9900 \\
+\,5.6290 \\
\hline
608.4363
\end{array}
$$

Copyright © 2018 Pearson Education, Inc.

31. $12 - 11.725$
An estimate is $12 - 12 = 0$, so 0.275 is the most reasonable answer.

33. $6.5 + 0.007$
An estimate is $7 + 0 = 7$, so 6.507 is the most reasonable answer.

35. $456.71 - 454.9$
An estimate is $457 - 455 = 2$, so 1.81 is the most reasonable answer.

37. $6004.003 + 52.7172$
An estimate is $6000 + 50 = 6050$, so 6056.7202 is the most reasonable answer.

39. Mexico: 50.9 million; Nigeria: 67 million

Estimate:	Exact:
70	67.0
$-$ 50	$-$ 50.9
20	16.1

There are 16.1 million fewer Internet users in Mexico than there are in Nigeria.

41. Add the number of users from all the countries.

Estimate:	Exact:
600	642.00
300	279.84
200	243.20
70	67.00
50	50.90
40	39.47
$+$ 20	$+$ 22.20
1280 million people	1344.61

There are 1344.61 million Internet users in all the countries listed in the table.

43. First, add the three players' heights.

Estimate:	Exact:
2	2.18
2	1.93
$+$ 2	$+$ 2.03
6 meters	6.14 meters

Now subtract the players' combined height of 6.14 m from the rhino's height of 6.40 m.

$$\begin{array}{r} 6.40 \\ -\ 6.14 \\ \hline 0.26 \end{array}$$

The NBA stars' combined height is 0.26 meter less than the rhino's height of 6.40 meters.

45. Subtract the weight of the lighter box from the weight of the heavier box.

Estimate:	Exact:
11	10.50
$-$ 10	$-$ 9.85
1 ounce	0.65 ounce

The difference in weight is 0.65 ounce.

47. Add the lengths of the four sides.

Estimate:	Exact:
20	19.75
20	19.75
6	6.30
$+$ 6	$+$ 6.30
52 inches	52.10 inches

The perimeter is 52.1 inches.

49. Subtract the price of the fluorescent fishing line, $6.99, from the price of the clear fishing line, $7.49.

Estimate:	Exact:
$7	$7.49
$-$ 7	$-$ 6.99
$0	$0.50

The clear fishing line costs $0.50 more than the fluorescent fishing line.

51. Add the price of the four items and the sales tax.

Estimate:	Exact:	
$30	$29.99	*second-lowest-priced spinning reel*
3	2.99	*environmentally safe split shot*
3	2.99	*environmentally safe split shot*
20	20.24	*three-tray tackle box*
$+$ 4	$+$ 3.93	*sales tax*
$60	$60.14	

The total cost of the four items and the sales tax is $60.14.

Copyright © 2018 Pearson Education, Inc.

53. Add the monthly expenses.

$1022.00
330.44
202.00
39.99
29.99
65.43
73.75
258.70
159.51
+ 68.00
—————
$2249.81

Olivia's total expenses are $2249.81 per month.

55. Subtract to find the difference.

$330.44
− 258.70
————
$71.74

The difference in the amounts spent for groceries and for the car payment is $71.74.

57. Add the car expenses; then subtract this amount from the rent.

330.44   1022.00
+ 202.00   − 532.44
————   ————
$532.44 $489.56

She spends $489.56 more on rent than on all her car expenses.

59. Subtract the two inside measurements from the total length.

3.00 *Total length*
− 0.91 *Leftmost measurement*
————
2.09
− 0.70 *Middle measurement*
————
1.39

$b = 1.39$ cm.

61. Add the given lengths and subtract the sum from the total length.

2.981   13.905
+ 2.981   − 5.962
————   ————
5.962 feet   7.943 feet

$q = 7.943$ feet

4.4   Multiplying Decimal Numbers

4.4   Margin Exercises

1.   (a)   2.6   ← 1 *decimal place*
        × 0.4   ← 1 *decimal place*
        ————
        1.04   ← 2 *decimal places*

(b)   45.2   ← 1 *decimal place*
        × 0.25 ← 2 *decimal places*
        ————
        2 2 60
        9 0 4
        ————
        11.3 0 0 ← 3 *decimal places*

(c)   0.104 ← 3 *decimal places*
        ×   7 ← 0 *decimal places*
        ————
        0.728 ← 3 *decimal places*

(d)   3.18   ← 2 *decimal places*
        × 2.23 ← 2 *decimal places*
        ————
        9 54
        63 6
        6 36
        ————
        7.09 14 ← 4 *decimal places*

(e)   611   ← 0 *decimal places*
        × 3.7   ← 1 *decimal place*
        ————
        427 7
        1833
        ————
        2260.7 ← 1 *decimal place*

2.   (a) 0.04×0.09

        0.04   ← 2 *decimal places*
        × 0.09 ← 2 *decimal places*
        ————
        36   ← 4 *decimal places needed*

Count 4 places from the right. Write in the decimal point and zeros.   0.0036

(b) $(0.2)(0.008)$

        0.008 ← 3 *decimal places*
        × 0.2   ← 1 *decimal place*
        ————
        16   ← 4 *decimal places needed*

Count 4 places from the right. Write in the decimal point and zeros.   0.0016

(c) $(0.003)^2$

        0.003 ← 3 *decimal places*
        × 0.003 ← 3 *decimal places*
        ————
        9   ← 6 *decimal places needed*

Count 6 places from the right. Write in the decimal point and zeros.   0.000009

Copyright © 2018 Pearson Education, Inc.

(d) $(0.0081)(0.003)$

$$
\begin{array}{r}
0.0081 \\
\times\ 0.003 \\
\hline
243
\end{array}
$$
← 4 *decimal places*
← 3 *decimal places*
← 7 *decimal places needed*

Count 7 places from the right. Write in the decimal point and zeros. 0.0000243

(e) $(0.11)(0.0005)$

$$
\begin{array}{r}
0.11 \\
\times\ 0.0005 \\
\hline
55
\end{array}
$$
← 2 *decimal places*
← 4 *decimal places*
← 6 *decimal places needed*

Count 6 places from the right. Write in the decimal point and zeros. 0.000055

3. **(a)** $(11.62)(4.01)$

Estimate: *Exact:*

$$
\begin{array}{r}
10 \\
\times\ 4 \\
\hline
40
\end{array}
\qquad
\begin{array}{r}
11.62 \\
\times\ 4.01 \\
\hline
1162 \\
46\,480 \\
\hline
46.5962
\end{array}
$$
← 2 *decimal places*
← 2 *decimal places*

← 4 *decimal places*

(b) $(5.986)(33)$

Estimate: *Exact:*

$$
\begin{array}{r}
6 \\
\times\ 30 \\
\hline
180
\end{array}
\qquad
\begin{array}{r}
5.986 \\
\times\ 33 \\
\hline
17\,958 \\
179\,58 \\
\hline
197.538
\end{array}
$$
← 3 *decimal places*
← 0 *decimal places*

← 3 *decimal places*

(c) $8.31(4.2)$

Estimate: *Exact:*

$$
\begin{array}{r}
8 \\
\times\ 4 \\
\hline
32
\end{array}
\qquad
\begin{array}{r}
8.31 \\
\times\ 4.2 \\
\hline
1\,662 \\
33\,24 \\
\hline
34.902
\end{array}
$$
← 2 *decimal places*
← 1 *decimal place*

← 3 *decimal places*

(d) 58.6×17.4

Estimate: *Exact:*

$$
\begin{array}{r}
60 \\
\times\ 20 \\
\hline
1200
\end{array}
\qquad
\begin{array}{r}
58.6 \\
\times\ 17.4 \\
\hline
23\,44 \\
4\,10\,2 \\
5\,86 \\
\hline
10\,19.64
\end{array}
$$
← 1 *decimal place*
← 1 *decimal place*

← 2 *decimal places*

4.4 Section Exercises

1. There is one decimal place in 4.2 and there are two decimal places in 3.46, so there will be $1 + 2 = \underline{3}$ decimal places in the product.

3.
$$
\begin{array}{r}
0.12 \\
\times\ 0.03 \\
\hline
0.0036
\end{array}
$$
← 2 *decimal places*
← 2 *decimal places*
← 4 *decimal places*

We must write $\underline{2}$ zeros in the product as placeholders: 0.$\underline{00}$36.

5.
$$
\begin{array}{r}
0.042 \\
\times\ 3.2 \\
\hline
84 \\
126 \\
\hline
0.1344
\end{array}
$$
← 3 *decimal places*
← 1 *decimal place*

← 4 *decimal places*

7.
$$
\begin{array}{r}
21.5 \\
\times\ 7.4 \\
\hline
8\,60 \\
150\,5 \\
\hline
159.10
\end{array}
$$
← 1 *decimal place*
← 1 *decimal place*

← 2 *decimal places*

9. $(0.666)(23.4)$
$$
\begin{array}{r}
0.666 \\
\times\ 23.4 \\
\hline
2664 \\
1\,998 \\
13\,32 \\
\hline
15.5844
\end{array}
$$
← 3 *decimal places*
← 1 *decimal place*

← 4 *decimal places*

11.
$$
\begin{array}{r}
\$51.88 \\
\times\ 6\,65 \\
\hline
259\,40 \\
3\,112\,8 \\
31\,128 \\
\hline
\$34,500.20
\end{array}
$$
← 2 *decimal places*
← 0 *decimal places*

← 2 *decimal places*

13. 72×0.6
$$
\begin{array}{r}
72 \\
\times\ 6 \\
\hline
432
\end{array}
\qquad
\begin{array}{r}
72 \\
\times\ 0.6 \\
\hline
43.2
\end{array}
$$
← 0 *decimal places*
← 1 *decimal place*
← 1 *decimal place*

15. $(7.2)(0.06)$
$$
\begin{array}{r}
72 \\
\times\ 6 \\
\hline
432
\end{array}
\qquad
\begin{array}{r}
7.2 \\
\times\ 0.06 \\
\hline
0.432
\end{array}
$$
← 1 *decimal place*
← 2 *decimal places*
← 3 *decimal places*

Copyright © 2018 Pearson Education, Inc.

17. 0.72(0.06)

$$
\begin{array}{r}
72 \\
\times\, 6 \\
\hline
432
\end{array}
\qquad
\begin{array}{r}
0.72 \leftarrow 2 \text{ decimal places} \\
\times\, 0.06 \leftarrow 2 \text{ decimal places} \\
\hline
0.0432 \leftarrow 4 \text{ decimal places}
\end{array}
$$

19. (0.006)(0.0052)

$$
\begin{array}{r}
0.0052 \leftarrow 4 \text{ decimal places} \\
\times\, 0.006 \leftarrow 3 \text{ decimal places} \\
\hline
0.0000312 \leftarrow 7 \text{ decimal places}
\end{array}
$$

21. $(0.005)^2 = 0.005 \cdot 0.005$

$$
\begin{array}{r}
0.005 \leftarrow 3 \text{ decimal places} \\
\times\, 0.005 \leftarrow 3 \text{ decimal places} \\
\hline
0.000025 \leftarrow 6 \text{ decimal places}
\end{array}
$$

23. *Estimate:*   *Exact:*

$$
\begin{array}{r}
40 \\
\times\, 5 \\
\hline
200
\end{array}
\qquad
\begin{array}{r}
39.6 \leftarrow 1 \text{ decimal place} \\
\times\,\ 4.8 \leftarrow 1 \text{ decimal place} \\
\hline
31\,68 \\
158\,4 \\
\hline
190.08 \leftarrow 2 \text{ decimal places}
\end{array}
$$

25. *Estimate:*   *Exact:*

$$
\begin{array}{r}
40 \\
\times\, 40 \\
\hline
1600
\end{array}
\qquad
\begin{array}{r}
37.1 \leftarrow 1 \text{ decimal place} \\
\times\,\ 42 \leftarrow 0 \text{ decimal places} \\
\hline
74\,2 \\
148\,4 \\
\hline
155\,8.2 \leftarrow 1 \text{ decimal place}
\end{array}
$$

27. *Estimate:*   *Exact:*

$$
\begin{array}{r}
7 \\
\times\, 5 \\
\hline
35
\end{array}
\qquad
\begin{array}{r}
6.53 \leftarrow 2 \text{ decimal places} \\
\times\,\ 4.6 \leftarrow 1 \text{ decimal place} \\
\hline
3\,918 \\
26\,12 \\
\hline
30.038 \leftarrow 3 \text{ decimal places}
\end{array}
$$

29. *Estimate:*   *Exact:*

$$
\begin{array}{r}
3 \\
\times\, 7 \\
\hline
21
\end{array}
\qquad
\begin{array}{r}
2.809 \leftarrow 3 \text{ decimal places} \\
\times\,\ 6.85 \leftarrow 2 \text{ decimal places} \\
\hline
14045 \\
2\,2472 \\
16\,854 \\
\hline
19.24165 \leftarrow 5 \text{ decimal places}
\end{array}
$$

31. A $28.90 car payment is *unreasonable.* A reasonable answer would be $289.00.

33. A height of 60.5 inches (about 5 feet) is *reasonable.*

35. A gallon of milk for $419 is *unreasonable.* A reasonable answer would be $4.19.

37. 0.095 pound for a baby's weight is *unreasonable.* A reasonable answer would be 9.5 pounds.

39. Multiply her pay per hour times the number of hours she worked.

Estimate: *Exact:*

$$
\begin{array}{r}
\$20 \\
\times\, 50 \\
\hline
\$1000
\end{array}
\qquad
\begin{array}{r}
\$18.73 \leftarrow 2 \text{ decimal places} \\
\times\,\ 50.5 \leftarrow 1 \text{ decimal place} \\
\hline
9\,365 \\
936\,50 \\
\hline
\$945.865 \leftarrow 3 \text{ decimal places}
\end{array}
$$

Round $945.865 to the nearest cent. LaTasha made $945.87.

41. Multiply the cost of one meter of canvas by the number of meters needed.

Estimate: *Exact:*

$$
\begin{array}{r}
\$5 \\
\times\, 1 \\
\hline
\$5
\end{array}
\qquad
\begin{array}{r}
\$5.09 \\
\times\, 0.8 \\
\hline
\$4.072
\end{array}
$$

$4.072 rounds to $4.07. Sid will spend $4.07 on the canvas.

43. Multiply the number of gallons that needed to fill her car's gas tank by the price per gallon for each type of gas. Use a calculator. Round answers to the nearest cent.

Regular unleaded: $15.5 \times \$3.199 \approx \$49.58$
Premium unleaded $15.5 \times \$3.599 \approx \$55.78$

Subtract the cost of regular gas from the cost of premium gas.

$$
\begin{array}{r}
\$55.78 \\
-\ \ 49.58 \\
\hline
\$6.20
\end{array}
$$

Michelle would save $6.20 filling her tank with regular unleaded.

45. Octavio forgot to write three zeros on the far left side when placing the decimal point in the product. The correct product is 0.000510.

47. **(a) Area before 1929:**

$$
\begin{array}{r}
7.4218 \\
\times\ \ \ \ \ 3.1250 \\
\hline
371090 \\
148436 \\
74218 \\
22\,2654 \\
\hline
23.1931250
\end{array}
$$

Rounding to the nearest tenth gives us 23.2 in.^{2}

(continued)

Copyright © 2018 Pearson Education, Inc.

Area after 1929:

$$
\begin{array}{r}
6.14 \\
\times\ 2.61 \\
\hline
614 \\
3\,684 \\
12\,28 \\
\hline
16.0254
\end{array}
$$

Rounding to the nearest tenth gives us 16.0 in.²

(b) Subtract to find the difference.

$$
\begin{array}{r}
23.2 \\
-\ 16.0 \\
\hline
7.2
\end{array}
$$

The difference in rounded areas is 7.2 in.²

49. **(a)** Multiply the thickness of one bill times the number of bills.

$$
\begin{array}{r}
0.0043 \\
\times\ \ 100 \\
\hline
0.4300
\end{array}
$$

A pile of 100 bills would be 0.43 inch high.

(b)
$$
\begin{array}{r}
0.0043 \\
\times\ 1000 \\
\hline
4.3000
\end{array}
$$

A pile of 1000 bills would be 4.3 inches high.

51. **(a)** The first 12 months cost $69.99 per month. Multiply the monthly cost for cable times 12 months.

$$
\begin{array}{r}
\$69.99 \leftarrow \textit{2 decimal places} \\
\times\ \ \ \ 12 \leftarrow \textit{0 decimal places} \\
\hline
139\,98 \\
699\,9 \\
\hline
\$839.88 \leftarrow \textit{2 decimal places}
\end{array}
$$

Then add the $90 installation fee.

$$
\begin{array}{r}
\$839.88 \\
+\ \ 90.00 \\
\hline
\$929.88 \leftarrow \textit{Amount}
\end{array}
$$

Judy pays $929.88 for the first year .

The second 12 months cost $89.99 per month. Multiply.

$$
\begin{array}{r}
\$89.99 \leftarrow \textit{2 decimal places} \\
\times\ \ \ \ 12 \leftarrow \textit{0 decimal places} \\
\hline
179\,98 \\
899\,9 \\
\hline
\$1079.98 \leftarrow \textit{2 decimal places}
\end{array}
$$

Judy pays $1079.98 for the second year .

(b) After the 2-year agreement is over, Judy pays $119.94 per month. Multiply by 12 months.

$$
\begin{array}{r}
\$119.94 \leftarrow \textit{2 decimal places} \\
\times\ \ \ \ 12 \leftarrow \textit{0 decimal places} \\
\hline
239\,88 \\
1199\,4 \\
\hline
\$1439.28 \leftarrow \textit{2 decimal places}
\end{array}
$$

After the two-year agreement is over, Judy will pay $1439.28 per year.

53. Multiply to find the cost of the rope, then multiply to find the cost of the wire. Add the results to find Barry's total purchases. Subtract the purchases from $15 (three $5 bills is $15).

Cost of rope	Cost of wire
16.5	$0.72
× $0.65	× 3
$10.725	$2.16

The cost of the rope rounds to $10.73.

Purchases	Change
$10.73 *rope*	$15.00
+ 2.16 *wire*	− 12.89
$12.89	$2.11

Barry received $2.11 in change.

55. **(a)** Multiply the current minimum wage, $7.75, by 25.

$$
\begin{array}{r}
\$7.75 \\
\times\ \ 25 \\
\hline
\$193.75
\end{array}
$$

Monique made $193.75 with the current minimum wage.

(b) Multiply the new minimum wage, $8.65, by 25.

$$
\begin{array}{r}
\$8.65 \\
\times\ \ 25 \\
\hline
\$216.25
\end{array}
$$

Monique will make $216.25 with the new minimum wage.

57. **(a)**
$$
\begin{aligned}
(5.386)(10) &= \underline{53.86} \\
(10.5)(10) &= \underline{105} \\
(7)(10) &= \underline{70} \\
(4.95)(10) &= \underline{49.5} \\
(0.9)(10) &= \underline{9} \\
(14.34)(10) &= \underline{143.4}
\end{aligned}
$$

(b) When you multiply a number by 10, the decimal point moves one place to the right.

Copyright © 2018 Pearson Education, Inc.

(c) Multiplying by 100 moves the decimal point two places to the right. Multiplying by 1000 moves the decimal point three places to the right.

58. **(a)** $(0.538)(0.1) = \underline{0.0538}$
$(10.5)(0.1) = \underline{1.05}$
$(58)(0.1) = \underline{5.8}$
$(4.95)(0.1) = \underline{0.495}$
$(0.9)(0.1) = \underline{0.09}$
$(300)(0.1) = \underline{30}$

(b) Multiplying by 0.1 moves the decimal point one place to the left.

(c) Multiplying by 0.01 moves the decimal point two places to the left. Multiplying by 0.001 moves the decimal point three places to the left.

Summary Exercises *Adding, Subtracting, and Multiplying Decimal Numbers*

1. $0.8 = \dfrac{8}{10} = \dfrac{8 \div 2}{10 \div 2} = \dfrac{4}{5}$

3. $0.35 = \dfrac{35}{100} = \dfrac{35 \div 5}{100 \div 5} = \dfrac{7}{20}$

5. 2.0003 is two and three ten-thousandths.

7. five hundredths: $\frac{5}{100} = 0.05$

9. ten and seven tenths: $10\frac{7}{10} = 10.7$

11. 0.95 to the nearest tenth
Draw a cut-off line: 0.9|5
The first digit cut is 5, which is 5 or more, so round up.

$\begin{array}{r} 0.9 \\ + 0.1 \\ \hline 1.0 \end{array}$

Answer: ≈ 1.0

13. $0.893 to the nearest cent
Draw a cut-off line: $0.89|3
The first digit cut is 3, which is 4 or less. The part you keep stays the same.
Answer: $\approx \$0.89$

15. $99.64 to the nearest dollar
Draw a cut-off line: $99|.64
The first digit cut is 6, which is 5 or more, so round up.

$\begin{array}{r} \$99 \\ + 1 \\ \hline \$100 \end{array}$

Answer: $\approx \$100$

17. $50 - 0.3801$

$\begin{array}{r} {\scriptstyle 4\,9\ 99\,9\,10} \\ \cancel{5}\,\cancel{0}\,.\,\cancel{0}\,\cancel{0}\,\cancel{0}\,\cancel{0} \quad \textit{Write four zeros.} \\ -\ 0.3\,8\,0\,1 \\ \hline 4\,9\,.6\,1\,9\,9 \end{array}$

Check: $\begin{array}{r} 0.3801 \\ + 49.6199 \\ \hline 50.0000 \end{array}$

19. $25(\$3.74)$

$\begin{array}{r} \$3.74 \quad \leftarrow \textit{2 decimal places} \\ \times \quad 25 \quad \leftarrow \textit{0 decimal places} \\ \hline 18\ 70 \\ 74\ 8 \\ \hline \$93.50 \quad \leftarrow \textit{2 decimal places} \end{array}$

21. $3.7 - 1.55$

$\begin{array}{r} {\scriptstyle 6\,10} \\ 3.7\cancel{0} \\ -\ 1.5\,5 \\ \hline 2.1\,5 \end{array}$ Check: $\begin{array}{r} 1.55 \\ + 2.15 \\ \hline 3.70 \end{array}$

23. $3.6 + 0.718 + 9 + 5.0829$

$\begin{array}{r} {\scriptstyle 1\ 11} \\ 3.6000 \quad \textit{Line up decimal points.} \\ 0.7180 \quad \textit{Write in zeros.} \\ 9.0000 \\ + 5.0829 \\ \hline 18.4009 \end{array}$

25. $8.9 - 0.4 - 0.03 - 7.6$ *Subtract.*
$8.5 - 0.03 - 7.6$ *Subtract.*
$8.47 - 7.6 = 0.87$ *Subtract.*

27. $(0.99)(83.672) = 82.83528$
Rounded to the nearest hundredth, we get 82.84.

29. Perimeter $= 0.875 + 0.75 + 0.875 + 0.75$ *Add.*
$= 1.625 + 0.875 + 0.75$ *Add.*
$= 2.5 + 0.75 = 3.25$ inches   *Add.*

Area $= (0.875)(0.75)$ *Multiply.*
$= 0.65625 \approx 0.66$ inch^2

31.
$\begin{array}{r} {\scriptstyle 22\ 1} \\ 68.564 \quad \textit{radish} \\ 7.750 \quad \textit{tomato} \\ + 18.740 \quad \textit{onion} \\ \hline 95.054 \quad \textit{combined weight} \end{array}$

The combined weight is 95.1 pounds (rounded).

Copyright © 2018 Pearson Education, Inc.

33. **(a)**
$$
\begin{array}{r}
\overset{6\ 14\ 9\ 10}{7.7\ \cancel{3}\ \cancel{0}\ \cancel{0}} \quad tomato \\
-\ 4.0\ 7\ 6\ 4 \quad apple \\
\hline
3.6\ 7\ 3\ 6 \quad difference
\end{array}
$$

The tomato is 3.6736 pounds heavier than the apple.

(b)
$$
\begin{array}{r}
\overset{3\ 10}{4.\cancel{0}\ 764} \quad apple \\
-\ 1.6\ 000 \quad peach \\
\hline
2.4\ 764 \quad difference
\end{array}
$$

The peach weighs 2.4764 pounds less than the apple.

35. First, find the cost to buy the pumpkin by multiplying the weight of the pumpkin, 2324 pounds, by the price per pound, $0.50.

$$
\begin{array}{r}
2324 \quad number\ of\ pounds \\
\times\ \ \ \$0.5 \quad cost\ per\ pound \\
\hline
\$1162 \quad total\ cost
\end{array}
$$

It cost $1162 to buy the pumpkin.
Now find how much they earned by multiplying the number of pies, 775, by the cost per pie, $4.50.

$$
\begin{array}{r}
775 \\
\times\ \ \ \$4.5 \\
\hline
387\ 5 \\
3100 \\
\hline
\$3487.5
\end{array}
$$

The total revenue from the pies was $3487.50

$$
\begin{array}{r}
\$3487.50 \quad total\ revenue \\
-\ 1162.00 \quad cost \\
\hline
\$2325.50 \quad profit
\end{array}
$$

They earned $2325.50 for the charity.

4.5 Dividing Decimal Numbers

4.5 Margin Exercises

1. **(a)**
$$
\begin{array}{r}
2\ 3.\ \underline{4} \\
4\overline{)9\ 3.\ 6} \\
\underline{8} \\
1\ 3 \\
\underline{1\ 2} \\
1\ 6 \\
\underline{1\ 6} \\
0
\end{array}
$$
Check:
$$
\begin{array}{r}
23.4 \\
\times\ \ 4 \\
\hline
93.6
\end{array}
$$

(b)
$$
\begin{array}{r}
1.\ 1\ 3\ 4 \\
6\overline{)6.\ 8\ 0\ 4} \\
\underline{6} \\
0\ 8 \\
\underline{6} \\
2\ 0 \\
\underline{1\ 8} \\
2\ 4 \\
\underline{2\ 4} \\
0
\end{array}
$$
Check:
$$
\begin{array}{r}
1.134 \\
\times\ \ 6 \\
\hline
6.804
\end{array}
$$

(c)
$$
\begin{array}{r}
2\ 5.\ 3 \\
11\overline{)2\ 7\ 8.\ 3} \\
\underline{2\ 2} \\
5\ 8 \\
\underline{5\ 5} \\
3\ 3 \\
\underline{3\ 3} \\
0
\end{array}
$$
Check:
$$
\begin{array}{r}
25.3 \\
\times\ 11 \\
\hline
25\ 3 \\
253 \\
\hline
278.3
\end{array}
$$

(d) $0.51835 \div 5$
$$
\begin{array}{r}
0.\ 1\ 0\ 3\ 6\ 7 \\
5\overline{)0.\ 5\ 1\ 8\ 3\ 5} \\
\underline{5} \\
1 \\
\underline{0} \\
1\ 8 \\
\underline{1\ 5} \\
3\ 3 \\
\underline{3\ 0} \\
3\ 5 \\
\underline{3\ 5} \\
0
\end{array}
$$
Check:
$$
\begin{array}{r}
0.10367 \\
\times\ \ \ \ \ 5 \\
\hline
0.51835
\end{array}
$$

(e) $213.45 \div 15$
$$
\begin{array}{r}
1\ 4.\ 2\ 3 \\
15\overline{)2\ 1\ 3.\ 4\ 5} \\
\underline{1\ 5} \\
6\ 3 \\
\underline{6\ 0} \\
3\ 4 \\
\underline{3\ 0} \\
4\ 5 \\
\underline{4\ 5} \\
0
\end{array}
$$
Check:
$$
\begin{array}{r}
14.23 \\
\times\ \ 15 \\
\hline
71\ 15 \\
142\ 3 \\
\hline
213.45
\end{array}
$$

Copyright © 2018 Pearson Education, Inc.

2.   (a)

$$
\begin{array}{r}
1.\,2\,8 \\
5\overline{\smash)6.\,4\,0} \quad \leftarrow \textit{Write one zero.} \\
\underline{5} \\
1\ 4 \\
\underline{1\ 0} \\
4\ 0 \\
\underline{4\ 0} \\
0
\end{array}
$$

Check: 1.28
$$
\begin{array}{r}
1.28 \\
\times\ 5 \\
\hline
6.40 \text{ or } 6.4
\end{array}
$$

(b) $30.87 \div 14$

$$
\begin{array}{r}
2.\,2\,0\,5 \\
14\overline{\smash)3\,0.\,8\,7\,0} \quad \leftarrow \textit{Write one zero.} \\
\underline{2\ 8} \\
2\ 8 \\
\underline{2\ 8} \\
0\ 7 \\
\underline{0} \\
7\ 0 \\
\underline{7\ 0} \\
0
\end{array}
$$

Check: 2.205
$$
\begin{array}{r}
2.205 \\
\times\ \ \ 14 \\
\hline
8\,820 \\
22\,05 \\
\hline
30.870 \text{ or } 30.87
\end{array}
$$

(c) $\dfrac{259.5}{30}$

$$
\begin{array}{r}
8.\,6\,5 \\
30\overline{\smash)2\,5\,9.\,5\,0} \quad \leftarrow \textit{Write one zero.} \\
\underline{2\ 4\ 0} \\
1\ 9\ 5 \\
\underline{1\ 8\ 0} \\
1\ 5\ 0 \\
\underline{1\ 5\ 0} \\
0
\end{array}
$$

Check: 8.65
$$
\begin{array}{r}
8.65 \\
\times\ 30 \\
\hline
259.50 \text{ or } 259.5
\end{array}
$$

(d) $0.3 \div 8$

$$
\begin{array}{r}
0.\,0\,3\,7\,5 \\
8\overline{\smash)0.\,3\,0\,0\,0} \quad \leftarrow \textit{Write three zeros.} \\
\underline{0} \\
3\ 0 \\
\underline{2\ 4} \\
6\ 0 \\
\underline{5\ 6} \\
4\ 0 \\
\underline{4\ 0} \\
0
\end{array}
$$

Check: 0.0375
$$
\begin{array}{r}
0.0375 \\
\times\ \ \ \ 8 \\
\hline
0.3000 \text{ or } 0.3
\end{array}
$$

3.   (a) $13\overline{\smash)2\,6\,7.\,0\,1}$

$267.01 \div 13 \approx \underline{20.53923}$

There are no repeating digits visible on the calculator.

20.539|23 rounds to 20.539.

Check: $(20.539)(13) = 267.007 \approx 267.01$

(b) $6\overline{\smash)2\,0.\,5}$

$20.5 \div 6 \approx 3.416666$

There is a repeating decimal: $3.41\overline{6}$.

3.416|666 rounds to 3.417.

Check: $(3.417)(6) = 20.502 \approx 20.5$

(c) $\dfrac{10.22}{9} = 10.22 \div 9 \approx 1.135555$

There is a repeating decimal: $1.13\overline{5}$.

1.135|555 rounds to 1.136.

Check: $(1.136)(9) = 10.224 \approx 10.22$

(d) $16.15 \div 3 \approx 5.383333$

There is a repeating decimal: $5.38\overline{3}$.

5.383|333 rounds to 5.383.

Check: $(5.383)(3) = 16.149 \approx 16.15$

(e) $116.3 \div 11 \approx 10.5727272$

The answer has a repeating decimal that starts repeating in the fourth place: $10.5\overline{72}$.

10.572|72 rounds to 10.573.

Check: $(10.573)(11) = 116.303 \approx 116.3$

Copyright © 2018 Pearson Education, Inc.

4.   (a)

$$
\begin{array}{r}
5.\,2 \\
0.2_\wedge\overline{\smash{\big)}1.\,0_\wedge\,4} \\
\underline{1\ 0} \\
0\ \ 4 \\
\underline{4} \\
0
\end{array}
$$

(b)

$$
\begin{array}{r}
3\,0.\,1\,2 \\
0.06_\wedge\overline{\smash{\big)}1.\,8\,0_\wedge\,7\,2} \\
\underline{1\ 8} \\
0\ 0 \\
\underline{0} \\
0\ \ 7 \\
\underline{6} \\
1\ 2 \\
\underline{1\ 2} \\
0
\end{array}
$$

(c)

$$
\begin{array}{r}
6\ 4\ 0\ 0 \\
0.005_\wedge\overline{\smash{\big)}3\,2.\,0\,0\,0\,0_\wedge} \\
\underline{3\ 0} \\
2\ 0 \\
\underline{2\ 0} \\
0\ 0 \\
\underline{0} \\
0\ 0 \\
\underline{0} \\
0
\end{array}
$$

(d) $8.1 \div 0.025$

$$
\begin{array}{r}
3\ 2\ 4 \\
0.025_\wedge\overline{\smash{\big)}8.\,1\,0\,0_\wedge} \\
\underline{7\ 5} \\
6\ 0 \\
\underline{5\ 0} \\
1\ 0\ 0 \\
\underline{1\ 0\ 0} \\
0
\end{array}
$$

(e) $\dfrac{7}{1.3}$

$$
\begin{array}{r}
5.\,3\,8\,4 \approx 5.38 \\
1.3_\wedge\overline{\smash{\big)}7.\,0_\wedge\,0\,0\,0} \\
\underline{6\ 5} \\
5\ 0 \\
\underline{3\ 9} \\
1\ 1\ 0 \\
\underline{1\ 0\ 4} \\
6\ 0 \\
\underline{5\ 2} \\
8
\end{array}
$$

(f) $5.3091 \div 6.2$

$$
\begin{array}{r}
0.\,8\,5\,6 \approx 0.86 \\
6.2_\wedge\overline{\smash{\big)}5.\,3_\wedge\,0\,9\,1} \\
\underline{4\ 9\ 6} \\
3\ 4\ 9 \\
\underline{3\ 1\ 0} \\
3\ 9\ 1 \\
\underline{3\ 7\ 2} \\
1\ 9
\end{array}
$$

5.   (a) $42.75 \div 3.8 = 1.125$

Estimate: $40 \div 4 = 10$

The answer is not reasonable.

$$
\begin{array}{r}
1\ 1.\,2\,5 \\
3.8_\wedge\overline{\smash{\big)}4\,2.\,7_\wedge\,5\,0} \\
\underline{3\ 8} \\
4\ 7 \\
\underline{3\ 8} \\
9\ 5 \\
\underline{7\ 6} \\
1\ 9\ 0 \\
\underline{1\ 9\ 0} \\
0
\end{array}
$$

The answer should be 11.25.

(b) $807.1 \div 1.76 = 458.580$

Estimate: $800 \div 2 = 400$

The answer is reasonable.

(c) $48.63 \div 52 = 93.519$

Estimate: $50 \div 50 = 1$

The answer is not reasonable.

$$
\begin{array}{r}
0.\,9\,3\,5\,1 \approx 0.935 \\
52\overline{\smash{\big)}4\,8.\,6\,3\,0\,0} \\
\underline{4\ 6\ 8} \\
1\ 8\ 3 \\
\underline{1\ 5\ 6} \\
2\ 7\ 0 \\
\underline{2\ 6\ 0} \\
1\ 0\ 0 \\
\underline{5\ 2} \\
4\ 8
\end{array}
$$

The answer should be 0.935.

(d) $9.0584 \div 2.68 = 0.338$

Estimate: $9 \div 3 = 3$

The answer is not reasonable.

$$
\begin{array}{r}
3.\ 3\ 8 \\
2.68_\wedge \overline{)9.\ 0\ 5_\wedge 8\ 4} \\
8\ 0\ 4 \\
\overline{1\ 0\ 1\ 8} \\
8\ 0\ 4 \\
\overline{2\ 1\ 4\ 4} \\
2\ 1\ 4\ 4 \\
\overline{0}
\end{array}
$$

The answer should be 3.38.

6.   **(a)** $4.6 - 0.79 + \underset{\smile}{1.5^2}$ *Exponent*

   $4.6 - 0.79 + 2.25$ *Subtract.*

   $3.81 + 2.25 = 6.06$ *Add.*

   (b) $\underset{\smile}{3.64 \div 1.3} \times 3.6$ *Divide.*

   $2.8 \times 3.6 = 10.08$ *Multiply.*

   (c) $0.08 + 0.6(\underset{\smile}{3 - 2.99})$ *Parentheses*

   $0.08 + 0.6(0.01)$ *Multiply.*

   $0.08 + 0.006 = 0.086$ *Add.*

   (d) $10.85 - \underset{\smile}{2.3(5.2)} \div 3.2$ *Multiply.*

   $10.85 - 11.96 \div 3.2$ *Divide.*

   $10.85 - 3.7375 = 7.1125$ *Subtract.*

4.5 Section Exercises

1.   Problem (c), $25.5 \div 5$, has a whole number for the divisor. It can be rewritten as $5\overline{)2\ 5.\ 5}$.

3.   To make 2.2 a whole number, move the decimal point one digit to the right.

$$2.2_\wedge \overline{)8.\ 2_\wedge 4}$$

5.
$$
\begin{array}{r}
3.\ 9 \quad \textit{Line up decimal points.} \\
7\overline{)2\ 7.\ 3} \\
2\ 1 \\
\overline{6\ 3} \\
6\ 3 \\
\overline{0}
\end{array}
$$

7. $\dfrac{4.23}{9}$

$$
\begin{array}{r}
0.\ 4\ 7 \quad \textit{Line up decimal points.} \\
9\overline{)4.\ 2\ 3} \\
3\ 6 \\
\overline{6\ 3} \\
6\ 3 \\
\overline{0}
\end{array}
$$

9.
$$
\begin{array}{r}
4\ 0\ 0.\ 2 \\
0.05_\wedge \overline{)2\ 0.\ 0\ 1_\wedge 0} \quad \textit{Move decimal point} \\
2\ 0 \qquad\qquad\quad \textit{in divisor and dividend} \\
\overline{0\ 0\ 1\ 0} \quad \textit{2 places; write one zero.} \\
1\ 0 \\
\overline{0}
\end{array}
$$

11.
$$
\begin{array}{r}
3\ 6. \\
1.5_\wedge \overline{)5\ 4.\ 0_\wedge} \quad \textit{Move decimal point} \\
4\ 5 \qquad\quad \textit{in divisor and dividend} \\
\overline{9\ 0} \quad \textit{1 place; write 0 in the dividend.} \\
9\ 0 \\
\overline{0}
\end{array}
$$

13.   Given: $108 \div 18 = 6$

Find: $0.108 \div 1.8$ by moving the decimal points.

$$
\begin{array}{r}
0\ .0\ 6 \\
1.8_\wedge \overline{)0.\ 1_\wedge 0\ 8}
\end{array}
$$

$\mathbf{0.108 \div 1.8 = 0.06}$

15.   Given: $108 \div 18 = 6$

Find: $0.018\overline{)108}$ or $108 \div 0.018$

The new divisor, 0.018, has the effect of moving the decimal place in the answer *right* three places.

$\mathbf{108 \div 0.018 = 6000}$

17.
$$
\begin{array}{r}
2\ 5.\ 3 \\
4.6_\wedge \overline{)1\ 1\ 6.\ 3_\wedge 8} \quad \textit{Move decimal point} \\
9\ 2 \qquad\qquad \textit{in divisor and dividend} \\
\overline{2\ 4\ 3} \quad \textit{1 place.} \\
2\ 3\ 0 \\
\overline{1\ 3\ 8} \\
1\ 3\ 8 \\
\overline{0}
\end{array}
$$

Copyright © 2018 Pearson Education, Inc.

19. $\dfrac{3.1}{0.006}$

```
                5 1 6. 6 6 6    Line up decimal points.
0.006∧ 3. 1 0 0∧0 0 0
        3 0                     Move decimal point in
          1 0                   divisor and dividend 3
           6                    places. Write 000
           4 0                  in dividend.
           3 6
             4 0                Write 0 in dividend.
             3 6
               4 0             Write 0 in dividend.
               3 6
                 4 0  Write 0 in dividend.
                 3 6
                   4  Stop and round
                      answer to the nearest
                      hundredth.
```

516.666 rounds to 516.67.

21. $240.8 \div 9$

```
        2 6. 7 5 5 5    Line up decimal points.
9 2 4 0. 8 0 0 0
  1 8
    6 0
    5 4
      6 8
      6 3
        5 0            Write 0 in dividend.
        4 5
          5 0          Write 0 in dividend.
          4 5
            5 0  Write 0 in dividend.
            4 5
              5  Stop and round answer
                 to the nearest thousandth.
```

26.7555 rounds to 26.756.

23. $0.034\overline{\smash{)}342.81}$

Enter on calculator:

342.81 ÷ 0.034 =

Round 10,082.64706 to 10,082.647.

25. $37.8 \div 8 = 47.25$

Estimate: $40 \div 8 = 5$

The answer 47.25 is *unreasonable*.

**

```
        4. 7 2 5
8 3 7. 8 0 0
  3 2
    5 8
    5 6
      2 0
      1 6
        4 0
        4 0
          0
```

The correct answer is 4.725.

27. $54.6 \div 48.1 = 1.135$

Estimate: $50 \div 50 = 1$

The answer 1.135 is *reasonable*.

29. $307.02 \div 5.1 = 6.2$

Estimate: $300 \div 5 = 60$

The answer 6.2 is *unreasonable* because it is so much less than 60.

```
          6 0. 2
5.1∧ 3 0 7. 0∧2
      3 0 6
          1 0
            0
          1 0 2
          1 0 2
              0
```

The correct answer is 60.2, which is close to the estimate of 60.

31. Divide the cost of 6 pairs of tights by 6.

```
      3. 9 9 6
6 2 3. 9 8 0
  1 8
    5 9
    5 4
      5 8
      5 4
        4 0
        3 6
          4
```

$3.996 rounds to $4.00.

One pair costs $4.00.

Copyright © 2018 Pearson Education, Inc.

33. Divide the balance by the number of months.

$$
\begin{array}{r}
67.08 \\
21\overline{\smash{)}1408.68} \\
\underline{126} \\
148 \\
\underline{147} \\
168 \\
\underline{168} \\
0
\end{array}
$$

Aimee is paying \$67.08 per month.

35. Divide the total earnings by the number of hours.

$$
\begin{array}{r}
11.92 \\
40\overline{\smash{)}476.80} \\
\underline{40} \\
76 \\
\underline{40} \\
368 \\
\underline{360} \\
80 \\
\underline{80} \\
0
\end{array}
$$

Darren earns \$11.92 per hour.

37. Divide the miles driven by the gallons of gas.

$$344.1 \div 12.3 \approx 27.975$$

She got 28.0 miles per gallon (rounded).

39. Multiply the price per can by the number of cans.

$$
\begin{array}{r}
\$0.67 \\
\times\ \ 6 \\
\hline
\$4.02
\end{array}
$$

Subtract to find total savings.

$$
\begin{array}{r}
\$4.02 \\
-\ 3.85 \\
\hline
\$0.17
\end{array}
$$

There are six cans, so divide by 6 to find the savings per can.

$$
\begin{array}{r}
0.028 \\
6\overline{\smash{)}0.170} \\
\underline{12} \\
50 \\
\underline{48} \\
2
\end{array}
$$

\$0.028 rounds to \$0.03.

You will save \$0.03 per can (rounded).

41. First find the sum of lengths for Jackie Joyner-Kersee.

$$
\begin{array}{r}
7.49 \\
7.45 \\
7.40 \\
7.32 \\
+\ 7.20 \\
\hline
36.86
\end{array}
$$

Then divide by the number of jumps, namely 5.

$$
\begin{array}{r}
7.372 \\
5\overline{\smash{)}36.860} \\
\underline{35} \\
18 \\
\underline{15} \\
36 \\
\underline{35} \\
10 \\
\underline{10} \\
0
\end{array}
$$

7.372 rounds to 7.37.

The average length of the long jumps made by Jackie Joyner-Kersee is 7.37 meters (rounded).

43. Subtract to find the difference.

$$
\begin{array}{rl}
7.40 & \textit{length of fifth-longest jump} \\
-\ 7.32 & \textit{length of sixth-longest jump} \\
\hline
0.08 & \textit{difference}
\end{array}
$$

The fifth-longest jump was 0.08 meter longer than the sixth-longest jump.

45. Add the lengths of the longest jump made by the top three athletes.

$$
\begin{array}{r}
7.52 \\
7.49 \\
+\ 7.48 \\
\hline
22.49
\end{array}
$$

The total length jumped by the top three athletes was 22.49 meters.

47. The first mistake was multiplying 4.4 by 2 instead of multiplying 4.4 by 4.4. The second mistake was ignoring the parentheses instead of doing the work inside the parenthese before subtracting. The correct solution is

$$
\begin{array}{ll}
4.4^2 - (2.5 + 3.1) & \textit{Exponent} \\
19.36 - (2.5 + 3.1) & \textit{Parentheses} \\
19.36 - 5.6 = 13.76 & \textit{Subtract.}
\end{array}
$$

49.
$$
\begin{array}{ll}
7.2 - 5.2 + 3.5^2 & \textit{Exponent} \\
7.2 - 5.2 + 12.25 & \textit{Subtract.} \\
2 + 12.25 = 14.25 & \textit{Add.}
\end{array}
$$

Copyright © 2018 Pearson Education, Inc.

51. $38.6 + 11.6(13.4 - 10.4)$ *Parentheses*
$38.6 + 11.6(3)$ *Multiply.*
$38.6 + 34.8 = 73.4$ *Add.*

53. $8.68 - 4.6(10.4) \div 6.4$ *Multiply.*
$8.68 - 47.84 \div 6.4$ *Divide.*
$8.68 - 7.475 = 1.205$ *Subtract.*

55. $33 - 3.2(0.68 + 9) - 1.3^2$ *Parentheses; Exponent*
$33 - 3.2(9.68) - 1.69$ *Multiply.*
$33 - 30.976 - 1.69$ *Subtract.*
$2.024 - 1.69 = 0.334$ *Subtract.*

57. **(a)** $\frac{38,000,000}{24} \approx 1,583,333$ pieces each hour

(b) There are $24 \times 60 = 1440$ minutes in a day.

$\frac{38,000,000}{1440} \approx 26,389$ pieces each minute

(c) There are $24 \times 60 \times 60 = 86,400$ seconds in a day.

$\frac{38,000,000}{86,400} \approx 440$ pieces each second

59. Divide \$20,000 by 10¢.

$$
\begin{array}{r}
2\,0\,0,0\,0\,0. \\
0.10_\wedge \overline{)2\,0,0\,0\,0.\,0\,0_\wedge} \\
\underline{2\,0} \\
0\;0\,0\,0\,0\,0
\end{array}
$$

The school would need to collect 200,000 box tops.

61. Divide 200,000 (the answer from Exercise 59) by 38 weeks.

$$
\begin{array}{r}
5\,2\,6\,3.\,1 \\
38\,\overline{)2\,0\,0,0\,0\,0.\,0} \\
\underline{1\,9\,0} \\
1\,0\,0 \\
\underline{7\,6} \\
2\,4\,0 \\
\underline{2\,2\,8} \\
1\,2\,0 \\
\underline{1\,1\,4} \\
6\,0 \\
\underline{3\,8} \\
2\,2
\end{array}
$$

5263.1 rounds to 5263.
The school needs to collect 5263 box tops (rounded) during each of the 38 weeks.

63. $3.77 \div 10 = \underline{0.377}$ $\quad 9.1 \div 10 = \underline{0.91}$
$0.886 \div 10 = \underline{0.0886}$ $\quad 30.19 \div 10 = \underline{3.019}$
$406.5 \div 10 = \underline{40.65}$ $\quad 6625.7 \div 10 = \underline{662.57}$

(a) Dividing by 10, decimal point moves one place to the left; by 100, two places to the left; by 1000, three places to the left.

(b) The decimal point moved to the *right* when *multiplying* by 10 or 100 or 1000. Here the decimal point moves to the *left* when *dividing* by those numbers.

64. $40.2 \div 0.1 = \underline{402}$ $\quad 7.1 \div 0.1 = \underline{71}$
$0.339 \div 0.1 = \underline{3.39}$ $\quad 15.77 \div 0.1 = \underline{157.7}$
$46 \div 0.1 = \underline{460}$ $\quad 873 \div 0.1 = \underline{8730}$

(a) Dividing by 0.1, decimal point moves one place to the right; by 0.01, two places to the right; by 0.001, three places to the right.

(b) The decimal point moved to the *left* when *multiplying* by 0.1 or 0.01 or 0.001. Here the decimal point moves to the *right* when *dividing* by those numbers.

4.6 Fractions and Decimals

4.6 Margin Exercises

1. **(a)** $\frac{1}{9}$ is written $9\overline{)1.}$

(b) $\frac{2}{3}$ is written $3\overline{)2.}$

(c) $\frac{5}{4}$ is written $4\overline{)5.}$

(d) $\frac{3}{10}$ is written $10\overline{)3.}$

(e) $\frac{21}{16}$ is written $16\overline{)21.}$

(f) $\frac{1}{50}$ is written $50\overline{)1.}$

2. **(a)** $\frac{1}{4}$

$$
\begin{array}{r}
0.\,2\,5 \\
4\,\overline{)1.\,0\,0} \\
\underline{8} \\
2\,0 \\
\underline{2\,0} \\
0
\end{array}
$$

$\frac{1}{4} = 0.25$

(b) $2\frac{1}{2} = \frac{5}{2}$

$$
\begin{array}{r}
2.\,5 \\
2\,\overline{)5.\,0} \\
\underline{4} \\
1\,0 \\
\underline{1\,0} \\
0
\end{array}
$$

$2\frac{1}{2} = 2.5$

Copyright © 2018 Pearson Education, Inc.

(c) $\dfrac{5}{8}$

$$\begin{array}{r} 0.625 \\ 8\overline{)5.000} \\ 48 \\ \hline 20 \\ 16 \\ \hline 40 \\ 40 \\ \hline 0 \end{array}$$

$$\dfrac{5}{8} = 0.625$$

(d) $4\frac{3}{5}$

$$\begin{array}{r} 0.6 \\ 5\overline{)3.0} \\ 30 \\ \hline 0 \end{array}$$

$$4 + 0.6 = 4.6$$
$$4\tfrac{3}{5} = 4.6$$

(e) $\dfrac{7}{8}$

$$\begin{array}{r} 0.875 \\ 8\overline{)7.000} \\ 64 \\ \hline 60 \\ 56 \\ \hline 40 \\ 40 \\ \hline 0 \end{array}$$

$$\dfrac{7}{8} = 0.875$$

3. **(a)** $\frac{1}{3} = 1 \div 3$ Use a calculator.
Rounded to the nearest thousandth, $\frac{1}{3} \approx 0.333$.

(b) $2\frac{7}{9} = \frac{25}{9} = 25 \div 9$ Use a calculator.
Rounded to the nearest thousandth, $2\frac{7}{9} \approx 2.778$.

(c) $\frac{10}{11} = 10 \div 11$ Use a calculator.
Rounded to the nearest thousandth, $\frac{10}{11} \approx 0.909$.

(d) $\frac{3}{7} = 3 \div 7$ Use a calculator.
Rounded to the nearest thousandth, $\frac{3}{7} \approx 0.429$.

(e) $3\frac{5}{6} = \frac{23}{6} = 23 \div 6$ Use a calculator.
Rounded to the nearest thousandth, $3\frac{5}{6} \approx 3.833$.

4. **(a)** On the first number line, 0.4375 is to the *left* of 0.5, so use the $\boxed{<}$ symbol: $0.4375 < 0.5$

(b) On the first number line, 0.75 is to the *right* of 0.6875, so use the $\boxed{>}$ symbol:
$0.75 > 0.6875$

(c) $0.625 > 0.0625$

(d) $\dfrac{2}{8} = \dfrac{1}{4} = 0.25$

$0.25 < 0.375$, so $\dfrac{2}{8} < 0.375$.

(e) On the second number line, $0.8\overline{3}$ and $\frac{5}{6}$ are at the *same point*, so use the $\boxed{=}$ symbol:
$$0.8\overline{3} = \dfrac{5}{6}$$

(f) $\dfrac{1}{2} < 0.\overline{5}$ $\left(\frac{1}{2} = 0.5\right)$

(g) $0.\overline{1} < 0.1\overline{6}$

(h) $\dfrac{8}{9} = 0.\overline{8}$

(i) $\dfrac{4}{6} = \dfrac{2}{3} = 0.\overline{6}$

$0.\overline{7} > 0.\overline{6}$, so $0.\overline{7} > \dfrac{4}{6}$.

(j) $\dfrac{1}{4} = 0.25$

5. **(a)** 0.7   0.703   0.7029
         $\downarrow$ $\downarrow$ $\downarrow$
       0.7000   0.7030   0.7029

From least to greatest:
0.7000, 0.7029, 0.7030
or
0.7, 0.7029, 0.703

(b) 6.39   6.309   6.401   6.4
        $\downarrow$ $\downarrow$ $\downarrow$ $\downarrow$
       6.390   6.309   6.401   6.400

From least to greatest:
6.309, 6.390, 6.400, 6.401
or
6.309, 6.39, 6.4, 6.401

(c) 1.085 $1\frac{3}{4}$ 0.9
        $\downarrow$ $\downarrow$ $\downarrow$
       1.085   1.750   0.900

From least to greatest:
0.900, 1.085, 1.750
or
0.9, 1.085, $1\frac{3}{4}$

(d) $\frac{1}{4}, \frac{2}{5}, \frac{3}{7}, 0.428$

To compare, change fractions to decimals.

$$\dfrac{1}{4} = 0.250 \qquad \dfrac{2}{5} = 0.400 \qquad \dfrac{3}{7} \approx 0.429$$

From least to greatest:
0.250, 0.400, 0.428, 0.429
or
$\frac{1}{4}, \frac{2}{5}, 0.428, \frac{3}{7}$

Copyright © 2018 Pearson Education, Inc.

4.6 Section Exercises

1. (a) incorrect, $\frac{2}{5}$ is written $5\overline{)2}$.

  (b) correct

  (c) correct

3. $\frac{3}{4} \rightarrow 4\overline{)3.}^{\,0.}$

To continue dividing, write zeros in the dividend.

5. (a) $\frac{2}{5}\ (=0.4)$ is <u>less than</u> 0.5.

  (b) $\frac{3}{4}\ (=0.75)$ is <u>greater than</u> 0.6.

  (c) 0.2 is <u>less than</u> $\frac{5}{8}\ (=0.625)$.

7. $\frac{1}{2} = 0.5$
$$2\overline{)1.0}\;\;0.5$$
$$\underline{1\ 0}$$
$$0$$

9. $\frac{3}{4} = 0.75$
$$4\overline{)3.00}\;\;0.75$$
$$\underline{2\ 8}$$
$$2\ 0$$
$$\underline{2\ 0}$$
$$0$$

11. $\frac{3}{10} = 0.3$
$$10\overline{)3.0}\;\;0.3$$
$$\underline{3\ 0}$$
$$0$$

13. $\frac{9}{10} = 0.9$
$$10\overline{)9.0}\;\;0.9$$
$$\underline{9\ 0}$$
$$0$$

15. $\frac{3}{5} = 0.6$
$$5\overline{)3.0}\;\;0.6$$
$$\underline{3\ 0}$$
$$0$$

17. $\frac{7}{8} = 0.875$
$$8\overline{)7.000}\;\;0.875$$
$$\underline{6\ 4}$$
$$6\ 0$$
$$\underline{5\ 6}$$
$$4\ 0$$
$$\underline{4\ 0}$$
$$0$$

19. $2\frac{1}{4} = \frac{9}{4} = 2.25$
$$4\overline{)9.00}\;\;2.25$$
$$\underline{8}$$
$$1\ 0$$
$$\underline{8}$$
$$2\ 0$$
$$\underline{2\ 0}$$
$$0$$

21. $14\frac{7}{10} = 14 + \frac{7}{10} = 14 + 0.7 = 14.7$

23. $3\frac{5}{8} = \frac{29}{8} = 3.625$
$$8\overline{)29.000}\;\;3.625$$
$$\underline{2\ 4}$$
$$5\ 0$$
$$\underline{4\ 8}$$
$$2\ 0$$
$$\underline{1\ 6}$$
$$4\ 0$$
$$\underline{4\ 0}$$
$$0$$

25. $6\frac{1}{3} = \frac{19}{3} \approx 6.333$
$$3\overline{)19.000}\;\;6.333$$
$$\underline{1\ 8}$$
$$1\ 0$$
$$\underline{9}$$
$$1\ 0$$
$$\underline{9}$$
$$1\ 0$$
$$\underline{9}$$
$$1$$

27. $\frac{5}{6} \approx 0.8333333 \approx 0.833$

29. $1\frac{8}{9} = \frac{17}{9} \approx 1.8888889 \approx 1.889$

31. $0.4 = \frac{4}{10} = \frac{4 \div 2}{10 \div 2} = \frac{2}{5}$

33. $0.625 = \frac{625}{1000} = \frac{625 \div 125}{1000 \div 125} = \frac{5}{8}$

35. $0.35 = \frac{35}{100} = \frac{35 \div 5}{100 \div 5} = \frac{7}{20}$

37. $\frac{7}{20} = 0.35$
$$20\overline{)7.00}\;\;0.35$$
$$\underline{6\ 0}$$
$$1\ 00$$
$$\underline{1\ 00}$$
$$0$$

39. $0.04 = \frac{4}{100} = \frac{4 \div 4}{100 \div 4} = \frac{1}{25}$

41. $\dfrac{1}{5} = 0.2$

$$5\overline{\smash{\big)}1.0}$$
$$\underline{1\ 0}$$
$$0$$

43. $0.09 = \dfrac{9}{100}$

45. Compare the two lengths.
average length $\rightarrow$ 20.80 *longer*
Charlene's baby $\rightarrow$ 20.08 *shorter*

$$\begin{array}{r} 20.80 \\ -\ 20.08 \\ \hline 0.72 \end{array}$$

Her baby is 0.72 inch *shorter* than the average length.

47. Compare the two amounts.

$3\frac{3}{4} \rightarrow 3.75$ *less*
$3.8 \rightarrow 3.80$ *more*

$$\begin{array}{r} 3.80 \\ -\ 3.75 \\ \hline 0.05 \end{array}$$

3.8 inches is 0.05 inch *more* than Ginny hoped for.

49. Write two zeros to the right of 0.5 so it has the same number of decimal places as 0.505. Then you can compare the numbers: $0.505 > 0.500$. There was too much calcium in each capsule. Subtract to find the difference.

$$\begin{array}{r} 0.505 \\ -\ 0.500 \\ \hline 0.005 \end{array}$$

There was 0.005 gram too much.

51. Write zeros so that all the numbers have four decimal places. The acceptable lengths must be greater than 0.9980 cm and less than 1.0020 cm.

$1.0100 > 1.0020$ *unacceptable*
$0.9991 > 0.9980$ and $0.9991 < 1.0020$
acceptable
$1.0007 > 0.9980$ and $1.0007 < 1.0020$
acceptable
$0.9900 < 0.9980$ *unacceptable*

The lengths of 0.9991 cm and 1.0007 cm are acceptable.

53. 0.54, 0.5455, 0.5399

$0.54 = 0.5400$
$0.5455 = 0.5455$ *greatest*
$0.5399 = 0.5399$ *least*

From least to greatest: 0.5399, 0.54, 0.5455

55. 5.8, 5.79, 5.0079, 5.804

$5.8 = 5.8000$
$5.79 = 5.7900$
$5.0079 = 5.0079$ *least*
$5.804 = 5.8040$ *greatest*

From least to greatest: 5.0079, 5.79, 5.8, 5.804

57. 0.628, 0.62812, 0.609, 0.6009

$0.628 = 0.62800$
$0.62812 = 0.62812$ *greatest*
$0.609 = 0.60900$
$0.6009 = 0.60090$ *least*

From least to greatest:
0.6009, 0.609, 0.628, 0.62812

59. 5.8751, 4.876, 2.8902, 3.88

The numbers after the decimal places are irrelevant since the whole number parts are all different.

From least to greatest:
2.8902, 3.88, 4.876, 5.8751

61. 0.043, 0.051, 0.006, $\frac{1}{20}$

$0.043 = 0.043$
$0.051 = 0.051$ *greatest*
$0.006 = 0.006$ *least*
$\frac{1}{20} = 0.050$

From least to greatest: 0.006, 0.043, $\frac{1}{20}$, 0.051

63. $\frac{3}{8}$, $\frac{2}{5}$, 0.37, 0.4001

$\frac{3}{8} = 0.3750$
$\frac{2}{5} = 0.4000$
$0.37 = 0.3700$ *least*
$0.4001 = 0.4001$ *greatest*

From least to greatest: 0.37, $\frac{3}{8}$, $\frac{2}{5}$, 0.4001

65. Find the greatest of:
0.018, 0.01, 0.008, 0.010

$0.018 = 0.018$ *greatest*
$0.01 = 0.010$
$0.008 = 0.008$ *least*
$0.010 = 0.010$

(continued)

Copyright © 2018 Pearson Education, Inc.

List from least to greatest:

$0.008, 0.01 = 0.010, 0.018$

The red box, labeled 0.018 in. diameter, has the strongest line.

67. Convert $\frac{1}{125}$ to a decimal.

$$
\begin{array}{r}
0.008 \\
125\overline{\smash)1.000} \\
\underline{1\,000} \leftarrow 8 \times 125 \\
0
\end{array}
$$

Since $\frac{1}{125} = 0.008$, the box having line $\frac{1}{125}$ inch in diameter is the green box.

69. $1\frac{7}{16} = \frac{23}{16}$

$$
\begin{array}{r}
1.43 \\
16\overline{\smash)23.00} \\
\underline{1\,6} \\
7\,0 \\
\underline{6\,4} \\
6\,0 \\
\underline{4\,8} \\
1\,2
\end{array}
$$

1.43 rounded to the nearest tenth is 1.4.
Length (a) is 1.4 inches (rounded).

71. $\dfrac{1}{4} = 0.25$

0.25 rounded to the nearest tenth is 0.3.
Length (c) is 0.3 inch (rounded).

73. $\dfrac{3}{8}$

$$
\begin{array}{r}
0.37 \\
8\overline{\smash)3.00} \\
\underline{2\,4} \\
6\,0 \\
\underline{5\,6} \\
4
\end{array}
$$

0.37 rounded to the nearest tenth is 0.4.
Length (e) is 0.4 inch (rounded).

75. **(a)** A proper fraction like $\frac{5}{9}$ is less than 1, so it cannot be equivalent to a decimal number that is greater than 1.

(b) $\frac{5}{9}$ means $5 \div 9$ or $9\overline{\smash)5}$, so the correct answer is 0.556 (rounded). This makes sense because both the fraction and the decimal are less than 1.

$$
\begin{array}{r}
0.5555 \\
9\overline{\smash)5.0000} \\
\underline{4\,5} \\
5\,0 \\
\underline{4\,5} \\
5\,0 \\
\underline{4\,5} \\
5\,0 \\
\underline{4\,5} \\
5
\end{array}
$$

76. **(a)** She changed 0.35 to 0.035 when adding to 2.

(b) Adding the whole number part gives $2 + 0.35$, which is 2.35, not 2.035. To check, $2.35 = 2\frac{35}{100} = 2\frac{7}{20}$, but $2.035 = 2\frac{35}{1000} = 2\frac{7}{200}$.

77. Just add the whole number part to 0.375. So $1\frac{3}{8} = 1.375$; $3\frac{3}{8} = 3.375$; $295\frac{3}{8} = 295.375$.

78. It works only when the fraction part has a one-digit numerator and a denominator of 10, a two-digit numerator and a denominator of 100, and so on.

Chapter 4 Review Exercises

1. 2 4 3 . 0 5 9
(tenths over the 0, hundredths over the 5)

2. 0 . 6 8 1 7
(ones over the 0, tenths over the 6)

3. $5 8 2 4 . 3 9
(hundreds over the 8, hundredths over the 9)

4. 8 9 6 . 5 0 3
(tens over the 9, tenths over the 5)

5. 2 0 . 7 3 8 6 1
(tenths over the 7, ten-thousandths over the 6)

Copyright © 2018 Pearson Education, Inc.

6. $0.5 = \dfrac{5}{10} = \dfrac{1}{2}$

7. $0.75 = \dfrac{75}{100} = \dfrac{75 \div 25}{100 \div 25} = \dfrac{3}{4}$

8. $4.05 = 4\dfrac{5}{100} = 4\dfrac{5 \div 5}{100 \div 5} = 4\dfrac{1}{20}$

9. $0.875 = \dfrac{875}{1000} = \dfrac{875 \div 125}{1000 \div 125} = \dfrac{7}{8}$

10. $0.027 = \dfrac{27}{1000}$

11. $27.8 = 27\dfrac{8}{10} = 27\dfrac{8 \div 2}{10 \div 2} = 27\dfrac{4}{5}$

12.   0.8 is eight tenths.

13.   400.29 is four hundred and twenty-nine hundredths.

14.   12.007 is twelve and seven thousandths.

15.   0.0306 is three hundred six ten-thousandths.

16.   eight and three tenths:

$8\dfrac{3}{10} = 8.3$

17.   two hundred five thousandths:

$\dfrac{205}{1000} = 0.205$

18.   seventy and sixty-six ten-thousandths:

$70\dfrac{66}{10,000} = 70.0066$

19.   thirty hundredths:

$\dfrac{30}{100} = 0.30$

20.   275.635 to the nearest tenth
Draw a cut-off line: 275.6|35
The first digit cut is 3, which is 4 or less. The part you keep stays the same.
Answer: ≈ 275.6

21.   72.789 to the nearest hundredth
Draw a cut-off line: 72.78|9
The first digit cut is 9, which is 5 or more, so round up.
Answer: ≈ 72.79

22.   0.1604 to the nearest thousandth
Draw a cut-off line: 0.160|4
The first digit cut is 4, which is 4 or less. The part you keep stays the same.
Answer: ≈ 0.160

23.   0.0905 to the nearest thousandth
Draw a cut-off line: 0.090|5
The first digit cut is 5, which is 5 or more, so round up.
Answer: ≈ 0.091

24.   0.98 to the nearest tenth
Draw a cut-off line: 0.9|8
The first digit cut is 8, which is 5 or more, so round up.
Answer: ≈ 1.0

25. $15.8333 to the nearest cent
Draw a cut-off line: $15.83|33
The first digit cut is 3, which is 4 or less. The part you keep stays the same.
Answer: $\approx \$15.83$

26. $0.698 to the nearest cent
Draw a cut-off line: $0.69|8
The first digit cut is 8, which is 5 or more, so round up.
Answer: $\approx \$0.70$

27. $17,625.7906 to the nearest cent
Draw a cut-off line: $17,625.79|06
The first digit cut is 0, which is 4 or less. The part you keep stays the same.
Answer: $\approx \$17,625.79$

28. $350.48 to the nearest dollar
Draw a cut-off line: $350|.48
The first digit cut is 4, which is 4 or less. The part you keep stays the same.
Answer: $\approx \$350$

29. $129.50 to the nearest dollar
Draw a cut-off line: $129|.50
The first digit cut is 5, which is 5 or more, so round up.
Answer: $\approx \$130$

30. $99.61 to the nearest dollar
Draw a cut-off line: $99|.61
The first digit cut is 6, which is 5 or more, so round up.
Answer: $\approx \$100$

31. $29.37 to the nearest dollar
Draw a cut-off line: $29|.37
The first digit cut is 3, which is 4 or less. The part you keep stays the same.
Answer: $\approx \$29$

Copyright © 2018 Pearson Education, Inc.

32.

Estimate:	*Exact:*
6	5.81
400	423.96
+ 20	+ 15.09
426	444.86

33.

Estimate:	*Exact:*
80	75.600
1	1.290
100	122.045
1	0.880
+ 30	33.700
212	233.515

34.

Estimate:	*Exact:*
300	308.5
− 20	− 17.8
280	290.7

35.

Estimate:	*Exact:*
9	9.2000
− 8	− 7.9316
1	1.2684

36.

Estimate:		*Exact:*	
90	*cats*	85.9	*cats*
− 80	*dogs*	− 77.9	*dogs*
10	*difference*	8.0	*difference*

There are 8.0 million more pet cats than there are pet dogs in America.

37. Add the amounts of the two payments.

Estimate:	*Exact:*
$300	$315.53
+ 70	+ 74.67
$370	$390.20

The total amount of the two payments was $390.20.
Now subtract from her balance of $406.

Estimate:	*Exact:*
$400	$406.00
− 370	− 390.20
$30	$15.80

The new balance is $15.80.

38. First total the money that Joey spent.

Estimate:	*Exact:*
$2	$1.59
5	5.33
+ 20	+ 18.94
$27	$25.86

Then subtract to find the change.

Estimate:	*Exact:*
$30	$30.00
− 27	− 25.86
$3	$4.14

Joey's change was $4.14.

39. Add the kilometers that she raced each day.

Estimate:	*Exact:*
2	2.30
4	4.00
+ 5	+ 5.25
11 km	11.55 km

Roseanne raced 11.55 kilometers.

40.

Estimate:	*Exact:*
6	6.138
× 4	× 3.7
24	4 2966
	18 414
	22.7106

41.

Estimate:	*Exact:*
40	42.9
× 3	× 3.3
120	12 87
	128 7
	141.57

42. $(5.6)(0.002)$

5.6	← 1 *decimal place*
× 0.002	← 3 *decimal places*
0.0112	← 4 *decimal places*

43. $0.071(0.005)$

0.071	← 3 *decimal places*
× 0.005	← 3 *decimal places*
0.000 355	← 6 *decimal places*

44. $706.2 \div 12 = 58.85$

Estimate: $700 \div 10 = 70$

58.85 is *reasonable.*

Copyright © 2018 Pearson Education, Inc.

45. $26.6 \div 2.8 = 0.95$

Estimate: $30 \div 3 = 10$

0.95 is *not reasonable.*

$$
\begin{array}{r}
9.\;5 \\
2.8_\wedge\overline{)2\;6.\;6_\wedge\;0} \\
2\;5\;2 \\
\hline
1\;4\;\;\;0 \\
1\;4\;\;\;0 \\
\hline
0
\end{array}
$$

Move decimal point 1 place in divisor and dividend; write one zero.

The correct answer is 9.5.

46.
$$
\begin{array}{r}
1\;4.\;4\;6\;6\;6 \\
3\overline{)4\;3.\;4\;0\;0\;0} \\
3 \\
\hline
1\;3 \\
1\;2 \\
\hline
1\;4 \\
1\;2 \\
\hline
2\;0 \\
1\;8 \\
\hline
2\;0 \\
1\;8 \\
\hline
2\;0 \\
1\;8 \\
\hline
2
\end{array}
$$

Write one zero.
Write one zero.
Write one zero.

14.4666 rounds to 14.467.

47. $\dfrac{72}{0.06} = \dfrac{7200}{6} = 1200$

48. $0.00048 \div 0.0012 = 4.8 \div 12$

$$
\begin{array}{r}
0.\;4 \\
12\overline{)4.\;8} \\
4\;8 \\
\hline
0
\end{array}
$$

49. Multiply the hourly wage times the hours worked.

Pay for first 40 hours	Pay rate after 40 hours	Pay for over 40 hours
14.24	14.24	21.36
× 40	× 1.5	× 6.5
569.60	7 120	10 680
	14 24	128 16
	21.360	138.840

Add the two amounts.

$569.60
 138.84
─────────
$708.44

Adrienne's total earnings were $708 (rounded).

50. Divide the cost of the book by the number of tickets in the book.

$$
\begin{array}{r}
1.\;9\;9 \\
30\overline{)\$5\;9.\;7\;5} \\
3\;0 \\
\hline
2\;9\;\;\;7 \\
2\;7\;\;\;0 \\
\hline
2\;\;\;7\;0 \\
2\;\;\;7\;0 \\
\hline
0
\end{array}
$$

Each ticket costs $1.99.

51. Divide the amount of the investment by the price per share.

$$
\begin{array}{r}
1\;3\;3.\;3 \\
3.75_\wedge\overline{)5\;0\;0.\;0\;0\;0_\wedge\;0} \\
3\;7\;5 \\
\hline
1\;2\;5\;\;0 \\
1\;1\;2\;\;5 \\
\hline
1\;2\;\;5\;0 \\
1\;1\;\;2\;5 \\
\hline
1\;\;2\;5\;0 \\
1\;\;1\;2\;5 \\
\hline
1\;2\;5
\end{array}
$$

Round 133.3 to the nearest whole number. Kenneth could buy 133 shares.

52. Multiply the price per pound by the amount to be purchased.

$0.99
 × 3.5
─────
  495
 2 97
─────
$3.465

$3.465 rounds to $3.47.
Ms. Lee will pay $3.47.

53. $3.5^2 + 8.7(1.95)$ *Exponent*
      $12.25 + 8.7(1.95)$ *Multiply.*
      $12.25 + 16.965 = 29.215$ *Add.*

54. $11 - 3.06 \div (3.95 - 0.35)$ *Parentheses*
      $11 - 3.06 \div 3.6$ *Divide.*
      $11 - 0.85 = 10.15$ *Subtract.*

55. $3\dfrac{4}{5} = \dfrac{19}{5} = 3.8$

$$
\begin{array}{r}
3.\;8 \\
5\overline{)1\;9.\;0} \\
1\;5 \\
\hline
4\;\;0 \\
4\;\;0 \\
\hline
0
\end{array}
$$

Copyright © 2018 Pearson Education, Inc.

56. $\dfrac{16}{25} = 0.64$

$$\begin{array}{r} 0.64 \\ 25\overline{)16.00} \\ \underline{15\ 0} \\ 1\ 00 \\ \underline{1\ 00} \\ 0 \end{array}$$

57. $1\dfrac{7}{8} = \dfrac{15}{8} = 1.875$

$$\begin{array}{r} 1.875 \\ 8\overline{)15.000} \\ \underline{8} \\ 7\ 0 \\ \underline{6\ 4} \\ 6\ 0 \\ \underline{5\ 6} \\ 4\ 0 \\ \underline{4\ 0} \\ 0 \end{array}$$

58. $\dfrac{1}{9}$

$$\begin{array}{r} 0.1111 \\ 9\overline{)1.0000} \\ \underline{9} \\ 1\ 0 \\ \underline{9} \\ 1\ 0 \\ \underline{9} \\ 1\ 0 \\ \underline{9} \\ 1 \end{array}$$

0.1111 rounds to 0.111.

$\dfrac{1}{9} \approx 0.111$

59. 3.68, 3.806, 3.6008

$3.68 = 3.6800$
$3.806 = 3.8060$ *greatest*
$3.6008 = 3.6008$ *least*

From least to greatest:
3.6008, 3.68, 3.806

60. 0.215, 0.22, 0.209, 0.2102

$0.215 = 0.2150$
$0.22 = 0.2200$ *greatest*
$0.209 = 0.2090$ *least*
$0.2102 = 0.2102$

From least to greatest:
0.209, 0.2102, 0.215, 0.22

61. $0.17, \dfrac{3}{20}, \dfrac{1}{8}, 0.159$

$0.17 = 0.170$ *greatest*

$\dfrac{3}{20} = 0.150$

$\dfrac{1}{8} = 0.125$ *least*

$0.159 = 0.159$

From least to greatest:

$\dfrac{1}{8}, \dfrac{3}{20}, 0.159, 0.17$

Chapter 4 Mixed Review Exercises

1. $89.19 + 0.075 + 310.6 + 5$

$$\begin{array}{r} \overset{11\ \ 1}{} \\ 89.190 \quad \textit{Line up decimal points.} \\ 0.075 \quad \textit{Write in zeros.} \\ 310.600 \\ \underline{+\ 5.000} \\ 404.865 \end{array}$$

2. 72.8×3.5

$$\begin{array}{r} 72.8 \quad \leftarrow \textit{1 decimal place} \\ \times\ 3.5 \quad \leftarrow \textit{1 decimal place} \\ \hline 36\ 40 \\ 218\ 4 \\ \hline 254.80 \quad \leftarrow \textit{2 decimal places} \end{array}$$

We may omit the last zero in this case.

3. $1648.3 \div 0.46 \approx 3583.2609$
3583.2609 rounds to 3583.261.

4.
$$\begin{array}{r} \overset{2\ \ 9\ \ 9\ \ 9\ \ 9\ \ 10}{\cancel{3}\,\cancel{0}.\cancel{0}\,\cancel{0}\,\cancel{0}\,\cancel{0}} \\ -\ 0.9102 \\ \hline 29.0898 \end{array}$$

5. $4.38(0.007)$

$$\begin{array}{r} 4.38 \quad \leftarrow \textit{2 decimal places} \\ \times\ 0.007 \quad \leftarrow \textit{3 decimal places} \\ \hline 0.03066 \quad \leftarrow \textit{5 decimal places} \end{array}$$

6.
$$\begin{array}{r} 9.4 \\ 0.005_\wedge\overline{)0.047_\wedge 0} \\ \underline{45} \\ 2\ 0 \\ \underline{2\ 0} \\ 0 \end{array}$$
Move decimal point 3 places in divisor and dividend; write 0.

Copyright © 2018 Pearson Education, Inc.

7. $72.105 + 8.2 + 95.37$

$$
\begin{array}{r}
72.105 \\
8.200 \\
+\,95.370 \\
\hline
175.675
\end{array}
$$

Line up decimal points.
Write in zeros.

8. $81.36 \div 9$

$$
\begin{array}{r}
9.04 \\
9\overline{)81.36} \\
81 \\
\hline
03 \\
0 \\
\hline
36 \\
36 \\
\hline
0
\end{array}
$$

9. $(5.6 - 1.22) + 4.8(3.15)$ *Parentheses*
$4.38 + 4.8(3.15)$ *Multiply.*
$4.38 + 15.12 = 19.50$ *Add.*

10. 0.455×18

$$
\begin{array}{r}
0.455 \\
\times\ \ 18 \\
\hline
3\,640 \\
4\,55 \\
\hline
8.190
\end{array}
$$

$\leftarrow$ *3 decimal places*
$\leftarrow$ *0 decimal places*

$\leftarrow$ *3 decimal places*

We may omit the last zero in this case.

11. $(1.6)(0.58)$

$$
\begin{array}{r}
0.58 \\
\times\ 1.6 \\
\hline
348 \\
58 \\
\hline
0.928
\end{array}
$$

$\leftarrow$ *2 decimal places*
$\leftarrow$ *1 decimal place*

$\leftarrow$ *3 decimal places*

12. $0.218\overline{)7.63}$

$7.63 \div 0.218 = 35$

13. $21.059 - 20.8$

$$
\begin{array}{r}
21.059 \\
-\,20.800 \\
\hline
0.259
\end{array}
$$

Line up decimal points.
Write in zeros.

14. $18.3 - 3^2 \div 0.5$ *Exponent*
$18.3 - 9 \div 0.5$ *Divide.*
$18.3 - 18 = 0.3$ *Subtract.*

15. Men's socks are 3 pairs for $8.99. Divide the price by 3 to find the cost of one pair.

$$
\begin{array}{r}
2.996 \\
3\overline{)8.990} \\
6 \\
\hline
29 \\
27 \\
\hline
29 \\
27 \\
\hline
20 \\
18 \\
\hline
2
\end{array}
$$

$2.996 rounds to $3.00.

One pair of men's socks costs $3.00 (rounded).

16. Children's socks cost 6 pairs for $5.00. Find the cost for one pair.

$$
\begin{array}{r}
0.833 \\
6\overline{)5.000} \\
48 \\
\hline
20 \\
18 \\
\hline
20 \\
18 \\
\hline
2
\end{array}
$$

$0.833 rounds to $0.83.

Subtract to find how much more men's socks cost.

$$
\begin{array}{r}
\$3.00 \\
-\,0.83 \\
\hline
\$2.17
\end{array}
$$

Men's socks cost $2.17 (rounded) more.

17. A dozen pair of socks is $4 \cdot 3 = 12$ pairs of socks. So multiply 4 times $8.99.

$$
\begin{array}{r}
\$8.99 \\
\times\ 4 \\
\hline
\$35.96
\end{array}
$$

The cost for a dozen pair of men's socks is $35.96.

18. Five pairs of teen jeans cost $19.95 times 5.

$$
\begin{array}{r}
\$19.95 \\
\times\ 5 \\
\hline
\$99.75
\end{array}
$$

(continued)

Copyright © 2018 Pearson Education, Inc.

Four pairs of women's jeans cost $24.99 times 4.

$24.99
$\underline{\times\ \ \ \ \ 4}$
$99.96

Add the two amounts.

$99.75
$\underline{+\ 99.96}$
$199.71

Akiko would pay $199.71.

19. The highest regular price for athletic shoes is $149.50. The cheapest sale price is $71.

Subtract to find the difference.

$149.50
$\underline{-\ 71.00}$
$78.50

The difference in price is $78.50.

Chapter 4 Test

1. $18.4 = 18\dfrac{4}{10} = 18\dfrac{4 \div 2}{10 \div 2} = 18\dfrac{2}{5}$

2. $0.075 = \dfrac{75}{1000} = \dfrac{75 \div 25}{1000 \div 25} = \dfrac{3}{40}$

3. 60.007 is sixty and seven thousandths.

4. 0.0208 is two hundred eight ten-thousandths.

5. 725.6089 to the nearest tenth
Draw a cut-off line: 725.6|089
The first digit cut is 0, which is 4 or less. The part you keep stays the same.

Answer: ≈ 725.6

6. 0.62951 to the nearest thousandth
Draw a cut-off line: 0.629|51
The first digit cut is 5, which is 5 or more, so round up.

Answer: ≈ 0.630

7. $1.4945 to the nearest cent
Draw a cut-off line: $1.49|45
The first digit cut is 4, which is 4 or less. The part you keep stays the same.

Answer: $\approx \$1.49$

8. $7859.51 to the nearest dollar
Draw a cut-off line: $7859|.51
The first digit cut is 5, which is 5 or more, so round up.

Answer: $\approx \$7860$

9. $7.6 + 82.0128 + 39.59$

Estimate:	Exact:	
8	7.6000	*Line up decimals.*
80	82.0128	*Write in zeros.*
$\underline{+\ 40}$	$\underline{+\ 39.5900}$	
128	129.2028	

10. $79.1 - 3.602$

Estimate:	Exact:	
80	79.100	*Line up decimals.*
$\underline{-\ \ 4}$	$\underline{-\ 3.602}$	*Write in zeros.*
76	75.498	

11. $5.79(1.2)$

Estimate:	Exact:
6	5.79
$\underline{\times\ 1}$	$\underline{\times\ 1.2}$
6	1 158
	5 79
	6.948

12. $20.04 \div 4.8$

Estimate: Exact:

$$4.8_\wedge \overline{)20.0_\wedge 400}$$
$$\begin{array}{r} 4.175 \\ \hline 19\ 2 \\ \hline 8\ 4 \\ 4\ 8 \\ \hline 3\ 60 \\ 3\ 36 \\ \hline 2\ 40 \\ 2\ 40 \\ \hline 0 \end{array}$$

Estimate: $5\overline{)20}$ → 4

13. $53.1 + 4.631 + 782 + 0.031$

53.100	*Line up decimals.*
4.631	*Write in zeros.*
782.000	
$\underline{+\ 0.031}$	
839.762	

14. $670 - 0.996$

670.000	*Line up decimals.*
$\underline{-\ 0.996}$	*Write in zeros.*
669.004	

15. $(0.0069)(0.007)$

0.0069	$\leftarrow$	4 *decimal places*
$\underline{\times\ \ \ 0.007}$	$\leftarrow$	3 *decimal places*
0.0000483	$\leftarrow$	7 *decimal places*

Copyright © 2018 Pearson Education, Inc.

16.

$$
\begin{array}{r}
4\ 8\ 0.\\
0.15_\wedge\overline{)7\ 2.\ 0\ 0_\wedge}\\
\underline{6\ 0}\\
1\ 2\ 0\\
\underline{1\ 2\ 0}\\
0\ 0\\
\underline{0}\\
0
\end{array}
$$

Move decimal point 2 places in dividend and divisor; write two zeros.

17. $2\frac{5}{8} = \frac{21}{8}$

$$
\begin{array}{r}
2.\ 6\ 2\ 5\\
8\overline{)2\ 1.\ 0\ 0\ 0}\\
\underline{1\ 6}\\
5\ 0\\
\underline{4\ 8}\\
2\ 0\\
\underline{1\ 6}\\
4\ 0\\
\underline{4\ 0}\\
0
\end{array}
$$

$2\frac{5}{8} = 2.625$

18. $0.44, 0.451, \frac{9}{20}, 0.4506$

Change to ten-thousandths, if necessary, and compare.

$0.44 = 0.4400$ *least*

$0.451 = 0.4510$ *greatest*

$\frac{9}{20} = 0.4500$

$0.4506 = 0.4506$

From least to greatest: $0.44, \frac{9}{20}, 0.4506, 0.451$

19.
$6.3^2 - 5.9 + 3.4(0.5)$	*Exponent*
$39.69 - 5.9 + 3.4(0.5)$	*Multiply.*
$39.69 - 5.9 + 1.7$	*Subtract.*
$33.79 + 1.7 = 35.49$	*Add.*

20.   To determine how much she paid, add the cost of the PlayStation, the two extra controllers, and the sales tax, and then subtract the discount.

Estimate:
$300 + (2 \times \$50) + \$20 - \$80 = \$340$

The exact cost of the two controllers is
$2 \times \$49.99 = 99.98$.

Exact:

$$
\begin{array}{rl}
\$299.99 & \textit{original price}\\
+\ \$99.98 & \textit{extra controllers}\\
\hline
\$399.97 & \\
+\ 22.75 & \textit{sales tax}\\
\hline
\$422.72 & \\
-\ 75.00 & \textit{discount}\\
\hline
\$347.72 & \textit{new price}
\end{array}
$$

She paid $347.72.

21.   2.500 million *gadwalls*
2.551 million *wigeons*
2.560 million *pintails*
From the greatest number to the least:
pintails, wigeons, gadwalls

22.   Multiply the price per pound by the amount purchased.

$$
\begin{array}{r}
\$2.89\\
\times\ \ 1.85\\
\hline
1445\\
2\ 312\\
2\ 89\\
\hline
\$5.3465
\end{array}
$$

Round $5.3465 to the nearest cent.
Draw a cut-off line: $5.34|65$
The first digit cut is 6, which is 5 or more, so round up.

Mr. Yamamoto paid $5.35 for the cheese.

23.   Subtract the two temperatures.

$$
\begin{array}{r}
102.7\\
-\ 99.9\\
\hline
2.8
\end{array}
$$

The temperature dropped 2.8 degrees.

24.   Divide to find the cost per foot.

$$
\begin{array}{r}
3.\ 7\ 9\\
6.5_\wedge\overline{)\$2\ 4.\ 6_\wedge\ 4\ 0}\\
\underline{1\ 9\ 5}\\
5\ 1\ 4\\
\underline{4\ 5\ 5}\\
5\ 9\ 0\\
\underline{5\ 8\ 5}\\
5
\end{array}
$$

Move decimal point 1 place in dividend and divisor; write one zero.

The chain costs $3.79 per foot (rounded).

25.   Answers will vary.

Copyright © 2018 Pearson Education, Inc.

Cumulative Review Exercises (Chapters 1–4)

1. 499,501 to the nearest thousand
    Draw a cut-off line: 499|501
    The first digit cut is 5, which is 5 or more, so round up.

    Answer: $\approx 500{,}000$

2. 602.4937 to the nearest hundredth
    Draw a cut-off line: 602.49|37
    The first digit cut is 3, which is 4 or less. The part you keep stays the same.

    Answer: ≈ 602.49

3. $709.60 to the nearest dollar
    Draw a cut-off line: $709|.60
    The first digit cut is 6, which is 5 or more, so round up.

    Answer: $\approx \$710$

4. $0.0528 to the nearest cent
    Draw a cut-off line: $0.05|28
    The first digit cut is 2, which is 4 or less. The part you keep stays the same.

    Answer: $\approx \$0.05$

5. $10 - 0.329$

    $$\begin{array}{r} 10.000 \\ -\ 0.329 \\ \hline 9.671 \end{array}\quad \begin{array}{l}\textit{Line up decimals.}\\ \textit{Write in zeros.}\end{array}$$

6. $2\dfrac{3}{5} \cdot \dfrac{5}{9} = \dfrac{13}{5} \cdot \dfrac{5}{9} = \dfrac{13}{\cancel{5}} \cdot \dfrac{\overset{1}{\cancel{5}}}{9} = \dfrac{13}{9} = 1\dfrac{4}{9}$

7. $11\dfrac{1}{5} \div 8 = \dfrac{56}{5} \div \dfrac{8}{1} = \dfrac{56}{5} \cdot \dfrac{1}{\cancel{8}} = \dfrac{7}{5} = 1\dfrac{2}{5}$

8. $5006 - 92$

    $$\begin{array}{r} 5006 \\ -\ 92 \\ \hline 4914 \end{array}$$

9. $0.7 + 85 + 7.903$

    $$\begin{array}{r} 0.700 \\ 85.000 \\ +\ 7.903 \\ \hline 93.603 \end{array}\quad \begin{array}{l}\textit{Line up decimals.}\\ \textit{Write in zeros.}\end{array}$$

10.
$$\begin{array}{r} 4\ 0\ 4\ \ \textbf{R3} \\ 7\,\overline{)2\ 8\ 3\ 1} \\ \underline{2\ 8}\ \ \ \ \ \ \\ 0\ 3\ \ \ \\ \underline{0}\ \ \ \ \\ 3\ 1 \\ \underline{2\ 8} \\ 3 \end{array}$$

11. $\dfrac{5}{6} + \dfrac{7}{8}$

    $$\begin{array}{r} \dfrac{5}{6} = \dfrac{20}{24} \\ +\ \dfrac{7}{8} = \dfrac{21}{24} \\ \hline \dfrac{41}{24} = 1\dfrac{17}{24} \end{array}$$

12. 332×704

    $$\begin{array}{r} 332 \\ \times\ 704 \\ \hline 1\,328 \\ 232\,40\ \ \\ \hline 233{,}728 \end{array}$$

13. $(0.006)(5.44)$

    $$\begin{array}{r} 0.006 \quad \leftarrow\ \textit{3 decimal places} \\ \times\ 5.44 \quad \leftarrow\ \textit{2 decimal places} \\ \hline 24 \\ 24 \\ 30 \\ \hline 0.03264 \quad \leftarrow\ \textit{5 decimal places needed} \end{array}$$

14. $3.2(2.5)$

    $$\begin{array}{r} 3.2 \quad \leftarrow\ \textit{1 decimal place} \\ \times\ 2.5 \quad \leftarrow\ \textit{1 decimal place} \\ \hline 1\,60 \\ 6\,4\ \ \\ \hline 8.00 \quad \leftarrow\ \textit{2 decimal places} \end{array}$$

    We may omit the last two zeros in this case.

15. $25.2 \div 0.56$

    $$\begin{array}{r} 4\ 5 \\ 0.56_\wedge\,\overline{)2\ 5.\ 2\ 0_\wedge} \\ \underline{2\ 2\ 4}\ \ \ \\ 2\ 8\ 0 \\ \underline{2\ 8\ 0} \\ 0 \end{array}$$

16. $\dfrac{2}{3} \div 5\dfrac{1}{6} = \dfrac{2}{3} \div \dfrac{31}{6} = \dfrac{2}{\cancel{3}} \cdot \dfrac{\overset{2}{\cancel{6}}}{31} = \dfrac{4}{31}$

Copyright © 2018 Pearson Education, Inc.

17. $5\frac{1}{4} - 4\frac{7}{12}$

$$5\frac{1}{4} \;=\; 5\frac{3}{12} \;=\; 4\frac{15}{12}$$
$$-4\frac{7}{12} \;=\; 4\frac{7}{12} \;=\; 4\frac{7}{12}$$
$$\frac{8}{12} = \frac{2}{3}$$

18. $4.7 \div 9.3$

$$
\begin{array}{r}
0.\,5\,0\,5 \\
9.3_\wedge\overline{)4.\,7_\wedge 0\,0\,0} \\
4\;6\;5 \\
\hline
5\;0\;0 \\
4\;6\;5 \\
\hline
3\;5
\end{array}
$$

Round 0.505 to the nearest hundredth.
Draw a cut-off line: 0.50|5
The first digit cut is 5, which is 5 or more, so round up.

Answer: ≈ 0.51

19. $10 - 4 \div 2 \cdot 3$ *Divide.*
$\;\;\;\;10 - 2 \cdot 3$ *Multiply.*
$\;\;\;\;10 - 6 = 4$ *Subtract.*

20. $\sqrt{36} + 3(8) - 4^2$ *Exponent*
$\;\;\;\;\sqrt{36} + 3(8) - 16$ *Square Root*
$\;\;\;\;6 + 3(8) - 16$ *Multiply.*
$\;\;\;\;6 + 24 - 16$ *Add.*
$\;\;\;\;30 - 16 = 14$ *Subtract.*

21.   40.035 is forty and thirty-five thousandths.

22.   Three hundred six ten-thousandths is 0.0306.

23.   7.005, 7.5005, 7.5, 7.505

$\;\;\;\;7.005 = 7.0050$ *least*
$\;\;\;\;7.5005 = 7.5005$
$\;\;\;\;\;\;\;7.5 = 7.5000$
$\;\;\;\;7.505 = 7.5050$ *greatest*

From least to greatest:
7.005, 7.5, 7.5005, 7.505

24. $\frac{7}{8}, 0.8, \frac{21}{25}, 0.8015$

$\;\;\;\;\dfrac{7}{8} = 0.8750$ *greatest*

$\;\;\;\;0.8 = 0.8000$ *least*

$\;\;\;\;\dfrac{21}{25} = 0.8400$

$\;\;\;\;0.8015 = 0.8015$

From least to greatest:
0.8, 0.8015, $\frac{21}{25}$, $\frac{7}{8}$

25.   He started with three \$20 bills, or \$60. He spent \$47.96 and \$0.87.

Estimate: \$60 − \$50 − \$1 = \$9

First find how much money Lameck spent.

$$
\begin{array}{r}
\$47.96 \\
+\,0.87 \\
\hline
\$48.83
\end{array}
$$

Now subtract from \$60.

$$
\begin{array}{r}
\$60.00 \\
-\,48.83 \\
\hline
\$11.17
\end{array}
$$

He has \$11.17 left.

26.   Subtract the two measurements.

Estimate: 50 − 47 = 3 inches

$$
\begin{array}{rcl}
\textit{Exact:} \quad 50 & = & 49\frac{8}{8} \\
-46\frac{5}{8} & = & 46\frac{5}{8} \\
\hline
& & 3\frac{3}{8}
\end{array}
$$

She has grown $3\frac{3}{8}$ inches.

27.   Multiply the hourly wage, \$11.63, by the hours worked, 16.5.

Estimate: \$10 × 20 = \$200
Exact: \$11.63

$$
\begin{array}{r}
\$11.63 \\
\times\;16.5 \\
\hline
5\;815 \\
69\;78\;\; \\
116\;3\;\;\; \\
\hline
191.895
\end{array}
$$

\$191.895 rounded to the nearest cent is \$191.90. Sharon earned \$191.90 (rounded).

28.   Find the number of students in each size of classroom (8 have 22 students and 12 have 26 students). Then add the two sums.

Estimate: $(8 \times 20) + (10 \times 30) = 160 + 300$
$\; = 460$ students

$$
\textit{Exact:}\quad
\begin{array}{r}
22 \\
\times\;8 \\
\hline
176
\end{array}
\qquad
\begin{array}{r}
26 \\
\times\;12 \\
\hline
52 \\
26\;\; \\
\hline
312
\end{array}
\qquad
\begin{array}{r}
176 \\
+\,312 \\
\hline
488
\end{array}
$$

There are 488 students attending the school.

Copyright © 2018 Pearson Education, Inc.

29. Add the number of yards of fabric in two purchases.

Estimate: $2 + 4 = 6$ yards

Exact:
$$2\frac{1}{3} = 2\frac{8}{24}$$
$$+3\frac{7}{8} = 3\frac{21}{24}$$
$$5\frac{29}{24} = 5 + 1 + \frac{5}{24}$$
$$= 6\frac{5}{24}$$

Toshihiro bought $6\frac{5}{24}$ yards of fabric.

30. Add the deposit to the amount Kimberly had in her account. Then subtract the payment amount for gas and the charge for insufficient funds.

Estimate: $\$30 + \$200 - \$40 - \$40 = \$150$

Exact:

$\$29.44$	*Balance on hand*
$+ \$220.06$	*Deposit*
$\$249.50$	
$- 40.00$	*Gas purchase*
$\$209.50$	
$- 35.00$	*Insuff. Funds charge*
$\$174.50$	*New Balance*

The new balance in her account is $174.50.

31. Divide the price of the apples, $6.18, by the number of pounds of apples, 2.7, that Paulette bought.

Estimate: $\$6 \div \$3 = \$2$ per pound

Exact:
```
          2. 2 8 8
2.7⌐$6. 1⌐8 0 0    Move decimal point
     5  4           one place in divisor
     7  8           and dividend.
     5  4
     2  4 0
     2  1 6
        2 4 0
        2 1 6
          2 4
```

2.288 rounds to 2.29.

The cost of the apples is $2.29 per pound (rounded).

32. Divide the amount of the grant by the number of students.

Estimate: $\$80,000 \div 100 = \$800$

Exact:
```
           7 2 8. 9
107⌐$7 8, 0 0 0. 0
    7 4 9
    3 1 0
    2 1 4
      9 6 0
      8 5 6
      1 0 4 0
        9 6 3
          7 7
```

728.9 rounds to 729.

Each student could be given $729 (rounded).

33. $20\frac{1}{4} < 21\frac{1}{16} < 21\frac{1}{8}$

Order size M (medium).

34. **Size XXS :**
$$18 = 17\frac{2}{2}$$
$$-16\frac{1}{2} = 16\frac{1}{2}$$
$$1\frac{1}{2}$$

The difference for size XXS is $1\frac{1}{2}$ inches.

Size XS:
$$19 = 18\frac{4}{4}$$
$$-18\frac{1}{4} = 18\frac{1}{4}$$
$$\frac{3}{4}$$

The difference for size XS is $\frac{3}{4}$ inch.

Size S:
$$20 = 19\frac{4}{4}$$
$$-19\frac{1}{4} = 19\frac{1}{4}$$
$$\frac{3}{4}$$

The difference for size S is $\frac{3}{4}$ inch.

Size M:
$$21\frac{1}{8} = 21\frac{1}{8} = 20\frac{9}{8}$$
$$-20\frac{1}{4} = 20\frac{2}{8} = 20\frac{2}{8}$$
$$\frac{7}{8}$$

The difference for size M is $\frac{7}{8}$ inch.

Size L:
$$22\frac{1}{4} = 22\frac{1}{4} = 21\frac{5}{4}$$
$$-21\frac{1}{2} = 21\frac{2}{4} = 21\frac{2}{4}$$
$$\frac{3}{4}$$

The difference for size L is $\frac{3}{4}$ inch.

Copyright © 2018 Pearson Education, Inc.

35. $\frac{1}{8} = 0.125$

$$
\begin{array}{r}
0.\ 1\ 2\ 5 \\
8\overline{)1.\ 0\ 0\ 0} \\
\underline{8} \\
2\ 0 \\
\underline{1\ 6} \\
4\ 0 \\
\underline{4\ 0} \\
0
\end{array}
$$

$\frac{1}{4} = 0.25$

$$
\begin{array}{r}
0.\ 2\ 5 \\
4\overline{)1.\ 0\ 0} \\
\underline{8} \\
2\ 0 \\
\underline{2\ 0} \\
0
\end{array}
$$

$$
\begin{array}{rcl}
21\frac{1}{8} & = & 21.125 \\
-20\frac{1}{4} & = & 20.250 \\
\hline
& & 0.875 \text{ in.}
\end{array}
$$

$0.875 = \dfrac{875}{1000} = \dfrac{875 \div 125}{1000 \div 125} = \dfrac{7}{8}$ inch,

which is the same as the answer in Exercise 34.

36. Answers will vary. Many people prefer using decimals because you do not need to find a common denominator or rewrite answers in lowest terms.

Copyright © 2018 Pearson Education, Inc.

CHAPTER 5   RATIO AND PROPORTION

5.1 Ratios

5.1 Margin Exercises

1. **(a)** $7 spent on fresh fruit to $5 spent on milk

Spent on fruit $\rightarrow \quad \dfrac{7}{5}$
Spent on milk $\rightarrow$

(b) $5 spent on milk to $14 spent on meat

Spent on milk $\rightarrow \quad \dfrac{5}{14}$
Spent on meat $\rightarrow$

(c) $14 spent on meat to $5 spent on milk

Spent on meat $\rightarrow \quad \dfrac{14}{5}$
Spent on milk $\rightarrow$

2. **(a)** 9 hours to 12 hours

$$\frac{9 \text{ hours}}{12 \text{ hours}} = \frac{9}{12} = \frac{9 \div 3}{12 \div 3} = \frac{3}{4}$$

(b) 100 meters to 50 meters

$$\frac{100 \text{ meters}}{50 \text{ meters}} = \frac{100}{50} = \frac{100 \div 50}{50 \div 50} = \frac{2}{1}$$

(c) width to length or 24 feet to 48 feet

$$\frac{24 \text{ feet}}{48 \text{ feet}} = \frac{24}{48} = \frac{24 \div 24}{48 \div 24} = \frac{1}{2}$$

3. **(a)** The increase in price is
$7.00 - $5.50 = $1.50.

The ratio of increase in price to original price is

$$\frac{1.50}{5.50}.$$

Rewrite as two whole numbers.

$$\frac{1.50}{5.50} = \frac{1.50 \times 100}{5.50 \times 100} = \frac{150}{550}$$

Write in lowest terms.

$$\frac{150 \div 50}{550 \div 50} = \frac{3}{11}$$

(b) The decrease in hours is
4.5 hours $-$ 3 hours = 1.5 hours.

The ratio of decrease in hours to the original number of hours is

$$\frac{1.5}{4.5}.$$

Rewrite as two whole numbers.

$$\frac{1.5 \cdot 10}{4.5 \cdot 10} = \frac{15}{45}$$

Write in lowest terms.

$$\frac{15 \div 15}{45 \div 15} = \frac{1}{3}$$

4. **(a)** $3\frac{1}{2}$ to 4

$$\frac{3\frac{1}{2}}{4} = \frac{\frac{7}{2}}{\frac{4}{1}} = \frac{7}{2} \div \frac{4}{1} = \frac{7}{2} \cdot \frac{1}{4} = \frac{7}{8}$$

(b) $5\frac{5}{8}$ pounds to $3\frac{3}{4}$ pounds

$$\frac{5\frac{5}{8}}{3\frac{3}{4}} = \frac{\frac{45}{8}}{\frac{15}{4}} = \frac{45}{8} \div \frac{15}{4} = \frac{\overset{3}{\cancel{45}}}{\underset{2}{\cancel{8}}} \cdot \frac{\overset{1}{\cancel{4}}}{\cancel{15}} = \frac{3}{2}$$

(c) $3\frac{1}{2}$ inches to $\frac{7}{8}$ inch

$$\frac{3\frac{1}{2}}{\frac{7}{8}} = \frac{\frac{7}{2}}{\frac{7}{8}} = \frac{7}{2} \div \frac{7}{8} = \frac{\overset{1}{\cancel{7}}}{\underset{1}{\cancel{2}}} \cdot \frac{\overset{4}{\cancel{8}}}{\underset{1}{\cancel{7}}} = \frac{4}{1}$$

5. **(a)** 9 inches to 6 feet

Change 6 feet to inches.

$$6 \text{ feet} = 6 \cdot 12 \text{ inches}$$
$$= 72 \text{ inches}$$

$$\frac{9 \text{ in.}}{6 \text{ ft}} = \frac{9 \text{ in.}}{72 \text{ in.}} = \frac{9}{72} = \frac{9 \div 9}{72 \div 9} = \frac{1}{8}$$

(b) 2 days to 8 hours

$2 \text{ days} = 2 \cdot 24 \text{ hours} = 48 \text{ hours}$

$$\frac{48 \text{ hours}}{8 \text{ hours}} = \frac{48}{8} = \frac{48 \div 8}{8 \div 8} = \frac{6}{1}$$

(c) 7 yards to 14 feet

$7 \text{ yards} = 7 \cdot 3 \text{ feet} = 21 \text{ feet}$

$$\frac{21 \text{ feet}}{14 \text{ feet}} = \frac{21}{14} = \frac{21 \div 7}{14 \div 7} = \frac{3}{2}$$

(d) 3 quarts to 3 gallons

$3 \text{ gallons} = 3 \cdot 4 \text{ quarts} = 12 \text{ quarts}$

$$\frac{3 \text{ quarts}}{12 \text{ quarts}} = \frac{3}{12} = \frac{3 \div 3}{12 \div 3} = \frac{1}{4}$$

(e) 25 minutes to 2 hours

$2 \text{ hours} = 2 \cdot 60 \text{ minutes}$
$= 120 \text{ minutes}$

$$\frac{25 \text{ minutes}}{120 \text{ minutes}} = \frac{25}{120} = \frac{25 \div 5}{120 \div 5} = \frac{5}{24}$$

Copyright © 2018 Pearson Education, Inc.

(f) 4 pounds to 12 ounces

$$4 \text{ pounds} = 4 \cdot 16 \text{ ounces}$$
$$= 64 \text{ ounces}$$

$$\frac{64 \text{ ounces}}{12 \text{ ounces}} = \frac{64}{12} = \frac{64 \div 4}{12 \div 4} = \frac{16}{3}$$

5.1 Section Exercises

1. Answers will vary. One possibility is: A ratio compares two quantities with the same units. Examples will vary.

3. To rewrite the ratio $\frac{80}{20}$ in lowest terms, divide both the numerator and the denominator by <u>20</u>. The ratio in lowest terms is

$$\frac{80}{20} = \frac{80 \div 20}{120 \div 20} = \frac{4}{1}.$$

5. 8 days to 9 days: $\dfrac{8 \text{ days}}{9 \text{ days}} = \dfrac{8}{9}$

7. $100 to $50

$$\frac{\$100}{\$50} = \frac{100}{50} = \frac{100 \div 50}{50 \div 50} = \frac{2}{1}$$

9. 30 minutes to 90 minutes

$$\frac{30 \text{ minutes}}{90 \text{ minutes}} = \frac{30}{90} = \frac{30 \div 30}{90 \div 30} = \frac{1}{3}$$

11. 80 miles to 50 miles

$$\frac{80 \text{ miles}}{50 \text{ miles}} = \frac{80}{50} = \frac{80 \div 10}{50 \div 10} = \frac{8}{5}$$

13. 6 hours to 16 hours

$$\frac{6 \text{ hours}}{16 \text{ hours}} = \frac{6 \div 2}{16 \div 2} = \frac{3}{8}$$

15. $4.50 to $3.50

$$\frac{\$4.50}{\$3.50} = \frac{4.50}{3.50} = \frac{4.50 \cdot 10}{3.50 \cdot 10} = \frac{45}{35}$$
$$= \frac{45 \div 5}{35 \div 5} = \frac{9}{7}$$

17. 15 to $2\frac{1}{2}$

$$\frac{15}{2\frac{1}{2}} = \frac{\frac{15}{1}}{\frac{5}{2}} = \frac{15}{1} \div \frac{5}{2} = \frac{\overset{3}{\cancel{15}}}{1} \cdot \frac{2}{\cancel{5}} = \frac{6}{1}$$

19. $1\frac{1}{4}$ to $1\frac{1}{2}$

$$\frac{1\frac{1}{4}}{1\frac{1}{2}} = \frac{\frac{5}{4}}{\frac{3}{2}} = \frac{5}{4} \div \frac{3}{2} = \frac{5}{\underset{2}{\cancel{4}}} \cdot \frac{\overset{1}{\cancel{2}}}{3} = \frac{5}{6}$$

21. Raji forgot to convert 2 feet to 24 inches.

$$\frac{18 \text{ in.}}{2 \text{ ft}} = \frac{18 \text{ in.}}{24 \text{ in.}} = \frac{18}{24} = \frac{18 \div 6}{24 \div 6} = \frac{3}{4}$$

The correct answer is $\frac{3}{4}$.

23. 4 feet to 30 inches

4 feet $= 4 \cdot 12$ inches $= 48$ inches

$$\frac{4 \text{ feet}}{30 \text{ inches}} = \frac{48 \text{ inches}}{30 \text{ inches}} = \frac{48}{30} = \frac{48 \div 6}{30 \div 6} = \frac{8}{5}$$

(Do not write the ratio as $1\frac{3}{5}$.)

25. 5 minutes to 1 hour

1 hour $= 60$ minutes

$$\frac{5 \text{ minutes}}{1 \text{ hour}} = \frac{5 \text{ minutes}}{60 \text{ minutes}} = \frac{5}{60} = \frac{5 \div 5}{60 \div 5} = \frac{1}{12}$$

27. 15 hours to 2 days

2 days $= 2 \cdot 24$ hours $= 48$ hours

$$\frac{15 \text{ hours}}{2 \text{ days}} = \frac{15 \text{ hours}}{48 \text{ hours}} = \frac{15}{48} = \frac{15 \div 3}{48 \div 3} = \frac{5}{16}$$

29. 5 gallons to 5 quarts

5 gallons $= 5 \cdot 4$ quarts $= 20$ quarts

$$\frac{5 \text{ gallons}}{5 \text{ quarts}} = \frac{20 \text{ quarts}}{5 \text{ quarts}} = \frac{20}{5} = \frac{20 \div 5}{5 \div 5} = \frac{4}{1}$$

31. Thanksgiving cards to graduation cards

$$\frac{10 \text{ million}}{70 \text{ million}} = \frac{10}{70} = \frac{10 \div 10}{70 \div 10} = \frac{1}{7}$$

33. Valentine's Day cards to Halloween cards

$$\frac{145 \text{ million}}{10 \text{ million}} = \frac{145}{10} = \frac{145 \div 5}{10 \div 5} = \frac{29}{2}$$

35. Answers will vary. One possibility is stocking cards of various types in the same ratios as those in the table.

37. *Candle in the Wind* to *...Baby One More Time*

$$\frac{33 \text{ million}}{10 \text{ million}} = \frac{33}{10}$$

Silent Night to *I Want to Hold Your Hand*

$$\frac{30 \text{ million}}{12 \text{ million}} = \frac{30}{12} = \frac{30 \div 6}{12 \div 6} = \frac{5}{2}$$

39. To get a ratio of $\frac{2}{1}$, we can start with the smallest value in the table and see if there is a value that is two times the smallest. $2 \times 10 = 20$ and 20 million is in the table twice, so there are at least 2 pairs of songs that give a ratio of $\frac{2}{1}$: *We are the World* to *...Baby One More Time* and *I Will Always Love You* to *...Baby One More Time* $2 \times 12 = 24$, which is not in the table.

(continued)

Copyright © 2018 Pearson Education, Inc.

$2 \times 15 = 30$, which in the table: *Silent Night* to *My Heart Will Go On*

$2 \times 16 = 32$, which is not in the table.

$2 \times 20 = 40$, which is greater than the largest value in the table, so we do not need to look for any more pairs.

41. $\dfrac{\text{longest side}}{\text{shortest side}} = \dfrac{7 \text{ feet}}{5 \text{ feet}} = \dfrac{7}{5}$

The ratio of the length of the longest side to the length of the shortest side is $\frac{7}{5}$.

43. $\dfrac{\text{longest side}}{\text{shortest side}} = \dfrac{1.8 \text{ meters}}{0.3 \text{ meters}} = \dfrac{1.8}{0.3}$

$= \dfrac{1.8 \cdot 10}{0.3 \cdot 10} = \dfrac{18}{3} = \dfrac{18 \div 3}{3 \div 3} = \dfrac{6}{1}$

The ratio of the length of the longest side to the length of the shortest side is $\frac{6}{1}$.

45. $\dfrac{\text{longest side}}{\text{shortest side}} = \dfrac{9\frac{1}{2} \text{ inches}}{4\frac{1}{4} \text{ inches}} = \dfrac{9\frac{1}{2}}{4\frac{1}{4}} = \dfrac{\frac{19}{2}}{\frac{17}{4}}$

$= \dfrac{19}{2} \div \dfrac{17}{4} = \dfrac{19}{\overset{}{2}} \cdot \dfrac{\overset{2}{4}}{17} = \dfrac{38}{17}$

The ratio of the length of the longest side to the length of the shortest side is $\frac{38}{17}$.

47. The increase in price is

$\$12.50 - \$10.00 = \$2.50.$

$\dfrac{\$2.50}{\$10.00} = \dfrac{2.50}{10.00} = \dfrac{2.50 \cdot 10}{10.00 \cdot 10} = \dfrac{25}{100}$

$= \dfrac{25 \div 25}{100 \div 25} = \dfrac{1}{4}$

The ratio of the increase in price to the original price is $\frac{1}{4}$.

49. $59\frac{1}{2} \text{ days} \div 7 = \dfrac{\overset{17}{\cancel{119}}}{2} \cdot \dfrac{1}{\underset{1}{\cancel{7}}} = \dfrac{17}{2} \text{ weeks}$

$\dfrac{\frac{17}{2} \text{ weeks}}{8\frac{3}{4} \text{ weeks}} = \dfrac{\frac{17}{2}}{\frac{35}{4}} = \dfrac{17}{2} \div \dfrac{35}{4} = \dfrac{17}{\underset{1}{\cancel{2}}} \cdot \dfrac{\overset{2}{\cancel{4}}}{35} = \dfrac{34}{35}$

The ratio of $59\frac{1}{2}$ days to $8\frac{3}{4}$ weeks is $\frac{34}{35}$.

51. $\dfrac{\text{longest side}}{\text{shortest side}} = \dfrac{2\frac{7}{12} \text{ feet}}{2\frac{7}{12} \text{ feet}} = \dfrac{1}{1}$

The ratio of the length of the longest side to the length of the shortest side is $\frac{1}{1}$.

As long as the sides all have the same length, any measurement you choose will maintain the ratio.

52. Answers will vary. Some possibilities are:

$\dfrac{4}{5} = \dfrac{8}{10} = \dfrac{12}{15} = \dfrac{16}{20} = \dfrac{20}{25} = \dfrac{24}{30} = \dfrac{28}{35}.$

53. It is not possible. Amelia would have to be older than her mother to have a ratio of 5 to 3.

54. Answers will vary, but a ratio of 3 to 1 means your income is 3 times your friend's income.

5.2 Rates

5.2 Margin Exercises

1. **(a)** $6 for 30 packets

To write the rate in lowest terms, divide both numerator and denominator by $\underline{6}$.

$\dfrac{\$6 \div 6}{30 \text{ packets} \div 6} = \dfrac{\$\underline{1}}{\underline{5} \text{ packets}}$

(b) 500 miles in 10 hours

$\dfrac{500 \text{ miles} \div 10}{10 \text{ hours} \div 10} = \dfrac{50 \text{ miles}}{1 \text{ hour}}$

(c) 4 teachers for 90 students

$\dfrac{4 \text{ teachers} \div 2}{90 \text{ students} \div 2} = \dfrac{2 \text{ teachers}}{45 \text{ students}}$

2. **(a)** $4.35 for 3 pounds of cheese

$\dfrac{\$4.35 \div 3}{3 \text{ pounds} \div 3} = \dfrac{\$1.45}{1 \text{ pound}}$

The unit rate is $1.45/pound.

(b) 304 miles on 9.5 gallons of gas

$\dfrac{304 \text{ miles} \div 9.5}{9.5 \text{ gallons} \div 9.5} = \dfrac{32 \text{ miles}}{1 \text{ gallon}}$

The unit rate is 32 miles/gallon.

(c) $850 in 5 days

$\dfrac{\$850 \div 5}{5 \text{ days} \div 5} = \dfrac{\$170}{1 \text{ day}}$

The unit rate is $170/day.

(d) 24-pound turkey for 15 people

$\dfrac{24 \text{ pounds} \div 15}{15 \text{ people} \div 15} = \dfrac{1.6 \text{ pounds}}{1 \text{ person}}$

The unit rate is 1.6 pounds/person.

3.   **(a)** Divide to find each unit cost.

Size	Cost per Unit
2 quarts	$\begin{array}{c}\text{cost}\rightarrow\\\text{per}\rightarrow\\\text{quart}\rightarrow\end{array}\dfrac{\$3.25}{2\text{ quarts}}$

The unit cost is $\underline{1.625}$ per quart. Similarly,

3 quarts	$\dfrac{\$4.95}{3\text{ quarts}} = \$1.65/\text{qt}$

The unit cost is $\underline{1.65}$ per quart.

Now find the cost for 4 quarts.

4 quarts	$\dfrac{\$6.48}{4\text{ quarts}} = \$1.62/\text{qt}$

The unit cost is $\underline{1.62}$ per quart.

The lowest cost per quart is $1.62, so the best buy is 4 quarts for $6.48.

(b)

Size	Cost per Unit
6 cans	$\dfrac{\$1.99}{6\text{ cans}} \approx \$0.332/\text{can}$
12 cans	$\dfrac{\$3.49}{12\text{ cans}} \approx \$0.291/\text{can}$
24 cans	$\dfrac{\$7}{24\text{ cans}} \approx \$0.292/\text{can}$

The lowest price per can is about $0.291, so the best buy is 12 cans of cola for $3.49.

4.   **(a)** One battery that lasts twice as long is like getting two batteries.

$$\dfrac{\$1.19 \div 2}{2\text{ batteries} \div 2} = \$0.595/\text{battery}.$$

The cost of the four-pack is

$$\dfrac{\$2.79 \div 4}{4\text{ batteries} \div 4} = \$0.6975/\text{battery}.$$

The better buy is the single AA-size battery.

(b) Brand C:

$3.89 - $0.85 = $3.04

$$\dfrac{\$3.04 \div 6}{6\text{ ounces} \div 6} \approx \$0.507/\text{ounce}$$

Brand D:

$1.89 - $0.50 = $1.39

$$\dfrac{\$1.39 \div 2.5}{2.5\text{ ounces} \div 2.5} = \$0.556/\text{ounce}$$

Brand C with the 85¢ coupon is the better buy at $0.507 per ounce (rounded).

5.2   Section Exercises

1.   10 cups for 6 people

$$\dfrac{10\text{ cups} \div 2}{6\text{ people} \div 2} = \dfrac{5\text{ cups}}{3\text{ people}}$$

3.   15 feet in 35 seconds

$$\dfrac{15\text{ feet} \div 5}{35\text{ seconds} \div 5} = \dfrac{3\text{ feet}}{7\text{ seconds}}$$

5.   72 miles on 4 gallons

$$\dfrac{72\text{ miles} \div 4}{4\text{ gallons} \div 4} = \dfrac{18\text{ miles}}{1\text{ gallon}}$$

7.   To find a unit rate, you <u>divide</u>. The correct set-up to find the unit rate of $5.85 for 3 boxes is

$$\dfrac{\$5.85}{3\text{ boxes}}, \text{ or, equivalently, } 3\overline{)5\,.\,8\,5}$$

9. $60 in 5 hours

$$\dfrac{\$60 \div 5}{5\text{ hours} \div 5} = \dfrac{\$12}{1\text{ hour}}$$

The unit rate is $12 per hour or $12/hour.

11.   7.5 pounds for 6 people

$$\dfrac{7.5\text{ pounds} \div 6}{6\text{ people} \div 6} = \dfrac{1.25\text{ pounds}}{1\text{ person}}$$

The unit rate is 1.25 pounds/person.

13.   Miles traveled:

$$27{,}758.2 - 27{,}432.3 = 325.9$$

Miles per gallon:

$$\dfrac{325.9\text{ miles} \div 15.5}{15.5\text{ gallons} \div 15.5} \approx 21.03 \approx 21.0$$

15.   Miles traveled:

$$28{,}396.7 - 28{,}058.1 = 338.6$$

Miles per gallon:

$$\dfrac{338.6\text{ miles} \div 16.2}{16.2\text{ gallons} \div 16.2} \approx 20.90 \approx 20.9$$

17.

Size	Cost per Unit
2 ounces	$\dfrac{\$2.25}{2\text{ ounces}} = \$1.125/\text{oz}$
4 ounces	$\dfrac{\$3.65}{4\text{ ounces}} = \$0.9125/\text{oz }(*)$

The best buy is 4 ounces for $3.65, about $0.913/ounce.

Copyright © 2018 Pearson Education, Inc.

19.

Size	Cost per Unit
12 ounces	$\dfrac{\$2.49}{12 \text{ ounces}} = \$0.2075/\text{oz}$
14 ounces	$\dfrac{\$2.89}{14 \text{ ounces}} \approx \$0.2064/\text{oz}\ (*)$
18 ounces	$\dfrac{\$3.96}{18 \text{ ounces}} = \$0.22/\text{oz}$

Because $0.206 is the lowest cost per ounce, the best buy is 14 ounces for $2.89.

21.

Size	Cost per Unit
12 ounces	$\dfrac{\$1.29}{12 \text{ ounces}} \approx \$0.108/\text{oz}$
18 ounces	$\dfrac{\$1.79}{18 \text{ ounces}} \approx \$0.099/\text{oz}\ (*)$
28 ounces	$\dfrac{\$3.39}{28 \text{ ounces}} \approx \$0.121/\text{oz}$
40 ounces	$\dfrac{\$4.39}{40 \text{ ounces}} \approx \$0.110/\text{oz}$

The best buy is 18 ounces for $1.79, about $0.099/ounce.

23. Answers will vary. For example, you might choose Brand B because you prefer to buy organic or your doctor told you to lower your salt intake. So the additional cost may be worth the added health benefits.

25. 10.5 pounds in 6 weeks

$$\frac{10.5 \text{ pounds} \div 6}{6 \text{ weeks} \div 6} = \frac{1.75 \text{ pounds}}{1 \text{ week}}$$

Her rate of loss was 1.75 pounds/week.

27. 7 hours to earn $85.82

$$\frac{\$85.82 \div 7}{7 \text{ hours} \div 7} = \frac{\$12.26}{1 \text{ hour}}$$

His pay rate is $12.26/hour.

29. **(a)** Add the initial fare and five times the cost per mile.

Uber:
$1.25 + 5(\$2.89) = \$1.25 + \$14.45 = \$15.70$

Taxi:
$4.75 + 5(\$2.55) = \$4.75 + \$12.75 = \$17.50$

Lyft:
$2.60 + 5(\$2.76) = \$2.60 + \$13.80 = \$16.40$

(b) Divide the total cost for each ride by five miles.

Uber: $\dfrac{\$15.70 \div 5}{5 \text{ miles} \div 5} = \$3.14/\text{mile}$

Taxi: $\dfrac{\$17.50 \div 5}{5 \text{ miles} \div 5} = \$3.50/\text{mile}$

Lyft: $\dfrac{\$16.40 \div 5}{5 \text{ miles} \div 5} = \$3.28/\text{mile}$

Uber is the best buy.

31. Add the initial fare and 10 times the cost per mile. Then divide the total cost by 10.

Uber:
$1.25 + 10(\$2.89) = \$1.25 + \$28.90 = \$30.15$

$$\frac{\$30.15 \div 10}{10 \text{ miles} \div 10} \approx \$3.02/\text{mile}$$

Taxi:
$4.75 + 10(\$2.55) = \$4.75 + \$25.50 = \$30.25$

$$\frac{\$30.25 \div 10}{10 \text{ miles} \div 10} \approx \$3.03/\text{mile}$$

Lyft:
$2.60 + 10(\$2.76) = \$2.60 + \$27.60 = \$30.20$

$$\frac{\$30.20 \div 10}{10 \text{ miles} \div 10} \approx \$3.02/\text{mile}$$

For the 10-mile ride, the unit rates are all about $3.00 per mile

33. One battery that lasts 3 times longer is like getting 3 batteries, so $1.79 \div 3 \approx \$0.597$ per battery.

An eight-pack of AAA batteries for $4.99 is $4.99 \div 8 \approx \$0.624$ per battery.

Because $0.597 is the lower cost per battery, the better buy is the package with one battery.

35. **Brand G:** $2.39 - \$0.60$ coupon $= \$1.79$

$$\frac{\$1.79}{10 \text{ ounces}} = \$0.179 \text{ per ounce}$$

Brand K: $\dfrac{\$3.99}{20.3 \text{ ounces}} \approx \$0.197 \text{ per ounce}$

Brand P: $3.39 - \$0.50$ coupon $= \$2.89$

$$\frac{\$2.89}{16.5 \text{ ounces}} \approx \$0.175 \text{ per ounce}$$

Brand P with the 50¢ coupon has the lowest cost per ounce and is the best buy.

37. **Plan A:** $\dfrac{\$45}{1500 \text{ MB}} = \$0.03/\text{MB}$

Plan B: $\dfrac{\$30}{500 \text{ MB}} = \$0.06/\text{MB}$

Plan C: $\dfrac{\$60}{2500 \text{ MB}} \approx \$0.02/\text{MB}$

Plan C is the best buy.

Copyright © 2018 Pearson Education, Inc.

38. Divide the number of megabytes by 30, the number of days in a month.

(a) Plan A: $\dfrac{1500 \text{ MB}}{30 \text{ days}} = 50 \text{ MB/day}$

(b) Plan B: $\dfrac{500 \text{ MB}}{30 \text{ days}} \approx 17 \text{ MB/day}$

(c) Plan C: $\dfrac{2500 \text{ MB}}{30 \text{ days}} \approx 83 \text{ MB/day}$

39. 4 extra MB per day times 30 days gives us 120 extra MB. 120 MB overage means you need to pay for 100 MB plus another 100 MB to cover the entire amount, so $2(\$5) = \$10$. (You cannot buy less than 100 MB.)

$\$30 + \$10 = \$40$

$\dfrac{\text{Total charge}}{\text{Total MB}} = \dfrac{\$40}{500 \text{ MB} + 130 \text{ MB}}$

$= \dfrac{\$40}{630 \text{ MB}} \approx \$0.06/\text{MB}$

40. 3 extra MB per day times 30 days gives us 90 extra MB. 90 extra MB times $1.50 per extra MB gives us $90 \times \$1.50 = \$135$ in overage charges.

$\dfrac{\text{Total charge}}{\text{Total MB}} = \dfrac{\$60 + \$135}{2500 + 90}$

$= \dfrac{\$195}{2590 \text{ MB}} \approx \$0.08/\text{MB}$

41. Plan A:
$2250 \text{ MB} - 1500 \text{ MB} = 750 \text{ MB}$ overage
You need to pay for 100 MB plus another 100 MB to cover the entire amount, so $2(\$10) = \$20$.

$\dfrac{\text{Total charge}}{\text{Total MB}} = \dfrac{\$45 + \$20}{2250 \text{ MB}}$

$= \dfrac{\$65}{2250 \text{ MB}} \approx \$0.03/\text{MB}$

Plan B:
$2250 \text{ MB} - 500 \text{ MB} = 1750 \text{ MB}$ overage

$\dfrac{1750 \text{ MB}}{100 \text{ MB}} \approx 18$

You need to pay for 18(100 MB), so $18(\$5) = \$90$.

$\dfrac{\text{Total charge}}{\text{Total MB}} = \dfrac{\$30 + \$90}{2250 \text{ MB}}$

$= \dfrac{\$120}{2250 \text{ MB}} \approx \$0.05/\text{MB}$

Plan C:
There is no overage charge because you use less data each month than is allowed on the plan.

$\dfrac{\text{Total charge}}{\text{Total MB}} = \dfrac{\$60}{2250 \text{ MB}} \approx \$0.03/\text{MB}$

42. Plan A:
$3000 \text{ MB} - 1500 \text{ MB} = 1500 \text{ MB}$ overage

$\dfrac{1500 \text{ MB}}{500 \text{ MB}} \approx 3$

You need to pay for 3(500 MB), so $3(\$10) = \$30$.

Total charge $= \$45 + \$30 = \$75$

Plan B:
$3000 \text{ MB} - 500 \text{ MB} = 2500 \text{ MB}$ overage

$\dfrac{2500 \text{ MB}}{100 \text{ MB}} = 25$

You need to pay for 25(100 MB), so $25(\$5) = \$125$.

Total charge $= \$30 + \$125 = \$155$

Plan C:
$3000 \text{ MB} - 2500 \text{ MB} = 500 \text{ MB}$ overage

You need to pay for 500(1 MB), so $500(\$1.50) = \$750$.

Total charge $= \$60 + \$750 = \$810$

5.3 Proportions

5.3 Margin Exercises

1. **(a)** $7 is to 3 cans as $28 is to 12 cans

$$\dfrac{\$7}{3 \text{ cans}} = \dfrac{\$28}{12 \text{ cans}}$$

(b) 9 meters is to 16 meters as 18 meters is to 32 meters

$\dfrac{9 \text{ m}}{16 \text{ m}} = \dfrac{18 \text{ m}}{32 \text{ m}}$, so $\dfrac{9}{16} = \dfrac{18}{32}$ *Common units cancel*

2. **(a)** $\dfrac{6}{12} = \dfrac{15}{30}$

$\dfrac{6 \div 6}{12 \div 6} = \dfrac{1}{2}$ and $\dfrac{15 \div 15}{30 \div 15} = \dfrac{1}{2}$

Both ratios are equivalent to $\frac{1}{2}$, so the proportion is *true*.

(b) $\dfrac{20}{24} = \dfrac{3}{4}$

$\dfrac{20}{24} = \dfrac{20 \div 4}{24 \div 4} = \dfrac{5}{6}$ and $\dfrac{3}{4}$

Because $\frac{5}{6}$ is *not* equivalent to $\frac{3}{4}$, the proportion is *false*.

Copyright © 2018 Pearson Education, Inc.

(c) $\dfrac{35}{45} = \dfrac{12}{18}$

$$\dfrac{35 \div 5}{45 \div 5} = \dfrac{7}{9} \text{ and } \dfrac{12 \div 6}{18 \div 6} = \dfrac{2}{3}$$

Because $\frac{7}{9}$ is *not* equivalent to $\frac{2}{3}$, the proportion is *false*.

3. **(a)** $\dfrac{5}{9} = \dfrac{10}{18}$ Cross products: $\begin{array}{l}(9)(10) = \underline{90}\\(5)(18) = \underline{90}\end{array}$

The cross products are *equal*, so the proportion is *true*.

(b) $\dfrac{32}{15} = \dfrac{16}{8}$ Cross products: $\begin{array}{l}(15)(16) = 240\\(32)(8) = 256\end{array}$

The cross products are *unequal*, so the proportion is *false*.

(c) $\dfrac{10}{17} = \dfrac{20}{34}$ Cross products: $\begin{array}{l}(17)(20) = 340\\(10)(34) = 340\end{array}$

The cross products are *equal*, so the proportion is *true*.

(d) $\dfrac{2.4}{6} = \dfrac{5}{12}$

Cross products: $\begin{array}{l}(6)(5) = \underline{30}\\(2.4)(12) = \underline{28.8}\end{array}$

The cross products are *unequal*, so the proportion is *false*.

(e) $\dfrac{3}{4.25} = \dfrac{24}{34}$

Cross products: $\begin{array}{l}(4.25)(24) = 102\\(3)(34) = 102\end{array}$

The cross products are *equal*, so the proportion is *true*.

(f) $\dfrac{1\frac{1}{6}}{2\frac{1}{3}} = \dfrac{4}{8}$

Cross products: $2\dfrac{1}{3} \cdot 4 = \dfrac{7}{3} \cdot \dfrac{4}{1} = \dfrac{28}{3}$

$1\dfrac{1}{6} \cdot 8 = \dfrac{7}{\cancel{6}_{3}} \cdot \dfrac{\cancel{8}^{4}}{1} = \dfrac{28}{3}$

The cross products are *equal*, so the proportion is *true*.

5.3 Section Exercises

1. $9 is to 12 cans as $18 is to 24 cans.

$$\dfrac{\$9}{12 \text{ cans}} = \dfrac{\$18}{24 \text{ cans}}$$

3. 200 adults is to 450 children as 4 adults is to 9 children.

$$\dfrac{200 \text{ adults}}{450 \text{ children}} = \dfrac{4 \text{ adults}}{9 \text{ children}}$$

5. 120 feet is to 150 feet as 8 feet is to 10 feet.

$$\dfrac{120}{150} = \dfrac{8}{10} \quad \textit{The common units (feet) cancel.}$$

7. $\dfrac{6}{10} = \dfrac{3}{5}$

$$\dfrac{6 \div 2}{10 \div 2} = \dfrac{3}{5} \text{ and } \dfrac{3}{5}$$

Both ratios are equivalent to $\frac{3}{5}$, so the proportion is *true*.

9. $\dfrac{5}{8} = \dfrac{25}{40}$

$$\dfrac{5}{8} \text{ and } \dfrac{25 \div 5}{40 \div 5} = \dfrac{5}{8}$$

Both ratios are equivalent to $\frac{5}{8}$, so the proportion is *true*.

11. $\dfrac{150}{200} = \dfrac{200}{300}$

$$\dfrac{150 \div 50}{200 \div 50} = \dfrac{3}{4} \text{ and } \dfrac{200 \div 100}{300 \div 100} = \dfrac{2}{3}$$

Because $\frac{3}{4}$ is *not* equivalent to $\frac{2}{3}$, the proportion is *false*.

13. $\dfrac{42}{15} = \dfrac{28}{10}$

$$\dfrac{42 \div 3}{15 \div 3} = \dfrac{14}{5} \text{ and } \dfrac{28 \div 2}{10 \div 2} = \dfrac{14}{5}$$

Both ratios are equivalent to $\frac{14}{5}$, so the proportion is *true*.

15. $\dfrac{32}{18} = \dfrac{48}{27}$

$$\dfrac{32 \div 2}{18 \div 2} = \dfrac{16}{9} \text{ and } \dfrac{48 \div 3}{27 \div 3} = \dfrac{16}{9}$$

Both ratios are equivalent to $\frac{16}{9}$, so the proportion is *true*.

17. $\dfrac{7}{6} = \dfrac{54}{48}$

$$\dfrac{7}{6} \text{ and } \dfrac{54 \div 6}{48 \div 6} = \dfrac{9}{8}$$

Because $\frac{7}{6}$ is *not* equivalent to $\frac{9}{8}$, the proportion is *false*.

Copyright © 2018 Pearson Education, Inc.

19. Irene did not simplify the second ratio correctly; the numerator and denominator must be divided by the same number. Both ratios are already in lowest terms and are not equal, so the proportion is false.

21. $\dfrac{2}{9} = \dfrac{6}{27}$ Cross products: $\begin{array}{l} 9 \cdot 6 = 54 \\ 2 \cdot 27 = 54 \end{array}$

The cross products are *equal*, so the proportion is *true.*

23. $\dfrac{20}{28} = \dfrac{12}{16}$ Cross products: $\begin{array}{l} 28 \cdot 12 = 336 \\ 20 \cdot 16 = 320 \end{array}$

The cross products are *unequal*, so the proportion is *false.*

25. $\dfrac{110}{18} = \dfrac{160}{27}$ Cross products: $\begin{array}{l} 18 \cdot 160 = 2880 \\ 110 \cdot 27 = 2970 \end{array}$

The cross products are *unequal*, so the proportion is *false.*

27. $\dfrac{3.5}{4} = \dfrac{7}{8}$ Cross products: $\begin{array}{l} 4 \cdot 7 = 28 \\ (3.5)(8) = 28 \end{array}$

The cross products are *equal*, so the proportion is *true.*

29. $\dfrac{18}{16} = \dfrac{2.8}{2.5}$ Cross products: $\begin{array}{l} 16(2.8) = 44.8 \\ 18(2.5) = 45 \end{array}$

The cross products are *unequal*, so the proportion is *false.*

31. $\dfrac{6}{3\frac{2}{3}} = \dfrac{18}{11}$

Cross products: $3\dfrac{2}{3} \cdot 18 = \dfrac{11}{\overset{1}{\cancel{3}}} \cdot \dfrac{\overset{6}{\cancel{18}}}{1} = 66$

$6 \cdot 11 = 66$

The cross products are *equal*, so the proportion is *true.*

33. $\dfrac{2\frac{5}{8}}{3\frac{1}{4}} = \dfrac{21}{26}$

Cross products: $3\dfrac{1}{4} \cdot 21 = \dfrac{13}{4} \cdot \dfrac{21}{1} = \dfrac{273}{4}$

$2\dfrac{5}{8} \cdot 26 = \dfrac{21}{\overset{}{\cancel{8}}} \cdot \dfrac{\overset{13}{\cancel{26}}}{1} = \dfrac{273}{4}$

The cross products are *equal*, so the proportion is *true.*

35. $\dfrac{\frac{2}{3}}{2} = \dfrac{2.7}{8}$

Cross products: $2 \cdot 2\dfrac{7}{10} = \dfrac{2}{1} \cdot \dfrac{27}{10} = \dfrac{54}{10} = 5.4$

$\dfrac{2}{3} \cdot 8 = \dfrac{2}{3} \cdot \dfrac{8}{1} = \dfrac{16}{3} \approx 5.3$

The cross products are *unequal*, so the proportion is *false.*

37. $\dfrac{2\frac{3}{10}}{8.05} = \dfrac{\frac{1}{4}}{0.9}$

Cross products:

$2\dfrac{3}{10}(0.9) = \dfrac{23}{10} \cdot \dfrac{9}{10} = \dfrac{207}{100} = 2.07$

or $(2.3)(0.9) = 2.07$

$8\dfrac{5}{100} \cdot \dfrac{1}{4} = \dfrac{805}{100} \cdot \dfrac{1}{4} = 2\dfrac{1}{80}$

or $(8.05)\dfrac{1}{4} = (8.05)(0.25) = 2.0125$

The cross products are *unequal*, so the proportion is *false.*

$\left(2\dfrac{7}{100} \neq 2\dfrac{1}{80} \text{ or } 2.07 \neq 2.0125 \right)$

39. (Answers may vary: proportion may also be set up with "at bats" in both numerators and "hits" in both denominators. Cross product will be the same.)

F. Freeman	K. Suzuki
$\dfrac{168 \text{ hits}}{600 \text{ at bats}}$	$= \dfrac{126 \text{ hits}}{450 \text{ at bats}}$

Cross products:
$168 \cdot 450 = 75{,}600; \quad 126 \cdot 600 = 75{,}600$

The cross products are *equal*, so the proportion is *true.* Paula is correct; the two men hit equally well.

5.4 Solving Proportions

5.4 Margin Exercises

1. **(a)** $\dfrac{1}{2} = \dfrac{x}{12}$

$2 \cdot x = 1 \cdot 12$ *Cross products are equal*

$\dfrac{\overset{1}{\cancel{2}} \cdot x}{\cancel{2}} = \dfrac{1 \cdot \overset{6}{\cancel{12}}}{\cancel{2}}$ *Divide both sides by 2.*

$x = 6$

We will check our answers by showing that the cross products are equal.

Check: $1 \cdot 12 = 12 = 2 \cdot 6$

Copyright © 2018 Pearson Education, Inc.

(b) $\dfrac{6}{10} = \dfrac{15}{x}$

$6 \cdot x = 10 \cdot 15$ *Cross products are equal*

$\dfrac{\overset{1}{\cancel{6}} \cdot x}{\underset{1}{\cancel{6}}} = \dfrac{\overset{25}{\cancel{150}}}{\underset{1}{\cancel{6}}}$ *Divide by 6.*

$x = 25$

Check: $6 \cdot 25 = 150 = 10 \cdot 15$

(c) $\dfrac{28}{x} = \dfrac{21}{9}$

$x \cdot 21 = 28 \cdot 9$ *Cross products are equal*

$\dfrac{x \cdot \overset{1}{\cancel{21}}}{\underset{1}{\cancel{21}}} = \dfrac{\overset{4}{\cancel{28}} \cdot 9}{\underset{3}{\cancel{21}}}$ *Divide by 21.*

$x = 12$

Check: $28 \cdot 9 = 252 = 12 \cdot 21$

(d) $\dfrac{x}{8} = \dfrac{3}{5}$

$x \cdot 5 = 8 \cdot 3$ *Cross products are equal*

$\dfrac{x \cdot \overset{1}{\cancel{5}}}{\underset{1}{\cancel{5}}} = \dfrac{24}{5}$ *Divide by 5.*

$x = \dfrac{24}{5} = 4.8$

Check: $(4.8)(5) = 24 = 8 \cdot 3$

(e) $\dfrac{14}{11} = \dfrac{x}{3}$

$11 \cdot x = 14 \cdot 3$ *Cross products are equal*

$\dfrac{\overset{1}{\cancel{11}} \cdot x}{\underset{1}{\cancel{11}}} = \dfrac{14 \cdot 3}{11}$ *Divide by 11.*

$x = \dfrac{42}{11} \approx 3.82$ *(rounded to the nearest hundredth)*

Check: $14 \cdot 3 = 42 = 11 \cdot \frac{42}{11}$

2. (a) $\dfrac{3\frac{1}{4}}{2} = \dfrac{x}{8}$

$2 \cdot x = 3\dfrac{1}{4} \cdot 8$ *Cross products are equal*

$2 \cdot x = \dfrac{13}{\cancel{4}} \cdot \dfrac{\overset{2}{\cancel{8}}}{1}$

$\dfrac{\overset{1}{\cancel{2}} \cdot x}{\underset{1}{\cancel{2}}} = \dfrac{\overset{13}{\cancel{26}}}{\underset{1}{\cancel{2}}}$ *Divide by 2.*

$x = 13$

Check: $3\frac{1}{4} \cdot 8 = 26 = 2 \cdot 13$

(b) $\dfrac{x}{3} = \dfrac{1\frac{2}{3}}{5}$

$5 \cdot x = 3 \cdot 1\dfrac{2}{3}$ *Cross products are equal*

$5 \cdot x = \overset{1}{\cancel{3}} \cdot \dfrac{5}{\underset{1}{\cancel{3}}}$

$\dfrac{\overset{1}{\cancel{5}} \cdot x}{\underset{1}{\cancel{5}}} = \dfrac{1 \cdot \overset{1}{\cancel{5}}}{\underset{1}{\cancel{5}}}$ *Divide by 5.*

$x = 1$

Check: $1 \cdot 5 = 5 = 3 \cdot 1\frac{2}{3}$

(c) $\dfrac{0.06}{x} = \dfrac{0.3}{0.4}$

$0.3 \cdot x = 0.06 \cdot 0.4$ *Cross products are equal*

$\dfrac{\overset{1}{\cancel{0.3}} \cdot x}{\underset{1}{\cancel{0.3}}} = \dfrac{\overset{0.08}{\cancel{0.024}}}{\underset{1}{\cancel{0.3}}}$ *Divide by 0.3.*

$x = 0.08$

Check: $(0.06)(0.4) = 0.024 = (0.08)(0.3)$

(d) $\dfrac{2.2}{5} = \dfrac{13}{x}$

$2.2 \cdot x = 5 \cdot 13$ *Cross products are equal*

$\dfrac{\overset{1}{\cancel{2.2}} \cdot x}{\underset{1}{\cancel{2.2}}} = \dfrac{65}{2.2}$ *Divide by 2.2.*

$x = \dfrac{65}{2.2} \approx 29.55$ *(rounded to the nearest hundredth)*

Check: $2.2\left(\frac{65}{2.2}\right) = 65 = 5 \cdot 13$

Copyright © 2018 Pearson Education, Inc.

(e) $\dfrac{x}{6} = \dfrac{0.5}{1.2}$

$1.2 \cdot x = 6(0.5)$ *Cross products are equal*

$\dfrac{\cancel{1.2} \cdot x}{\cancel{1.2}} = \dfrac{3}{1.2} = \dfrac{30}{12} = \dfrac{5}{2}$ *Divide by 1.2.*

$x = 2.5$

Check: $(2.5)(1.2) = 3 = 6(0.5)$

(f) $\dfrac{0}{2} = \dfrac{x}{7.092}$ Since $\frac{0}{2} = 0$, x *must* equal 0.

Note: The denominator 7.092 could be any number except 0.

5.4 Section Exercises

1. The first step is to find the cross products, as shown below.

$$\frac{7}{10} = \frac{5}{x}$$

$10 \cdot 5 = 7 \cdot x$ *Find the cross products.*

$50 = 7x$ *Cross products are equal.*

3. $\dfrac{1}{3} = \dfrac{x}{12}$

$3 \cdot x = 1 \cdot 12$ *Cross products are equal*

$3 \cdot x = 12$

$\dfrac{\cancel{3} \cdot x}{\cancel{3}} = \dfrac{12}{3}$ *Divide by 3.*

$x = \underline{4}$

Check the answer by showing that the cross products are equal.

Check: $1 \cdot 12 = 12 = 3 \cdot 4$

5. It is a good idea to first reduce any fractions. We will do this when possible, and work with the reduced fraction.

$$\frac{15}{10} = \frac{3}{x} \quad \text{OR} \quad \frac{3}{2} = \frac{3}{x}$$

Since the numerators are equal, the denominators must be equal, so $x = 2$.

7. $\dfrac{x}{11} = \dfrac{32}{4}$ **OR** $\dfrac{x}{11} = \dfrac{8}{1}$

$1 \cdot x = 11 \cdot 8$ *Cross products are equal*

$x = 88$

Check: $88 \cdot 4 = 352 = 11 \cdot 32$

9. $\dfrac{42}{x} = \dfrac{18}{39}$

$x \cdot 18 = 42 \cdot 39$ *Cross products are equal*

$\dfrac{x \cdot \cancel{18}}{\cancel{18}} = \dfrac{\overset{7}{\cancel{42}} \cdot \overset{13}{\cancel{39}}}{\underset{1}{\cancel{18}}}$ *Divide by 18.*

$x = 91$

Check: $42 \cdot 39 = 1638 = 91 \cdot 18$

11. $\dfrac{x}{25} = \dfrac{4}{20}$ **OR** $\dfrac{x}{25} = \dfrac{1}{5}$

$5 \cdot x = 25 \cdot 1$ *Cross products are equal*

$\dfrac{\cancel{5} \cdot x}{\cancel{5}} = \dfrac{\overset{5}{\cancel{25}}}{\cancel{5}}$ *Divide by 5.*

$x = 5$

Check: $5 \cdot 20 = 100 = 25 \cdot 4$

13. $\dfrac{8}{x} = \dfrac{24}{30}$ **OR** $\dfrac{8}{x} = \dfrac{4}{5}$

$4 \cdot x = 8 \cdot 5$ *Cross products are equal*

$\dfrac{\cancel{4} \cdot x}{\cancel{4}} = \dfrac{\overset{2}{\cancel{8}} \cdot 5}{\cancel{4}}$ *Divide by 4.*

$x = 10$

Check: $8 \cdot 30 = 240 = 10 \cdot 24$

15. $\dfrac{99}{55} = \dfrac{44}{x}$ **OR** $\dfrac{9}{5} = \dfrac{44}{x}$

$9 \cdot x = 5 \cdot 44$ *Cross products are equal*

$\dfrac{\cancel{9} \cdot x}{\cancel{9}} = \dfrac{220}{9}$ *Divide by 9.*

$x = \dfrac{220}{9}$

≈ 24.44 *rounded*

Check: $99 \cdot \frac{220}{9} = 2420 = 55 \cdot 44$

17. $\dfrac{0.7}{9.8} = \dfrac{3.6}{x}$

$0.7 \cdot x = 9.8\,(3.6)$ *Cross products are equal.*

$\dfrac{\cancel{0.7} \cdot x}{\cancel{0.7}} = \dfrac{\overset{50.4}{\cancel{35.28}}}{\cancel{0.7}}$ *Divide by 0.7.*

$x = 50.4$

Check: $0.7(50.4) = 35.28 = 9.8(3.6)$

Copyright © 2018 Pearson Education, Inc.

19. $$\frac{250}{24.8} = \frac{x}{1.75}$$

$24.8 \cdot x = 250(1.75)$ *Cross products are equal*

$$\frac{\overset{1}{\cancel{24.8}} \cdot x}{\underset{1}{\cancel{24.8}}} = \frac{437.5}{24.8}$$ *Divide by 24.8.*

$$\approx 17.64$$ *Approximation*

Check: $250(1.75) = 437.5 = 24.8 \cdot \frac{437.5}{24.8}$

21. $$\frac{15}{1\frac{2}{3}} = \frac{9}{x}$$

$15 \cdot x = 1\frac{2}{3} \cdot 9$ *Cross products are equal*

$$15 \cdot x = \frac{5}{\underset{1}{\cancel{3}}} \cdot \frac{\overset{3}{\cancel{9}}}{1}$$

$$\frac{\overset{1}{\cancel{15}} \cdot x}{\underset{1}{\cancel{15}}} = \frac{15}{15}$$ *Divide by 15.*

$$x = 1$$

23. $$\frac{2\frac{1}{3}}{1\frac{1}{2}} = \frac{x}{2\frac{1}{4}}$$

$$1\frac{1}{2} \cdot x = 2\frac{1}{3} \cdot 2\frac{1}{4}$$

$$\frac{3}{2} \cdot x = \frac{7}{\underset{1}{\cancel{3}}} \cdot \frac{\overset{3}{\cancel{9}}}{4}$$

$$\frac{3}{2} \cdot x = \frac{21}{4}$$

$$\frac{\overset{1}{\cancel{\frac{3}{2}}} \cdot x}{\underset{1}{\cancel{\frac{3}{2}}}} = \frac{\frac{21}{4}}{\frac{3}{2}}$$

$$x = \frac{21}{4} \div \frac{3}{2} = \frac{\overset{7}{\cancel{21}}}{\underset{2}{\cancel{4}}} \cdot \frac{\overset{1}{\cancel{2}}}{\underset{1}{\cancel{3}}} = \frac{7}{2} = 3\frac{1}{2}$$

25. Change to decimals:

$$\frac{\frac{1}{2}}{x} = \frac{2}{0.8}$$

$$\frac{0.5}{x} = \frac{2}{0.8}$$

$$x \cdot 2 = 0.5(0.8)$$

$$x \cdot 2 = 0.4$$

$$\frac{x \cdot \overset{1}{\cancel{2}}}{\underset{1}{\cancel{2}}} = \frac{0.4}{2}$$

$$x = 0.2$$

Change to fractions:

$$\frac{\frac{1}{2}}{x} = \frac{2}{\frac{4}{5}} \quad \left(0.8 \text{ equals } \frac{4}{5} \right)$$

$$x \cdot 2 = \frac{1}{\underset{1}{\cancel{2}}} \cdot \frac{\overset{2}{\cancel{4}}}{5}$$

$$x \cdot 2 = \frac{2}{5}$$

$$\frac{x \cdot \overset{1}{\cancel{2}}}{\underset{1}{\cancel{2}}} = \frac{\frac{2}{5}}{2}$$

$$x = \frac{2}{5} \div 2 = \frac{2}{5} \cdot \frac{1}{\underset{1}{\cancel{2}}} = \frac{1}{5}$$

27. $$\frac{x}{\frac{3}{50}} = \frac{0.15}{1\frac{4}{5}}$$

Change to decimals:

$$\frac{3}{50} = 3 \div 50 = 0.06$$

$$1\frac{4}{5} = \frac{9}{5} \text{ and } 9 \div 5 = 1.8$$

$$\frac{x}{0.06} = \frac{0.15}{1.8}$$

$$x \cdot 1.8 = 0.15(0.06)$$

$$\frac{x \cdot 1.8}{1.8} = \frac{0.009}{1.8}$$

$$x = 0.005$$

(continued)

Copyright © 2018 Pearson Education, Inc.

Change to fractions:

$$0.15 = \frac{15 \div 5}{100 \div 5} = \frac{3}{20}$$

$$\frac{x}{\frac{3}{50}} = \frac{\frac{3}{20}}{1\frac{4}{5}}$$

$$1\frac{4}{5} \cdot x = \frac{3}{50} \cdot \frac{3}{20}$$

$$\frac{1\frac{4}{5} \cdot x}{1\frac{4}{5}} = \frac{\frac{9}{1000}}{1\frac{4}{5}}$$

$$x = \frac{9}{1000} \div 1\frac{4}{5} = \frac{\overset{1}{\cancel{9}}}{\underset{200}{\cancel{1000}}} \cdot \frac{\overset{1}{\cancel{5}}}{\underset{1}{\cancel{9}}} = \frac{1}{200}$$

Compare answers.

$$0.005 = \frac{5 \div 5}{1000 \div 5} = \frac{1}{200} \quad \textit{Matches}$$

29. $\dfrac{10}{4} = \dfrac{5}{3}$

Find the cross products.

$$10 \cdot 3 = 30$$
$$4 \cdot 5 = 20$$

The cross products are *unequal,* so the proportion is *not* true.

Replace 10 with x.

$$\frac{x}{4} = \frac{5}{3}$$
$$3 \cdot x = 4 \cdot 5$$
$$x = \frac{20}{3} = 6\frac{2}{3}, \text{ so}$$
$$\frac{6\frac{2}{3}}{4} = \frac{5}{3} \text{ is a true proportion.}$$

Replace 4 with x.

$$\frac{10}{x} = \frac{5}{3}$$
$$5 \cdot x = 10 \cdot 3$$
$$x = \frac{30}{5} = 6, \text{ so}$$
$$\frac{10}{6} = \frac{5}{3} \text{ is a true proportion.}$$

Replace 5 with x.

$$\frac{10}{4} = \frac{x}{3}$$
$$4 \cdot x = 10 \cdot 3$$
$$x = \frac{30}{4} = \frac{15}{2} = 7.5, \text{ so}$$
$$\frac{10}{4} = \frac{7.5}{3} \text{ is a true proportion.}$$

Replace 3 with x.

$$\frac{10}{4} = \frac{5}{x}$$
$$10 \cdot x = 4 \cdot 5$$
$$x = \frac{20}{10} = 2, \text{ so}$$
$$\frac{10}{4} = \frac{5}{2} \text{ is a true proportion.}$$

30. $\dfrac{6}{8} = \dfrac{24}{30}$

Find the cross products.

$$6 \cdot 30 = 180$$
$$8 \cdot 24 = 192$$

The cross products are *unequal,* so the proportion is *not* true.

Replace 6 with x.

$$\frac{x}{8} = \frac{24}{30}$$
$$30 \cdot x = 8 \cdot 24$$
$$x = \frac{8 \cdot \overset{4}{\cancel{24}}}{\underset{5}{\cancel{30}}} = \frac{32}{5} = 6.4, \text{ so}$$
$$\frac{6.4}{8} = \frac{24}{30} \text{ is a true proportion.}$$

Replace 8 with x.

$$\frac{6}{x} = \frac{24}{30}$$
$$24 \cdot x = 6 \cdot 30$$
$$x = \frac{6 \cdot \overset{5}{\cancel{30}}}{\underset{4}{\cancel{24}}} = \frac{30}{4} = 7.5, \text{ so}$$
$$\frac{6}{7.5} = \frac{24}{30} \text{ is a true proportion.}$$

Replace 24 with x.

$$\frac{6}{8} = \frac{x}{30}$$
$$8 \cdot x = 6 \cdot 30$$
$$x = \frac{\overset{3}{\cancel{6}} \cdot \overset{15}{\cancel{30}}}{\underset{\underset{2}{\cancel{4}}}{\cancel{8}}} = \frac{45}{2} = 22.5, \text{ so}$$
$$\frac{6}{8} = \frac{22.5}{30} \text{ is a true proportion.}$$

(continued)

Copyright © 2018 Pearson Education, Inc.

Replace 30 with x.

$$\frac{6}{8} = \frac{24}{x}$$

$$6 \cdot x = 8 \cdot 24$$

$$x = \frac{8 \cdot \overset{4}{\cancel{24}}}{\underset{1}{\cancel{6}}} = 32, \text{ so}$$

$\frac{6}{8} = \frac{24}{32}$ is a true proportion.

Summary Exercises *Ratios, Rates, and Proportions*

1. 60 million websites in Japan to 20 million websites in China

$$\frac{60 \text{ million}}{20 \text{ million}} = \frac{60 \div 20}{20 \div 20} = \frac{3}{1}$$

3. Combined number of websites in Italy and China (25 million + 20 million = 45 million) to 60 million websites in Japan

$$\frac{45 \text{ million}}{60 \text{ million}} = \frac{45 \div 15}{60 \div 15} = \frac{3}{4}$$

5. **(i)** 5 million bass players to 100 million piano players

$$\frac{5 \text{ million}}{100 \text{ million}} = \frac{5 \div 5}{100 \div 5} = \frac{1}{20}$$

   (ii) 5 million bass players to 50 million guitar players

$$\frac{5 \text{ million}}{50 \text{ million}} = \frac{5 \div 5}{50 \div 5} = \frac{1}{10}$$

   (iii) 5 million bass players to 40 million woodwind players

$$\frac{5 \text{ million}}{40 \text{ million}} = \frac{5 \div 5}{40 \div 5} = \frac{1}{8}$$

   (iv) 5 million bass players to 30 million drum players

$$\frac{5 \text{ million}}{30 \text{ million}} = \frac{5 \div 5}{30 \div 5} = \frac{1}{6}$$

   (v) 5 million bass to 15 million string players

$$\frac{5 \text{ million}}{15 \text{ million}} = \frac{5 \div 5}{15 \div 5} = \frac{1}{3}$$

7. $$\frac{100 \text{ points}}{48 \text{ minutes}} = \frac{100 \div 48 \text{ points}}{48 \div 48 \text{ minutes}}$$
$$= 2.08\overline{3} \approx 2.1 \text{ points/minute}$$

$$\frac{48 \text{ minutes}}{100 \text{ points}} = \frac{48 \div 100 \text{ minutes}}{100 \div 100 \text{ points}}$$
$$= 0.48 \approx 0.5 \text{ minute/point}$$

9. $$\frac{\$652.80}{40 \text{ hours}} = \frac{\$652.80 \div 40}{40 \div 40 \text{ hours}} = \$16.32/\text{hour}$$

Her regular hourly pay rate is $16.32/hour.

$$\frac{\$195.84}{8 \text{ hours}} = \frac{\$195.84 \div 8}{8 \div 8 \text{ hours}} = \$24.48/\text{hour}$$

Her overtime hourly pay rate is $24.48/hour.

11.

Size	Cost per Unit	
11 ounces	$\dfrac{\$6.79}{11 \text{ ounces}} \approx \$0.62/\text{oz}$	
12 ounces	$\dfrac{\$7.24}{12 \text{ ounces}} = \$0.60/\text{oz}$	$(*)$
16 ounces	$\dfrac{\$10.99}{16 \text{ ounces}} \approx \$0.69/\text{oz}$	

The best buy is 12 ounces for $7.24, about $0.60 per ounce.

13. $$\frac{28}{21} = \frac{44}{33}$$

$$\frac{28 \div 7}{21 \div 7} = \frac{4}{3} \text{ and } \frac{44 \div 11}{33 \div 11} = \frac{4}{3}$$

Both ratios are equivalent to $\frac{4}{3}$, so the proportion is *true*.

Note: We could also use the method of finding cross products and show that both cross products equal 924.

15. $$\frac{2\frac{5}{8}}{3\frac{1}{4}} = \frac{21}{26}$$

Simplify fractions:

$$\frac{2\frac{5}{8}}{3\frac{1}{4}} = \frac{\frac{21}{8}}{\frac{13}{4}} = \frac{21}{\underset{2}{\cancel{8}}} \cdot \frac{\overset{1}{\cancel{4}}}{13} = \frac{21}{26}$$

$$\frac{21}{26} = \frac{21}{26}$$

Or find cross products:

$$3\frac{1}{4} \cdot 21 = \frac{13}{4} \cdot \frac{21}{1} = \frac{273}{4}$$

$$2\frac{5}{8} \cdot 26 = \frac{21}{\underset{4}{\cancel{8}}} \cdot \frac{\overset{13}{\cancel{26}}}{1} = \frac{273}{4}$$

The cross products are *equal*, so the proportion is *true*.

Copyright © 2018 Pearson Education, Inc.

17.
$$\frac{15}{8} = \frac{6}{x}$$

$15 \cdot x = 8 \cdot 6$ *Cross products are equal*

$$\frac{\overset{1}{\cancel{15}} \cdot x}{\underset{1}{\cancel{15}}} = \frac{8 \cdot \overset{2}{\cancel{6}}}{\underset{5}{\cancel{15}}} = \frac{16}{5} = 3.2$$

Check: $15(3.2) = 48 = 8 \cdot 6$

19.
$$\frac{10}{11} = \frac{x}{4}$$

$11 \cdot x = 10 \cdot 4$ *Cross products are equal*

$$\frac{\overset{1}{\cancel{11}} \cdot x}{\underset{1}{\cancel{11}}} = \frac{40}{11} \approx 3.64$$

Check: $10 \cdot 4 = 40 = 11 \cdot \frac{40}{11}$

21.
$$\frac{2.6}{x} = \frac{13}{7.8}$$

$13 \cdot x = 2.6(7.8)$ *Cross products are equal*

$$\frac{\overset{1}{\cancel{13}} \cdot x}{\underset{1}{\cancel{13}}} = \frac{20.28}{13} = 1.56$$

Check: $2.6(7.8) = 20.28 = 1.56(13)$

23.
$$\frac{\frac{1}{3}}{8} = \frac{x}{24}$$

$8 \cdot x = \frac{1}{3} \cdot 24$ *Cross products are equal*

$$\frac{\overset{1}{\cancel{8}} \cdot x}{\underset{1}{\cancel{8}}} = \frac{8}{8} = 1$$

Check: $\frac{1}{3}(24) = 8 = 8 \cdot 1$

5.5 Solving Application Problems with Proportions

5.5 Margin Exercises

1. **(a)**
$$\frac{2 \text{ pounds}}{50 \text{ square feet}} = \frac{x \text{ pounds}}{\underline{225} \text{ square feet}}$$

$50 \cdot x = 2 \cdot 225$ *Cross products are equal*

$$\frac{50 \cdot x}{50} = \frac{450}{50}$$ *Divide both sides by 50*

$$x = 9$$

9 pounds of fertilizer are needed for 225 square feet.

(b)
$$\frac{1 \text{ inch}}{75 \text{ miles}} = \frac{4.75 \text{ inches}}{x \text{ miles}}$$

$1 \cdot x = 75(4.75)$ *Cross products are equal*

$x = 356.25 \approx 356 \text{ miles}$

The lake's actual length is about 356 miles.

(c)
$$\frac{30 \text{ milliliters}}{100 \text{ pounds}} = \frac{x \text{ milliliters}}{34 \text{ pounds}}$$

$100 \cdot x = 30 \cdot 34$ *Cross products are equal.*

$$\frac{100 \cdot x}{100} = \frac{1020}{100}$$ *Divide both sides by 100*

$x = 10.2 \approx 10 \text{ milliliters}$

A 34-pound child should be given about 10 milliliters of cough syrup.

2. **(a)** *Correct set-up*
$$\frac{\text{(lose weight) 2 people}}{\text{(surveyed) 3 people}} = \frac{x \text{ (lose weight)}}{\underline{150} \text{ people (in group)}}$$

$3 \cdot x = 2 \cdot 150$ *Cross products are equal*

$$\frac{3 \cdot x}{3} = \frac{300}{3}$$ *Divide both sides by 3*

$$x = 100$$

In a group of 150 people, 100 people want to lose weight. This is a reasonable answer because it's more than half the people, but less than all the people.

Incorrect set-up
$$\frac{\text{(lose weight) 2 people}}{\text{(surveyed) 3 people}} = \frac{150 \text{ people (in group)}}{x \text{ (lose weight)}}$$

$$2 \cdot x = 3 \cdot 150$$

$$\frac{2 \cdot x}{2} = \frac{450}{2}$$

$$x = 225$$

The incorrect set-up gives an answer of 225 people, which is unreasonable since there are only 150 people in the group.

(b)
$$\frac{3 \text{ FA students}}{5 \text{ students}} = \frac{x \text{ students}}{4500 \text{ students}}$$

$5 \cdot x = 3 \cdot 4500$ *Cross products are equal*

$$\frac{5 \cdot x}{5} = \frac{13,500}{5}$$ *Divide both sides by 5*

$$x = 2700$$

At Central Community College 2700 students receive financial aid. This is a reasonable answer because it's more than half the people, but less than all the people.

(continued)

Copyright © 2018 Pearson Education, Inc.

Incorrect set-up

$$\frac{5 \text{ students}}{3 \text{ FA students}} = \frac{x \text{ students}}{4500 \text{ students}}$$

$$3 \cdot x = 5 \cdot 4500$$

$$\frac{3 \cdot x}{3} = \frac{22{,}500}{3}$$

$$x = 7500$$

The incorrect set-up gives an answer of 7500 students, which is unreasonable since there are only 4500 students at the college.

5.5 Section Exercises

1. $\dfrac{6 \text{ potatoes}}{4 \text{ eggs}} = \dfrac{12 \text{ potatoes}}{x \text{ eggs}}$

3. Here's one set-up:

$$\frac{4 \text{ cartoon strips}}{5 \text{ hours}} = \frac{18 \text{ cartoon strips}}{x \text{ hours}}$$

Here's another set-up and solution:

$$\frac{5 \text{ hours}}{4 \text{ cartoon strips}} = \frac{x \text{ hours}}{18 \text{ cartoon strips}}$$

$$\frac{5}{4} = \frac{x}{18}$$

$$4 \cdot x = 5 \cdot 18$$

$$4 \cdot x = 90$$

$$\frac{\overset{1}{\cancel{4}} \cdot x}{\underset{1}{\cancel{4}}} = \frac{90}{4}$$

$$x = 22.5$$

It will take 22.5 hours to sketch 18 cartoon strips.

5. $\dfrac{4 \text{ cookies}}{150 \text{ calories}} = \dfrac{10 \text{ cookies}}{x \text{ calories}}$

$$4 \cdot x = 150 \cdot 10$$

$$\frac{4 \cdot x}{4} = \frac{1500}{4}$$

$$x = 375$$

There are 375 calories in 10 cookies.

7. $\dfrac{3 \text{ pounds}}{350 \text{ square feet}} = \dfrac{x \text{ pounds}}{4900 \text{ square feet}}$

$$350 \cdot x = 3 \cdot 4900$$

$$\frac{350 \cdot x}{350} = \frac{14{,}700}{350}$$

$$x = 42$$

42 pounds are needed to cover 4900 square feet.

9. $\dfrac{\$672.80}{5 \text{ days}} = \dfrac{\$x}{3 \text{ days}}$

$$5 \cdot x = (672.80)(3)$$

$$\frac{5 \cdot x}{5} = \frac{2018.40}{5}$$

$$x = 403.68$$

In 3 days Tom makes \$403.68.

11. $\dfrac{6 \text{ ounces}}{7 \text{ servings}} = \dfrac{x \text{ ounces}}{12 \text{ servings}}$

$$7 \cdot x = 6 \cdot 12$$

$$\frac{7 \cdot x}{7} = \frac{72}{7}$$

$$x = 10.3 \approx 10 \text{ (rounded)}$$

You need about 10 ounces for 12 servings.

13. $\dfrac{3 \text{ quarts}}{270 \text{ square feet}} = \dfrac{x \text{ quarts}}{(350 + 100) \text{ square feet}}$

$$\frac{3}{270} = \frac{x}{450}$$

$$270 \cdot x = 3 \cdot 450$$

$$\frac{270 \cdot x}{270} = \frac{1350}{270}$$

$$x = 5$$

You will need 5 quarts.

15. First find the length.

$$\frac{1 \text{ inch}}{4 \text{ feet}} = \frac{3.5 \text{ inches}}{x \text{ feet}}$$

$$1 \cdot x = 4(3.5)$$

$$x = 14$$

The kitchen is 14 feet long.

Then find the width.

$$\frac{1 \text{ inch}}{4 \text{ feet}} = \frac{2.5 \text{ inches}}{x \text{ feet}}$$

$$1 \cdot x = 4(2.5)$$

$$x = 10$$

The kitchen is 10 feet wide.

17. The length of the dining area is the same as the length of the kitchen, which is 14 feet from Exercise 15.

Find the width of the dining area.
$4.5 - 2.5$ inches $= 2$ inches on the floor plan.

$$\frac{1 \text{ inch}}{4 \text{ feet}} = \frac{2 \text{ inches}}{x \text{ feet}}$$

$$1 \cdot x = 4 \cdot 2$$

$$x = 8$$

The dining area is 8 feet wide.

Copyright © 2018 Pearson Education, Inc.

19. Set up and solve a proportion with pieces of chicken and number of guests.

$$\frac{40 \text{ pieces}}{25 \text{ guests}} = \frac{x \text{ pieces}}{60 \text{ guests}}$$
$$25 \cdot x = 40 \cdot 60$$
$$\frac{25 \cdot x}{25} = \frac{40 \cdot 60}{25}$$
$$x = \frac{40 \cdot 60}{25} = 96 \text{ pieces of chicken}$$

For the other food items, the proportions are similar, so we simply replace "40" in the last step with the appropriate value.

$$\frac{14 \cdot 60}{25} = 33.6 \text{ pounds of lasagna}$$

$$\frac{4.5 \cdot 60}{25} = 10.8 \text{ pounds of deli meats}$$

$$\frac{\frac{7}{3} \cdot 60}{25} = \frac{28}{5} = 5\frac{3}{5} \text{ pounds of cheese}$$

$$\frac{3 \text{ dozen} \cdot 60}{25} = 7.2 \text{ dozen (about 86) buns}$$

$$\frac{6 \cdot 60}{25} = 14.4 \text{ pounds of salad}$$

21. 44 students is an *unreasonable* answer because there are only 35 students in the class.

23.
$$\frac{7 \text{ refresher}}{10 \text{ entering}} = \frac{x \text{ refresher}}{2950 \text{ entering}}$$
$$10 \cdot x = 7 \cdot 2950$$
$$\frac{10 \cdot x}{10} = \frac{20,650}{10}$$
$$x = 2065$$

2065 students will probably need a refresher course. This is a reasonable answer because it's more than half the students, but not all the students.

Incorrect set-up

$$\frac{10 \text{ entering}}{7 \text{ refresher}} = \frac{x \text{ refresher}}{2950 \text{ entering}}$$
$$7 \cdot x = 10 \cdot 2950$$
$$\frac{7 \cdot x}{7} = \frac{29,500}{7}$$
$$x \approx 4214$$

The *incorrect* set-up gives an estimate of 4214 entering students, which is *unreasonable* since there are only 2950 entering students.

25.
$$\frac{3 \text{ chooses vanilla}}{4 \text{ people}} = \frac{x \text{ choose vanilla}}{250 \text{ people}}$$
$$4 \cdot x = 3 \cdot 250$$
$$\frac{4 \cdot x}{4} = \frac{750}{4}$$
$$x = 187.5 \approx 188$$

You would expect about 188 people to choose vanilla ice cream. This is a reasonable answer.

Incorrect set-up

$$\frac{4 \text{ people}}{3 \text{ chooses vanilla}} = \frac{x \text{ choose vanilla}}{250 \text{ people}}$$
$$3 \cdot x = 4 \cdot 250$$
$$\frac{3 \cdot x}{3} = \frac{1000}{3}$$
$$x = 333.3 \approx 333$$

With an incorrect set-up, 333 people choose vanilla ice cream. This is unreasonable because only 250 people attended the ice cream social.

27.
$$\frac{5 \text{ stocks up}}{6 \text{ stocks down}} = \frac{x \text{ stocks up}}{750 \text{ stocks down}}$$
$$\frac{5}{6} = \frac{x}{750}$$
$$6 \cdot x = 5 \cdot 750$$
$$\frac{6 \cdot x}{6} = \frac{3750}{6}$$
$$x = 625$$

625 stocks went up.

29.
$$\frac{8 \text{ length}}{1 \text{ width}} = \frac{32.5 \text{ meters length}}{x \text{ meters width}}$$
$$8 \cdot x = 1 \cdot 32.5$$
$$\frac{\overset{1}{\cancel{8}} \cdot x}{\underset{1}{\cancel{8}}} = \frac{32.5}{8}$$
$$x = 4.0625 \approx 4.06$$

The wing must be about 4.06 meters wide.

31. Tali did not use the correct numbers in the proportion. The problem compares a person's weight to the number of calories burned and does not involve time. The correct proportion and solution is

$$\frac{150 \text{ pounds}}{210 \text{ calories}} = \frac{180 \text{ pounds}}{x \text{ calories}}$$
$$150 \cdot x = 210 \cdot 180$$
$$\frac{150 \cdot x}{150} = \frac{37,800}{150}$$
$$x = 252$$

A 210-pound person would burn 252 calories.

Copyright © 2018 Pearson Education, Inc.

33.
$$\frac{1.05 \text{ meters (shadow)}}{1.68 \text{ meters (height)}} = \frac{6.58 \text{ meters (shadow)}}{x \text{ meters (height)}}$$
$$1.05 \cdot x = 1.68(6.58)$$
$$\frac{1.05 \cdot x}{1.05} = \frac{11.0544}{1.05}$$
$$x = 10.528 \approx 10.53$$

The height of the tree is about 10.53 meters.

35. You cannot solve this problem using a proportion because the ratio of age to weight is not constant. As Jim's age increases from 25 to 50 years old, his weight may decrease, stay the same, or increase.

37. First find the number of coffee drinkers.
$$\frac{4}{5} = \frac{x}{56,100}$$
$$5 \cdot x = 4 \cdot 56,100$$
$$\frac{5 \cdot x}{5} = \frac{224,400}{5}$$
$$x = 44,880$$

The survey showed that 44,880 students drink coffee. Now find the number of coffee drinkers who use cream.
$$\frac{1}{8} = \frac{x}{44,880}$$
$$8 \cdot x = 44,880$$
$$\frac{8 \cdot x}{8} = \frac{44,880}{8}$$
$$x = 5610$$

According to the survey, 5610 students use cream.

39. First find the number of calories in a $\frac{1}{2}$-cup serving.
$$\frac{\frac{1}{3} \text{ cup}}{80 \text{ calories}} = \frac{\frac{1}{2} \text{ cup}}{x \text{ calories}}$$
$$\frac{1}{3} \cdot x = 80 \cdot \frac{1}{2}$$
$$\frac{\frac{1}{3} \cdot x}{\frac{1}{3}} = \frac{40}{\frac{1}{3}} = \frac{40}{1} \cdot \frac{3}{1}$$
$$x = 120 \text{ calories}$$

Then find the number of grams of sugar in a $\frac{1}{2}$-cup serving.
$$\frac{\frac{1}{3} \text{ cup}}{8 \text{ grams of sugar}} = \frac{\frac{1}{2} \text{ cup}}{x \text{ grams of sugar}}$$
$$\frac{1}{3} \cdot x = 8 \cdot \frac{1}{2}$$

$$\frac{\frac{1}{3} \cdot x}{\frac{1}{3}} = \frac{4}{\frac{1}{3}} = \frac{4}{1} \cdot \frac{3}{1}$$
$$x = 12 \text{ grams of sugar}$$

A $\frac{1}{2}$-cup serving of diet ice cream has 120 calories and 12 grams of sugar.

41. *Use proportions.*

Water:
$$\frac{1\frac{1}{3} \text{ cups}}{4 \text{ servings}} = \frac{x \text{ cups}}{2 \text{ servings}}$$
$$4 \cdot x = 1\frac{1}{3} \cdot 2$$
$$4 \cdot x = \frac{4}{3} \cdot 2$$
$$\frac{4 \cdot x}{4} = \frac{8}{3} \div 4$$
$$x = \frac{8}{3} \cdot \frac{1}{4}$$
$$x = \frac{2}{3} \text{ cup}$$

Margarine/butter:
$$\frac{2 \text{ Tbsp}}{4 \text{ servings}} = \frac{x \text{ Tbsp}}{2 \text{ servings}}$$
$$4 \cdot x = 2 \cdot 2$$
$$\frac{4 \cdot x}{4} = \frac{4}{4}$$
$$x = 1 \text{ Tbsp}$$

Milk:
$$\frac{\frac{2}{3} \text{ cups}}{4 \text{ servings}} = \frac{x \text{ cups}}{2 \text{ servings}}$$
$$12 \cdot x = \frac{2}{3} \cdot 2$$
$$4 \cdot x = \frac{4}{3}$$
$$\frac{4 \cdot x}{4} = \frac{4}{3} \div 4$$
$$x = \frac{4}{3} \cdot \frac{1}{4}$$
$$x = \frac{1}{3} \text{ cup}$$

Potato flakes:
$$\frac{1\frac{1}{2} \text{ cups}}{4 \text{ servings}} = \frac{x \text{ cups}}{2 \text{ servings}}$$
$$4 \cdot x = 1\frac{1}{2} \cdot 2$$
$$4 \cdot x = \frac{3}{2} \cdot 2 = 3$$
$$\frac{4 \cdot x}{4} = \frac{3}{4}$$
$$x = \frac{3}{4} \text{ cup}$$

(continued)

Copyright © 2018 Pearson Education, Inc.

Alternatively, multiply the quantities by $\frac{1}{2}$ (or divide by 2), since *2 servings is $\frac{1}{2}$ of 4 servings.*

Water: $\qquad \dfrac{4}{3} \div 2 = \dfrac{4}{3} \cdot \dfrac{1}{2} = \dfrac{2}{3}$

Margarine/butter: $\dfrac{2}{2} = 1$

Milk: $\qquad \dfrac{2}{3} \div 2 = \dfrac{2}{3} \cdot \dfrac{1}{2} = \dfrac{1}{3}$

Potato flakes: $\dfrac{3}{2} \div 2 = \dfrac{3}{2} \cdot \dfrac{1}{2} = \dfrac{3}{4}$

For 2 servings, use $\frac{2}{3}$ cup water, 1 Tbsp margarine or butter, $\frac{1}{3}$ cup milk, and $\frac{3}{4}$ cup potato flakes.

42.   *Use proportions.*

Water: $\qquad \dfrac{1\frac{1}{3} \text{ cups}}{4 \text{ servings}} = \dfrac{x \text{ cups}}{6 \text{ servings}}$

$$4 \cdot x = 1\dfrac{1}{3} \cdot 6$$
$$4 \cdot x = \dfrac{4}{3} \cdot 6$$
$$\dfrac{4 \cdot x}{4} = \dfrac{8}{4}$$
$$x = 2 \text{ cup}$$

Margarine/butter: $\dfrac{2 \text{ Tbsp}}{4 \text{ servings}} = \dfrac{x \text{ Tbsp}}{6 \text{ servings}}$

$$4 \cdot x = 2 \cdot 6$$
$$\dfrac{4 \cdot x}{4} = \dfrac{12}{4}$$
$$x = 3 \text{ Tbsp}$$

Milk: $\qquad \dfrac{\frac{2}{3} \text{ cups}}{4 \text{ servings}} = \dfrac{x \text{ cups}}{6 \text{ servings}}$

$$4 \cdot x = \dfrac{2}{3} \cdot 6$$
$$4 \cdot x = 4$$
$$\dfrac{4 \cdot x}{4} = \dfrac{4}{4}$$
$$x = 1 \text{ cup}$$

Potato flakes: $\dfrac{1\frac{1}{2} \text{ cups}}{4 \text{ servings}} = \dfrac{x \text{ cups}}{6 \text{ servings}}$

$$4 \cdot x = 1\dfrac{1}{2} \cdot 6$$
$$4 \cdot x = \dfrac{3}{2} \cdot 6$$
$$\dfrac{4 \cdot x}{4} = \dfrac{9}{4}$$
$$x = 2\dfrac{1}{4} \text{ cups}$$

Alternatively, multiply the quantities in Exercise 41 by 3, *since 6 servings is 3 times 2 servings.*

Water: $\qquad \dfrac{2}{3} \cdot 3 = 2$

Margarine/butter: $\quad 1 \cdot 3 = 3$

Milk: $\qquad \dfrac{1}{3} \cdot 3 = 1$

Potato flakes: $\quad \dfrac{3}{4} \cdot 3 = \dfrac{9}{4} = 2\dfrac{1}{4}$

For 6 servings, use 2 cups water, 3 Tbsp margarine or butter, 1 cup milk, and $2\frac{1}{4}$ cups potato flakes.

Chapter 5 Review Exercises

1.   orca whale's length to whale shark's length

$$\dfrac{30 \text{ ft}}{40 \text{ ft}} = \dfrac{30}{40} = \dfrac{30 \div 10}{40 \div 10} = \dfrac{3}{4}$$

2.   blue whale's length to great white shark's length

$$\dfrac{80 \text{ ft}}{20 \text{ ft}} = \dfrac{80}{20} = \dfrac{80 \div 20}{20 \div 20} = \dfrac{4}{1}$$

3.   To get a ratio of $\frac{1}{2}$, we can start with the smallest value in the table and see if there is a value that is two times the smallest. $2 \times 20 = 40$, so the ratio of the *great white shark's* length to the *whale shark's* length is $\frac{20}{40} = \frac{1}{2}$. The length of the orca whale is 30 ft, but $2 \times 30 = 60$ is not in the table. $2 \times 40 = 80$, so the ratio of the *whale shark's* length to the *blue whale's* length is $\frac{40}{80} = \frac{1}{2}$.

4.   \$2.50 to \$1.25

$$\dfrac{\$2.50}{\$1.25} = \dfrac{2.50}{1.25} = \dfrac{2.50 \div 1.25}{1.25 \div 1.25} = \dfrac{2}{1}$$

5.   \$0.30 to \$0.45

$$\dfrac{\$0.30}{\$0.45} = \dfrac{0.30}{0.45} = \dfrac{0.30 \div 0.15}{0.45 \div 0.15} = \dfrac{2}{3}$$

6. $1\frac{2}{3}$ cups to $\frac{2}{3}$ cup

$$\dfrac{1\frac{2}{3} \text{ cups}}{\frac{2}{3} \text{ cup}} = \dfrac{1\frac{2}{3}}{\frac{2}{3}} = \dfrac{5}{3} \div \dfrac{2}{3}$$
$$= \dfrac{5}{\overset{1}{\cancel{3}}} \cdot \dfrac{\overset{1}{\cancel{3}}}{2} = \dfrac{5}{2}$$

7. $2\frac{3}{4}$ miles to $16\frac{1}{2}$ miles

$$\dfrac{2\frac{3}{4} \text{ miles}}{16\frac{1}{2} \text{ miles}} = \dfrac{2\frac{3}{4}}{16\frac{1}{2}} = \dfrac{\frac{11}{4}}{\frac{33}{2}} = \dfrac{11}{4} \div \dfrac{33}{2}$$
$$= \dfrac{\overset{1}{\cancel{11}}}{\underset{2}{\cancel{4}}} \cdot \dfrac{\overset{1}{\cancel{2}}}{\underset{3}{\cancel{33}}} = \dfrac{1}{6}$$

Copyright © 2018 Pearson Education, Inc.

8. 5 hours to 100 minutes

5 hours $= 5 \cdot 60$ minutes $= 300$ minutes

$$\frac{300 \text{ minutes}}{100 \text{ minutes}} = \frac{300}{100} = \frac{300 \div 100}{100 \div 100} = \frac{3}{1}$$

9. 9 inches to 2 feet

2 feet $= 24$ inches

$$\frac{9 \text{ inches}}{24 \text{ inches}} = \frac{9}{24} = \frac{9 \div 3}{24 \div 3} = \frac{3}{8}$$

10. 1 ton to 1500 pounds

1 ton $= 2000$ pounds

$$\frac{2000 \text{ pounds}}{1500 \text{ pounds}} = \frac{2000}{1500} = \frac{2000 \div 500}{1500 \div 500} = \frac{4}{3}$$

11. 8 hours to 3 days

3 days $= 3 \cdot 24$ hours $= 72$ hours

$$\frac{8 \text{ hours}}{72 \text{ hours}} = \frac{8}{72} = \frac{8 \div 8}{72 \div 8} = \frac{1}{9}$$

12. Ramona's $500 to Jake's $350

$$\frac{\$500}{\$350} = \frac{500 \div 50}{350 \div 50} = \frac{10}{7}$$

The ratio of Ramona's sales to Jake's sales is $\frac{10}{7}$.

13. $\dfrac{35 \text{ miles per gallon}}{25 \text{ miles per gallon}} = \dfrac{35}{25} = \dfrac{35 \div 5}{25 \div 5} = \dfrac{7}{5}$

The ratio of the new car's mileage to the old car's mileage is $\frac{7}{5}$.

14. $\dfrac{6000 \text{ students}}{7200 \text{ students}} = \dfrac{6000}{7200} = \dfrac{6000 \div 1200}{7200 \div 1200} = \dfrac{5}{6}$

The ratio of the math students to the English students is $\frac{5}{6}$.

15. $88 for 8 dozen

$$\frac{\$88 \div 8}{8 \text{ dozen} \div 8} = \frac{\$11}{1 \text{ dozen}}$$

16. 96 children in 40 families

$$\frac{96 \text{ children} \div 8}{40 \text{ families} \div 8} = \frac{12 \text{ children}}{5 \text{ families}}$$

17. 4 pages in 20 minutes

(i) $\dfrac{4 \text{ pages} \div 20}{20 \text{ minutes} \div 20} = \dfrac{0.2 \text{ page}}{1 \text{ minute}}$

$= 0.2$ page/minute

or $\dfrac{1}{5}$ page/minute

(ii) $\dfrac{20 \text{ minutes} \div 4}{4 \text{ pages} \div 4} = \dfrac{5 \text{ minutes}}{1 \text{ page}}$

$= 5$ minutes/page

18. $60 in 3 hours

(i) $\dfrac{\$60 \div 3}{3 \text{ hours} \div 3} = \dfrac{\$20}{1 \text{ hour}} = \$20/\text{hour}$

(ii) $\dfrac{3 \text{ hours} \div 60}{\$60 \div 60} = \dfrac{0.05 \text{ hour}}{\$1}$

$= 0.05$ hour/dollar

or $\dfrac{1}{20}$ hour/dollar

19.

Size	Cost per Unit	
8 ounces	$\dfrac{\$4.98}{8 \text{ ounces}} \approx \$0.623/\text{oz}$	$(*)$
3 ounces	$\dfrac{\$2.49}{3 \text{ ounces}} = \$0.83/\text{oz}$	
2 ounces	$\dfrac{\$1.89}{2 \text{ ounces}} = \$0.945/\text{oz}$	

The best buy is 8 ounces for $4.98, about $0.623/ounce.

20. 35.2 pounds for $36.96 $- \$1$ (coupon) $= \$35.96$

$$\frac{\$35.96}{35.2 \text{ pounds}} \approx \$1.022/\text{pound}$$

17.6 pounds for $18.69 $- \$1$ (coupon) $= \$17.69$

$$\frac{\$17.69}{17.6 \text{ pounds}} \approx \$1.005/\text{pound} \ (*)$$

3.5 pounds for $4.25

$$\frac{\$4.25}{3.5 \text{ pounds}} \approx \$1.214/\text{pound}$$

The best buy is 17.6 pounds for $18.69 with the $1 coupon, about $1.005/pound.

21. $\dfrac{6}{10} = \dfrac{9}{15}$

$$\frac{6 \div 2}{10 \div 2} = \frac{3}{5} \text{ and } \frac{9 \div 3}{15 \div 3} = \frac{3}{5}$$

Both ratios are equal to $\frac{3}{5}$, so the proportion is *true*.

Cross products: $\begin{aligned} 6 \cdot 15 &= 90 \\ 10 \cdot 9 &= 90 \end{aligned}$

The cross products are *equal,* so the proportion is *true.*

Copyright © 2018 Pearson Education, Inc.

22. $\dfrac{6}{48} = \dfrac{9}{36}$

$$\dfrac{6 \div 6}{48 \div 6} = \dfrac{1}{8} \text{ and } \dfrac{9 \div 9}{36 \div 9} = \dfrac{1}{4}$$

The ratios are not equal, so the proportion is *false.*

Cross products: $\begin{aligned} 48 \cdot 9 &= 432 \\ 6 \cdot 36 &= 216 \end{aligned}$

The cross products are *unequal,* so the proportion is *false.*

23. $\dfrac{47}{10} = \dfrac{98}{20}$

$$\dfrac{98 \div 2}{20 \div 2} = \dfrac{49}{10}$$

The ratios are not equal, so the proportion is *false.*

Cross products: $\begin{aligned} 10 \cdot 98 &= 980 \\ 47 \cdot 20 &= 940 \end{aligned}$

The cross products are *unequal,* so the proportion is *false.*

24. $\dfrac{64}{36} = \dfrac{96}{54}$

$$\dfrac{64 \div 4}{36 \div 4} = \dfrac{16}{9} \text{ and } \dfrac{96 \div 6}{54 \div 6} = \dfrac{16}{9}$$

Both ratios are equal to $\frac{16}{9}$, so the proportion is *true.*

Cross products: $\begin{aligned} 64 \cdot 54 &= 3456 \\ 36 \cdot 96 &= 3456 \end{aligned}$

The cross products are *equal,* so the proportion is *true.*

25. $\dfrac{1.5}{2.4} = \dfrac{2}{3.2}$

$$\dfrac{1.5 \div 0.3}{2.4 \div 0.3} = \dfrac{5}{8} \text{ and }$$

$$\dfrac{2 \div 2}{3.2 \div 2} = \dfrac{1}{1.6} = \dfrac{1}{\frac{16}{10}} = \dfrac{10}{16} = \dfrac{5}{8}$$

Both ratios are equal to $\frac{5}{8}$, so the proportion is *true.*

Cross products: $\begin{aligned} 2.4 \cdot 2 &= 4.8 \\ 1.5(3.2) &= 4.8 \end{aligned}$

The cross products are *equal,* so the proportion is *true.*

26. $\dfrac{3\frac{1}{2}}{2\frac{1}{3}} = \dfrac{6}{4}$

$$\dfrac{3\frac{1}{2}}{2\frac{1}{3}} = \dfrac{\frac{7}{2}}{\frac{7}{3}} = \dfrac{7}{2} \cdot \dfrac{3}{7} = \dfrac{3}{2} \text{ and } \dfrac{6}{4} = \dfrac{3}{2}$$

Both ratios are equal to $\frac{3}{2}$, so the proportion is *true.*

Cross products:

$$2\dfrac{1}{3} \cdot 6 = \dfrac{7}{\cancel{3}} \cdot \dfrac{\overset{2}{\cancel{6}}}{1} = 14$$

$$3\dfrac{1}{2} \cdot 4 = \dfrac{7}{\cancel{2}} \cdot \dfrac{\overset{2}{\cancel{4}}}{1} = 14$$

The cross products are *equal,* so the proportion is *true.*

27. $\dfrac{4}{42} = \dfrac{150}{x}$ **OR** $\dfrac{2}{21} = \dfrac{150}{x}$ *Cross products are equal*

$$2 \cdot x = 21 \cdot 150$$

$$\dfrac{2 \cdot x}{2} = \dfrac{3150}{2}$$

$$x = 1575$$

Check: $4 \cdot 1575 = 6300 = 42 \cdot 150$

28. $\dfrac{16}{x} = \dfrac{12}{15}$ **OR** $\dfrac{16}{x} = \dfrac{4}{5}$ *Cross products are equal*

$$4 \cdot x = 5 \cdot 16$$

$$\dfrac{4 \cdot x}{4} = \dfrac{80}{4}$$

$$x = 20$$

Check: $16 \cdot 15 = 240 = 20 \cdot 12$

29. $\dfrac{100}{14} = \dfrac{x}{56}$ **OR** $\dfrac{50}{7} = \dfrac{x}{56}$ *Cross products are equal*

$$7 \cdot x = 50 \cdot 56$$

$$\dfrac{7 \cdot x}{7} = \dfrac{2800}{7}$$

$$x = 400$$

Check: $100 \cdot 56 = 5600 = 14 \cdot 400$

30. $\dfrac{5}{8} = \dfrac{x}{20}$

$8 \cdot x = 5 \cdot 20$ *Cross products are equal*

$$\dfrac{8 \cdot x}{8} = \dfrac{100}{8}$$

$$x = 12.5$$

Check: $5 \cdot 20 = 100 = 8(12.5)$

Copyright © 2018 Pearson Education, Inc.

31. $\dfrac{x}{24} = \dfrac{11}{18}$

$18 \cdot x = 24 \cdot 11$ *Cross products are equal*

$\dfrac{18 \cdot x}{18} = \dfrac{264}{18}$

$x = \dfrac{44}{3} \approx 14.67$

Check: $\frac{44}{3} \cdot 18 = 264 = 24 \cdot 11$

32. $\dfrac{7}{x} = \dfrac{18}{21}$ **OR** $\dfrac{7}{x} = \dfrac{6}{7}$

$6 \cdot x = 7 \cdot 7$ *Cross products are equal*

$\dfrac{6 \cdot x}{6} = \dfrac{49}{6}$

$x = \dfrac{49}{6} \approx 8.17$

Check: $7 \cdot 21 = 147 = \frac{49}{6} \cdot 18$

33. $\dfrac{x}{3.6} = \dfrac{9.8}{0.7}$

$0.7 \cdot x = 9.8(3.6)$ *Cross products are equal*

$\dfrac{0.7 \cdot x}{0.7} = \dfrac{35.28}{0.7}$

$x = 50.4$

Check: $50.4(0.7) = 35.28 = 3.6(9.8)$

34. $\dfrac{13.5}{1.7} = \dfrac{4.5}{x}$

$13.5 \cdot x = 1.7(4.5)$ *Cross products are equal*

$\dfrac{13.5 \cdot x}{13.5} = \dfrac{7.65}{13.5}$

$x \approx 0.57$

Check: $13.5\left(\frac{7.65}{13.5}\right) = 7.65 = 1.7(4.5)$

35. $\dfrac{0.82}{1.89} = \dfrac{x}{5.7}$

$1.89 \cdot x = 0.82(5.7)$ *Cross products are equal*

$\dfrac{1.89 \cdot x}{1.89} = \dfrac{4.674}{1.89}$

$x \approx 2.47$

Check: $0.82(5.7) = 4.674 = 1.89\left(\frac{4.674}{1.89}\right)$

36. $\dfrac{3 \text{ cats}}{5 \text{ dogs}} = \dfrac{x}{45 \text{ dogs}}$

$5 \cdot x = 3 \cdot 45$

$\dfrac{5 \cdot x}{5} = \dfrac{135}{5}$

$x = 27$

There are 27 cats.

37. $\dfrac{8 \text{ hits}}{28 \text{ at bats}} = \dfrac{x}{161 \text{ at bats}}$

$28 \cdot x = 8 \cdot 161$

$\dfrac{28 \cdot x}{28} = \dfrac{1288}{28}$

$x = 46$

She will get 46 hits.

38. $\dfrac{3.5 \text{ pounds}}{\$14.84} = \dfrac{5.6 \text{ pounds}}{x}$

$3.5 \cdot x = 14.84(5.6)$

$\dfrac{3.5 \cdot x}{3.5} = \dfrac{83.101}{3.5}$

$x \approx 23.74$

The cost for 5.6 pounds of ground beef is $23.74 (rounded).

39. $\dfrac{4 \text{ voting students}}{10 \text{ students}} = \dfrac{x}{8247 \text{ students}}$

$10 \cdot x = 4 \cdot 8247$

$\dfrac{10 \cdot x}{10} = \dfrac{32{,}988}{10}$

$x \approx 3299$

They should expect about 3299 students to vote.

40. $\dfrac{1 \text{ inch}}{16 \text{ feet}} = \dfrac{4.25 \text{ inches}}{x}$

$1 \cdot x = 16(4.25)$

$x = 68$

The length of the real boxcar is 68 feet.

41.   2 dozen necklaces $= 2 \cdot 12 = 24$ necklaces

$\dfrac{24 \text{ necklaces}}{16\frac{1}{2} \text{ hours}} = \dfrac{40 \text{ necklaces}}{x}$

$24 \cdot x = 16\dfrac{1}{2} \cdot 40 = \dfrac{33}{2} \cdot \dfrac{40}{1}$

$\dfrac{24 \cdot x}{24} = \dfrac{660}{24}$

$x = 27.5 = 27\dfrac{1}{2}$

It will take Marvette $27\frac{1}{2}$ hours or 27.5 hours to make 40 necklaces.

42. $\dfrac{272 \text{ calories}}{25 \text{ minutes}} = \dfrac{x}{45 \text{ minutes}}$

$25 \cdot x = 272 \cdot 45$

$\dfrac{25 \cdot x}{25} = \dfrac{12{,}240}{25}$

$x = 489.6 \approx 490$

A 180-pound person would burn about 490 calories in 45 minutes.

Copyright © 2018 Pearson Education, Inc.

43.
$$\frac{3.5 \text{ milligrams}}{50 \text{ pounds}} = \frac{x}{210 \text{ pounds}}$$
$$50 \cdot x = 3.5 \cdot 210$$
$$\frac{50 \cdot x}{50} = \frac{735}{50}$$
$$x = 14.7$$

A patient who weighs 210 pounds should be given 14.7 milligrams of the medicine.

Chapter 5 Mixed Review Exercises

1.
$$\frac{x}{45} = \frac{70}{30} \text{ OR } \frac{x}{45} = \frac{7}{3}$$
$$3 \cdot x = 7 \cdot 45 \quad \begin{array}{l}\textit{Cross products}\\ \textit{are equal}\end{array}$$
$$\frac{3 \cdot x}{3} = \frac{315}{3}$$
$$x = 105$$

Check: $105 \cdot 30 = 3150 = 45 \cdot 70$

2. $\dfrac{x}{52} = \dfrac{0}{20}$ Since $\frac{0}{20} = 0$, x *must* equal 0.

Note: The denominator 52 could be any number (except 0).

3.
$$\frac{64}{10} = \frac{x}{20} \text{ OR } \frac{32}{5} = \frac{x}{20}$$
$$5 \cdot x = 32 \cdot 20 \quad \begin{array}{l}\textit{Cross products}\\ \textit{are equal}\end{array}$$
$$\frac{5 \cdot x}{5} = \frac{640}{5}$$
$$x = 128$$

Check: $64 \cdot 20 = 1280 = 10 \cdot 128$

4.
$$\frac{15}{x} = \frac{65}{100} \text{ OR } \frac{15}{x} = \frac{13}{20}$$
$$13 \cdot x = 15 \cdot 20 \quad \begin{array}{l}\textit{Cross products}\\ \textit{are equal}\end{array}$$
$$\frac{13 \cdot x}{13} = \frac{300}{13}$$
$$x = \frac{300}{13} \approx 23.08$$

Check: $15 \cdot 100 = 1500 = \frac{300}{13} \cdot 65$

5.
$$\frac{7.8}{3.9} = \frac{13}{x} \text{ OR } \frac{2}{1} = \frac{13}{x}$$
$$2 \cdot x = 1 \cdot 13$$
$$\frac{2 \cdot x}{2} = \frac{13}{2}$$
$$x = 6.5$$

Check: $7.8(6.5) = 50.7 = 3.9 \cdot 13$

6.
$$\frac{34.1}{x} = \frac{0.77}{2.65}$$
$$0.77 \cdot x = 34.1(2.65)$$
$$\frac{0.77 \cdot x}{0.77} = \frac{90.365}{0.77}$$
$$x \approx 117.36$$

Check: $34.1(2.65) = 90.365 = \left(\frac{90.365}{0.77}\right)(0.77)$

7. $\dfrac{55}{18} = \dfrac{80}{27}$

Cross products: $\begin{array}{l}18 \cdot 80 = 1440\\ 55 \cdot 27 = 1485\end{array}$

The cross products are *unequal,* so the proportion is *false.*

8. $\dfrac{5.6}{0.6} = \dfrac{18}{1.94}$

Cross products: $\begin{array}{l}0.6 \cdot 18 = 10.8\\ 5.6(1.94) = 10.864\end{array}$

The cross products are *unequal,* so the proportion is *false.*

9. $\dfrac{\frac{1}{5}}{2} = \dfrac{1\frac{1}{6}}{11\frac{2}{3}}$

Cross products:
$$2 \cdot 1\frac{1}{6} = \frac{\overset{1}{\cancel{2}}}{1} \cdot \frac{7}{\underset{3}{\cancel{6}}} = \frac{7}{3} = 2\frac{1}{3}$$
$$\frac{1}{5} \cdot 11\frac{2}{3} = \frac{1}{\cancel{5}} \cdot \frac{\overset{7}{\cancel{35}}}{3} = \frac{7}{3} = 2\frac{1}{3}$$

The cross products are *equal,* so the proportion is *true.*

10. 4 dollars to 10 quarters

4 dollars $= 4 \cdot 4 = 16$ quarters
$$\frac{16 \text{ quarters} \div 2}{10 \text{ quarters} \div 2} = \frac{8}{5}$$

11. $4\frac{1}{8}$ inches to 10 inches
$$\frac{4\frac{1}{8} \text{ inches}}{10 \text{ inches}} = \frac{\frac{33}{8}}{10} = \frac{33}{8} \div 10$$
$$= \frac{33}{8} \cdot \frac{1}{10} = \frac{33}{80}$$

12. 10 yards to 8 feet

10 yards $= 10 \cdot 3 = 30$ feet
$$\frac{30 \text{ feet}}{8 \text{ feet}} = \frac{30 \div 2}{8 \div 2} = \frac{15}{4}$$

Copyright © 2018 Pearson Education, Inc.

13. $3.60 to $0.90

$$\frac{\$3.60}{\$0.90} = \frac{3.60 \div 0.90}{0.90 \div 0.90} = \frac{4}{1}$$

14. 12 eggs to 15 eggs

$$\frac{12 \text{ eggs}}{15 \text{ eggs}} = \frac{12 \div 3}{15 \div 3} = \frac{4}{5}$$

15. 37 meters to 7 meters

$$\frac{37 \text{ meters}}{7 \text{ meters}} = \frac{37}{7}$$

16. 3 pints to 4 quarts

$$4 \text{ quarts} = 4 \cdot 2 = 8 \text{ pints}$$

$$\frac{3 \text{ pints}}{8 \text{ pints}} = \frac{3}{8}$$

17. 15 minutes to 3 hours

$$3 \text{ hours} = 3 \cdot 60 = 180 \text{ minutes}$$

$$\frac{15 \text{ minutes}}{180 \text{ minutes}} = \frac{15 \div 15}{180 \div 15} = \frac{1}{12}$$

18. $4\frac{1}{2}$ miles to $1\frac{3}{10}$ miles

$$\frac{4\frac{1}{2} \text{ miles}}{1\frac{3}{10} \text{ miles}} = \frac{4\frac{1}{2}}{1\frac{3}{10}} = 4\frac{1}{2} \div 1\frac{3}{10}$$

$$= \frac{9}{2} \div \frac{13}{10} = \frac{9}{\overset{}{2}} \cdot \frac{\overset{5}{\cancel{10}}}{13}$$
$$\phantom{= \frac{9}{2} \div \frac{13}{10}}_{1}$$

$$= \frac{45}{13}$$

19. $\dfrac{7 \text{ buying fans}}{8 \text{ fans}} = \dfrac{x \text{ buying fans}}{28{,}500 \text{ fans}}$

$$8 \cdot x = 7 \cdot 28{,}500$$
$$\frac{8 \cdot x}{8} = \frac{199{,}500}{8}$$
$$x \approx 24{,}937.5 \text{ or } 24{,}900$$
(rounded to the nearest hundred)

At today's concert, about 24,900 fans can be expected to buy a beverage.

20. $\dfrac{\$400 \text{ spent on car insurance}}{\$150 \text{ spent on repairs}} = \dfrac{400 \div 50}{150 \div 50}$

$$= \frac{8}{3}$$

The ratio of the amount spent on insurance to the amount spent on repairs is $\frac{8}{3}$.

21. 25 feet for $0.78

$$\frac{\$0.78}{25 \text{ feet}} \approx \$0.031/\text{foot}$$

75 feet for $1.99 − $0.50 (coupon) = $1.49

$$\frac{\$1.49}{75 \text{ feet}} \approx \$0.020/\text{foot} (*)$$

100 feet for $2.59 − $0.50 (coupon) = $2.09

$$\frac{\$2.09}{100 \text{ feet}} \approx \$0.021/\text{foot}$$

The best buy is 75 feet for $1.99 with a 50¢ coupon, about $0.020/foot.

22. First find the length.

$$\frac{0.5 \text{ inch}}{6 \text{ feet}} = \frac{1.75 \text{ inches}}{x \text{ feet}}$$
$$0.5 \cdot x = 6(1.75)$$
$$\frac{0.5 \cdot x}{0.5} = \frac{10.5}{0.5}$$
$$x = 21$$

When it is built, the actual length of the patio will be 21 feet.

Then find the width.

$$\frac{0.5 \text{ inch}}{6 \text{ feet}} = \frac{1.25 \text{ inches}}{x \text{ feet}}$$
$$0.5 \cdot x = 6(1.25)$$
$$\frac{0.5 \cdot x}{0.5} = \frac{7.5}{0.5}$$
$$x = 15$$

When it is built, the actual width of the patio will be 15 feet.

23. $\dfrac{0.8 \text{ gallons}}{3 \text{ hours}} = \dfrac{2 \text{ gallons}}{x \text{ hours}}$

$$0.8 \cdot x = 3 \cdot 2$$
$$\frac{0.8 \cdot x}{0.8} = \frac{6}{0.8}$$
$$x = 7.5 \text{ or } 7\frac{1}{2}$$

The lawn mower should run 7.5 or $7\frac{1}{2}$ hours on a full tank.

24. $\dfrac{1\frac{1}{2} \text{ teaspoons}}{24 \text{ pounds}} = \dfrac{x \text{ teaspoons}}{8 \text{ pounds}}$

$$24 \cdot x = 1\frac{1}{2} \cdot 8 = \frac{3}{2} \cdot \frac{8}{1} = \frac{24}{2} = 12$$
$$\frac{24 \cdot x}{24} = \frac{12}{24}$$
$$x = \frac{1}{2} \text{ or } 0.5$$

The infant should be given $\frac{1}{2}$ or 0.5 teaspoon.

Copyright © 2018 Pearson Education, Inc.

25. $\dfrac{251 \text{ points}}{169 \text{ minutes}} = \dfrac{x \text{ points}}{14 \text{ minutes}}$

$$169 \cdot x = 251 \cdot 14$$

$$\dfrac{169 \cdot x}{169} = \dfrac{3514}{169}$$

$$x \approx 20.792899$$

Charles should score 21 points (rounded).

Chapter 5 Test

1. 16 fish to 20 fish

$$\dfrac{16 \text{ fish}}{20 \text{ fish}} = \dfrac{16 \div 4}{20 \div 4} = \dfrac{4}{5}$$

2. 300 miles on 15 gallons

$$\dfrac{300 \text{ miles} \div 15}{15 \text{ gallons} \div 15} = \dfrac{20 \text{ miles}}{1 \text{ gallon}}$$

3. $15 for 75 minutes

$$\dfrac{\$15 \div 15}{75 \text{ minutes} \div 15} = \dfrac{\$1}{5 \text{ minutes}}$$

4. 3 hours to 40 minutes

$$3 \text{ hours} = 3 \cdot 60 \text{ minutes} = 180 \text{ minutes}$$

$$\dfrac{180 \text{ minutes}}{40 \text{ minutes}} = \dfrac{180 \div 20}{40 \div 20} = \dfrac{9}{2}$$

5. $\dfrac{1200 \text{ seats}}{320 \text{ seats}} = \dfrac{1200 \div 80}{320 \div 80} = \dfrac{15}{4}$

6.

Size	Cost per Unit
Small	$\dfrac{\$5.99}{5 \text{ inches}} = \$1.198/\text{inch}$
Regular	$\dfrac{\$6.89}{8 \text{ inches}} \approx \$0.861/\text{inch} \; (*)$
Large	$\dfrac{\$10}{11 \text{ inches}} \approx \$0.909/\text{inch}$

The best buy is the 8-inch sub, which costs about $0.861 per inch.

7. 26 ounces of Brand X for
$3.89 − $0.75 (coupon) = $3.14

$$\dfrac{\$3.14}{26 \text{ ounces}} \approx \$0.121/\text{ounce}$$

16 ounces of Brand Y for
$1.89 − $0.50 (coupon) = $1.39

$$\dfrac{\$1.39}{16 \text{ ounces}} \approx \$0.087/\text{ounce} \; (*)$$

14 ounces of Brand Z for $1.29.

$$\dfrac{\$1.29}{14 \text{ ounces}} \approx \$0.092/\text{ounce}$$

The best buy is 16 ounces of Brand Y for $1.89 with a 50¢ coupon, about $0.087/ounce.

8. You earned less this year.

An example is:

$$\begin{array}{l} \text{Last year} \rightarrow \\ \text{This year} \rightarrow \end{array} \dfrac{\$30,000}{\$20,000} = \dfrac{3}{2}$$

9. $\dfrac{6}{14} = \dfrac{18}{45}$ Cross products: $\begin{array}{l} 14 \cdot 18 = 252 \\ 6 \cdot 45 = 270 \end{array}$

The cross products are *unequal*, so the proportion is *false*. We could also write the fractions in lowest terms to see that they are not equal $\left(\frac{3}{7} \neq \frac{2}{5}\right)$.

10. $\dfrac{8.4}{2.8} = \dfrac{2.1}{0.7}$

$$\dfrac{3}{1} = \dfrac{3}{1} \qquad \textit{Write in lowest terms}$$

The proportion is *true*. We could also show that the cross products are equal (5.88).

11. $\dfrac{5}{9} = \dfrac{x}{45}$

$$9 \cdot x = 5 \cdot 45 \qquad \begin{array}{l}\textit{Cross products} \\ \textit{are equal}\end{array}$$

$$\dfrac{9 \cdot x}{9} = \dfrac{225}{9}$$

$$x = 25$$

Check: $5 \cdot 45 = 225 = 9 \cdot 25$

12. $\dfrac{3}{1} = \dfrac{8}{x}$

$$3 \cdot x = 8 \cdot 1 \qquad \begin{array}{l}\textit{Cross products} \\ \textit{are equal}\end{array}$$

$$\dfrac{3 \cdot x}{3} = \dfrac{8}{3}$$

$$x \approx 2.67$$

Check: $3\left(\frac{8}{3}\right) = 8 = 1 \cdot 8$

13. $\dfrac{x}{20} = \dfrac{6.5}{0.4}$

$$0.4 \cdot x = 20(6.5) \qquad \begin{array}{l}\textit{Cross products} \\ \textit{are equal}\end{array}$$

$$\dfrac{0.4 \cdot x}{0.4} = \dfrac{130}{0.4}$$

$$x = 325$$

Check: $325(0.4) = 130 = 20(6.5)$

Copyright © 2018 Pearson Education, Inc.

14.
$$\frac{2\frac{1}{3}}{x} = \frac{\frac{8}{9}}{4}$$

$$\frac{8}{9} \cdot x = 2\frac{1}{3} \cdot 4 = \frac{7}{3} \cdot \frac{4}{1} = \frac{28}{3}$$

$$\frac{\frac{8}{9} \cdot x}{\frac{8}{9}} = \frac{\frac{28}{3}}{\frac{8}{9}}$$

$$x = \frac{28}{3} \div \frac{8}{9} = \frac{\overset{7}{\cancel{28}}}{\underset{1}{\cancel{3}}} \cdot \frac{\overset{3}{\cancel{9}}}{\underset{2}{\cancel{8}}} = \frac{21}{2} = 10\frac{1}{2}$$

Check: $2\frac{1}{3} \cdot 4 = \frac{28}{3} = \frac{21}{2} \cdot \frac{8}{9}$

15.
$$\frac{18 \text{ orders}}{30 \text{ minutes}} = \frac{x \text{ orders}}{40 \text{ minutes}}$$

$$30 \cdot x = 18 \cdot 40$$

$$\frac{30 \cdot x}{30} = \frac{720}{30}$$

$$x = 24$$

Pedro could enter 24 orders in 40 minutes.

16.
$$\frac{0.8 \text{ ounce}}{50 \text{ square feet}} = \frac{x \text{ ounces}}{225 \text{ square feet}}$$

$$50 \cdot x = 0.8(225)$$

$$\frac{50 \cdot x}{50} = \frac{180}{50}$$

$$x = 3.6$$

3.6 ounces of wildflower seeds are needed for a garden with 225 square feet.

17.
$$\frac{2 \text{ left-handed people}}{15 \text{ people}} = \frac{x \text{ left-handed people}}{650 \text{ students}}$$

$$15 \cdot x = 2 \cdot 650$$

$$\frac{15 \cdot x}{15} = \frac{1300}{15}$$

$$x = \frac{260}{3} \approx 87$$

You could expect about 87 students to be left-handed.

18. No, 4875 cannot be correct because there are only 650 students in the whole school.

19.
$$\frac{8.2 \text{ grams}}{50 \text{ pounds}} = \frac{x \text{ grams}}{145 \text{ pounds}}$$

$$50 \cdot x = 8.2(145)$$

$$\frac{50 \cdot x}{50} = \frac{1189}{50}$$

$$x = 23.78 \approx 23.8$$

A 145-pound person should be given 23.8 grams (rounded).

20.
$$\frac{1 \text{ inch}}{8 \text{ feet}} = \frac{7.5 \text{ inches}}{x \text{ feet}}$$

$$1 \cdot x = 8(7.5)$$

$$x = 60$$

The actual height of the building is 60 feet.

Cumulative Review Exercises (Chapters 1–5)

1. 9903 to the nearest hundred
Draw a cut-off line: 99|03
The first digit cut is 0, which is 4 or less. The part you keep stays the same. **9900**

2. 617.0519 to the nearest tenth:
Draw a cut-off line: 617.0|519
The first digit cut is 5 or more, so round up.
617.1

3. $99.81 to the nearest dollar:
Draw a cut-off line: $99|.81
The first digit cut is 5 or more, so round up.
$100

4. $3.0555 to the nearest cent:
Draw a cut-off line: $3.05|55
The first digit cut is 5 or more, so round up.
$3.06

5. $30 - 0.66$

$$
\begin{array}{r}
9 \\
29\ \overset{10}{\cancel{1}}\overset{10}{0} \\
\cancel{30.0}\ \cancel{0} \\
-\ 0.6\ 6 \\
\hline
29.3\ 4
\end{array}
$$

6.
$$
\begin{array}{r}
6\ 1\ 0 \quad \mathbf{R}27 \\
33\overline{)2\ 0,\ 1\ 5\ 7} \\
\underline{1\ 9\ 8} \\
3\ 5 \\
\underline{3\ 3} \\
2\ 7 \\
\underline{0} \\
2\ 7
\end{array}
$$

7. $(1.9)(0.004)$

$$
\begin{array}{rl}
1.9 & \leftarrow \text{ 1 } \textit{decimal place} \\
\times\ 0.004 & \leftarrow \text{ 3 } \textit{decimal places} \\
\hline
0.0076 & \leftarrow \text{ 4 } \textit{decimal places needed}
\end{array}
$$

8. $3020 - 708$

$$
\begin{array}{r}
{\scriptstyle 2\,10\,1\,10}\\
\cancel{3\,0\,2\,0}\\
-\ \ \ 7\,0\,8\\
\hline
2\,3\,1\,2
\end{array}
$$

9. $0.401 + 62.98 + 5$

$$
\begin{array}{r}
0.401\\
62.980\\
+\ 5.000 \quad \textit{Write in 0's}\\
\hline
68.381
\end{array}
$$

10. $1.39 \div 0.025$

$$
\begin{array}{r}
5\,5.\ 6\\
0.025_\wedge \overline{)1.\,3\,9\,0_\wedge\,0}\ \textit{Move decimal point in}\\
\underline{1\ \ 2\ 5}\quad\quad \textit{divisor and dividend}\\
1\ 4\ 0\\
\underline{1\ 2\ 5}\\
1\ 5\ \ 0\ \textit{Write 0 in}\\
\underline{1\ 5\ \ 0}\ \textit{the dividend}\\
0
\end{array}
$$

11. $36 + 18 \div 6 = 36 + (18 \div 6) = 36 + 3 = 39$

12. $8 \div 4 + \left(10 - 3^2\right) \cdot 4^2 = 8 \div 4 + (10 - 9) \cdot 16$
$$
\begin{aligned}
&= 8 \div 4 + 1 \cdot 16\\
&= 2 + 1 \cdot 16\\
&= 2 + 16 = 18
\end{aligned}
$$

13. $88 \div \sqrt{121} \cdot 2^3 = 88 \div \sqrt{121} \cdot 8$
$$
\begin{aligned}
&= 88 \div 11 \cdot 8\\
&= 8 \cdot 8 = 64
\end{aligned}
$$

14. $(16.2 - 5.85) - 2.35(4) = 10.35 - 2.35(4)$
$$
= 10.35 - 9.4 = 0.95
$$

15. 0.0105 is one hundred five ten-thousandths.

16. Sixty and seventy-one thousandths is 60.071.

17. don't drink coffee to total
$$
\frac{200}{1000} = \frac{200 \div 200}{1000 \div 200} = \frac{1}{5}
$$

18. drink coffee to don't drink coffee
$$
\frac{550 + 250}{200} = \frac{800}{200} = \frac{800 \div 200}{200 \div 200} = \frac{4}{1}
$$

19. First get a decimal representation for each mixed number.

$1\frac{9}{16} = \frac{25}{16}$

$$
\begin{array}{r}
1.\ 5\ 6\ 2\ 5\quad 1\frac{9}{16} \approx 1.563\\
16\overline{)2\,5.\,0\,0\,0\,0}\\
\underline{1\ 6}\\
9\ 0\\
\underline{8\ 0}\\
1\ 0\ 0\\
\underline{9\ 6}\\
4\ 0\\
\underline{3\ 2}\\
8\ 0\\
\underline{8\ 0}\\
0
\end{array}
$$

$1\frac{3}{8} = \frac{11}{8}$

$$
\begin{array}{r}
1.\ 3\ 7\ 5\quad 1\frac{3}{8} = 1.375\\
8\overline{)1\,1.\,0\,0\,0}\\
\underline{8}\\
3\ 0\\
\underline{2\ 4}\\
6\ 0\\
\underline{5\ 6}\\
4\ 0\\
\underline{4\ 0}\\
0
\end{array}
$$

$1\frac{1}{8} = \frac{9}{8}$

$$
\begin{array}{r}
1.\ 1\ 2\ 5\quad 1\frac{1}{8} = 1.125\\
8\overline{)9.\,0\,0\,0}\\
\underline{8}\\
1\ 0\\
\underline{8}\\
2\ 0\\
\underline{1\ 6}\\
4\ 0\\
\underline{4\ 0}\\
0
\end{array}
$$

Now compare all the numbers.

$$1.5 = 1.500$$
$$1\frac{9}{16} \approx 1.563 \quad \textit{largest}$$
$$1.25 = 1.250$$
$$1\frac{3}{8} = 1.375$$
$$1\frac{1}{8} = 1.125 \quad \textit{smallest}$$

The nest box openings (in inches) in order from smallest to largest are:

$$1\frac{1}{8}, \quad 1.25, \quad 1\frac{3}{8}, \quad 1.5, \quad 1\frac{9}{16}$$

Copyright © 2018 Pearson Education, Inc.

20. $1\dfrac{9}{16} - 1.5 = 1\dfrac{9}{16} - 1\dfrac{1}{2} = 1\dfrac{9}{16} - 1\dfrac{8}{16}$

$\qquad = \dfrac{25}{16} - \dfrac{24}{16} = \dfrac{25-24}{16} = \dfrac{1}{16}$

$$1\dfrac{9}{16} = 1.5625$$
$$-1.5 = 1.5000$$
$$\overline{0.0625}$$

The difference is $\frac{1}{16}$ inch or about 0.063 inch.

21. $\quad \dfrac{9}{12} = \dfrac{x}{28}$ **OR** $\dfrac{3}{4} = \dfrac{x}{28}$

$4 \cdot x = 3 \cdot 28$ *Cross products*
are equal

$\dfrac{4 \cdot x}{4} = \dfrac{84}{4}$ *Divide both*
sides by 4

$x = 21$

Check: $9 \cdot 28 = 252 = 12 \cdot 21$

22. $\quad \dfrac{7}{12} = \dfrac{10}{x}$

$7 \cdot x = 10 \cdot 12$ *Cross products*
are equal

$\dfrac{7 \cdot x}{7} = \dfrac{120}{7}$ *Divide both*
sides by 7

$x \approx 17.14$

Check: $7\left(\dfrac{120}{7}\right) = 120 = 12 \cdot 10$

23. $\quad \dfrac{6.7}{x} = \dfrac{62.8}{9.15}$

$62.8 \cdot x = 6.7(9.15)$

$\dfrac{62.8 \cdot x}{62.8} = \dfrac{61.305}{62.8}$

$x \approx 0.98$

Check: $6.7(9.15) = 61.305 = \left(\dfrac{61.305}{62.8}\right)62.8$

24. Find the amount already collected.

$$\dfrac{5}{\cancel{6}_1} \cdot \dfrac{\cancel{1500}^{250}}{1} = 1250$$

They need to collect

$1500 - 1250 = 250$ more pounds.

25. $\quad \dfrac{10 \text{ cm wide}}{15 \text{ cm long}} = \dfrac{x \text{ cm wide}}{40 \text{ cm long}}$

$15 \cdot x = 10 \cdot 40$

$\dfrac{15 \cdot x}{15} = \dfrac{400}{15} = \dfrac{80}{3}$

$x \approx 26.7$

The enlarged width of the photo is 26.7 centimeters (rounded).

26. First add the number of times Norma ran in the morning and in the afternoon.

$$4 + 2\dfrac{1}{2} = 6\dfrac{1}{2}$$

Then multiply the sum by the distance around Dunning Pond.

$$1\dfrac{1}{10} \cdot 6\dfrac{1}{2} = \dfrac{11}{10} \cdot \dfrac{13}{2} = \dfrac{143}{20} = 7\dfrac{3}{20}$$

Norma ran $7\frac{3}{20}$ miles.

27. 20 servings for $2.15

$$\dfrac{\$2.15}{20 \text{ servings}} \approx \$0.108/\text{serving}$$

34 servings for $3.35 − $0.50 (coupon) = $2.85

$$\dfrac{\$2.85}{34 \text{ servings}} \approx \$0.084/\text{serving} \; (*)$$

65 servings for $10.41 − $0.50 (coupon) = $9.91

$$\dfrac{\$9.91}{65 \text{ servings}} \approx \$0.152/\text{serving}$$

The best buy is 34 servings for $3.35 with a 50¢ coupon, about $0.084/serving.

28. $\quad \dfrac{5 \text{ bothered residents}}{8 \text{ residents}} = \dfrac{x \text{ bothered residents}}{224 \text{ residents}}$

$8 \cdot x = 5 \cdot 224$

$\dfrac{8 \cdot x}{8} = \dfrac{1120}{8}$

$x = 140$

You would expect 140 residents to be bothered by noise.

29. $\quad \dfrac{\frac{1}{2} \text{ teaspoon}}{2 \text{ quarts water}} = \dfrac{x \text{ teaspoon}}{5 \text{ quarts water}}$

$2 \cdot x = \dfrac{1}{2} \cdot 5$

$\dfrac{2 \cdot x}{2} = \dfrac{\frac{5}{2}}{2}$

$x = \dfrac{5}{2} \div 2 = \dfrac{5}{2} \cdot \dfrac{1}{2} = \dfrac{5}{4} = 1\dfrac{1}{4}$

Use $1\frac{1}{4}$ teaspoons of plant food.

30. The cost per minute rates (in dollars) from least to greatest are:
$0.005, $0.018, $0.02, $0.084, $0.099, $0.10, $0.106

Copyright © 2018 Pearson Education, Inc.

31. $\dfrac{\$20}{\$0.099/\text{minute}} \approx 202$ minutes

You could make a call to a cell phone in Bolivia for 202 minutes (rounded).

32. $\dfrac{\$20}{\$0.106/\text{minute}} \approx 189$ minutes

You get 189 minutes (rounded) per $20 card, so buy 2 cards (378 minutes) to cover 360 minutes (6 hours).

33. For Argentina:

$$\dfrac{\$20}{\$0.02/\text{minute}} = 1000 \text{ minutes}$$

For Yemen:

$$\dfrac{\$20}{\$0.10/\text{minute}} = 200 \text{ minutes}$$

Ratio of minutes you can call Argentina for $20 to minutes you can call Yemen for $20:

$$\dfrac{1000 \div 200}{200 \div 200} = \dfrac{5}{1}$$

The ratio of minutes for a $20 call to Argentina to minutes for a $20 call to Yemen is $\frac{5}{1}$.

34. It would cost $90 \times \$0.005 \times 24 = \$10.80$ for the 24 90-minute calls to Mexico, leaving $\$20 - \$10.80 = \$9.20$ for the call to the Philippines.

$$\dfrac{\$9.20}{\$0.084/\text{minute}} \approx 109.5 \text{ minutes}$$

109.5 minutes rounds to 110 minutes, but the call would be cut off after 109 minutes.

Copyright © 2018 Pearson Education, Inc.

CHAPTER 6   PERCENT

6.1 Basics of Percent

6.1 Margin Exercises

1.   **(a)** 74 out of 100 is $\frac{74}{100}$ or 74%, so 74% of the adults keep fit by walking.

(b) The tax rate is $6 per $100. The ratio is $\frac{6}{100}$ and the percent of tax is 6%.

(c) 32 out of 100 is $\frac{32}{100}$ or 32%, so 32% of the Americans picked football as their favorite sport.

2.   **(a)** $68\% = 68 \div \underline{100} = \underline{0.68}$

(b) $34\% = 34 \div 100 = 0.34$

(c) $58.5\% = 58.5 \div 100 = 0.585$

(d) $175\% = 175 \div 100 = 1.75$

(e) $200\% = 200 \div 100 = 2.00$ or 2

3.   **(a)** $96\% = 96.\% = \underline{0.96}$

Drop the percent sign and move the decimal point two places to the left.

(b) $6\% = 6.\% = 0.06$

Drop the percent sign. Attach one zero so the decimal point can be moved two places to the left.

(c) $24.8\% = 0.248$

Drop the percent sign and move the decimal point two places to the left.

(d) $0.9\% = 0.\underline{009}$

Drop the percent sign. Attach two zeros so the decimal point can be moved two places to the left.

4.   **(a)** $0.74 = 74\%$
Move the decimal point two places to the right. (The decimal point is not written with whole number percents.) Attach a percent sign.

(b) $0.15 = 15\%$
Move the decimal point two places to the right and attach a percent sign.

(c) $0.09 = 9\%$
Move the decimal point two places to the right and attach a percent sign.

(d) $0.617 = \underline{61.7}\%$
Move the decimal point two places to the right and attach a percent sign.

(e) $0.834 = 83.4\%$
Move the decimal point two places to the right and attach a percent sign.

(f) $5.34 = 534\%$
Move the decimal point two places to the right and attach a percent sign.

(g) $2.8 = 280\%$
One zero is attached so the decimal point can be moved two places to the right. Attach a percent sign.

(h) $4 = 400\%$
Two zeros are attached so the decimal point can be moved two places to the right. Attach a percent sign.

5.   **(a)** 100% is all of the money.
So, 100% of $7.80 is $7.80.

(b) 100% is all of the workers.
So, 100% of 1850 workers is 1850 workers.

(c) 200% is twice ($\underline{2}$ times) as many photographs.
So, 200% of 24 photographs is $2 \cdot 24 = \underline{48}$ photographs.

(d) 300% is 3 times as many miles.
So, 300% of 8 miles is $3 \cdot 8 = 24$ miles.

6.   **(a)** 50% is half of the patients.
So, 50% of 200 patients is

$$\tfrac{1}{2} \text{ of } 200 = \underline{100} \text{ patients.}$$

(b) 50% is half of the tweets.
So, 50% of 64 tweets is 32 tweets.

(c) 10% is $\frac{1}{10}$ of the elm trees.
Move the decimal point <u>one</u> place to the <u>left</u>.
So, 10% of 3850. elm trees is $\underline{385}$ elm trees.

(d) 10% is $\frac{1}{10}$ of the pounds.
Move the decimal point *one* place to the left.
So, 10% of 7. pounds is 0.7 pound.

(e) 1% is $\frac{1}{100}$ of the length.
Move the decimal point <u>two</u> places to the <u>left</u>.
So, 1% of 240. feet is $\underline{2.4}$ feet.

(f) 1% is $\frac{1}{100}$ of the money.
Move the decimal point *two* places to the left.
So, 1% of $3000. is $30.

Copyright © 2018 Pearson Education, Inc.

6.1   Section Exercises

1.   To write a percent as a decimal, drop the <u>%</u> symbol and then divide by <u>100</u>.

3. $12\% = 12.\% = \underline{0}.12$
Drop the percent symbol and move the decimal point two places to the left.

5. $70\% = 70.\% = 0.70$ or 0.7
Drop the percent symbol and move the decimal point two places to the left.

7. $25\% = 25.\% = 0.25$
Drop the percent symbol and move the decimal point two places to the left.

9. $140\% = 140.\% = 1.40$ or 1.4
Drop the percent symbol and move the decimal point two places to the left.

11. $5.5\% = 0.055$
Drop the percent symbol. Attach one zero so the decimal point can be moved two places to the left.

13. $100\% = 100.\% = 1.00$ or 1
Drop the percent symbol and move the decimal point two places to the left.

15. $0.5\% = 0.005$
Drop the percent symbol. Attach two zeros so the decimal point can be moved two places to the left.

17. $0.35\% = 0.0035$
Drop the percent symbol. Attach two zeros so the decimal point can be moved two places to the left.

19.   To write a decimal as a percent, multiply by <u>100</u> and then attach a <u>%</u> symbol.

21. $0.6 = 60\%$
0 is attached so the decimal point can be moved two places to the right. Attach a percent sign. (The decimal point is not written with whole numbers.)

23. $0.01 = 1\%$
Move the decimal point two places to the right and attach a percent symbol.

25. $0.375 = 37.5\%$
Move the decimal point two places to the right and attach a percent symbol.

27. $2 = 200\%$
Two zeros are attached so the decimal point can be moved two places to the right. Attach a percent symbol.

29. $3.7 = 370\%$
One zero is attached so the decimal point can be moved two places to the right. Attach a percent symbol.

31. $0.0312 = 3.12\%$
Move the decimal point two places to the right and attach a percent symbol.

33. $4.162 = 416.2\%$
Move the decimal point two places to the right and attach a percent symbol.

35. $0.0028 = 0.28\%$
Move the decimal point two places to the right and attach a percent symbol.

37.   Answers will vary. Some possibilities are: No common denominators are needed with percents. The denominator is always 100 with percent, which makes comparisons easier to understand.

39.   Drop the percent sign and move the decimal point two places to the left.

Truck drivers: $5.5\% = 0.055$

41.   Move the decimal point two places to the right and attach a percent sign.

Computer systems analysts: $0.209 = 20.9\%$

43.   Drop the percent sign and move the decimal point two places to the left.

Accountants and auditors: $10.7\% = 0.107$

45.   Move the decimal point two places to the right and attach a percent sign.

Customer service representatives: $0.098 = 9.8\%$

47.   Drop the percent sign and move the decimal point two places to the left.

Cooks: $14.3\% = 0.143$

49.   Move the decimal point two places to the right and attach a percent sign.

0.08 of the population $= 8\%$

51.   Drop the percent sign and move the decimal point two places to the left.

30% above normal $= 0.30$ or 0.3

53.   100% is all of the children.
So, 100% of 12 children is 12 children.
12 children are present.

55.   200% is two times as many employees.
So, 200% of 210 employees is $2 \cdot 210 = 420$ employees. There are 420 employees this year.

Copyright © 2018 Pearson Education, Inc.

57. 300% is three times as many chairs.
So, 300% of 90 chairs is $3 \cdot 90 = 270$ chairs.
We'll need 270 chairs.

59. 50% is half of the tuition.
So, 50% of \$755 is $\frac{1}{2} \cdot \$755 = \$377.50$.
Financial aid will pay \$377.50.

61. 10% is found by moving the decimal point one place to the left.
So, 10% of 8200 commuters is 820 commuters.
820 commuters carpool.

63. 1% is found by moving the decimal point two places to the left.
So 1% of 2600 plants is 26 plants.
26 plants are poisonous.

65. **(a)** Since 100% means 100 parts out of 100 parts, 100% is all of the number.

(b) Answers will vary. For example, 100% of \$72 is \$72.

67. **(a)** Since 200% is two times a number, find 200% of the number by multiplying the number by 2 (double it).

(b) Answers will vary. For example, 200% of \$20 is $2 \cdot \$20 = \$40$.

69. **(a)** Since 10% means 10 parts out of 100 parts or $\frac{1}{10}$, the shortcut for finding 10% of a number is to move the decimal point in the number one place to the left.

(b) Answers will vary. For example, 10% of \$90 is \$9.

71. 12% or 0.12 of the pet owners selected the devil costume for their pet.

73. **(a)** The third longest bar, Witch, is the third-most-popular costume.

(b) Witch was selected by 4.5% or 0.045 of the pet owners.

75. 9% or 0.09 of the motorists chose "Tailgating."

77. **(a)** "Speeding" was the violation that was least often chosen for the biggest fine.

(b) 5% or 0.05 of the motorists chose "Speeding."

79. 21% of the children eat french fries. To write 21% as a decimal, drop the percent symbol and move the decimal point two places to the left, resulting in 0.21. So, 0.21 of the children eat french fries.

81. **(a)** The food eaten by the lowest portion of children is candy.

(b) 10% or 0.10 of the children eat candy.

83. Ninety-five parts of the one hundred parts are shaded.

$$\frac{95}{100} = 0.95 = \underline{95}\%$$

Five parts of the one hundred parts are unshaded.

$$\frac{5}{100} = 0.05 = \underline{5}\%$$

84. Twenty parts of the one hundred parts are shaded.

$$\frac{20}{100} = 0.20 = \underline{20}\%$$

Eighty parts of the one hundred parts are unshaded.

$$\frac{80}{100} = 0.80 = \underline{80}\%$$

85. Three parts of the ten parts are shaded.

$$\frac{3}{10} = 0.3 = 30\%$$

Seven parts of the ten parts are unshaded.

$$\frac{7}{10} = 0.7 = 70\%$$

86. Eight parts of the ten parts are shaded.

$$\frac{8}{10} = 0.8 = 80\%$$

Two parts of the ten parts are unshaded.

$$\frac{2}{10} = 0.2 = 20\%$$

87. Fifty-five parts of the one hundred parts are shaded.

$$\frac{55}{100} = 0.55 = 55\%$$

Forty-five parts of the one hundred parts are unshaded.

$$\frac{45}{100} = 0.45 = 45\%$$

88. Thirty-seven parts of the one hundred parts are shaded.

$$\frac{37}{100} = 0.37 = 37\%$$

Sixty-three parts of the one hundred parts are unshaded.

$$\frac{63}{100} = 0.63 = 63\%$$

Copyright © 2018 Pearson Education, Inc.

89. The shaded portions contain 16 squares, so four of them give us $4 \cdot 16 = 64$ shaded squares.

$$\frac{64}{100} = 0.64 = 64\%$$

$100 - 64 = 36$ squares of the one hundred squares are unshaded.

$$\frac{36}{100} = 0.36 = 36\%$$

90. The corner-shaded portions contain 4 squares, so four of them give us $4 \cdot 4 = 16$ shaded squares. The middle-shaded portion contains 16 squares, so the total number of shaded squares is $16 + 16 = 32$.

$$\frac{32}{100} = 0.32 = 32\%$$

$100 - 32 = 68$ squares of the one hundred squares are unshaded.

$$\frac{68}{100} = 0.68 = 68\%$$

91. Exercise 83: $95\% + 5\% = 100\%$
Exercise 84: $20\% + 80\% = 100\%$
Exercise 85: $30\% + 70\% = 100\%$
Exercise 86: $80\% + 20\% = 100\%$
Exercise 87: $55\% + 45\% = 100\%$
Excrcise 88: $37\% + 63\% = 100\%$
Exercise 89: $64\% + 36\% = 100\%$
Exercise 90: $32\% + 68\% = 100\%$
The sum is always 100%, or all of the parts.

6.2   Percents and Fractions

6.2   Margin Exercises

1. **(a)** $50\% = \dfrac{50}{100} = \dfrac{50 \div 50}{100 \div 50} = \dfrac{1}{2}$

(b) $48\% = \dfrac{48}{100} = \dfrac{48 \div 4}{100 \div 4} = \dfrac{12}{25}$

(c) $23\% = \dfrac{23}{100}$

(d) $125\% = \dfrac{125}{100} = \dfrac{125 \div 25}{100 \div 25} = \dfrac{5}{4} = 1\dfrac{1}{4}$

(e) $250\% = \dfrac{250}{100} = \dfrac{250 \div 50}{100 \div 50} = \dfrac{5}{2} = 2\dfrac{1}{2}$

2. **(a)** $37.5\% = \dfrac{37.5}{100} = \dfrac{37.5(10)}{100(10)} = \dfrac{375 \div 125}{1000 \div 125} = \dfrac{3}{8}$

(b) $62.5\% = \dfrac{62.5}{100} = \dfrac{62.5(10)}{100(10)} = \dfrac{625 \div 125}{1000 \div 125} = \dfrac{5}{8}$

(c) $4.5\% = \dfrac{4.5}{100} = \dfrac{4.5(10)}{100(10)} = \dfrac{45 \div 5}{1000 \div 5} = \dfrac{9}{200}$

(d) $66\dfrac{2}{3}\% = \dfrac{66\frac{2}{3}}{100} = \dfrac{\frac{200}{3}}{100} = \dfrac{200}{3} \div \dfrac{100}{1}$

$= \dfrac{200}{3} \cdot \dfrac{1}{100} = \dfrac{2}{3}$

(e) $10\dfrac{1}{3}\% = \dfrac{10\frac{1}{3}}{100} = \dfrac{\frac{31}{3}}{100} = \dfrac{31}{3} \div \dfrac{100}{1}$

$= \dfrac{31}{3} \cdot \dfrac{1}{100} = \dfrac{31}{300}$

(f) $87\dfrac{1}{2}\% = \dfrac{87\frac{1}{2}}{100} = \dfrac{\frac{175}{2}}{100} = \dfrac{175}{2} \div \dfrac{100}{1}$

$= \dfrac{\overset{7}{\cancel{175}}}{2} \cdot \dfrac{1}{\underset{4}{\cancel{100}}} = \dfrac{7}{8}$

3. **(a)** $\dfrac{1}{4} = \dfrac{p}{100}$

$4 \cdot p = 1 \cdot \underline{100}$

$\dfrac{4 \cdot p}{4} = \dfrac{100}{4}$

$p = 25$

Thus, $\frac{1}{4} = 25\%$.

(b) $\dfrac{3}{10} = \dfrac{p}{100}$

$10 \cdot p = 3 \cdot 100$

$\dfrac{10 \cdot p}{10} = \dfrac{300}{10}$

$p = 30$

Thus, $\frac{3}{10} = 30\%$.

(c) $\dfrac{6}{25} = \dfrac{p}{100}$

$25 \cdot p = 6 \cdot 100$

$\dfrac{25 \cdot p}{25} = \dfrac{600}{25}$

$p = 24$

Thus, $\frac{6}{25} = 24\%$.

(d) $\dfrac{5}{8} = \dfrac{p}{100}$

$8 \cdot p = 5 \cdot 100$

$\dfrac{8 \cdot p}{8} = \dfrac{500}{8}$

$p = 62.5$

Thus, $\frac{5}{8} = 62.5\%$.

(e) $\dfrac{1}{6} = \dfrac{p}{100}$

$6 \cdot p = 1 \cdot 100$

$\dfrac{6 \cdot p}{6} = \dfrac{100}{6}$

$p = 16.\overline{6} \approx 16.7$

Thus, $\frac{1}{6} \approx 16.7\%$ or exactly $16\frac{2}{3}\%$.

Copyright © 2018 Pearson Education, Inc.

(f) $\dfrac{2}{9} = \dfrac{p}{100}$

$9 \cdot p = 2 \cdot 100$

$\dfrac{9 \cdot p}{9} = \dfrac{200}{9}$

$p = 22.\overline{2} \approx 22.2$

Thus, $\dfrac{2}{9} \approx 22.2\%$ or exactly $22\frac{2}{9}\%$.

4. (a) $\dfrac{3}{4}$ as a percent

$\dfrac{3}{4} = 75\%$

(b) 10% as a fraction

$10\% = \dfrac{1}{10}$

(c) $0.\overline{6} \approx 0.667$ as a fraction

$0.\overline{6} \approx 0.667 = \dfrac{2}{3}$

(d) $37\frac{1}{2}\%$ as a fraction

$37\dfrac{1}{2}\% = \dfrac{3}{8}$

(e) $\dfrac{7}{8}$ as a percent

$\dfrac{7}{8} = 87\dfrac{1}{2}\%$ or 87.5%

(f) $\dfrac{1}{2}$ as a percent

$\dfrac{1}{2} = 50\%$

(g) $33\frac{1}{3}\%$ as a fraction

$33\dfrac{1}{3}\% = \dfrac{1}{3}$

(h) $1\frac{1}{2}$ as a percent

$1\dfrac{1}{2} = 150\%$

6.2 Section Exercises

1. $25\% = \dfrac{25}{100} = \dfrac{25 \div 25}{100 \div 25} = \dfrac{1}{4}$

Thus, $25\% = \frac{1}{2}$ is *false*.

3. $75\% = \dfrac{75}{100} = \dfrac{75 \div 25}{100 \div 25} = \dfrac{3}{4}$

Thus, $75\% = \frac{3}{4}$ is *true*.

5. $85\% = 0.85 = \dfrac{85}{100} = \dfrac{85 \div 5}{100 \div 5} = \dfrac{17}{20}$

7. $62.5\% = \dfrac{62.5}{100} = \dfrac{62.5(10)}{100(10)} = \dfrac{625 \div 125}{1000 \div 125} = \dfrac{5}{8}$

9. First write 6.25 over 100. Then, to get a whole number in the numerator, multiply the numerator and denominaotr by 100. Finally, write the fraction in lowest terms.

$6.25\% = \dfrac{6.25}{100} = \dfrac{6.25(100)}{100(100)}$

$= \dfrac{625 \div 625}{10{,}000 \div 625} = \dfrac{1}{16}$

11. $16\dfrac{2}{3}\% = \dfrac{16\frac{2}{3}}{100} = 16\dfrac{2}{3} \div \dfrac{100}{1} = \dfrac{\overset{50}{\cancel{50}}}{3} \cdot \dfrac{1}{\underset{2}{\cancel{100}}} = \dfrac{1}{6}$

13. $6\dfrac{2}{3}\% = \dfrac{6\frac{2}{3}}{100} = \dfrac{\frac{20}{3}}{100} = \dfrac{20}{3} \div \dfrac{100}{1}$

$= \dfrac{\overset{1}{\cancel{20}}}{3} \cdot \dfrac{1}{\underset{5}{\cancel{100}}} = \dfrac{1}{15}$

15. $0.5\% = \dfrac{0.5}{100} = \dfrac{0.5(10)}{100(10)} = \dfrac{5 \div 5}{1000 \div 5} = \dfrac{1}{200}$

17. $180\% = \dfrac{180}{100} = \dfrac{180 \div 20}{100 \div 20} = \dfrac{9}{5} = 1\dfrac{4}{5}$

19. $375\% = \dfrac{375}{100} = \dfrac{375 \div 25}{100 \div 25} = \dfrac{15}{4} = 3\dfrac{3}{4}$

21. $\dfrac{1}{2} = \dfrac{p}{100}$

$2 \cdot p = 1 \cdot 100$

$\dfrac{2 \cdot p}{2} = \dfrac{100}{2}$

$p = 50$

Thus, $\frac{1}{2} = 50\%$ is *true*.

23. $\dfrac{4}{5} = \dfrac{p}{100}$

$5 \cdot p = 4 \cdot 100$

$\dfrac{5 \cdot p}{5} = \dfrac{400}{5}$

$p = 80$

Thus, $\frac{4}{5} = 75\%$ is *false*.

25. $\dfrac{7}{10} = \dfrac{p}{100}$

$10 \cdot p = 7 \cdot 100$

$\dfrac{10 \cdot p}{10} = \dfrac{700}{10}$

$p = 70$

Thus, $\frac{7}{10} = 70\%$.

Copyright © 2018 Pearson Education, Inc.

27. The denominator is already 100.

$$\frac{37}{100} = 37\%$$

29.
$$\frac{5}{8} = \frac{p}{100}$$
$$8 \cdot p = 5 \cdot 100$$
$$\frac{8 \cdot p}{8} = \frac{500}{8} = \frac{125}{2}$$
$$p = 62.5$$

Thus, $\frac{5}{8} = 62.5\%$.

31.
$$\frac{7}{8} = \frac{p}{100}$$
$$8 \cdot p = 7 \cdot 100$$
$$\frac{8 \cdot p}{8} = \frac{700}{8} = \frac{175}{2}$$
$$p = 87.5$$

Thus, $\frac{7}{8} = 87.5\%$.

33.
$$\frac{12}{25} = \frac{p}{100}$$
$$25 \cdot p = 12 \cdot 100$$
$$\frac{25 \cdot p}{25} = \frac{1200}{25}$$
$$p = 48$$

Thus, $\frac{12}{25} = 48\%$.

35. As a different approach, we see that the denominator, 50, can be multiplied by 2 to get a denominator of 100. We'll multiply both the numerator and denominator by 2.

$$\frac{23}{50} = \frac{23 \cdot 2}{50 \cdot 2} = \frac{46}{100} = 46\%$$

37.
$$\frac{7}{20} = \frac{p}{100}$$
$$20 \cdot p = 7 \cdot 100$$
$$\frac{20 \cdot p}{20} = \frac{700}{20}$$
$$p = 35$$

Thus, $\frac{7}{20} = 35\%$.

39.
$$\frac{5}{6} = \frac{p}{100}$$
$$6 \cdot p = 5 \cdot 100$$
$$\frac{6 \cdot p}{6} = \frac{500}{6} = \frac{250}{3}$$
$$p \approx 83.3$$

Thus, $\frac{5}{6} = 83.3\%$ (rounded).

41.
$$\frac{5}{9} = \frac{p}{100}$$
$$9 \cdot p = 5 \cdot 100$$
$$\frac{9 \cdot p}{9} = \frac{500}{9}$$
$$p = \frac{500}{9}$$
$$p = 55.\overline{5}$$

Thus, $\frac{5}{9} = 55.6\%$ (rounded).

43.
$$\frac{1}{7} = \frac{p}{100}$$
$$7 \cdot p = 1 \cdot 100$$
$$\frac{7 \cdot p}{7} = \frac{100}{7}$$
$$p \approx 14.3$$

Thus, $\frac{1}{7} = 14.3\%$ (rounded).

45. $\frac{3}{4} = \frac{3 \cdot 25}{4 \cdot 25} = \frac{75}{100} = 0.75$ *decimal*

So choice (b) is correct.

47. $0.5 = \frac{5}{10} = \frac{5 \div 5}{10 \div 5} = \frac{1}{2}$ *fraction*

$0.5 = 0.50 = 50\%$ *percent*

49. $87.5\% = 0.875$ *decimal*

$= \frac{875}{1000} = \frac{875 \div 125}{1000 \div 125} = \frac{7}{8}$ *fraction*

51. $\frac{1}{6} \approx 0.167$ *decimal*

$\approx 16.7\%$ *percent*

53. $0.7 = \frac{7}{10}$ *fraction*

$0.7 = 0.70 = 70\%$ *percent*

55. $12.5\% = 0.125$ *decimal*

$= \frac{125}{1000} = \frac{125 \div 125}{1000 \div 125} = \frac{1}{8}$ *fraction*

57. $\frac{2}{3} \approx 0.667$ *decimal*

$\approx 66.7\%$ *percent*

59. $\frac{3}{50} = \frac{3 \cdot 2}{50 \cdot 2} = \frac{6}{100} = 0.06$ *decimal*

$= 6\%$ *percent*

61. $\frac{8}{100} = 8\%$ *percent*

$= 0.08$ *decimal*

63. $\frac{1}{200} = 1 \div 200 = 0.005$ *decimal*

$= 0.5\%$ *percent*

Copyright © 2018 Pearson Education, Inc.

65. $2.5 = 2\frac{5}{10} = 2\frac{1}{2}$ *fraction*

$2.5 = 250\%$ *percent*

67. $3\frac{1}{4} = \frac{13}{4} = 13 \div 4 = 3.25$ *decimal*

$= 325\%$ *percent*

69. There are many possible answers. Examples 2 and 3 show the steps that students should include in their answers.

71. 90 out of 500 people who used the Internet to find pet information, said they used it when buying a pet.

$\frac{90}{500} = \frac{90 \div 10}{500 \div 10} = \frac{9}{50}$ *fraction*

$\frac{9}{50} = \frac{9 \cdot 2}{50 \cdot 2} = \frac{18}{100} = 0.18$ *decimal*

$0.18 = 18\%$ *percent*

73. 13 adults out of 100 adults consumes the recommended 1000 mg of calcium daily.

$\frac{13}{100}$ *fraction*

$\frac{13}{100} = 0.13$ *decimal*

$0.13 = 13\%$ *percent*

75. 150 out of 750 workers said no.

$\frac{150}{750} = \frac{150 \div 150}{750 \div 150} = \frac{1}{5}$ *fraction*

$\frac{1}{5} = \frac{1 \cdot 2}{5 \cdot 2} = \frac{2}{10} = 0.2$ *decimal*

$0.2 = 0.20 = 20\%$ *percent*

77. **(a)** 64 out of 80 employees have iPhones.

$\frac{64}{80} = \frac{64 \div 16}{80 \div 16} = \frac{4}{5}$ *fraction*

$\frac{4}{5} = \frac{4 \cdot 2}{5 \cdot 2} = \frac{8}{10} = 0.8$ *decimal*

$0.8 = 0.80 = 80\%$ *percent*

(b) If $\frac{4}{5}$ have cell phones, then $1 - \frac{4}{5} = \frac{1}{5}$ do *not* have iPhones.

$\frac{1}{5} = \frac{1 \cdot 2}{5 \cdot 2} = \frac{2}{10} = 0.2$ *decimal*

$0.2 = 0.20 = 20\%$ *percent*

79. **(a)** 21 million gallons out of 60 million gallons go toward milk that people drink.

$\frac{21}{60} = \frac{21 \div 3}{60 \div 3} = \frac{7}{20}$ *fraction*

$\frac{7}{20} = \frac{7 \cdot 5}{20 \cdot 5} = \frac{35}{100} = 0.35$ *decimal*

$0.35 = 35\%$ *percent*

(b) $60 - 21 = 39$ million gallons go toward other dairy products.

$\frac{39}{60} = \frac{39 \div 3}{60 \div 3} = \frac{13}{20}$ *fraction*

$\frac{13}{20} = \frac{13 \cdot 5}{20 \cdot 5} = \frac{65}{100} = 0.65$ *decimal*

$0.65 = 65\%$ *percent*

81. 100% of a number means all of the parts or <u>100</u> parts out of <u>100</u> parts. 200% means two times as many parts and 300% means three times as many parts.

83. 50% of a number is <u>50</u> parts out of <u>100</u> parts, which is half or $\frac{1}{2}$ of the parts.

85. 1% of a number is <u>1</u> part out of <u>100</u> parts and can be found quickly by moving the decimal point <u>2</u> places to the <u>left</u>.

87. 432 people out of 5400 people picked Winter as their favorite season.

$\frac{432}{5400} = \frac{432 \div 216}{5400 \div 216} = \frac{2}{25}$ *fraction*

$\frac{2}{25} = \frac{2 \cdot 4}{25 \cdot 4} = \frac{8}{100} = 0.08$ *decimal*

$0.08 = 8\%$ *percent*

88. 2052 people out of 5400 people picked Spring as their favorite season.

$\frac{2052}{5400} = \frac{2052 \div 108}{5400 \div 108} = \frac{19}{50}$ *fraction*

$\frac{19}{50} = \frac{19 \cdot 2}{50 \cdot 2} = \frac{38}{100} = 0.38$ *decimal*

$0.38 = 38\%$ *percent*

89. 1404 people out of 5400 people picked Summer as their favorite season.

$\frac{1404}{5400} = \frac{1404 \div 108}{5400 \div 108} = \frac{13}{50}$ *fraction*

$\frac{13}{50} = \frac{13 \cdot 2}{50 \cdot 2} = \frac{26}{100} = 0.26$ *decimal*

$0.26 = 26\%$ *percent*

90. 1512 people out of 5400 people picked Fall as their favorite season.

$\frac{1512}{5400} = \frac{1512 \div 216}{5400 \div 216} = \frac{7}{25}$ *fraction*

$\frac{7}{25} = \frac{7 \cdot 4}{25 \cdot 4} = \frac{28}{100} = 0.28$ *decimal*

$0.28 = 28\%$ *percent*

Copyright © 2018 Pearson Education, Inc.

91. Find 10% of $160, then add $\frac{1}{2}$ of the 10% amount.

$$10\% + 5\% = 15\%$$
$$\downarrow \qquad \downarrow \qquad \downarrow$$
$$\$16 + \$8 = \$24$$

92. Find 100% of $160, then add 50% of $160.

$$100\% + 50\% = 150\%$$
$$\downarrow \qquad \downarrow \qquad \downarrow$$
$$\$160 + \$80 = \$240$$

93. From 100% of $450, subtract 10% of $450.

$$100\% - 10\% = 90\%$$
$$\downarrow \qquad \downarrow \qquad \downarrow$$
$$\$450 - \$45 = \$405$$

94. To 100% of $800, add another 100% of $800, and then add 10% of $800.

$$100\% + 100\% + 10\% = 210\%$$
$$\downarrow \qquad \downarrow \qquad \downarrow \qquad \downarrow$$
$$\$800 + \$800 + \$80 = \$1680$$

95. To find 25% of a number, first find 10% of the number, add another 10%, and then add 5% of the number. For example, to find 25% of $400:

$$10\% \;+\; 10\% \;+\; 5\% \;=\; 25\%$$
$$\downarrow \qquad \downarrow \qquad \downarrow \qquad \downarrow$$
$$\$40 \;+\; \$40 \;+\; \$20 \;=\; \$100$$

Alternatively, find 50% of $400 and divide the result by 2.

50% of $400 = $\frac{1}{2} \cdot \$400 = \$200$
$\frac{1}{2} \cdot \$200 = \$100$

96. To find 104% of a number, add 5% of a number to 100% of the number, and then subtract 1% of the number. For example, to find 104% of $1500:

$$100\% \;+\; 5\% \;-\; 1\% \;=\; 104\%$$
$$\downarrow \qquad \downarrow \qquad \downarrow \qquad \downarrow$$
$$\$1500 \;+\; \$75 \;-\; \$15 \;=\; \$1560$$

Alternatively, find 1% of $1500 and multiply the result by 4. Then add the result to $1500.

1% of $1500 = $15
$4 \cdot \$15 = \$60$
$\$1500 + \$60 = \$1560$

6.3   Using the Percent Proportion and Identifying the Components in a Percent Problem

6.3   Margin Exercises

1.   **(a)** $\dfrac{1}{2} = \dfrac{25}{50}$

Cross product: $2 \cdot 25 = 50;\; 1 \cdot 50 = 50$

The cross products are equal, so the proportion is *true*.

(b) $\dfrac{7}{8} = \dfrac{180}{200}$

Cross product: $8 \cdot 180 = 1440;\; 7 \cdot 200 = 1400$

The cross products are *not* equal, so the proportion is *false*.

(c) $\dfrac{112}{41} = \dfrac{332}{123}$

Cross product:
$41 \cdot 332 = 13{,}612;\; 112 \cdot 123 = 13{,}776$

The cross products are *not* equal, so the proportion is *false*.

2.   **(a)** part = 12, percent = 16

$$\frac{\text{part}}{\text{whole}} = \frac{\text{percent}}{100}$$

$$\frac{12}{x} = \frac{16}{100} \quad \textbf{OR} \quad \frac{12}{x} = \frac{4}{25}$$
$$x \cdot 4 = 12 \cdot \underline{25}$$
$$x \cdot 4 = \underline{300}$$
$$\frac{x \cdot \overset{1}{\cancel{4}}}{\underset{1}{\cancel{4}}} = \frac{300}{4}$$
$$x = \underline{75}$$

The whole is 75.

(b) part = 30, whole = 120

$$\frac{30}{120} = \frac{x}{100} \quad \textbf{OR} \quad \frac{1}{4} = \frac{x}{100}$$
$$4 \cdot x = 1 \cdot 100$$
$$\frac{4 \cdot x}{4} = \frac{100}{4}$$
$$x = 25$$

The percent is 25, written as 25%.

Copyright © 2018 Pearson Education, Inc.

(c) whole = 210, percent = 20

$$\frac{x}{210} = \frac{20}{100} \quad \textbf{OR} \quad \frac{x}{210} = \frac{1}{5}$$
$$x \cdot 5 = 210 \cdot 1$$
$$\frac{x \cdot 5}{5} = \frac{210}{5}$$
$$x = 42$$

The part is 42.

(d) whole = 4000, percent = 32

$$\frac{x}{4000} = \frac{32}{100} \quad \textbf{OR} \quad \frac{x}{4000} = \frac{8}{25}$$
$$x \cdot 25 = 4000 \cdot 8$$
$$\frac{x \cdot 25}{25} = \frac{32,000}{25}$$
$$x = 1280$$

The part is 1280.

(e) part = 74, whole = 185

Note: If you don't see how to write the fraction in lowest terms, just proceed with the original fraction.

$$\frac{74}{185} = \frac{x}{100} \quad \textbf{OR} \quad \frac{2}{5} = \frac{x}{100}$$
$$5 \cdot x = 2 \cdot 100$$
$$\frac{5 \cdot x}{5} = \frac{200}{5}$$
$$x = 40$$

The percent is 40, written as 40%.

3. **(a)** The number preceding the <u>percent</u> symbol is the percent. So, <u>25</u> is the percent.

(b) The percent is 65. The number 65 appears with the symbol %.

(c) The percent is $6\frac{1}{2}$ because $6\frac{1}{2}$ appears with the word *percent*.

(d) The percent is 42. The number 42 appears with the symbol %.

(e) The word *percent* has no number with it, so the percent is the unknown part of the problem.

4. **(a)** The whole is 900 since there are 900 total tests.

(b) The whole is 620 since there are 620 total students.

(c) For "8500 tons of recyclables is 42% of what number of tons?," the whole is, "what number," the unknown part of the problem.

5. **(a)** The percent is 25 and the whole is 900. 225 is the remaining number, so the part is 225.

$$\text{Part} \rightarrow \frac{225}{900} = \frac{25}{100} \begin{array}{l} \leftarrow \text{Percent} \\ \end{array}$$
$$\text{Whole} \rightarrow \qquad\qquad \leftarrow \text{Always 100}$$

(b) The percent is 65 and the whole is 620. 403 is the remaining number, so the part is 403.

$$\text{Part} \rightarrow \frac{403}{620} = \frac{65}{100} \begin{array}{l} \leftarrow \text{Percent} \\ \leftarrow \text{Always 100} \end{array}$$
$$\text{Whole} \rightarrow$$

(c) The percent is 42 and the whole is unknown. 8500 is the part.

$$\text{Part} \rightarrow \frac{8500}{\text{unknown}} = \frac{42}{100} \begin{array}{l} \leftarrow \text{Percent} \\ \leftarrow \text{Always 100} \end{array}$$
$$\text{Whole} \rightarrow$$

6.3 Section Exercises

1. part = 5, percent = 10

$$\frac{\text{part}}{\text{whole}} = \frac{\text{percent}}{100}$$
$$\frac{5}{\text{unknown}} = \frac{10}{100}$$

The unknown component is the whole.

3. part = 36, whole = 24

$$\frac{\text{part}}{\text{whole}} = \frac{\text{percent}}{100}$$
$$\frac{36}{24} = \frac{\text{unknown}}{100}$$

The unknown component is the percent.

5. whole = 160, percent = 35

$$\frac{\text{part}}{\text{whole}} = \frac{\text{percent}}{100}$$
$$\frac{\text{unknown}}{160} = \frac{35}{100}$$

The unknown component is the part.

7. part = 30, percent = 20

$$\frac{30}{x} = \frac{20}{100} \quad \textbf{OR} \quad \frac{30}{x} = \frac{1}{5}$$
$$x \cdot 1 = 30 \cdot 5$$
$$x = 150$$

The whole is 150.

9. part = 28, percent = 40

$$\frac{28}{x} = \frac{40}{100} \quad \textbf{OR} \quad \frac{28}{x} = \frac{2}{5}$$
$$x \cdot 2 = 28 \cdot 5$$
$$\frac{x \cdot 2}{2} = \frac{140}{2}$$
$$x = 70$$

The whole is 70.

Copyright © 2018 Pearson Education, Inc.

11. part = 15, whole = 60

$$\frac{15}{60} = \frac{x}{100} \quad \textbf{OR} \quad \frac{1}{4} = \frac{x}{100}$$
$$4 \cdot x = 1 \cdot 100$$
$$\frac{4 \cdot x}{4} = \frac{100}{4}$$
$$x = 25$$

The percent is 25, written as 25%.

13. part = 9.25, whole = 27.75

Use a calculator.

$$\frac{9.25}{27.75} = \frac{x}{100}$$
$$27.75 \cdot x = 9.25 \cdot 100$$
$$\frac{27.75 \cdot x}{27.75} = \frac{925}{27.75}$$
$$x = 33.3 \text{ (rounded)}$$

The percent is about 33.3, written as 33.3%.

15. whole = 52, percent = 50

$$\frac{x}{52} = \frac{50}{100} \quad \textbf{OR} \quad \frac{x}{52} = \frac{1}{2}$$
$$x \cdot 2 = 52 \cdot 1$$
$$\frac{x \cdot 2}{2} = \frac{52}{2}$$
$$x = 26$$

The part is 26.

17. whole = 94.4, part = 25

$$\frac{25}{94.4} = \frac{x}{100}$$
$$94.4 \cdot x = 25 \cdot 100$$
$$\frac{94.4 \cdot x}{94.4} = \frac{2500}{94.4}$$
$$x = 26.5 \text{ (rounded)}$$

The percent is about 26.5, written as 26.5%.

19. part = 46, percent = 40

$$\frac{\text{part}}{\text{whole}} = \frac{\text{percent}}{100}$$
$$\frac{46}{x} = \frac{40}{100}$$
$$x \cdot 40 = 46 \cdot 100$$
$$\frac{x \cdot 40}{40} = \frac{4600}{40}$$
$$x = \underline{115}$$

The whole is 115.

21. whole = 5000, part = 20

$$\frac{20}{5000} = \frac{x}{100} \qquad \textit{Percent proportion}$$
$$5000 \cdot x = 20 \cdot 100 \qquad \textit{Find cross products.}$$
$$\frac{5000 \cdot x}{5000} = \frac{2000}{5000} \qquad \textit{Divide both sides by 5000.}$$
$$x = 0.4$$

The percent is 0.4, written as 0.4%.

23. whole = 4300, part = $107\frac{1}{2}$

$$\frac{107\frac{1}{2}}{4300} = \frac{x}{100}$$
$$4300 \cdot x = 107\frac{1}{2} \cdot 100$$
$$\frac{4300 \cdot x}{4300} = \frac{10{,}750}{4300}$$
$$x = 2.5$$

The percent is 2.5, written as 2.5%.

25. whole = 6480, part = 19.44

$$\frac{19.44}{6480} = \frac{x}{100}$$
$$6480 \cdot x = 19.44 \cdot 100$$
$$\frac{6480 \cdot x}{6480} = \frac{1944}{6480}$$
$$x = 0.3$$

The percent is 0.3, written as 0.3%.

27. In a percent problem, the percent can be identified because it appears with the word <u>percent</u> or with the <u>%</u> symbol after it.

29. In a percent problem, the part is the portion being compared with the <u>whole</u>.

31.

10%	of how many bicycles	is 60 bicycles?
↑	↑	↑
percent	whole	part
10	unknown	60

$$\text{Part} \rightarrow \frac{60}{\text{unknown}} = \frac{10}{100} \quad \begin{matrix}\leftarrow \text{Percent}\\ \leftarrow \text{Always 100}\end{matrix}$$
$$\text{Whole} \rightarrow$$

33. What percent of $800 is $600?

↑	↑	↑
percent	whole	part
unknown	800	600

$$\text{Part} \rightarrow \frac{600}{800} = \frac{\text{unknown}}{100} \quad \begin{matrix}\leftarrow \text{Percent}\\ \leftarrow \text{Always 100}\end{matrix}$$
$$\text{Whole} \rightarrow$$

Copyright © 2018 Pearson Education, Inc.

35. What is 25% of $970?

    ↑   ↑   ↑

  part percent whole

unknown 25   970

Part → $\dfrac{\text{unknown}}{970} = \dfrac{25}{100}$ ← Percent

Whole →   ← Always 100

37. 12 injections is 20% of what number
                        of injections?

     ↑   ↑   ↑

   part   percent   whole

   12   20   unknown

Part → $\dfrac{12}{\text{unknown}} = \dfrac{20}{100}$ ← Percent

Whole →   ← Always 100

39. 54.34 is 3.25% of what number?

  ↑   ↑   ↑

 part percent   whole

54.34 3.25   unknown

Part → $\dfrac{54.34}{\text{unknown}} = \dfrac{3.25}{100}$ ← Percent

Whole →   ← Always 100

41. 0.68% of $487 is what number?

   ↑   ↑   ↑

percent whole   part

0.68   487   unknown

Part → $\dfrac{\text{unknown}}{487} = \dfrac{0.68}{100}$ ← Percent

Whole →   ← Always 100

43. Percent—the ratio of the part to the whole. It appears with the word *percent* or "%" after it. Whole—the entire quantity. Often appears after the word *of*. Part—the part being compared with the whole.

45. 730 of 1262 computers is what percent?

 ↑   ↑   ↑

part   whole   percent

730   1262   unknown

Part → $\dfrac{730}{\underline{1262}} = \dfrac{\text{unknown}}{100}$ ← Percent

Whole →   ← Always 100

47. 86 of 142 people is what percent?

  ↑   ↑   ↑

part whole   percent

 86   142   unknown

Part → $\dfrac{86}{142} = \dfrac{\text{unknown}}{100}$ ← Percent

Whole →   ← Always 100

49. 23% live 610 is the total   the number of
in lower   number of   pilots who live in
48 states   pilots surveyed the lower 48 states

  ↑   ↑   ↑

percent   whole   part

  23   610   unknown

Part → $\dfrac{\text{unknown}}{610} = \dfrac{23}{100}$ ← Percent

Whole →   ← Always 100

51. 4300 students is what of 8600?
                      percent

  ↑   ↑   ↑

 part   percent   whole

4300   unknown 8600

Part → $\dfrac{4300}{8600} = \dfrac{\text{unknown}}{100}$ ← Percent

Whole →   ← Always 100

53. 55% of 480 adults is how many
                       adults?

  ↑   ↑   ↑

percent   whole   part

  55   480   unknown

Part → $\dfrac{\text{unknown}}{480} = \dfrac{55}{100}$ ← Percent

Whole →   ← Always 100

55. 49.5% of 822 people is how many
                     people?

  ↑   ↑   ↑

percent   whole   part

49.5   822   unknown

Part → $\dfrac{\text{unknown}}{822} = \dfrac{49.5}{100}$ ← Percent

Whole →   ← Always 100

57. 12%   total number is 168 people.
  of   surveyed

  ↑   ↑   ↑

percent   whole   part

  12   unknown 168

Part → $\dfrac{168}{\text{unknown}} = \dfrac{12}{100}$ ← Percent

Whole →   ← Always 100

59. 45% of 680 toys is what number?

  ↑   ↑   ↑

percent whole   part

 45   680   unknown

Part → $\dfrac{\text{unknown}}{680} = \dfrac{45}{100}$ ← Percent

Whole →   ← Always 100

Copyright © 2018 Pearson Education, Inc.

6.4   Using Proportions to Solve Percent Problems

6.4   Margin Exercises

1. **(a)** 8% of 400 patients

   percent is 8; whole is 400

   $$\frac{\text{part}}{\text{whole}} = \frac{\text{percent}}{100}$$

   $$\frac{x}{400} = \frac{8}{100} \quad \textbf{OR} \quad \frac{x}{400} = \frac{2}{25}$$

   $$x \cdot 25 = 400 \cdot 2$$

   $$x \cdot 25 = \underline{800}$$

   $$\frac{x \cdot \overset{1}{\cancel{25}}}{\underset{1}{\cancel{25}}} = \frac{800}{25}$$

   $$x = \underline{32}$$

   8% of 400 patients is 32 patients.

   (b) 15% of $3220

   percent is 15; whole is 3220

   $$\frac{\text{part}}{\text{whole}} = \frac{\text{percent}}{100}$$

   $$\frac{x}{3220} = \frac{15}{100} \quad \textbf{OR} \quad \frac{x}{3220} = \frac{3}{20}$$

   $$x \cdot 20 = 3220 \cdot 3$$

   $$\frac{x \cdot 20}{20} = \frac{9660}{20}$$

   $$x = 483$$

   15% of $3220 is $483.

   (c) 7% of 2700 miles

   percent is 7; whole is 2700

   $$\frac{\text{part}}{\text{whole}} = \frac{\text{percent}}{100}$$

   $$\frac{x}{2700} = \frac{7}{100}$$

   $$x \cdot 100 = 2700 \cdot 7$$

   $$\frac{x \cdot 100}{100} = \frac{18,900}{100}$$

   $$x = 189$$

   7% of 2700 miles is 189 miles.

   (d) 48% of 1580 kilowatts

   percent is 48; whole is 1580

   $$\frac{\text{part}}{\text{whole}} = \frac{\text{percent}}{100}$$

   $$\frac{x}{1580} = \frac{48}{100}$$

   $$x \cdot 100 = 1580 \cdot 48$$

   $$\frac{x \cdot 100}{100} = \frac{75,840}{100}$$

   $$x = 758.4$$

   48% of 1580 kilowatts is 758.4 kilowatts.

2. **(a)** 55% of 10,000 xrays
   55% = 0.55
   Write 55% in decimal form as 0.55.
   part = (0.55)(10,000) = 5500
   55% of 10,000 xrays is 5500 xrays.

   (b) 16% of 120 miles
   16% = 0.16
   part = (0.16)(120) = 19.2
   16% of 120 miles is 19.2 miles.

   (c) 135% of 60 dosages
   135% = 1.35
   part = (1.35)(60) = 81
   135% of 60 dosages is 81 dosages.

   (d) 0.5% of $238
   0.5% = 0.005
   part = (0.005)(238) = 1.19
   0.5% of $238 is $1.19.

3. **(a)** *Step 1*   The problem asks us to find the number of pieces of mail that are advertising pieces.

   Step 2   The total number of pieces is 2920, so the whole is 2920. The percent is 45. To find the number of advertising pieces, find the part.

   Step 3   2920 is about 3000 and 45% is about 50%, or $\frac{1}{2}$. To get an estimate, multiply 3000 by $\frac{1}{2}$ to get 1500.

   Step 4   Write 45% in decimal form as 0.45.

   $$\text{part} = (0.45) \cdot (2920)$$
   $$= 1314$$

   Step 5   There were 1314 advertising pieces of mail.

   Step 6   The exact answer, 1314, is close to the estimate, 1500.

   (b) *Step 1*   The problem asks us to find the number of students who wear glasses or contact lenses.

   Step 2   The total number of students is 9750, so the whole is 9750. The percent is 12. To find the number of students who wear glasses or contact lenses, find the part.

   (continued)

Copyright © 2018 Pearson Education, Inc.

Step 3 9750 is about 10,000 and 12% is about 10%, or $\frac{1}{10}$. To get an estimate, multiply 10,000 by $\frac{1}{10}$ to get 1000.

Step 4 part $= (0.12)(9750) = 1170$

Step 5 1170 students wear glasses or contact lenses.

Step 6 The exact answer, 1170, is close to the estimate, 1000.

4. **(a)** part is 75; percent is 5

$$\frac{75}{x} = \frac{5}{100} \quad \textbf{OR} \quad \frac{75}{x} = \frac{1}{20}$$
$$x \cdot 1 = 75 \cdot 20$$
$$x = \underline{1500}$$

75 applicants is 5% of 1500 applicants.

(b) part is 28; percent is 35

$$\frac{28}{x} = \frac{35}{100} \quad \textbf{OR} \quad \frac{28}{x} = \frac{7}{20}$$
$$x \cdot 7 = 28 \cdot 20$$
$$\frac{x \cdot 7}{7} = \frac{560}{7}$$
$$x = 80$$

28 antiques is 35% of 80 antiques.

(c) part is 387; percent is 36

$$\frac{387}{x} = \frac{36}{100} \quad \textbf{OR} \quad \frac{387}{x} = \frac{9}{25}$$
$$x \cdot 9 = 387 \cdot 25$$
$$\frac{x \cdot 9}{9} = \frac{9675}{9}$$
$$x = 1075$$

387 customers is 36% of 1075 customers.

(d) part is 292.5; percent is 37.5

$$\frac{292.5}{x} = \frac{37.5}{100}$$
$$x \cdot 37.5 = (292.5)(100)$$
$$\frac{x \cdot 37.5}{37.5} = \frac{29,250}{37.5}$$
$$x = 780$$

292.5 miles is 37.5% of 780 miles.

5. **(a)** *Step 1* The problem asks us to find the total number of tons in the crop.

Step 2 The percent is 52. The part is 182 tons. The whole is unknown.

Step 3 52% is about 50%, or $\frac{1}{2}$, and 182 is about 200. 200 is about $\frac{1}{2}$ of the total tons, so multiply 200 by 2 to get 400 as an estimate.

Step 4 $\quad \dfrac{182}{x} = \dfrac{52}{100} \quad \textbf{OR} \quad \dfrac{182}{x} = \dfrac{13}{25}$
$$x \cdot 13 = 182 \cdot 25$$
$$\frac{x \cdot \cancel{13}^{1}}{\cancel{13}_{1}} = \frac{4550}{13}$$
$$x = \underline{350}$$

Step 5 182 tons is 52% of 350 tons.

Step 6 The exact answer, 350, is close to the estimate, 400.

(b) *Step 1* The problem asks us to find the total weight of a batch of cake mix.

Step 2 The percent is 18. The part is 900 pounds. The whole is unknown.

Step 3 18% is close to 20%, or $\frac{1}{5}$, and the part 900 is close to 1000. 1000 is about $\frac{1}{5}$ of the total weight, so multiply 1000 by 5 to get 5000 as an estimate.

Step 4 part is 900; percent is 18

$$\frac{900}{x} = \frac{18}{100} \quad \textbf{OR} \quad \frac{900}{x} = \frac{9}{50}$$
$$x \cdot 9 = 900 \cdot 50$$
$$\frac{x \cdot 9}{9} = \frac{45,000}{9}$$
$$x = 5000$$

Step 5 900 pounds of sugar is 18% of 5000 pounds of cake mix.

Step 6 The exact answer matches the estimate.

6. **(a)** part is 21; whole is 105

$$\frac{21}{105} = \frac{x}{100} \quad \textbf{OR} \quad \frac{1}{5} = \frac{x}{100}$$
$$5 \cdot x = 1 \cdot 100$$
$$\frac{\cancel{5}^{1} \cdot x}{\cancel{5}_{1}} = \frac{100}{5}$$
$$x = \underline{20}$$

$21 is 20% of $105.

(b) part is 48; whole is 320

$$\frac{48}{320} = \frac{x}{100} \quad \textbf{OR} \quad \frac{3}{20} = \frac{x}{100}$$
$$20 \cdot x = 3 \cdot 100$$
$$\frac{20 \cdot x}{20} = \frac{300}{20}$$
$$x = 15$$

48 Internet companies is 15% of 320 Internet companies.

Copyright © 2018 Pearson Education, Inc.

(c) part is 1026; whole is 2280

$$\frac{1026}{2280} = \frac{x}{100} \quad \textbf{OR} \quad \frac{513}{1140} = \frac{x}{100}$$

$$1140 \cdot x = 513 \cdot 100$$

$$\frac{1140 \cdot x}{1140} = \frac{51{,}300}{1140}$$

$$x = 45$$

1026 trials is 45% of 2280 trials.

7. **(a)** part is 289; whole is 425

$$\frac{289}{425} = \frac{x}{100}$$

$$425 \cdot x = 289 \cdot 100$$

$$\frac{425 \cdot x}{425} = \frac{28{,}900}{425}$$

$$x = 68$$

The bid price is 68% of the minimum.

(b) part is 52; whole is 80

$$\frac{52}{80} = \frac{x}{100}$$

$$80 \cdot x = (52)(100)$$

$$\frac{80 \cdot x}{80} = \frac{5200}{80}$$

$$x = 65$$

65% of the tests were completed in the morning.

8. **(a)** part is 48; whole is 32

$$\frac{48}{32} = \frac{x}{100} \quad \textbf{OR} \quad \frac{3}{2} = \frac{x}{100}$$

$$2 \cdot x = 3 \cdot 100$$

$$\frac{2 \cdot x}{2} = \frac{300}{2}$$

$$x = \underline{150}$$

The miles per gallon around town is 150% of the miles per gallon on the highway.

(b) part is 432; whole is 360

$$\frac{432}{360} = \frac{x}{100} \quad \textbf{OR} \quad \frac{6}{5} = \frac{x}{100}$$

$$5 \cdot x = 6 \cdot 100$$

$$\frac{5 \cdot x}{5} = \frac{600}{5}$$

$$x = 120$$

The service calls made were 120% of their goal for the week.

6.4 Section Exercises

1. To find the part using the multiplication shortcut, use the formula

$$\text{part} = \underline{\text{percent}} \cdot \underline{\text{whole}}$$

3. 35% of 120 test tubes

$$\text{part} = (0.35)(120)$$
$$\text{part} = \underline{42 \text{ test tubes}}$$

5. 45% of 4080 military personnel

$$\text{part} = (0.45)(4080)$$
$$\text{part} = 1836 \text{ military personnel}$$

7. 4% of 120 ft

$$\text{part} = (0.04)(120)$$
$$\text{part} = 4.8 \text{ ft}$$

9. 150% of 210 files

$$\text{part} = (1.50)(210)$$
$$\text{part} = 315 \text{ files}$$

11. 52.5% of 1560 trucks

$$\text{part} = (0.525)(1560)$$
$$\text{part} = 819 \text{ trucks}$$

13. 2% of $164

$$\text{part} = (0.02)(164)$$
$$\text{part} = \$3.28$$

15. Write 225% as a decimal, 2.25.
225% of 680 tables

$$(2.25)(680) = 1530$$
$$\text{part} = 1530 \text{ tables}$$

17. 17.5% of 1040 cell phones

$$\text{part} = (0.175)(1040)$$
$$\text{part} = 182 \text{ cell phones}$$

19. 0.9% of $2400

$$\text{part} = (0.009)(2400)$$
$$\text{part} = \$21.60$$

21. part is 80; percent is 25

$$\frac{80}{\text{unknown}} = \frac{25}{100}$$

23. part is 48; percent is 30

$$\frac{48}{x} = \frac{30}{100} \quad \textbf{OR} \quad \frac{48}{x} = \frac{3}{10}$$

$$x \cdot 3 = 48 \cdot 10$$

$$\frac{x \cdot 3}{3} = \frac{480}{3}$$

$$x = 160$$

30% of <u>160 hay bales</u> is 48 hay bales.

Copyright © 2018 Pearson Education, Inc.

25. part is 495; percent is 90

$$\frac{495}{x} = \frac{90}{100} \quad \textbf{OR} \quad \frac{495}{x} = \frac{9}{10}$$
$$x \cdot 9 = 495 \cdot 10$$
$$\frac{x \cdot 9}{9} = \frac{4950}{9}$$
$$x = 550$$

495 successful students is 90% of 550 students.

27. part is 462; percent is 140

$$\frac{462}{x} = \frac{140}{100} \quad \textbf{OR} \quad \frac{462}{x} = \frac{7}{5}$$
$$x \cdot 7 = 462 \cdot 5$$
$$\frac{x \cdot 7}{7} = \frac{2310}{7}$$
$$x = 330$$

462 mountain bikes is 140% of 330 mountain bikes.

29. part is 350; percent is 12.5

$$\frac{\text{part}}{\text{whole}} = \frac{\text{percent}}{100}$$
$$\text{so} \quad \frac{350}{x} = \frac{12.5}{100}$$
$$12.5 \cdot x = 35{,}000 \quad \textit{Cross products}$$
$$\frac{12.5 \cdot x}{12.5} = \frac{35{,}000}{12.5} \quad \textit{Divide both sides by 12.5.}$$
$$x = 2800$$

$12\frac{1}{2}\%$ of 2800 is 350.

31. part is 18; whole is 36

$$\frac{18}{36} = \frac{\text{unknown}}{100}$$

33. part is 780; whole is 1500

$$\frac{780}{1500} = \frac{x}{100} \quad \textbf{OR} \quad \frac{13}{25} = \frac{x}{100}$$
$$25 \cdot x = 13 \cdot 100$$
$$\frac{25 \cdot x}{25} = \frac{1300}{25}$$
$$x = 52$$

780 hybrid cars is 52% of 1500 hybrid cars.

35. part is 27; whole is 1800

$$\frac{27}{1800} = \frac{x}{100} \quad \textbf{OR} \quad \frac{3}{200} = \frac{x}{100}$$
$$200 \cdot x = 3 \cdot 100$$
$$\frac{200 \cdot x}{200} = \frac{300}{200}$$
$$x = 1.5$$

27 downloaded songs is <u>1.5%</u> of 1800 downloaded songs.

37. part is 64; whole is 344

$$\frac{64}{344} = \frac{x}{100} \quad \textbf{OR} \quad \frac{8}{43} = \frac{x}{100}$$
Cross products $\qquad 43 \cdot x = 8 \cdot 100$
Divide both sides by 43. $\quad \dfrac{43 \cdot x}{43} = \dfrac{800}{43}$
$$x \approx 18.6$$

$64 is 18.6% (rounded) of $344.

39. part is 23; whole is 250

$$\frac{23}{250} = \frac{x}{100}$$
$$250 \cdot x = 23 \cdot 100$$
$$\frac{250 \cdot x}{250} = \frac{2300}{250}$$
$$x = 9.2$$

23 tires is 9.2% of 250 tires.

41. 50% of $84 is $42 and 100% of $217 is $217 are correct statements.
150% of $30 cannot be less than $30 because 150% is greater than 1 (100%). The answer must be greater than $30.
25% of $16 cannot be greater than $16 because 25% is less than 1 (100%). The answer must be less than $16.

43. **(a)** part is unknown; whole is 240; percent is 22

$$\frac{x}{240} = \frac{22}{100} \quad \textbf{OR} \quad \frac{x}{240} = \frac{11}{50}$$
Cross products $\qquad x \cdot 50 = 240 \cdot 11$
Divide both sides by 50. $\quad \dfrac{x \cdot 50}{50} = \dfrac{2640}{50}$
$$x = 52.8$$

The amount withheld is $52.80.

(b) $240 earnings − $52.80 withheld = $187.20 amount remaining.

45. part is unknown; whole is 3020; percent is 65

$$\frac{x}{3020} = \frac{65}{100}$$
$$x \cdot 100 = 3020 \cdot 65$$
$$\frac{x \cdot 100}{100} = \frac{196{,}300}{100}$$
$$x = 1963$$

The number of people who check their smartphones while working is 1963.

Copyright © 2018 Pearson Education, Inc.

47.   part is unknown; whole is 3020; percent is 71

$$\frac{x}{3020} = \frac{71}{100}$$
$$x \cdot 100 = 3020 \cdot 71$$
$$\frac{x \cdot 100}{100} = \frac{214{,}420}{100}$$
$$x = 2144.2 \approx 2144$$

The number of people who check their smartphones as soon as they wake up is 2144 (rounded).

49.   **(a)** part is unknown; whole is 48,000,000; percent is 2

$$\frac{x}{48{,}000{,}000} = \frac{2}{100}$$
OR $$\frac{x}{48{,}000{,}000} = \frac{1}{50}$$
$$x \cdot 50 = 48{,}000{,}000 \cdot 1$$
$$\frac{x \cdot 50}{50} = \frac{48{,}000{,}000}{50}$$
$$x = 960{,}000$$

The number of trips using public transportation is 960,000 or 0.96 million.

  (b) The number of trips not using public transportation is

$$48{,}000{,}000 - 960{,}000 = 47{,}040{,}000$$

or 48 million − 0.96 million = 47.04 million.

51.   15% of American youth use Tumblr as part of their social media, so 100% − 15% = 85% do not use Tumblr as part of their social media.

53.   74% of 3400 = (0.74)(3400) = 2516
2516 of the 3400 people who answered the survey use Facebook.

55.   part is 960; whole is 48,000; percent is unknown

$$\frac{960}{48{,}000} = \frac{x}{100}$$
OR $$\frac{1}{50} = \frac{x}{100}$$
$$50 \cdot x = 1 \cdot 100$$
$$\frac{50 \cdot x}{50} = \frac{100}{50}$$
$$x = 2$$

There are 2% of these jobs filled by women.

57.   part is 240; whole is unknown; percent is 7.5%

$$\frac{240}{x} = \frac{7.5}{100}$$
$$7.5 \cdot x = 240 \cdot 100$$
$$\frac{7.5 \cdot x}{7.5} = \frac{24{,}000}{7.5}$$
$$x = 3200$$

She has monthly earnings of $3200 and yearly earnings of 12($3200) = $38,400.

59.   After "Other" and "Private Label," the brand that was purchased most often was Blue Bell.

61.   10.1% of 2373 people

$$\text{part} = (0.101)(2373) \approx 239.7$$
$$\text{part} \approx 240 \text{ people (rounded)}$$

240 people said they purchase Häagen-Dazs.

63.   15.3% of the drivers in the 16–19 age group were pulled over by the police, so
100% − 15.3% = 84.7% were not.

65.   part is unknown; whole is 8000; percent is 13.4

$$\frac{x}{8000} = \frac{13.4}{100}$$
$$x \cdot 100 = 8000 \cdot 13.4$$
$$\frac{x \cdot 100}{100} = \frac{107{,}200}{100}$$
$$x = 1072$$

1072 of the drivers in the 20–29 age group were pulled over by the police.

67.   percent is 86; whole is 15,401; part is unknown

$$\frac{x}{15{,}401} = \frac{86}{100}$$ **OR** $$\frac{x}{15{,}401} = \frac{43}{50}$$
$$50 \cdot x = 15{,}401 \cdot 43$$
$$\frac{50 \cdot x}{50} = \frac{662{,}243}{50}$$
$$x \approx 13{,}245$$

15,401 − 13,245 = 2156
2156 products were successful.

69.   If 85% of her customers paid for their orders using Pay Pal, then 100% − 85% = 15%, used some other method of payment.
part is unknown; whole is 1540; percent is 15

$$\frac{x}{1540} = \frac{15}{100}$$ **OR** $$\frac{x}{1540} = \frac{3}{20}$$
$$\frac{20 \cdot x}{20} = \frac{3 \cdot 1540}{20}$$
$$\frac{20 \cdot x}{20} = \frac{4620}{20}$$
$$x = 231$$

231 customers used some other method of payment.

71.   In the percent proportion, part is to <u>whole</u> as percent is to <u>100</u>.

72.   All percent problems involve a comparison between a part of something and the <u>whole</u>.

Copyright © 2018 Pearson Education, Inc.

73. Since there are 27 grams of total carbohydrates per serving, and there are 4 calories per gram of carbohydrates, there are

$$27 \cdot 4 = 108$$

calories per serving from total carbohydrates.

74. 9% of what number is 27 grams?
part is 27; whole is unknown; percent is 9

$$\frac{27}{x} = \frac{9}{100}$$
$$x \cdot 9 = 27 \cdot 100$$
$$\frac{x \cdot 9}{9} = \frac{2700}{9}$$
$$x = 300$$

300 grams of total carbohydrates are needed to meet the recommended daily value.

75. 12% of what number is 8 grams?
part is 8; whole is unknown; percent is 12

$$\frac{8}{x} = \frac{12}{100} \quad \textbf{OR} \quad \frac{8}{x} = \frac{3}{25}$$
$$x \cdot 3 = 8 \cdot 25$$
$$\frac{x \cdot 3}{3} = \frac{200}{3}$$
$$x \approx 67$$

67 grams (rounded) of fat are needed to meet the recommended daily value.

76. 20% of what number is 4 grams?
part is 4; whole is unknown; percent is 20

$$\frac{4}{x} = \frac{20}{100} \quad \textbf{OR} \quad \frac{4}{x} = \frac{1}{5}$$
$$x \cdot 1 = 4 \cdot 5$$
$$x = 20$$

20 grams of saturated fat are needed to meet the recommended daily value.

77. Yes, since they would eat
28% × 4 servings = 112% of the daily value.

78. Each package has 2 servings and each serving has 3% of the recommended daily value of fiber.

$$\frac{100\%}{2 \cdot 3\%} = \frac{100\%}{6\%} \approx 16.7$$

17 packages (rounded) would have to be eaten. It may be possible but would result in a diet that is high in total fat, saturated fat, sodium, and total carbohydrates.

6.5   Using the Percent Equation

6.5   Margin Exercises

1. **(a)** 15% of 880 policyholders
   $$15\% = 0.15$$

   $$\text{part} = \text{percent} \cdot \text{whole}$$
   $$\text{part} = (\underline{0.15})(\underline{880})$$
   $$\text{part} = \underline{132}$$

   132 policy holders is 15% of 880 policyholders.

(b) 23% of 840 gallons
   $$23\% = 0.23$$

   $$\text{part} = \text{percent} \cdot \text{whole}$$
   $$x = (0.23)(840)$$
   $$x = 193.2$$

   193.2 gallons is 23% of 840 gallons.

(c) 120% of $220
   $$120\% = 1.2$$

   $$\text{part} = \text{percent} \cdot \text{whole}$$
   $$x = (1.2)(220)$$
   $$x = 264$$

   $264 is 120% of $220.

(d) 135% of $1080
   $$135\% = 1.35$$

   $$\text{part} = \text{percent} \cdot \text{whole}$$
   $$x = (\underline{1.35})(\underline{1080})$$
   $$x = 1458$$

   $1458 is 135% of $1080.

(e) 0.5% of 1200 fruit cups
   $$0.5\% = 0.005$$

   $$\text{part} = \text{percent} \cdot \text{whole}$$
   $$x = (0.005)(1200)$$
   $$x = 6$$

   6 fruit cups is 0.5% of 1200 fruit cups.

(f) 0.25% of 1600 lab tests
   $$0.25\% = 0.0025$$

   $$\text{part} = \text{percent} \cdot \text{whole}$$
   $$x = (0.0025)(1600)$$
   $$x = 4$$

   4 lab tests is 0.25% of 1600 lab tests.

2. **(a)** 18 is 45% of what number?

   $$\text{part} = \text{percent} \cdot \text{whole}$$
   $$18 = (\underline{0.45})(x) \qquad 45\% = 0.45$$
   $$\frac{18}{0.45} = \frac{(0.45)(x)}{0.45}$$
   $$\underline{40} = x$$

   18 dancers is 45% of 40 dancers.

Copyright © 2018 Pearson Education, Inc.

(b) 67.5 is 27% of what number?

part = percent • whole

$67.5 = (0.27)(x)$ $27\% = 0.27$

$\dfrac{67.5}{0.27} = \dfrac{(0.27)(x)}{0.27}$

$250 = x$

67.5 containers is 27% of 250 containers.

(c) 666 is 45% of what number?

part = percent • whole

$666 = (0.45)(x)$ $45\% = 0.45$

$\dfrac{666}{0.45} = \dfrac{(0.45)(x)}{0.45}$

$1480 = x$

666 inoculations is 45% of 1480 inoculations.

(d) $5\frac{1}{2}\%$ of what number is 66?

$5\frac{1}{2}\% = 5.5\% = 0.055$

part = percent • whole

$66 = (0.055)(x)$

$\dfrac{66}{0.055} = \dfrac{(0.055)(x)}{0.055}$

$1200 = x$

$5\frac{1}{2}\%$ of 1200 policies is 66 policies.

3.   (a) The whole is 350 and the part is 70.

part = percent • whole

$70 = x \cdot \underline{350}$

$\dfrac{70}{350} = \dfrac{x \cdot 35}{35}$

$0.20 = x$

0.20 is <u>20%</u>.
20% of 350 Facebook friends is 70 Facebook friends.

(b) The whole is 85 and the part is 34.

part = percent • whole

$34 = x \cdot 85$

$\dfrac{34}{85} = \dfrac{x \cdot 85}{85}$

$0.40 = x$

0.40 is 40%.
34 post office boxes is 40% of 85 post office boxes.

(c) The whole is 920 and the part is 1288.

part = percent • whole

$1288 = x \cdot 920$

$\dfrac{1288}{920} = \dfrac{x \cdot 920}{920}$

$1.40 = x$

1.40 is 140%.
140% of 920 invitations is 1288 invitations.

(d) The whole is 1125 and the part is 9.

part = percent • whole

$9 = x \cdot 1125$

$\dfrac{9}{1125} = \dfrac{x \cdot 1125}{1125}$

$0.008 = x$

0.008 is 0.8%.
9 world class runners is 0.8% of 1125 runners.

6.5   Section Exercises

1.   46% of 780 text messages
The percent sign follows 46, so write <u>percent</u> next to 46.
There are 780 total messages, so write <u>whole</u> next to 780 (also, 780 follows *of*).

3.   Write 25% as the decimal 0.25. The whole is 1080. Let x represent the unknown part.

part = percent • whole

$x = (0.25)(1080)$

$x = 270$

25% of 1080 blood donors is <u>270 blood donors</u>.

5.   part = percent • whole

$x = (0.45)(3000)$ $45\% = 0.45$

$x = 1350$

45% of 3000 bath towels is 1350 bath towels.

7.   part = percent • whole

$x = (0.32)(260)$ $32\% = 0.32$

$x = 83.2$

32% of 260 quarts is 83.2 quarts.

9.   part = percent • whole

$x = (1.4)(2500)$ $140\% = 1.4$

$x = 3500$

3500 air bags is 140% of 2500 air bags.

11.   part = percent • whole

$x = (0.124)(8300)$ $12.4\% = 0.124$

$x = 1029.2$

12.4% of 8300 meters is 1029.2 meters.

13.   part = percent • whole

$x = (0.008)(520)$ $0.8\% = 0.008$

$x = 4.16$

$4.16 is 0.8% of $520.

Copyright © 2018 Pearson Education, Inc.

15. 70% of what number of backpackers is 476 backpackers?

The percent sign follows 70, so write <u>percent</u> next to 70.

"What number" follows *of*, so it is the "whole." 476 is the given portion of the whole, so write <u>part</u> next to 476.

17. part = percent · whole
$$24 = (0.15)(x) \qquad 15\% = 0.15$$
$$\frac{24}{0.15} = \frac{(0.15)(x)}{0.15}$$
$$160 = x$$

24 patients is 15% of 160 patients.

19. part = percent · whole
$$130 = (0.40)(x) \qquad 40\% = 0.40$$
$$\frac{130}{0.40} = \frac{(0.40)(x)}{0.40}$$
$$325 = x$$

40% of 325 salads is 130 salads.

21. $12\frac{1}{2}\% = 12.5\% = 0.125$

part = percent · whole
$$135 = (0.125)(x)$$
$$\frac{135}{0.125} = \frac{(0.125)(x)}{0.125}$$
$$1080 = x$$

$12\frac{1}{2}\%$ of 1080 people is 135 people.

23. The part is 3.75 and the percent is $1\frac{1}{4}\% = 1.25\% = 0.0125$ as a decimal. The whole is unknown.

part = percent · whole
$$3.75 = (0.0125)(x)$$
$$\frac{3.75}{0.0125} = \frac{(0.0125)(x)}{0.0125}$$
$$300 = x$$

$1\frac{1}{4}\%$ of 300 gallons is 3.75 gallons.

25. When using the percent equation to solve for percent, you must always change the decimal answer to <u>percent</u> by moving the decimal point <u>two</u> places to the <u>right</u> and attaching a <u>%</u> symbol.

27. part = percent · whole
$$114 = x \cdot 150$$
$$\frac{114}{150} = \frac{x \cdot 150}{150}$$
$$0.76 = x$$

0.76 is 76%. 114 tuxedos is <u>76%</u> of 150 tuxedos.

29. Because 160 follows *of*, the whole is 160. The part is 2.4, and the percent is unknown.

part = percent · whole
$$2.4 = x \cdot 160$$
$$\frac{2.4}{160} = \frac{x \cdot 160}{160}$$
$$0.015 = x$$

0.015 is 1.5%. 1.5% of 160 liters is 2.4 liters.

31. part = percent · whole
$$170 = x \cdot 68$$
$$\frac{170}{68} = \frac{x \cdot 68}{68}$$
$$2.5 = x$$

2.5 is 250%. 170 cartons is 250% of 68 cartons.

33. You must first write the fraction in the percent as a decimal, then divide the percent by 100 to change it to a decimal.

$$2\frac{1}{2}\% = 2.5\% = 0.025$$

2.5% as a decimal

Write $2\frac{1}{2}$ as 2.5

35. 27% of 14 million

part = percent · whole
$$x = (0.27)(14)$$
$$x = 3.78$$

3.78 million or 3,780,000 office workers want more storage space.

37. **(a)** 77% of 12,500 people

part = percent · whole
$$x = (0.77)(12,500)$$
$$x = 9625$$

The number of people in the survey who would rather admit their age is 9625.

(b) 15% of 12,500 people

part = percent · whole
$$x = (0.15)(12,500)$$
$$x = 1875$$

The number of people in the survey who would rather admit their weight is 1875.

(c) 8% of 12,500 people

part = percent · whole
$$x = (0.08)(12,500)$$
$$x = 1000$$

The number of people in the survey who would rather admit their salary is 1000.

Copyright © 2018 Pearson Education, Inc.

39. (a) 98% of 18,000

$$\text{part} = \text{percent} \cdot \text{whole}$$
$$x = (0.98)(18,000)$$
$$x = 17,640$$

17,640 households are expected to have a refrigerator.

(b) $18,000 - 17,640 = 360$ households are expected to not have a refrigerator.

41. (a) 3 of the 50 states

$$\text{part} = \text{percent} \cdot \text{whole}$$
$$3 = x \cdot 50$$
$$\frac{3}{50} = \frac{x \cdot 50}{50}$$
$$0.06 = x$$
$$0.06 \text{ is } 6\%$$

6% of the states do not have drive-in movies.

(b) $100\% - 6\% = 94\%$ of the states do have drive-in movies.

43. Because 1250 follows *of*, the whole is 1250. The part is 461, and the percent is unknown.

$$\text{part} = \text{percent} \cdot \text{whole}$$
$$461 = x \cdot 1250$$
$$\frac{461}{1250} = \frac{x \cdot 1250}{1250}$$
$$0.369 \approx x$$

0.369 is 36.9%. 36.9% (rounded) of these Americans rate their health as excellent.

44.
$$\text{part} = \text{percent} \cdot \text{whole}$$
$$155 = x \cdot 3480$$
$$\frac{155}{3480} = \frac{x \cdot 3480}{3480}$$
$$0.0445 \approx x$$

0.0445 is 4.45%. They have planned 4.5% (rounded) additional stores.

45.
$$\text{part} = \text{percent} \cdot \text{whole}$$
$$x = (0.03)(35,051) \quad 3\% = 0.03$$
$$x = 1051.53$$

The average amount of loan debt increased by $1051.53. Therefore, the average amount of loan is $35,051 + \$1052$ (rounded) $= \$36,103$.

47. 220,917 is 46.2% of what number?

$$\text{part} = \text{percent} \cdot \text{whole}$$
$$220,917 = (0.462)(x)$$
$$\frac{220,917}{0.462} = \frac{(0.462)(x)}{0.462}$$
$$478,175 \approx x$$

In 1967, 478,175 Mustangs (rounded) were sold.

49.
$$\text{part} = \text{percent} \cdot \text{whole}$$
$$x = (0.325)(385,200) \quad 32\tfrac{1}{2}\% = 0.325$$
$$x = 125,190$$

There was an increase in sales of $125,190. Therefore, the amount of sales this year is $385,200 + \$125,190 = \$510,390$.

51. Find the sales tax on the new Polaris unit. The whole is 524. The percent is $7\tfrac{3}{4}\% = 7.75\%$, which is 0.0775 as a decimal.

$$\text{part} = \text{percent} \cdot \text{whole}$$
$$x = (0.0775)(524)$$
$$x = 40.61$$

Add the sales tax to the purchase price.

$$\$524 + \$40.61 = \$564.61$$

Subtract the trade-in.

$$\$564.61 - \$125 = \$439.61$$

The total cost to the customer is $439.61.

53. (a) The largest percent is 25%, so "Relatives" is the type of day care used most often.

(b)
$$\text{part} = \text{percent} \cdot \text{whole}$$
$$x = (0.25)(7800)$$
$$x = 1950$$

1950 families used "Relatives" for day care.

54. (a) The smallest percent is 3%, so "Mother while working" is the type of day care used least often.

(b)
$$\text{part} = \text{percent} \cdot \text{whole}$$
$$x = (0.03)(7800)$$
$$x = 234$$

234 families used "Mother while working" for day care.

55.
$$\text{part} = \text{percent} \cdot \text{whole}$$
$$x = (0.17)(7800)$$
$$x = 1326$$

1326 families used "Nanny or family day care home" for day care.

56.
$$\text{part} = \text{percent} \cdot \text{whole}$$
$$x = (0.24)(7800)$$
$$x = 1872$$

1872 families used "Child care center or nursery school" for day care.

Copyright © 2018 Pearson Education, Inc.

Summary Exercises *Using Percent Proportion and Percent Equation*

1. $6.25\% = 0.0625$
Drop the percent sign and move the decimal point two places to the left.

3. $0.375 = 37.5\%$
Move the decimal point two places to the right and attach a percent sign.

5. $87.5\% = \dfrac{87.5}{100} = \dfrac{87.5 \cdot 10}{100 \cdot 10}$
$= \dfrac{875}{1000} = \dfrac{875 \div 125}{1000 \div 125} = \dfrac{7}{8}$

7. $\dfrac{5}{8} = \dfrac{p}{100}$
$8 \cdot p = 5 \cdot 100$
$\dfrac{8 \cdot p}{8} = \dfrac{500}{8}$
$p = 62.5$

Thus, $\frac{5}{8} = 62.5\%$.

9. 286 of 1100 adults have a fear of putting on weight.
$\dfrac{286}{1100} = \dfrac{286 \div 22}{1100 \div 22} = \dfrac{13}{50}$ *fraction*
$\dfrac{13}{50} = \dfrac{13 \cdot 2}{50 \cdot 2} = \dfrac{26}{100} = 0.26$ *decimal*
$0.26 = 26\%$ *percent*

11. 66 of 1100 adults dread too much family time.
$\dfrac{66}{1100} = \dfrac{66 \div 22}{1100 \div 22} = \dfrac{3}{50}$ *fraction*
$\dfrac{3}{50} = \dfrac{3 \cdot 2}{50 \cdot 2} = \dfrac{6}{100} = 0.06$ *decimal*
$0.06 = 6\%$ *percent*

13. percent is 6; whole is 780
$$\text{part} = \text{percent} \cdot \text{whole}$$
$$x = (0.06)(780)$$
$$x = 46.8$$
6% of $780 is $46.80.

15. part is 176; whole is 320
$\dfrac{176}{320} = \dfrac{x}{100}$ **OR** $\dfrac{11}{20} = \dfrac{x}{100}$
$20 \cdot x = 11 \cdot 100$
$\dfrac{20 \cdot x}{20} = \dfrac{1100}{20}$
$x = 55$

55% of 320 policies is 176 policies.

17. part is 1016.4; percent is 280
$\dfrac{1016.4}{x} = \dfrac{280}{100}$ **OR** $\dfrac{1016.4}{x} = \dfrac{14}{5}$
$x \cdot 14 = (1016.4)(5)$
$\dfrac{x \cdot 14}{14} = \dfrac{5082}{14}$
$x = 363$

1016.4 acres is 280% of 363 acres.

19. 10.6% of 1582
$$\text{part} = \text{percent} \cdot \text{whole}$$
$$x = (0.106)(1582)$$
$$x = 167.692 \approx 168$$
168 people (rounded) prefer Dreyer's/Edy's brand ice cream.

21. 13.7% of 39,508,000
$$\text{part} = \text{percent} \cdot \text{whole}$$
$$x = (0.137)(39{,}508{,}000)$$
$$x = 5{,}412{,}596$$
To the nearest thousand, there were 5,413,000 student loans that were in default last year.

23. $65 - 39 = 26$ miles remain.
$\dfrac{26}{65} = \dfrac{x}{100}$ **OR** $\dfrac{2}{5} = \dfrac{x}{100}$
$x \cdot 5 = 2 \cdot 100$
$\dfrac{x \cdot 5}{5} = \dfrac{200}{5}$
$x = 40$

40% of the run remains.

25. part is 120.6 million; percent is 91%
$\dfrac{120.6 \text{ million}}{x} = \dfrac{91}{100}$
$x \cdot 91 = 120.6 \text{ million} \cdot 100$
$\dfrac{x \cdot 91}{91} = \dfrac{12{,}060 \text{ million}}{91}$
$x \approx 132.5 \text{ million}$

To the nearest tenth of a million, there were 132.5 million tax returns filed electronically last year.

Copyright © 2018 Pearson Education, Inc.

6.6 Solving Application Problems with Percent

6.6 Margin Exercises

1. Use the sales tax formula:

amount of sales tax = rate of tax · cost of item

The sales tax rate is 6%.

(a) The cost of the bat is $29.

$$amount\ of\ sales\ tax = (6\%)(\underline{\$29})$$
$$a = (0.06)(\$29)$$
$$a = \$\underline{1.74}$$

The total cost is $29 + \underline{\$1.74} = \$\underline{30.74}$.

(b) The cost of the leather chair and ottoman is $1287.

$$amount\ of\ sales\ tax = (6\%)(\$1287)$$
$$a = (0.06)(\$1287)$$
$$a = \$77.22$$

The total cost is $1287 + $77.22 = $1364.22.

(c) The cost of the pickup truck is $24,500.

$$amount\ of\ sales\ tax = (6\%)(\$24,500)$$
$$a = (0.06)(\$24,500)$$
$$a = \$1470$$

The total cost is $24,500 + $1470 = $25,970.

2. Use the sales tax formula:

sales tax = rate of tax · cost of item

(a) The tax on a $320 patio set is $25.60.

$$\$25.60 = r \cdot \underline{\$320}$$
$$\frac{25.60}{320} = \frac{r \cdot 320}{320}$$
$$0.08 = r$$

0.08 is $\underline{8}$%. The sales tax rate is 8%.

(b) The tax on a $96 park bench is $6.24.

$$\$6.24 = r \cdot \$96$$
$$\frac{6.24}{96} = \frac{r \cdot 96}{96}$$
$$0.065 = r$$

0.065 is 6.5% or $6\frac{1}{2}$%. The sales tax rate is 6.5% or $6\frac{1}{2}$%.

(c) The tax on a $28,990 Dodge Charger is $1159.60.

$$\$1159.60 = r \cdot \$28,990$$
$$\frac{1159.60}{28,990} = \frac{r \cdot 28,990}{28,990}$$
$$0.04 = r$$

0.04 is 4%. The sales tax rate is 4%.

3. Use the commission formula:

amount of commission (c)
= rate of commission · amount of sales

(a) Rate of commission is 11%; sales are $38,700.

$$c = (11\%)(\underline{\$38,700})$$
$$c = (0.11)(\$38,700)$$
$$c = \$\underline{4257}$$

Danielle earned a commission of $4257 for selling dental equipment.

(b) Rate of commission is 6%; sales are $189,500.

$$c = (6\%)(\$189,500)$$
$$c = (0.06)(\$189,500)$$
$$c = \$11,370$$

Alyssa earned a commission of $11,370 for selling a home.

4. **(a)** A commission of $450 and sales worth $22,500:

$$\frac{part}{whole} = \frac{x}{100}$$
$$\frac{450}{22,500} = \frac{x}{100}$$
$$22,500 \cdot x = 450 \cdot 100$$
$$\frac{22,500 \cdot x}{22,500} = \frac{45,000}{22,500}$$
$$x = \underline{2}$$

The rate of commission is 2%.

(b) A commission of $2898 and sales worth $32,200:

$$\frac{part}{whole} = \frac{x}{100}$$
$$\frac{2898}{32,200} = \frac{x}{100}$$
$$32,200 \cdot x = 2898 \cdot 100$$
$$\frac{32,200 \cdot x}{32,200} = \frac{289,800}{32,200}$$
$$x = 9$$

The rate of commission is 9%.

Copyright © 2018 Pearson Education, Inc.

5. **(a)** 42% discount on a leather recliner priced at $950:

amount of discount
$$= \text{rate of discount} \cdot \text{original price}$$
$$= (0.42)(\$950)$$
$$= \$399$$

sale price
$$= \text{original price} - \text{amount of discount}$$
$$= \$950 - \$399$$
$$= \$551$$

The sale price for the Easy-Boy leather recliner is $551.

(b) 35% discount on a sweater set priced at $30.

$$\text{amount of discount} = (0.35)(\$30)$$
$$= \$10.50$$

$$\text{sale price} = \$30 - \$10.50$$
$$= \$19.50$$

The sale price for the sweater set is $19.50.

6. **(a)** Production increased from 14,100 units to 19,035 units.

$$\text{increase} = 19{,}035 - 14{,}100 = \underline{4935}$$

$$\frac{\text{amount of increase (part)}}{\text{original value (whole)}} = \frac{\text{percent}}{100}$$
$$\frac{4935}{14{,}100} = \frac{x}{100}$$
$$14{,}100 \cdot x = 4935 \cdot 100$$
$$\frac{14{,}100 \cdot x}{14{,}100} = \frac{493{,}500}{14{,}100}$$
$$x = 35$$

The percent of increase is 35%.

(b) Flu cases rose from 496 cases to 620 cases.

$$\text{increase} = 620 - 496 = 124$$

$$\frac{124}{496} = \frac{x}{100} \quad \textbf{OR} \quad \frac{1}{4} = \frac{x}{100}$$
$$4 \cdot x = 1 \cdot 100$$
$$\frac{4 \cdot x}{4} = \frac{100}{4}$$
$$x = 25$$

The percent of increase is 25%.

7. **(a)** The number of service calls decreased from 380 to 285.

$$\text{decrease} = 380 - 285 = \underline{95}$$

$$\frac{\text{amount of decrease (part)}}{\text{original value (whole)}} = \frac{x}{100}$$
$$\frac{95}{380} = \frac{x}{100}$$
$$380 \cdot x = 95 \cdot 100$$
$$\frac{380 \cdot x}{380} = \frac{9500}{380}$$
$$x = 25$$

The percent of decrease is <u>25%</u>.

(b) The number of workers decreased from 4850 to 3977.

$$\text{decrease} = 4850 - 3977 = 873$$

$$\frac{873}{4850} = \frac{p}{100}$$
$$4850 \cdot p = 873 \cdot 100$$
$$\frac{4850 \cdot p}{4850} = \frac{87{,}300}{4850}$$
$$p = 18$$

The percent of decrease is 18%.

6.6 Section Exercises

1. When solving a sales tax problem, the cost of an item is the <u>whole</u>, the sales tax rate is the <u>percent</u>, and the amount of sales tax is the <u>part</u>.

3. $\text{sales tax} = \text{rate of tax} \cdot \text{cost of item}$
$$= (4\%)(\$6)$$
$$= (0.04)(\$6)$$
$$= \$0.24$$

The amount of sales tax is $0.24 and the total cost is $6 + \$0.24 = \$6.24$.

5. Find the sales tzx rate using the sales tzx formula. The sales tax is $12.75, and the cost of the item is $425.

$\text{sales tax} = \text{rate of tax} \cdot \text{cost of item}$
$$\$12.75 = r \cdot \$425$$
$$\frac{12.75}{425} = \frac{r \cdot 425}{425}$$
$$0.03 = r$$

0.03 is 3%. The tax rate is 3% and the total cost is $425 + \$12.75 = \$437.75$.

7. $\text{sales tax} = \text{rate of tax} \cdot \text{cost of item}$
$$= \left(5\tfrac{1}{2}\%\right)(\$12{,}229)$$
$$= (0.055)(\$12{,}229)$$
$$\approx \$672.60$$

The amount of sales tax is $672.60 (rounded) and the total cost is $12,229 + \$672.60 = \$12{,}901.60$.

Copyright © 2018 Pearson Education, Inc.

9. When solving commission problems, the whole is the <u>sales amount</u>, the percent is the <u>commission rate</u>, and the part is the <u>commission</u>.

11. The problem asks for the amount of commission. Use the commission formula. The reate of commission is 8%, and the sales are $280.

$$
\begin{aligned}
commission &= rate\ of\ commission \cdot sales \\
&= (8\%)(\$280) \\
&= (0.08)(\$280) \\
&= \$22.40
\end{aligned}
$$

The amount of commission is $22.40.

13. $\dfrac{part}{whole} = \dfrac{percent}{100}$

$$\frac{600}{3000} = \frac{x}{100} \quad \textbf{OR} \quad \frac{1}{5} = \frac{x}{100}$$
$$5 \cdot x = 1 \cdot 100$$
$$\frac{5 \cdot x}{5} = \frac{100}{5}$$
$$x = 20$$

The rate of commission is 20%.

15.
$$
\begin{aligned}
commission &= rate\ of\ commission \cdot sales \\
&= (3\%)(\$6183.50) \\
&= (0.03)(\$6183.50) \\
&\approx \$185.51
\end{aligned}
$$

The amount of commission is $185.51 (rounded).

17. When solving retail sales discount problems, the *original* price is always the <u>whole</u>, the rate of discount is the <u>percent</u>, and the amount of discount is the <u>part</u>.

19.
$$
\begin{aligned}
discount &= rate\ of\ discount \cdot original\ price \\
&= (10\%)(\$199.99) \\
&= (0.10)(\$199.99) \\
&\approx \$20.00
\end{aligned}
$$

The amount of discount is $20.00 (rounded) and the sale price is $199.99 − $20 = $179.99.

21.
$$
\begin{aligned}
discount &= rate\ of\ discount \cdot original\ price \\
\$54 &= r \cdot \$180 \\
\frac{54}{180} &= \frac{r \cdot 180}{180} \\
0.3 &= r
\end{aligned}
$$

0.3 is 30%. The rate of discount is 30% and the sale price is $180 − $54 = $126.

23. The problem asks for the amount of the discount and the sale price. First, find the amount of discount using the discount formula. The rate of discount is 15%, and the original price is $58.40.

$$
\begin{aligned}
sales\ tax &= rate\ of\ tax \cdot cost\ of\ item \\
&= (15\%)(\$58.40) \\
&= (0.15)(\$58.40) \\
&= \$8.76
\end{aligned}
$$

Now find the sale price by subtracting the amount of the discount ($8.76) from the original price: $58.40 − $8.76 = $49.64. The amount of discount is $8.76 and the sale price is $49.64.

25. Answers will vary. A sample answer follows: On the basis of commission alone you would choose Company A. Other considerations might be: reputation of the company; expense allowances; other employee benefits; travel; promotion and training (to name a few).

27. The cost of 6 skillets is

$$6(\$29.99) = \$179.94.$$

Since the cost of the skillets is "$100.00 or more," we need to add $14.99 for shipping and insurance. The sales tax is

$$(5\%)(\$179.94) = (0.05)(\$179.94) \approx \$9.00.$$

Thus, the total cost is

$179.94 + $14.99 + $9.00 = $203.93 (rounded).

29. The cost of 3 pop-up hampers is

$$3(\$9.99) = \$29.97.$$

The cost of 4 nonstick mini donut pans is

$$4(\$10.99) = \$43.96.$$

The cost of the hampers and pans is $29.97 + $43.96 = $73.93. Since this cost is "$70.01 to $99.99," we need to add $12.99 for shipping and insurance.
The sales tax is

$$(5\%)(\$73.93) = (0.05)(\$73.93) \approx \$3.70.$$

Thus, the total cost is

$73.93 + $12.99 + $3.70 = $90.62 (rounded).

31. The problem asks for the rate of sales tax. Use the sales tax formula. The sales tax is $99. The cost of the door is $1980. Use r to represent the unknown rate of tax.

$$
\begin{aligned}
sales\ tax &= rate\ of\ tax \cdot cost\ of\ item \\
\$99 &= r \cdot \$1980 \\
\frac{99}{1980} &= \frac{r \cdot 1980}{1980} \\
0.05 &= r
\end{aligned}
$$

0.05 is 5%. The rate of sales tax is 5%.

Copyright © 2018 Pearson Education, Inc.

33. increase $= 635{,}000 - 467{,}400 = 167{,}600$

$$\frac{\text{increase}}{\text{original}} = \frac{x}{100}$$

$$\frac{167{,}600}{467{,}400} = \frac{x}{100}$$

$$467{,}400 \cdot x = 167{,}600 \cdot 100$$

$$\frac{467{,}400 \cdot x}{467{,}400} = \frac{16{,}760{,}000}{467{,}400}$$

$$x \approx 35.9$$

The percent of increase is 35.9% (rounded).

35. decrease $= 48 - 36 = 12$

$$\frac{\text{decrease}}{\text{original}} = \frac{x}{100}$$

$$\frac{12}{48} = \frac{x}{100}$$

$$48 \cdot x = 12 \cdot 100$$

$$\frac{48 \cdot x}{48} = \frac{1200}{48}$$

$$x = 25$$

The percent of decrease is 25%.

37. *discount = rate of discount • original price*

$$= (60\%)(\$335)$$

$$= (0.60)(\$335)$$

$$= \$201$$

The sale price of the coat is
$335 - $201 = $134.

39. *commission = rate of commission • sales*

$$= (3\%)(\$18{,}960)$$

$$= (0.03)(\$18{,}960)$$

$$= \$568.80$$

Strong's commission is $568.80.

41. *commission = rate of commission • sales*

$$\$707.20 = r \cdot \$17{,}680$$

$$\frac{707.20}{17{,}680} = \frac{r \cdot 17{,}680}{17{,}680}$$

$$0.04 = r$$

$$0.04 \text{ is } 4\%$$

The rate of commission for Brasher is 4%.

43. *discount = rate of discount • original price*

$$= (22\%)(\$390)$$

$$= (0.22)(\$390)$$

$$= \$85.80$$

The discount is $85.80, and the sale price is
$390 - $85.80 = $304.20.

45. The price decreased $45.50 - $35.49 = $10.01.

$$\frac{\text{decrease}}{\text{original}} = \frac{x}{100}$$

$$\frac{10.01}{45.50} = \frac{x}{100}$$

$$45.50 \cdot x = 10.01 \cdot 100$$

$$\frac{45.50 \cdot x}{45.50} = \frac{1001}{45.50}$$

$$x = 22$$

The percent of decrease in price is 22%.

47. The problem asks for the cost of the dictionary. First find the amount of discount using the discount formula. The rate of discount is 6%, and the original price is $18.50.

discount = rate of discount • original price

$$= (6\%)(\$18.50)$$

$$= (0.06)(\$18.50)$$

$$= \$1.11$$

To find the sale price, subtract the amount of discount from the original price.

Sale price = original price − amount of discount

$$= \$18.50 - \$1.11 = \$17.39$$

To find the sales tax, use the sales tax formula. The rate of tax is 6% and the cost of the dictionary is $17.39.

sales tax = rate of tax • cost of item

$$= (6\%)(\$17.39)$$

$$= (0.06)(\$17.39)$$

$$\approx \$1.04$$

The cost of the dictionary is
$17.39 + $1.04 = $18.43.

49. *sales commission = rate of commission • sales*

$$= (2\%)(\$380{,}450)$$

$$= (0.02)(\$380{,}450)$$

$$= \$7609$$

The sales representative gets
$100\% - 20\% = 80\%$.

commission for agent $= (80\%)(\$7609)$

$$= (0.80)(\$7609)$$

$$= \$6087.20$$

The agent receives $6087.20.

51. *discount = rate of discount • original price*

$$= (15\%)(\$15{,}321)$$

$$= (0.15)(\$15{,}321)$$

$$= \$2298.15$$

The sale price is
$15,321 - $2298.15 = $13,022.85.

(continued)

Copyright © 2018 Pearson Education, Inc.

sales tax = rate of tax · cost of item
$$= (7\tfrac{3}{4}\%)(\$13{,}022.85)$$
$$= (0.0775)(\$13{,}022.85)$$
$$\approx \$1009.27$$

The total cost of the boat is
$\$13{,}022.85 + \$1009.27 = \$14{,}032.12$ (rounded).

53.   The percent equation is

$$part = \underline{percent} \cdot \underline{whole}$$

54.   The sales tax formula is

sales tax = $\underline{\text{rate (percent) of tax}}$ · $\underline{\text{cost of item}}$

55.   *number working = percent · number surveyed*
$$= (61\%)(4250)$$
$$= (0.61)(4250)$$
$$= 2592.5$$

Of those surveyed, 2593 (rounded) people planned to work while on vacation.

56.   Fielding requests is the least common activity among working vacationers, 2593 (from exercise 55.)
number = percent · number working
$$= (20\%)(2593)$$
$$= (0.2)(2593)$$
$$= 518.6$$

About 519 workers (rounded) said they will field requests.

57.   Emailing is the most common activity among working vacationers, 2593 (from exercise 55.)
number = percent · number working
$$= (38\%)(2593)$$
$$= (0.38)(2593)$$
$$= 985.34$$

About 985 workers (rounded) said they will email.

58.   30% will be calling work
number = percent · number working
$$= (30\%)(2593)$$
$$= (0.3)(2593)$$
$$= 777.9$$

About 778 workers (rounded) said they will be calling work.

6.7   Simple Interest

6.7   Margin Exercises

1.   **(a)** $1000 at 3% for 1 year
$$I = p \cdot r \cdot t$$
$$I = (1000)(\underline{0.03})(1)$$
$$I = \underline{30}$$

The interest is $30.

   (b) $3650 at 2% for 1 year
$$I = p \cdot r \cdot t$$
$$I = (3650)(0.02)(1)$$
$$I = 73$$

The interest is $73.

2.   **(a)** $820 at 1% for $3\tfrac{1}{2}$ years
$$I = p \cdot r \cdot t$$
$$I = (\underline{820})(\underline{0.01})(3.5)$$
$$I = \underline{28.70}$$

The interest is $28.70.

   (b) $16,800 at 3% for $2\tfrac{3}{4}$ years
$$I = p \cdot r \cdot t$$
$$I = (16{,}800)(0.03)(2.75)$$
$$I = 1386$$

The interest is $1386.

3.   **(a)** $1800 at 3% for 4 months
$$I = p \cdot r \cdot t$$
$$I = (1800)(0.03)\left(\tfrac{4}{12}\right) \quad \text{4 months}$$
$$I = 54\left(\tfrac{1}{3}\right) \qquad\qquad = \left(\tfrac{4}{12}\right) \text{ year}$$
$$I = \frac{54}{3} = \underline{18}$$

The interest is $18.

   (b) $28,000 at $7\tfrac{1}{2}$% for 3 months
$$I = p \cdot r \cdot t$$
$$I = (28{,}000)(0.075)\left(\tfrac{3}{12}\right)$$
$$I = (2100)\left(\tfrac{1}{4}\right)$$
$$I = \frac{2100}{4} = 525$$

The interest is $525.

Copyright © 2018 Pearson Education, Inc.

4. **(a)** $3800 at $6\frac{1}{2}$% for 6 months

$$I = p \cdot r \cdot t$$
$$I = (3800)(0.065)\left(\frac{6}{12}\right)$$
$$I = 247\left(\frac{1}{2}\right)$$
$$I = \frac{247}{2} = \underline{123.5}$$

The interest is $123.50.

total amount due = principal + interest
$$= \$3800 + \$123.50$$
$$= \underline{\$3923.50}$$

The total amount due is $3923.50.

(b) $12,400 at 5% for 5 years

$$I = p \cdot r \cdot t$$
$$I = (12,400)(0.05)(5)$$
$$I = 3100$$

The interest is $3100.

total amount due = principal + interest
$$= \$12,400 + \$3100$$
$$= \$15,500$$

The total amount due is $15,500.

(c) $2400 at $4\frac{1}{2}$% for $2\frac{3}{4}$ years

$$I = p \cdot r \cdot t$$
$$= (2400)(0.045)(2.75)$$
$$= 297$$

The interest is $297.

total amount due = principal + interest
$$= \$2400 + \$297$$
$$= \$2697$$

The total amount due is $2697.

6.7 Section Exercises

1. $3\frac{1}{2}$% is the same as 3.5%, and is written as a decimal as 0.035.

3. $2\frac{1}{4}$ years is written as a decimal as 2.25 years.

5. $100 at 6% for 1 year

$$I = p \cdot r \cdot t$$
$$I = (100)(0.06)(1)$$
$$I = 6$$

The interest is $6.

7. $700 at 5% for 3 years

$$I = p \cdot r \cdot t$$
$$I = (700)(0.05)(3)$$
$$I = 105$$

The interest is $105.

9. $2300 at $4\frac{1}{2}$% for $2\frac{1}{2}$ years
The principal (p) is $2300. The rate ($r$) is $4\frac{1}{2}$% or 0.045 as a decimal and the time (t) is $2\frac{1}{2}$ or 2.5 years.

$$I = p \cdot r \cdot t$$
$$I = (2300)(0.045)(2.5)$$
$$I = 258.75$$

The interest is $258.75.

11. $10,800 at $7\frac{1}{2}$% for $2\frac{3}{4}$ years

$$I = p \cdot r \cdot t$$
$$I = (10,800)(0.075)(2.75)$$
$$I = 2227.5$$

The interest is $2227.50.

13. 3 months

$$\frac{3}{12} = \frac{3 \div 3}{12 \div 3} = \frac{1}{4} \qquad \textit{fraction}$$
$$\frac{1}{4} = \frac{1 \cdot 25}{4 \cdot 25} = \frac{25}{100} = 0.25 \quad \textit{decimal}$$

15. 9 months

$$\frac{9}{12} = \frac{9 \div 3}{12 \div 3} = \frac{3}{4} \qquad \textit{fraction}$$
$$\frac{3}{4} = \frac{3 \cdot 25}{4 \cdot 25} = \frac{75}{100} = 0.75 \quad \textit{decimal}$$

17. $400 at 3% for 6 months

$$I = p \cdot r \cdot t$$
$$I = (400)(0.03)\left(\frac{6}{12}\right)$$
$$I = (12)\left(\frac{1}{2}\right)$$
$$I = \frac{12}{2} = 6$$

The interest is $6.

19. $940 at 3% for 18 months
The principal is $940. The rate is 3% or 0.03, and the time is $\frac{18}{12}$ of a year.

$$I = p \cdot r \cdot t$$
$$I = (940)(0.03)\left(\frac{18}{12}\right)$$
$$I = (28.2)(1.5)$$
$$I = 42.3$$

The interest is $42.30.

Copyright © 2018 Pearson Education, Inc.

21. $1225 at $5\frac{1}{2}$% for 3 months

$$I = p \cdot r \cdot t$$
$$I = (1225)(0.055)\left(\tfrac{3}{12}\right)$$
$$I = (67.375)\left(\tfrac{1}{4}\right)$$
$$I = 16.84 \text{ (rounded)}$$

The interest is $16.84.

23. $15,300 at $7\frac{1}{4}$% for 7 months

$$I = p \cdot r \cdot t$$
$$I = (15,300)(0.0725)\left(\tfrac{7}{12}\right)$$
$$I = (1109.25)\left(\tfrac{7}{12}\right)$$
$$I = 647.06 \text{ (rounded)}$$

The interest is $647.06.

25. $200 at 5% for 1 year

$$I = p \cdot r \cdot t$$
$$I = (200)(0.05)(1)$$
$$I = 10$$

The interest is $10.

total amount due = principal + interest
$$= \$200 + \$10$$
$$= \$210$$

The total amount due is $210.

27. $740 at 6% for 9 months

$$I = p \cdot r \cdot t$$
$$I = (740)(0.06)\left(\tfrac{9}{12}\right)$$
$$I = (44.4)\left(\tfrac{3}{4}\right)$$
$$I = \frac{133.2}{4} = 33.3$$

The interest is $33.30.

total amount due = principal + interest
$$= \$740 + \$33.30$$
$$= \$773.30$$

The total amount due is $773.30.

29. $1800 at 9% for 18 months

$$I = p \cdot r \cdot t$$
$$I = (1800)(0.09)\left(\tfrac{18}{12}\right)$$
$$I = (162)(1.5)$$
$$I = 243$$

The interest is $243.

total amount due = principal + interest
$$= \$1800 + \$243$$
$$= \$2043$$

The total amount due is $2043.

31. $3250 at $3\frac{1}{2}$% for 6 months

$$I = p \cdot r \cdot t$$
$$I = (3250)(0.035)\left(\tfrac{6}{12}\right)$$
$$I = (113.75)(0.5)$$
$$I = 56.88 \text{ (rounded)}$$

The interest is $56.88.

total amount due = principal + interest
$$= \$3250 + \$56.88$$
$$= \$3306.88$$

The total amount due is $3306.88 (rounded).

33. $16,850 at $7\frac{1}{2}$% for 9 months
First find the interest. The principal (p) is $16,850. The rate is $7\frac{1}{2}$% or 0.075 as a decimal; and the time (t) is $\frac{9}{12}$ of a year.

$$I = p \cdot r \cdot t$$
$$I = (16,850)(0.075)\left(\tfrac{9}{12}\right)$$
$$I = (1263.75)\left(\tfrac{3}{4}\right)$$
$$I = 947.81 \text{ (rounded)}$$

The interest is $947.81.

To find the total amount due, add the principal and the interest.

total amount due = principal + interest
$$= \$16,850 + \$947.81$$
$$= \$17,797.81$$

The total amount due is $17,797.81.

35. The answer should include:
Amount of principal—This is the amount of money borrowed or loaned.
Interest rate—This is the percent used to calculate the interest.
Time of loan—The length of time that money is loaned or borrowed is an important factor in determining interest.

37. $8642 at 2% for 2 years

$$I = p \cdot r \cdot t$$
$$I = (8642)(0.02)(2)$$
$$I = 345.68$$

He will earn $345.68 in interest.

39. $150,000 at 7% for 30 months

$$I = p \cdot r \cdot t$$
$$I = (150,000)(0.07)\left(\tfrac{30}{12}\right)$$
$$I = (10,500)(2.5)$$
$$I = 26,250$$

The amount of interest is $26,250.

Copyright © 2018 Pearson Education, Inc.

41. $4650 at 5% for 5 months

$$I = p \cdot r \cdot t$$
$$I = (4650)(0.05)\left(\tfrac{5}{12}\right)$$
$$I = (232.50)\left(\tfrac{5}{12}\right)$$
$$I = 96.88 \text{ (rounded)}$$

The interest is $96.88.

total amount due = principal + interest
$$= \$4650 + \$96.88$$
$$= \$4746.88$$

The total amount due is $4746.88 (rounded).

43. $14,800 at $2\tfrac{1}{4}$% for 10 months

The principal (p) is $14,800. The rate ($r$) is $2\tfrac{1}{4}$% or 0.0225 as a decimal and the time (t) is $\tfrac{10}{12}$ of a year.

$$I = p \cdot r \cdot t$$
$$I = (14,800)(0.0225)\left(\tfrac{10}{12}\right)$$
$$I = (333)\left(\tfrac{5}{6}\right)$$
$$I = 277.50$$

He will earn $277.50 in interest.

45. $8800 at $7\tfrac{1}{4}$% for 3 months

$$I = p \cdot r \cdot t$$
$$I = (8800)(0.0725)\left(\tfrac{3}{12}\right)$$
$$I = (638)\left(\tfrac{1}{4}\right)$$
$$I = 159.50$$

She will earn $159.50 in interest every 3 months.

47. (a) Bank: $5400 at $2\tfrac{3}{4}$% for $1\tfrac{1}{2}$ years

$$I = p \cdot r \cdot t$$
$$I = (5400)(0.0275)(1.5)$$
$$I = (148.50)(1.5)$$
$$I = 222.75$$

The interest earned at the bank is $222.75.

Credit union: $5400 at $3\tfrac{1}{2}$% for $1\tfrac{1}{2}$ years

$$I = p \cdot r \cdot t$$
$$I = (5400)(0.035)(1.5)$$
$$I = (189)(1.5)$$
$$I = 283.50$$

The interest earned at the credit union is $283.50.

(b) Bank: The total amount in the account is $5400 + $222.75 = $5622.75.

Credit union: The total amount in the account is $5400 + $283.50 = $5683.50.

48. (a) Dealer: $29,400 at $9\tfrac{1}{2}$% for $3\tfrac{1}{2}$ years

$$I = p \cdot r \cdot t$$
$$I = (29,400)(0.095)(3.5)$$
$$I = (2793)(3.5)$$
$$I = 9775.50$$

The interest will be $9775.50.
The total amount due will be
$29,400 + $9775.50 = $39,175.50

Credit union: $29,400 at $6\tfrac{1}{4}$% for $3\tfrac{1}{2}$ years

$$I = p \cdot r \cdot t$$
$$I = (29,400)(0.0625)(3.5)$$
$$I = (1837.5)(3.5)$$
$$I = 6431.25$$

The interest will be $6431.25.
The total amount due will be
$29,400 + $6431.25 = $35,831.25

(b) If the loan is made with the credit union, he will save $39,175.50 − $35,831.25 = $3344.25.

49. The total cost of the bird cages is
$$4 \cdot \$980 = \$3920.$$

The amount borrowed is 70% of $3920.
$$x = (0.70)(3920)$$
$$x = 2744$$

$2744 at $7\tfrac{1}{2}$% for $2\tfrac{1}{2}$ years

$$I = p \cdot r \cdot t$$
$$I = (2744)(0.075)(2.5)$$
$$I = (205.80)(2.5)$$
$$I = 514.50$$

The interest is $514.50.

total amount due = principal + interest
$$= \$2744 + \$514.50$$
$$= \$3258.50$$

The total amount due is $3258.50.

50. The total cost of the earth movers is
$$4 \cdot \$485,000 = \$1,940,000.$$

The amount borrowed is 80% of $1,940,000.
$$x = (0.80)(1,940,000)$$
$$x = 1,552,000$$

$1,552,000 at $10\tfrac{1}{2}$% for $2\tfrac{1}{2}$ years

$$I = p \cdot r \cdot t$$
$$I = (1,552,000)(0.105)(2.5)$$
$$I = 407,400$$

The interest is $407,400.

(continued)

Copyright © 2018 Pearson Education, Inc.

total amount due = principal + interest
$$= \$1,552,000 + \$407,400$$
$$= \$1,959,400$$

The total amount due is $1,959,400.

6.8 Compound Interest

6.8 Margin Exercises

1. (a) $400 at 4% for 2 years

Year	Interest	Compound Amount
1	($400)(0.04)(1)	$400 + $16
	= $16	= $416
2	($416)(0.04)(1)	$416 + $16.64
	= $16.64	= $432.64

The compound amount is $432.64.

(b) $2000 at 2% for 3 years

Year	Interest	Compound Amount
1	($2000)(0.02)(1)	$2000 + $40
	= $40	= $2040
2	($2040)(0.02)(1)	$2040 + $40.80
	= $40.80	= $2080.80
3	($2080.80)(0.02)(1)	$2080.80 + $41.62
	≈ $41.62	= $2122.42

The compound amount is $2122.42 (rounded).

2. (a) $1800 at 2% for 3 years

$$100\% + 2\% = 102\% = 1.02$$
$$(\$1800)\underbrace{(1.02)(1.02)(1.02)}_{3 \text{ times}} \approx \$1910.17$$

The compound amount is $1910.17 (rounded).

(b) $900 at 3% for 2 years

$$100\% + 3\% = 103\% = 1.03$$
$$(\$900)\underbrace{(1.03)(1.03)}_{2 \text{ times}} = \$954.81$$

The compound amount is $954.81.

(c) $2500 at 5% for 4 years

$$100\% + 5\% = 105\% = 1.05$$
$$(\$2500)\underbrace{(1.05)(1.05)(1.05)(1.05)}_{4 \text{ times}} \approx \$3038.77$$

The compound amount is $3038.77 (rounded).

3. Read from the compound interest table.

(a) $1 at 3% for 6 years: $1.1941 ≈ $1.19

(b) $1 at 2% for 8 years: $1.1717 ≈ $1.17

(c) $1 at $4\frac{1}{2}$% for 12 years: $1.6959 ≈ $1.70

4. Use the table and multiply to find the compound amount. Then subtract the original deposit from the compound amount to find the interest.

(a) $4000 at 3% for 10 years from the table = 1.3439

compound amount = ($4000)(1.3439)
$$= \$5375.60$$
$$interest = \$5375.60 - \$4000$$
$$= \$1375.60$$

(b) $12,600 at $3\frac{1}{2}$% for 8 years from the table = 1.3168

compound amount = ($12,600)(1.3168)
$$= \$16,591.68$$
$$interest = \$16,591.68 - \$12,600$$
$$= \$3991.68$$

(c) $32,700 at $4\frac{1}{2}$% for 12 years from the table = 1.6959

compound amount = ($32,700)(1.6959)
$$= \$55,455.93$$
$$interest = \$55,455.93 - \$32,700$$
$$= \$22,755.93$$

6.8 Section Exercises

1. The statement is *false*. Compound interest will yield more interest than simple interest.

3. $500 at 4% for 2 years

Year	Interest	Compound Amount
1	($500)(0.04)(1)	$500 + $20
	= $20	= $520
2	($520)(0.04)(1)	$520 + $20.80
	= $20.80	= $540.80

The compound amount is $540.80.

5. $1800 at 3% for 3 years

Year	Interest	Compound Amount
1	($1800)(0.03)(1)	$1800 + $54
	= $54	= $1854
2	($1854)(0.03)(1)	$1854 + $55.62
	= $55.62	= $1909.62
3	($1909.62)(0.03)(1)	$1909.62 + $57.29
	≈ $57.29	= $1966.91

The compound amount is $1966.91 (rounded).

Copyright © 2018 Pearson Education, Inc.

7. $3500 at 7% for 4 years

Year	Interest	Compound Amount
1	($3500)(0.07)(1) = $245	$3500 + $245 = $3745
2	($3745)(0.07)(1) = $262.15	$3745 + $262.15 = $4007.15
3	($4007.15)(0.07)(1) ≈ $280.50	$4007.15 + $280.50 = $4287.65
4	($4287.65)(0.07)(1) ≈ $300.14	$4287.65 + $300.14 = $4587.79

The compound amount is $4587.79 (rounded).

9.   ($2000)(1.02)(1.02)(1.02) = $2122.42 (rounded).
Adding the compound interest rate to 100% for each year, and then multiplying by the original amount, will give the compound amount at the end of the compound interest period.

11. $1000 at 5% for 2 years

100% + 5% = 105% = 1.05
($1000)(1.05)(1.05) = $1102.50

The compound amount is $1102.50.

13. $1400 at 6% for 5 years

100% + 6% = 106% = 1.06
($1400)(1.06)(1.06)(1.06)(1.06)(1.06) ≈ $1873.52

The compound amount is $1873.52 (rounded).

15. $1180 at 7% for 8 years

100% + 7% = 107% = 1.07
($1180)(1.07)(1.07)(1.07)(1.07)
        (1.07)(1.07)(1.07)(1.07) ≈ $2027.46

The compound amount is $2027.46 (rounded).

17. $10,940 at 4% for 6 years

100% + 4% = 104% = 1.04
($10,940)(1.04)(1.04)(1.04)(1.04)(1.04)(1.04)
        ≈ $13,842.59

The compound amount is $13,842.59 (rounded).

19. $1 at 3% for 6 years
3% column, row 6 of the table gives 1.1941.

21. $1 at 2% for 8 years
2% column, row 8 of the table gives 1.1717.

23. $1000 at 4% for 5 years
Look down the column headed 4%, and across to row 5 (because 5 years = 5 time periods.) At the intersection of the column and the row, read the compound amount, 1.2167.

compound amount = ($1000)(1.2167)
                = $1216.70

Find the interest by subtracting the principal ($1000) from the compound amount.

interest = $1216.70 − $1000
         = $216.70

25. $8000 at 2% for 10 years
2% column, row 10 of the table gives 1.2190.

compound amount = ($8000)(1.2190)
                = $9752.00
      interest = $9752.00 − $8000
               = $1752.00

27. $8428.17 at $4\frac{1}{2}$% for 6 years
4.50% column, row 6 of the table gives 1.3023.

compound amount = ($8428.17)(1.3023)
                ≈ $10,976.01
      interest = $10,976.01 − $8428.17
               = $2547.84 (rounded)

29.   Compound interest is interest paid on past interest as well as on the principal. Many people describe compound interest as "interest on interest."

31. $50,000 at 4% compounded annually for 5 years
4% column, row 5 of the table gives 1.2167.

compound amount = ($50,000)(1.2167)
                = $60,835.00

He will have $60,835.00 at the end of 5 years.

33.   (a) $76,000 at 6% compounded annually for 9 years
6% column, row 9 from the table is 1.6895.

compound amount = ($76,000)(1.6895)
                = $128,402

The total amount that should be repaid is $128,402.

(b) To find the amount of interest earned, subtract the principal ($76,000) from the compound amount.

The amount of interest earned is
    $128,402 − $76,000 = $52,402

35.   The formula for simple interest is
    Interest = principal · rate · time
    or I = p · r · t.

37.   Compound interest is interest calculated on principal plus past interest.

Copyright © 2018 Pearson Education, Inc.

39. **(a)** $30,000 at 6% compounded annually for 2 years

6% column, row 2 from the table is 1.1236.

compound amount $= (\$30{,}000)(1.1236)$
$$= \$33{,}708$$

After 2 years she will have $33,708. After the $40,000 deposit, she will have $73,708.

6% column, row 3 from the table is 1.1910.

compound amount $= (\$73{,}708)(1.1910)$
$$\approx \$87{,}786.23$$

The total amount will be $87,786.23 (rounded).

(b) The amount of interest earned is
$87,786.23 − $30,000 − $40,000 = $17,786.23.

40. **(a)** $25,000 at 4% compounded annually for 3 years

4% column, row 3 from the table is 1.1249.

compound amount $= (\$25{,}000)(1.1249)$
$$= \$28{,}122.50$$

After 3 years she will have $28,122.50. After the second $25,000 deposit, she will have $53,122.50.

4% column, row 2 from the table is 1.0816.

compound amount $= (\$53{,}122.50)(1.0816)$
$$\approx \$57{,}457.30$$

After 5 years she will have $57,457.30 (rounded).

(b) The amount of interest earned is

$57,457.30 − $25,000 − $25,000 = $7,457.30.

41. **(a)** $25,000 at 4% simple interest for 3 years:
$$(\$25{,}000)(0.04)(3) = \$3000$$

(b) $25,000 at $3\frac{1}{2}$% compounded annually for 3 years (use the table):

$$(\$25{,}000)(1.1087) = \$27{,}717.50$$

Interest $= \$27{,}717.50 - \$25{,}000$
$$= \$2717.50$$

(c) The loan to his sister will earn
$3000 − $2717.50 = $282.50 more.

42. **(a)** $35,000 at 6% simple interest for 5 years:
$$(\$35{,}000)(0.06)(5) = \$10{,}500$$

She would owe $35,000 + $10,500 = $45,500.

(b) $35,000 at $5\frac{1}{2}$% compounded annually for 5 years (use the table):

$$(\$35{,}000)(1.3070) = \$45{,}745$$

She would owe $45,745.

(c) The loan from her family results in
$45,745 − $45,500 = $245 less interest.

43. **(a)** $4350 at 6% simple interest for 6 years:
$$(\$4350)(0.06)(6) = \$1566$$

$4350 at 6% compounded annually for 6 years (use the table):

$$(\$4350)(1.4185) \approx \$6170.48$$

$6170.48 − $4350 = $1820.48

The difference in the amount of interest is
$1820.48 − $1566 = $254.48.

(b) $4350 at 6% simple interest for 12 years:
$$(\$4350)(0.06)(12) = \$3132$$

$4350 at 6% compounded annually for 12 years (use the table):

$$(\$4350)(2.0122) = \$8753.07$$

$8753.07 − $4350 = $4403.07

The difference is $4403.07 − $3132 = $1271.07.

No, it has more than doubled. It is about five times as much ($1271.07 ÷ $254.48 ≈ 5).

(c) Examples will vary.

44. **(a)** $10,000 at 5% simple interest for 10 years:
$$(\$10{,}000)(0.05)(10) = \$5000$$

$10,000 + $5000 = $15,000

$9500 at 5% compounded annually for 10 years (use the table):

$$(\$9500)(1.6289) = \$15{,}474.55$$

The compound interest account has the higher balance. It has $15,474.55 − $15,000 = $474.55 more.

(b) Greater amounts of interest are earned with compound interest than with simple interest because interest is earned on both principal and past interest.

Copyright © 2018 Pearson Education, Inc.

Chapter 6 Review Exercises

1. $35\% = 0.35$
Drop the percent sign and move the decimal point 2 places to the left.

2. $150\% = 1.5$
Drop the percent sign and move the decimal point 2 places to the left.

3. $99.44\% = 0.9944$
Drop the percent sign and move the decimal point 2 places to the left.

4. $0.085\% = 0.00085$
Drop the percent sign and move the decimal point 2 places to the left.

5. $3.15 = 315\%$
Move the decimal point 2 places to the right and attach a percent sign.

6. $0.02 = 2\%$
Move the decimal point 2 places to the right and attach a percent sign.

7. $0.875 = 87.5\%$
Move the decimal point 2 places to the right and attach a percent sign.

8. $0.002 = 0.2\%$
Move the decimal point 2 places to the right and attach a percent sign.

9. $15\% = \dfrac{15}{100} = \dfrac{15 \div 5}{100 \div 5} = \dfrac{3}{20}$

10. $37.5\% = \dfrac{37.5}{100} = \dfrac{37.5 \cdot 10}{100 \cdot 10} = \dfrac{375 \div 125}{1000 \div 125} = \dfrac{3}{8}$

11. $175\% = \dfrac{175}{100} = \dfrac{175 \div 25}{100 \div 25} = \dfrac{7}{4} = 1\dfrac{3}{4}$

12. $0.25\% = \dfrac{0.25}{100} = \dfrac{(0.25)(4)}{(100)(4)} = \dfrac{1}{400}$

13. $\dfrac{3}{4} = \dfrac{p}{100}$
$4 \cdot p = 3 \cdot 100$
$\dfrac{4 \cdot p}{4} = \dfrac{300}{4}$
$p = 75$
Thus, $\dfrac{3}{4} = 75\%$.

14. $\dfrac{5}{8} = \dfrac{p}{100}$
$8 \cdot p = 5 \cdot 100$
$\dfrac{8 \cdot p}{8} = \dfrac{500}{8}$
$p = 62.5$
Thus, $\dfrac{5}{8} = 62.5\%$ or $62\dfrac{1}{2}\%$.

15. $3\dfrac{1}{4} = 3\dfrac{25}{100} = 3.25 = 325\%$

16. $\dfrac{1}{200} = \dfrac{p}{100}$
$200 \cdot p = 1 \cdot 100$
$\dfrac{200 \cdot p}{200} = \dfrac{100}{200}$
$p = 0.5$
Thus, $\dfrac{1}{200} = 0.5\%$.

17. $\dfrac{1}{8}$
$$\begin{array}{r} 0.\,1\,2\,5 \\ 8\overline{\smash{)}1.\,0\,0\,0} \\ \underline{8} \\ 2\,0 \\ \underline{1\,6} \\ 4\,0 \\ \underline{4\,0} \\ 0 \end{array}$$
$\dfrac{1}{8} = 0.125$

18. From Exercise 17, $0.125 = 12.5\%$.

19. $0.25 = \dfrac{25}{100} = \dfrac{25 \div 25}{100 \div 25} = \dfrac{1}{4}$

20. $0.25 = 25\%$

21. $180\% = \dfrac{180}{100} = \dfrac{180 \div 20}{100 \div 20} = \dfrac{9}{5} = 1\dfrac{4}{5}$

22. $180\% = 1.80 = 1.8$

23. part $= 25$, percent $= 10$
$$\dfrac{25}{x} = \dfrac{10}{100} \quad \textbf{OR} \quad \dfrac{25}{x} = \dfrac{1}{10}$$
$$x \cdot 1 = 25 \cdot 10$$
$$x = 250$$
The whole is 250.

24. whole $= 480$, percent $= 5$
$$\dfrac{x}{480} = \dfrac{5}{100} \quad \textbf{OR} \quad \dfrac{x}{480} = \dfrac{1}{20}$$
$$x \cdot 20 = 480 \cdot 1$$
$$\dfrac{x \cdot 20}{20} = \dfrac{480}{20}$$
$$x = 24$$
The part is 24.

Copyright © 2018 Pearson Education, Inc.

25. 35% of 820 mailboxes is 287 mailboxes.

$$\text{Part} \rightarrow \frac{287}{820} = \frac{35}{100} \leftarrow \text{Percent}$$
$$\text{Whole} \rightarrow \phantom{\frac{287}{820}} \phantom{\frac{35}{100}} \leftarrow \text{Always 100}$$

26. 73 Blu-ray discs is what percent of 90 Blue-ray discs?

$$\text{Part} \rightarrow \frac{73}{90} = \frac{\text{unknown}}{100} \leftarrow \text{Percent}$$
$$\text{Whole} \rightarrow \phantom{\frac{73}{90}} \phantom{\frac{}{100}} \leftarrow \text{Always 100}$$

27. Find 14% of 160 mountain bikes.

$$\text{Part} \rightarrow \frac{\text{unknown}}{160} = \frac{14}{100} \leftarrow \text{Percent}$$
$$\text{Whole} \rightarrow \phantom{\frac{}{160}} \phantom{\frac{14}{100}} \leftarrow \text{Always 100}$$

28. 418 curtains is 16% of what number of curtains?

$$\text{Part} \rightarrow \frac{418}{\text{unknown}} = \frac{16}{100} \leftarrow \text{Percent}$$
$$\text{Whole} \rightarrow \phantom{\frac{418}{}} \phantom{\frac{16}{100}} \leftarrow \text{Always 100}$$

29. A golfer lost 3 of his 8 golf balls. What percent were lost?

$$\text{Part} \rightarrow \frac{3}{8} = \frac{\text{unknown}}{100} \leftarrow \text{Percent}$$
$$\text{Whole} \rightarrow \phantom{\frac{3}{8}} \phantom{\frac{}{100}} \leftarrow \text{Always 100}$$

30. What number is 88% of 1280 keys?

$$\text{Part} \rightarrow \frac{\text{unknown}}{1280} = \frac{88}{100} \leftarrow \text{Percent}$$
$$\text{Whole} \rightarrow \phantom{\frac{}{1280}} \phantom{\frac{88}{100}} \leftarrow \text{Always 100}$$

31. whole is 950; percent is 18; part is unknown.
part $= (0.18)(950) = 171$
18% of 950 programs is 171 programs.

32. whole is 1450; percent is 60; part is unknown.

$$\frac{x}{1450} = \frac{60}{100} \quad \textbf{OR} \quad \frac{x}{1450} = \frac{3}{5}$$
$$5 \cdot x = 1450 \cdot 3$$
$$\frac{5 \cdot x}{5} = \frac{4350}{5}$$
$$x = 870$$

60% of 1450 reference books is 870 reference books.

33. whole is 5200; percent is 0.6; part is unknown.
part $= (0.006)(5200) = 31.2$
0.6% of 5200 acres is 31.2 acres.

34. whole is 1400; percent is 0.2; part is unknown.

$$\frac{x}{1400} = \frac{0.2}{100}$$
$$100 \cdot x = (0.2)(1400)$$
$$\frac{100 \cdot x}{100} = \frac{280}{100}$$
$$x = 2.8$$

0.2% of 1400 kilograms is 2.8 kilograms.

35. part is 105; percent is 14; whole is unknown.

$$\frac{105}{x} = \frac{14}{100} \quad \textbf{OR} \quad \frac{105}{x} = \frac{7}{50}$$
$$x \cdot 7 = 105 \cdot 50$$
$$\frac{x \cdot 7}{7} = \frac{5250}{7}$$
$$x = 750$$

105 crates is 14% of 750 crates.

36. part is 348; percent is 15; whole is unknown.

$$\frac{348}{x} = \frac{15}{100} \quad \textbf{OR} \quad \frac{348}{x} = \frac{3}{20}$$
$$x \cdot 3 = 348 \cdot 20$$
$$\frac{x \cdot 3}{3} = \frac{6960}{3}$$
$$x = 2320$$

348 test tubes is 15% of 2320 test tubes.

37. part is 677.6; percent is 140; whole is unknown.

$$\frac{677.6}{x} = \frac{140}{100} \quad \textbf{OR} \quad \frac{677.6}{x} = \frac{7}{5}$$
$$x \cdot 7 = 677.6 \cdot 5$$
$$\frac{x \cdot 7}{7} = \frac{3388}{7}$$
$$x = 484$$

677.6 miles is 140% of 484 miles.

38. part is 425; percent is 2.5; whole is unknown.

$$\frac{425}{x} = \frac{2.5}{100}$$
$$(2.5)(x) = 425 \cdot 100$$
$$\frac{(2.5)(x)}{2.5} = \frac{42,500}{2.5}$$
$$x = 17,000$$

2.5% of 17,000 cases is 425 cases.

39. part is 649; whole is 1180; percent is unknown.

$$\frac{649}{1180} = \frac{x}{100}$$
$$1180 \cdot x = 649 \cdot 100$$
$$\frac{1180 \cdot x}{1180} = \frac{64,900}{1180}$$
$$x = 55$$

649 tulip bulbs is 55% of 1180 tulip bulbs.

Copyright © 2018 Pearson Education, Inc.

40. part is 85; whole is 1620; percent is unknown.

$$\frac{85}{1620} = \frac{x}{100} \quad \textbf{OR} \quad \frac{17}{324} = \frac{x}{100}$$
$$324 \cdot x = 17 \cdot 100$$
$$\frac{324 \cdot x}{324} = \frac{1700}{324}$$
$$x \approx 5.2$$

85 dinner rolls is 5.2% (rounded) of 1620 dinner rolls.

41. part is 36; whole is 380; percent is unknown.

$$\frac{36}{380} = \frac{x}{100} \quad \textbf{OR} \quad \frac{9}{95} = \frac{x}{100}$$
$$95 \cdot x = 9 \cdot 100$$
$$\frac{95 \cdot x}{95} = \frac{900}{95}$$
$$x \approx 9.5$$

36 pairs is 9.5% (rounded) of 380 pairs.

42. part is 200; whole is 650; percent is unknown.

$$\frac{200}{650} = \frac{x}{100} \quad \textbf{OR} \quad \frac{4}{13} = \frac{x}{100}$$
$$13 \cdot x = 4 \cdot 100$$
$$\frac{13 \cdot x}{13} = \frac{400}{13}$$
$$x \approx 30.8$$

200 cans is 30.8% (rounded) of 650 cans.

43. Add 25.4% of 63 to 63. part is unknown; whole is 63; percent is 25.4

$$\frac{x}{63} = \frac{25.4}{100}$$
$$100 \cdot x = 25.4 \cdot 63$$
$$\frac{100 \cdot x}{100} = \frac{1600.2}{100}$$
$$x = 16.002 \approx 16$$

The number of shark attacks on humans this year was $63 + 16 = 79$ (rounded).

44. part = 896; whole = 3200; percent is unknown.

$$\frac{896}{3200} = 0.28 \text{ (use a calculator)}$$

28% of the shoppers are seniors.

45. *part = percent · whole*
$$x = (0.32)(454) \qquad 32\% = 0.32$$
$$x = 145.28$$

32% of $454 is $145.28.

46. *part = percent · whole*
$$x = (1.55)(120) \qquad 155\% = 1.55$$
$$x = 186$$

155% of 120 trucks is 186 trucks.

47. *part = percent · whole*
$$0.128 = x \cdot 32$$
$$\frac{0.128}{32} = \frac{x \cdot 32}{32}$$
$$0.004 = x$$

0.004 is 0.4%. 0.128 ounces is 0.4% of 32 ounces.

48. *part = percent · whole*
$$304.5 = x \cdot 174$$
$$\frac{304.5}{174} = \frac{x \cdot 174}{174}$$
$$1.75 = x$$

1.75 is 175%. 304.5 meters is 175% of 174 meters.

49. *part = percent · whole*
$$33.6 = (0.28)(x)$$
$$\frac{33.6}{0.28} = \frac{(0.28)(x)}{0.28}$$
$$120 = x$$

33.6 miles is 28% of 120 miles.

50. *part = percent · whole*
$$92 = (0.16)(x)$$
$$\frac{92}{0.16} = \frac{(0.16)(x)}{0.16}$$
$$575 = x$$

$92 is 16% of $575.

51. *sales tax = rate of tax · cost of item*
$$= (5\%)(\$630)$$
$$= (0.05)(\$630)$$
$$= \$31.50$$

The sales tax is $31.50 and the total cost is $630 + $31.50 = $661.50.

52. *sales tax = rate of tax · cost of item*
$$\$58.50 = r \cdot \$780$$
$$\frac{58.50}{780} = \frac{r \cdot 780}{780}$$
$$0.075 = r$$
$$0.075 \text{ is } 7\tfrac{1}{2}\%$$

The tax rate is $7\tfrac{1}{2}\%$ and the total cost is $780 + $58.50 = $838.50.

53. *commission = rate of commission · sales*
$$= (8\%)(\$3450)$$
$$= (0.08)(\$3450)$$
$$= \$276$$

The amount of the commission is $276.

Copyright © 2018 Pearson Education, Inc.

54. $commission = rate\ of\ commission \cdot sales$

$$\$3265 = r \cdot \$65,300$$
$$\frac{3265}{65,300} = \frac{r \cdot 65,300}{65,300}$$
$$0.05 = r$$
$$0.05 \text{ is } 5\%$$

The rate of commission is 5%.

55. $discount = rate\ of\ discount \cdot original\ price$

$$= (30\%)(\$112.50)$$
$$= (0.30)(\$112.50)$$
$$= \$33.75$$

The amount of discount is $33.75 and the sale price is

$$\$112.50 - \$33.75 = \$78.75.$$

56. $discount = rate\ of\ discount \cdot original\ price$

$$\$63 = r \cdot \$252$$
$$\frac{63}{252} = \frac{r \cdot 252}{252}$$
$$0.25 = r$$
$$0.25 \text{ is } 25\%$$

The rate of discount is 25% and the sale price is

$$\$252 - \$63 = \$189.$$

57. $200 at 4% for 1 year

$$I = p \cdot r \cdot t$$
$$= (200)(0.04)(1)$$
$$= 8$$

The interest is $8.

58. $1080 at 5% for $1\frac{1}{4}$ years

$$I = p \cdot r \cdot t$$
$$= (1080)(0.05)(1.25) \quad 1\tfrac{1}{4} = 1.25$$
$$= (54)(1.25)$$
$$= 67.5$$

The interest is $67.50.

59. $400 at $3\frac{1}{2}$% for 3 months

$$I = p \cdot r \cdot t$$
$$= (400)(0.035)\left(\tfrac{3}{12}\right)$$
$$= (14)\left(\tfrac{3}{12}\right)$$
$$= 3.5$$

The interest is $3.50.

60. $1560 at $6\frac{1}{2}$% for 18 months

$$I = p \cdot r \cdot t$$
$$= (1560)(0.065)\left(\tfrac{18}{12}\right)$$
$$= (101.4)\left(\tfrac{18}{12}\right)$$
$$= 152.1$$

The interest is $152.10.

61. $750 at $5\frac{1}{2}$% for 2 years

$$I = p \cdot r \cdot t$$
$$= (750)(0.055)(2)$$
$$= 82.5$$

The total amount due is
$750 + $82.50 = $832.50.

62. $1560 at 3% for 9 months

$$I = p \cdot r \cdot t$$
$$= (1560)(0.03)\left(\tfrac{9}{12}\right)$$
$$= (46.8)\left(\tfrac{9}{12}\right)$$
$$= 35.10$$

The total amount due is
$1560 + $35.10 = $1595.10.

63. $4000 at 3% for 10 years

3% column, row 10 of the table gives 1.3439.

$$compound\ amount = (\$4000)(1.3439)$$
$$= \$5375.60$$
$$interest = \$5375.60 - \$4000$$
$$= \$1375.60$$

64. $1870 at 4% for 4 years

4% column, row 4 of the table gives 1.1699.

$$compound\ amount = (\$1870)(1.1699)$$
$$\approx \$2187.71$$
$$interest = \$2187.71 - \$1870$$
$$= \$317.71 \text{ (rounded)}$$

65. $3600 at $4\frac{1}{2}$% for 3 years

$4\frac{1}{2}$% column, row 3 of the table gives 1.1412.

$$compound\ amount = (\$3600)(1.1412)$$
$$= \$4108.32$$
$$interest = \$4108.32 - \$3600$$
$$= \$508.32$$

66. $12,500 at $5\frac{1}{2}$% for 5 years

$5\frac{1}{2}$% column, row 5 of the table gives 1.3070.

$$compound\ amount = (\$12,500)(1.3070)$$
$$= \$16,337.50$$
$$interest = \$16,337.50 - \$12,500$$
$$= \$3837.50$$

Copyright © 2018 Pearson Education, Inc.

Chapter 6 Mixed Review Exercises

1. whole $= 80$, percent $= 15$

$$\frac{\text{part}}{\text{whole}} = \frac{\text{percent}}{100}$$

$$\frac{x}{80} = \frac{15}{100} \quad \textbf{OR} \quad \frac{x}{80} = \frac{3}{20}$$

$$x \cdot 20 = 80 \cdot 3$$

$$\frac{x \cdot 20}{20} = \frac{240}{20}$$

$$x = 12$$

The part is 12.

2. part $= 738$, percent $= 45$

$$\frac{\text{part}}{\text{whole}} = \frac{\text{percent}}{100}$$

$$\frac{738}{x} = \frac{45}{100} \quad \textbf{OR} \quad \frac{738}{x} = \frac{9}{20}$$

$$x \cdot 9 = 738 \cdot 20$$

$$\frac{x \cdot 9}{9} = \frac{14,760}{9}$$

$$x = 1640$$

The whole is 1640.

3. 12% of 194 meters

$$\text{part} = \text{percent} \cdot \text{whole}$$

$$x = (0.12)(194)$$

$$x = 23.28$$

12% of 194 meters is 23.28 meters.

4. part is 327; whole is 218; percent is unknown.

$$\frac{327}{218} = \frac{x}{100}$$

$$218 \cdot x = 327 \cdot 100$$

$$\frac{218 \cdot x}{218} = \frac{32,700}{218}$$

$$x = 150$$

The percent is 150, written as 150%.
327 cars is 150% of 218 cars.

5. 0.6% of $85

$$\text{part} = \text{percent} \cdot \text{whole}$$

$$x = (0.006)(85)$$

$$x = 0.51$$

0.6% of $85 is $0.51.

6. part is 99; percent is 5; whole is unknown.

$$\frac{99}{x} = \frac{5}{100} \quad \textbf{OR} \quad \frac{99}{x} = \frac{1}{20}$$

$$x \cdot 1 = 99 \cdot 20$$

$$x = 1980$$

99 employees is 5% of 1980 employees.

7. part is 76; whole is 190; percent is unknown.

$$\frac{76}{190} = \frac{x}{100} \quad \textbf{OR} \quad \frac{38}{95} = \frac{x}{100}$$

$$95 \cdot x = 38 \cdot 100$$

$$\frac{95 \cdot x}{95} = \frac{3800}{95}$$

$$x = 40$$

The percent is 40, written as 40%.
76 chickens is 40% of 190 chickens.

8. part is 214.484; percent is 43; whole is unknown.

$$\frac{214.484}{x} = \frac{43}{100}$$

$$x \cdot 43 = (214.484)(100)$$

$$\frac{x \cdot 43}{43} = \frac{21,448.4}{43}$$

$$x = 498.8$$

214.484 liters is 43% of 498.8 liters.

9. $55\% = 0.55$
Drop the percent sign and move the decimal point two places to the left.

10. $300\% = 3$
Drop the percent sign and move the decimal point two places to the left.

11. $5 = 500\%$
Attach two zeros so the decimal point can be moved two places to the right. Attach a percent sign.

12. $4.71 = 471\%$
Move the decimal point two places to the right and attach a percent sign.

13. $8.6\% = 0.086$
Drop the percent sign and move the decimal point two places to the left.

14. $0.621 = 62.1\%$
Move the decimal point two places to the right and attach a percent sign.

15. $0.375\% = 0.00375$
Drop the percent sign and move the decimal point two places to the left.

16. $0.0006 = 0.06\%$
Move the decimal point two places to the right and attach a percent sign.

17. $\dfrac{3}{4} = \dfrac{3 \cdot 25}{4 \cdot 25} = \dfrac{75}{100} = 0.75 = 75\%$

18. $42\% = \dfrac{42}{100} = \dfrac{42 \div 2}{100 \div 2} = \dfrac{21}{50}$

Copyright © 2018 Pearson Education, Inc.

19. $87.5\% = \dfrac{87.5 \cdot 10}{100 \cdot 10} = \dfrac{875}{1000} = \dfrac{875 \div 125}{1000 \div 125} = \dfrac{7}{8}$

20.
$$\dfrac{3}{8} = \dfrac{p}{100}$$
$$8 \cdot p = 3 \cdot 100$$
$$8 \cdot p = 300$$
$$\dfrac{8 \cdot p}{8} = \dfrac{300}{8}$$
$$p = 37.5 \text{ or } 37\dfrac{1}{2}$$
$$\dfrac{3}{8} = 37.5\% \text{ or } 37\dfrac{1}{2}\%$$

21. $32\dfrac{1}{2}\% = \dfrac{32.5 \cdot 10}{100 \cdot 10} = \dfrac{325}{1000} = \dfrac{325 \div 25}{1000 \div 25} = \dfrac{13}{40}$

22.
$$\dfrac{3}{5} = \dfrac{p}{100}$$
$$5 \cdot p = 3 \cdot 100$$
$$5 \cdot p = 300$$
$$\dfrac{5 \cdot p}{5} = \dfrac{300}{5}$$
$$p = 60$$

Thus, $\dfrac{3}{5} = 60\%$.

23. $0.25\% = \dfrac{0.25 \cdot 100}{100 \cdot 100}$
$$= \dfrac{25}{10,000} = \dfrac{25 \div 25}{10,000 \div 25} = \dfrac{1}{400}$$

24. $3\dfrac{3}{4} = 3\dfrac{3 \cdot 25}{4 \cdot 25} = 3\dfrac{75}{100} = 3.75 = 375\%$

25. $20,500 at $2\frac{1}{2}\%$ for 30 months

$$I = p \cdot r \cdot t$$
$$= (20,500)\left(2\dfrac{1}{2}\%\right)\left(\dfrac{30}{12}\right)$$
$$= (20,500)(0.025)(2.5) \quad \dfrac{30}{12} = 2.5$$
$$= 1281.25$$

The interest earned is $1281.25.

26. $14,750 at 8% for 18 months
$$I = p \cdot r \cdot t$$
$$= (14,750)(8\%)\left(\dfrac{18}{12}\right)$$
$$= (14,750)(0.08)(1.5)$$
$$= 1770$$

The interest is $1770 and the total amount due is $14,750 + $1770 = $16,520.

27. (a) part is 965; whole is 1005; percent is unknown.

$$\dfrac{965}{1005} = \dfrac{x}{100}$$
$$1005 \cdot x = (965)(100)$$
$$\dfrac{1005 \cdot x}{1005} = \dfrac{96,500}{1005}$$
$$x \approx 96.0$$

The percent of adults who have a smoke alarm in their home is 96.0% (rounded).

(b) part is 461; whole is 1005; percent is unknown.

$$\dfrac{461}{1005} = \dfrac{x}{100}$$
$$1005 \cdot x = (461)(100)$$
$$\dfrac{1005 \cdot x}{1005} = \dfrac{46,100}{1005}$$
$$x \approx 45.9$$

The percent of adults who have a carbon monoxide detector in their home is 45.9% (rounded).

28. (a) $148,000 at 5% compounded annually for 4 years
$100\% + 5\% = 105\% = 1.05$

$(\$148,000)(1.05)(1.05)(1.05)(1.05) \approx$ $179,894.93
The compound amount is $179,894.93.

(b) The amount of interest is

$179,894.93 − $148,000 = $31,894.93.

29. Add the two sales.

$$\$125,000 + \$290,000 = \$415,000$$

amount of commission
$= $ *rate of commission · amount of sales*
$= \left(1\dfrac{1}{2}\%\right)(\$415,000)$
$= (0.015)(\$415,000)$
$= \$6225$

The commission that he earned was $6225.

30. increase $= 5107 − 4320 = 787$
Use the percent proportion.

$$\dfrac{787}{4320} = \dfrac{x}{100}$$
$$4320 \cdot x = 787 \cdot 100$$
$$\dfrac{4320 \cdot x}{4320} = \dfrac{78,700}{4320}$$
$$x \approx 18.2$$

The increase is 18.2% (rounded).

Copyright © 2018 Pearson Education, Inc.

31. *amount of discount*
   $= $ *rate of discount $\cdot$ original price*
   $= (18\%)(\$958)$
   $= (0.18)(\$958)$
   $= \$172.44$

The price is $\$958 - \$172.44 = \$785.56$.

sales tax $=$ rate of tax $\cdot$ cost of item
   $= (8\%)(\$785.56)$
   $= (0.08)(\$785.56)$
   $= \$62.84$ (rounded)

The cost of the washer/dryer set is

   $\$785.56 + \$62.84 = \$848.40.$

32. part $= 88$; percent $= 8$; whole is unknown

$$\frac{88}{x} = \frac{8}{100}$$
$$8 \cdot x = 88 \cdot 100$$
$$\frac{8 \cdot x}{8} = \frac{8800}{8}$$
$$x = 1100$$

There were 1100 boaters (rounded) in the survey.

33. $25\% + 20\% + 7\% + 5\% + 10\% + 12\%$
$+ 4\% = 83\%$ and $100\% - 83\% = 17\%$, so the couple saves 17% of their earnings.

Jack's yearly income is

   $\$2850 \cdot 12 = \$34,200.$

Their total yearly income is

   $\$34,200 + \$42,300 = \$76,500.$

The amount saved is

   $17\% \text{ of } \$76,500 = (0.17)(\$76,500)$
   $= \$13,005.$

34. decrease $= 42.8 - 28.5 = 14.3$
Use the percent proportion.

$$\frac{14.3}{42.8} = \frac{x}{100}$$
$$42.8 \cdot x = (14.3)(100)$$
$$\frac{42.8 \cdot x}{42.8} = \frac{1430}{42.8}$$
$$x \approx 33.4$$

The percent of decrease is 33.4% (rounded).

Chapter 6 Test

1. $65\% = 0.65$
Drop the percent sign and move the decimal point two places to the left.

2. $0.8 = 80\%$
0 is attached so the decimal point can be moved two places to the right. Attach a percent sign.

3. $1.75 = 175\%$
Move the decimal point two places to the right and attach a percent sign.

4. $0.875 = 87.5\%$ or $87\frac{1}{2}\%$
Move the decimal point two places to the right and attach a percent sign.

5. $300\% = 3.00$ or 3
Drop the percent sign and move the decimal point two places to the left.

6. $2\% = 0.02$
Drop the percent sign, attach two zeros on the left, and move the decimal point two places to the left.

7. $12.5\% = \frac{12.5 \cdot 10}{100 \cdot 10} = \frac{125}{1000} = \frac{125 \div 125}{1000 \div 125} = \frac{1}{8}$

8. $0.25\% = \frac{0.25}{100} = \frac{(0.25)(4)}{(100)(4)} = \frac{1}{400}$

9. $\frac{3}{5} = \frac{3 \cdot 20}{5 \cdot 20} = \frac{60}{100} = 60\%$

10. $\frac{5}{8} = \frac{p}{100}$
$8 \cdot p = 5 \cdot 100$
$\frac{8 \cdot p}{8} = \frac{500}{8}$
$p = 62.5$

Thus, $\frac{5}{8} = 62.5\%$ or $62\frac{1}{2}\%$.

11. $2\frac{1}{2} = 2\frac{1 \cdot 50}{2 \cdot 50} = 2\frac{50}{100} = 2.50 = 250\%$

12. part is 32; percent is 4; whole is unknown
Use the percent proportion.

$$\frac{32}{x} = \frac{4}{100} \quad \textbf{OR} \quad \frac{32}{x} = \frac{1}{25}$$
$$x \cdot 1 = 32 \cdot 25$$
$$x = 800$$

32 sacks is 4% of 800 sacks.

Copyright © 2018 Pearson Education, Inc.

13. part is 680; whole is 3400; percent is unknown.
Use the percent proportion.

$$\frac{680}{3400} = \frac{x}{100} \quad \textbf{OR} \quad \frac{1}{5} = \frac{x}{100}$$
$$5 \cdot x = 1 \cdot 100$$
$$\frac{5 \cdot x}{5} = \frac{100}{5}$$
$$x = 20$$

$680 is 20% of $3400.

14. part is 100,000; percent is 0.08; whole is unknown.
Use the percent proportion.

$$\frac{100,000}{x} = \frac{0.08}{100}$$
$$x \cdot 0.08 = 100,000 \cdot 100$$
$$\frac{x \cdot 0.08}{0.08} = \frac{10,000,000}{0.08}$$
$$x = 125,000,000$$

The total number of households is 125,000,000.

15. *sales tax = rate of tax · cost of item*
$$= \left(6\tfrac{1}{2}\%\right)(\$3240)$$
$$= (0.065)(\$3240)$$
$$= \$210.60$$

The total cost is

$$\$3240 + \$210.60 = \$3450.60.$$

16. *commission = rate of commission · sales*
$$\$628 = r \cdot \$7850$$
$$\frac{628}{7850} = \frac{r \cdot 7850}{7850}$$
$$0.08 = r$$

0.08 is 8%. The rate of commission paid is 8%.

17. decrease = 5760 − 4320 = 1440

$$\frac{1440}{5760} = \frac{x}{100} \quad \textbf{OR} \quad \frac{1}{4} = \frac{x}{100}$$
$$4 \cdot x = 1 \cdot 100$$
$$\frac{4 \cdot x}{4} = \frac{100}{4}$$
$$x = 25$$

The percent decrease is 25%.

18. A possible answer is:
Part is the increase in salary.
Whole is the last year's salary.
Percent of increase is unknown.

$$\frac{\text{amount of increase}}{\text{last year's salary}} = \frac{p}{100}$$

19. The interest formula is $I = p \cdot r \cdot t$.
If time is in months, it is expressed as a fraction with 12 as the denominator. If time is expressed in years, it is placed over 1 or shown as a decimal number. Problems will vary. Some possibilities are:

$$I = \quad p \quad \cdot \quad r \quad \cdot \quad t$$
$$I = (\$1000)(0.05)\left(\tfrac{9}{12}\right) = \$37.50$$
$$I = (\$1000)(0.05)(2.5) \ = \$125$$

20. *amount of discount*
$$= rate\ of\ discount \cdot original\ price$$
$$= (12\%)(\$96)$$
$$= (0.12)(\$96)$$
$$= \$11.52$$

The amount of discount is $11.52. The sale price is $96 − $11.52 = $84.48.

21. *amount of discount* $= (32.5\%)(\$280)$
$$= (0.325)(\$280)$$
$$= \$91$$

The amount of discount is $91. The sale price is $280 − $91 = $189.

22. $4200 at 6% for $1\tfrac{1}{2}$ years

$$I = p \cdot r \cdot t$$
$$= (4200)(0.06)(1.5)$$
$$= 378$$

The interest on the loan is $378.

23. $6400 at 9% for 4 months

$$I = p \cdot r \cdot t$$
$$= (6400)(0.09)\left(\tfrac{4}{12}\right)$$
$$= (576)\left(\tfrac{1}{3}\right)$$
$$= 192$$

The interest on the loan is $192.

24. $19,200 at 7% for 15 months

$$I = p \cdot r \cdot t$$
$$= (19,200)(0.07)\left(\tfrac{15}{12}\right) \quad \frac{15}{12} = \frac{5}{4} = 1.25$$
$$= (1344)(1.25)$$
$$= 1680$$

The interest is $1680 and the total amount due is $19,200 + $1680 = $20,880.

Copyright © 2018 Pearson Education, Inc.

25. **(a)** $4000 at 6% for 2 years
6% column, row 2 of the table gives 1.1236.

$$(\$4000)(1.1236) = \$4494.40$$

$9494.40 ($5000 + $4494.40) at 6% for 2 years
6% column, row 2 of the table gives 1.1236.

$$(\$9494.40)(1.1236) \approx \$10,667.91$$

Rounded to the nearest dollar, the total amount after four years is $10,668.

(b) The amount of interest earned is

$$\$10,668 - \$4000 - \$5000 = \$1668.$$

Cumulative Review Exercises (Chapters 1–6)

1. *Estimate:*   *Exact:*

$$\begin{array}{r} 80,000 \\ -\,50,000 \\ \hline 30,000 \end{array} \qquad \begin{array}{r} 75,078 \\ -\,46,090 \\ \hline 28,988 \end{array}$$

2. *Estimate:*   *Exact:*

$$\begin{array}{r} 8 \\ -\,4 \\ \hline 4 \end{array} \qquad \begin{array}{r} 7.8000 \\ -\,3.5029 \\ \hline 4.2971 \end{array}$$

3. *Estimate:*   *Exact:*

$$\begin{array}{r} 7000 \\ \times\ 700 \\ \hline 4,900,000 \end{array} \qquad \begin{array}{r} 6538 \\ \times\ 708 \\ \hline 52\ 304 \\ 4\ 576\ 60 \\ \hline 4,628,904 \end{array}$$

4. *Estimate:*   *Exact:*

$$\begin{array}{r} 70 \\ \times\ 9 \\ \hline 630 \end{array} \qquad \begin{array}{r} 65.3 \\ \times\ 8.7 \\ \hline 45\ 71 \\ 522\ 4 \\ \hline 568.11 \end{array} \begin{array}{l} \leftarrow 1\ decimal\ place \\ \leftarrow 1\ decimal\ place \\ \\ \\ \leftarrow 2\ decimal\ places \end{array}$$

5. *Estimate:*   *Exact:*

$$40\overline{)40,000}^{\,1\,000} \qquad 43\overline{)38,786}^{\,902} \\ \underline{38\ 7} \\ 8 \\ \underline{0} \\ 8\ 6 \\ \underline{8\ 6} \\ 0$$

6. *Estimate:*   *Exact:*

$$1\overline{)7}^{\,7} \qquad 0.8_\wedge\overline{)6.\,7_\wedge\,6\,0}^{\,8.\,4\,5} \\ \underline{6\ 4} \\ 3\ 6 \\ \underline{3\ 2} \\ 4\ 0 \\ \underline{4\ 0} \\ 0$$

7. $6^2 - 3(6)$

$$\begin{aligned} &= 36 - 3(6) && \textit{Exponent} \\ &= 36 - 18 && \textit{Multiply.} \\ &= 18 && \textit{Subtract.} \end{aligned}$$

8. $\sqrt{49} + 5 \cdot 4 - 8$

$$\begin{aligned} &= 7 + 5 \cdot 4 - 8 && \textit{Square root} \\ &= 7 + 20 - 8 && \textit{Multiply.} \\ &= 27 - 8 && \textit{Add.} \\ &= 19 && \textit{Subtract.} \end{aligned}$$

9. $9 + 6 \div 3 + 7(4)$

$$\begin{aligned} &= 9 + 2 + 7(4) && \textit{Divide.} \\ &= 9 + 2 + 28 && \textit{Multiply.} \\ &= 11 + 28 = 39 && \textit{Add.} \end{aligned}$$

10. 7,583,281 rounded to the nearest hundred thousand: 7,5̲83,281

Next digit is 5 or more. Hundred thousands place changes $(5 + 1 = 6)$. All digits to the right of the underlined place change to 0. **7,600,000**

11. $513.499 rounded to the nearest dollar:
Draw a cut-off line: $513|.499
First digit cut is less than 5 so the part you keep stays the same. **$513**

12. $362.735 rounded to the nearest cent:
Draw a cut-off line: $362.73|5
First digit cut is 5 or more, so the cents place changes $(3 + 1 = 4)$. **$362.74**

13.

$$\begin{aligned} 5\tfrac{3}{4} &= 5\tfrac{6}{8} \\ +\,7\tfrac{5}{8} &= 7\tfrac{5}{8} \\ \hline &\ 12\tfrac{11}{8} \end{aligned}$$

$$12\tfrac{11}{8} = 12 + \tfrac{8}{8} + \tfrac{3}{8} = 13\tfrac{3}{8}$$

Copyright © 2018 Pearson Education, Inc.

14.
$$8\frac{3}{8} = 8\frac{3}{8} = 7\frac{11}{8}$$
$$-4\frac{1}{2} = 4\frac{4}{8} = 4\frac{4}{8}$$
$$\overline{\phantom{-4\frac{1}{2}} = 4\frac{4}{8} = } 3\frac{7}{8}$$

15. $36 \cdot \frac{4}{5} = \frac{36}{1} \cdot \frac{4}{5} = \frac{36 \cdot 4}{1 \cdot 5} = \frac{144}{5} = 28\frac{4}{5}$

16. $12 \div \frac{3}{4} = \frac{12}{1} \div \frac{3}{4} = \frac{\overset{4}{\cancel{12}}}{1} \cdot \frac{4}{\underset{1}{\cancel{3}}} = \frac{16}{1} = 16$

17. Area (cell) = length $\cdot$ width
$$= (5 \text{ ft}) \cdot (9 \text{ ft}) = 45 \text{ ft}^2$$

Area (cage) = length $\cdot$ width
$$= (5 \text{ ft}) \cdot (6\tfrac{1}{2} \text{ ft}) = \frac{5}{1} \cdot \frac{13}{2} = \frac{65}{2} \text{ ft}^2$$

Subtract the areas.
$$45 - \frac{65}{2} = \frac{90}{2} - \frac{65}{2} = \frac{25}{2} = 12\tfrac{1}{2}$$

There are $12\tfrac{1}{2}$ more square feet in the floor of the prison cell than in the shark cage.

18. Add the hours together.
$$
\begin{array}{rcl}
5\frac{1}{2} & = & 5\frac{2}{4} \\
6\frac{1}{4} & = & 6\frac{1}{4} \\
3\frac{3}{4} & = & 3\frac{3}{4} \\
+7 & = & 7 \\
\hline
& & 21\frac{6}{4} = 21 + \frac{4}{4} + \frac{2}{4} = 22\frac{1}{2}
\end{array}
$$

Altogether, she studied $22\tfrac{1}{2}$ hours.

19. $\dfrac{5}{8} - \dfrac{2}{3}$

$\dfrac{5}{8} = \dfrac{15}{24}$ and $\dfrac{2}{3} = \dfrac{16}{24}$

$\dfrac{15}{24} < \dfrac{16}{24}$, so $\dfrac{5}{8} \boxed{<} \dfrac{2}{3}$.

20. $\dfrac{8}{15} - \dfrac{11}{20}$

$\dfrac{8}{15} = \dfrac{32}{60}$ and $\dfrac{11}{20} = \dfrac{33}{60}$

$\dfrac{32}{60} < \dfrac{33}{60}$, so $\dfrac{8}{15} \boxed{<} \dfrac{11}{20}$.

21. $\dfrac{2}{3} - \dfrac{7}{12}$

$\dfrac{2}{3} = \dfrac{8}{12}$

$\dfrac{8}{12} > \dfrac{7}{12}$, so $\dfrac{2}{3} \boxed{>} \dfrac{7}{12}$.

22. $\dfrac{2}{3}\left(\dfrac{7}{8} - \dfrac{1}{2}\right) = \dfrac{2}{3}\left(\dfrac{7}{8} - \dfrac{4}{8}\right)$

$= \dfrac{\overset{1}{\cancel{2}}}{\underset{1}{\cancel{3}}} \cdot \dfrac{\overset{1}{\cancel{3}}}{\underset{4}{\cancel{8}}}$

$= \dfrac{1}{4}$

23. $\dfrac{7}{8} \div \left(\dfrac{3}{4} + \dfrac{1}{8}\right) = \dfrac{7}{8} \div \left(\dfrac{6}{8} + \dfrac{1}{8}\right)$

$= \dfrac{7}{8} \div \dfrac{7}{8}$

$= \dfrac{\overset{1}{\cancel{7}}}{\underset{1}{\cancel{8}}} \cdot \dfrac{\overset{1}{\cancel{8}}}{\underset{1}{\cancel{7}}}$

$= 1$

24. $\left(\dfrac{5}{6} - \dfrac{5}{12}\right) - \left(\dfrac{1}{2}\right)^2 \cdot \dfrac{2}{3}$

$= \left(\dfrac{10}{12} - \dfrac{5}{12}\right) - \left(\dfrac{1}{2}\right)^2 \cdot \dfrac{2}{3}$

$= \dfrac{5}{12} - \left(\dfrac{1}{2}\right)^2 \cdot \dfrac{2}{3}$

$= \dfrac{5}{12} - \dfrac{1}{4} \cdot \dfrac{2}{3}$

$= \dfrac{5}{12} - \dfrac{2}{12} = \dfrac{3}{12} = \dfrac{1}{4}$

25. $\dfrac{3}{4} = 0.75$

$$
\begin{array}{r}
0.\,7\,5 \\
4\overline{)3.\,0\,0} \\
\underline{2\,8} \\
2\,0 \\
\underline{2\,0} \\
0
\end{array}
$$

26. $\dfrac{3}{8} = 0.375$

$$
\begin{array}{r}
0.\,3\,7\,5 \\
8\overline{)3.\,0\,0\,0} \\
\underline{2\,4} \\
6\,0 \\
\underline{5\,6} \\
4\,0 \\
\underline{4\,0} \\
0
\end{array}
$$

Copyright © 2018 Pearson Education, Inc.

27. $\dfrac{7}{12} = 0.583$ (rounded)

$$
\begin{array}{r}
0.\,5\,8\,3\,3 \\
12\,\overline{\big)\,7.\,0\,0\,0\,0} \\
6\;0 \\
\overline{1\;0\;0} \\
9\;6 \\
\overline{4\;0} \\
3\;6 \\
\overline{4\;0} \\
3\;6 \\
\overline{4}
\end{array}
$$

28. $\dfrac{11}{20} = 0.55$

$$
\begin{array}{r}
0.\,5\,5 \\
20\,\overline{\big)\,1\,1.\,0\,0} \\
1\;0\;0 \\
\overline{1\;0\;0} \\
1\;0\;0 \\
\overline{0}
\end{array}
$$

29. $\dfrac{1}{5} = \dfrac{x}{30}$

$5 \cdot x = 1 \cdot 30$ *Cross products*

$\dfrac{5 \cdot x}{5} = \dfrac{30}{5}$

$x = 6$

30. $\dfrac{224}{32} = \dfrac{28}{x}$ **OR** $\dfrac{7}{1} = \dfrac{28}{x}$

$7 \cdot x = 1 \cdot 28$

$x = 4$

31. $\dfrac{8}{x} = \dfrac{72}{144}$ **OR** $\dfrac{8}{x} = \dfrac{1}{2}$

$8 \cdot 2 = x \cdot 1$

$16 = x$

32. $\dfrac{x}{120} = \dfrac{7.5}{30}$

$30 \cdot x = 120(7.5)$

$\dfrac{30 \cdot x}{30} = \dfrac{900}{30}$

$x = 30$

33. $3\% = 0.03$

Drop the percent sign, attach one zero on the left, and move the decimal point two places to the left.

34. $200\% = 2.00$ or 2

Drop the percent sign and move the decimal point two places to the left.

35. $0.87 = 87\%$

Move the decimal point two places to the right and attach a percent sign.

36. $3.8 = 380\%$

0 is attached so the decimal point can be moved two places to the right. Attach a percent sign.

37. $8\% = 0.08 = \dfrac{8}{100} = \dfrac{8 \div 4}{100 \div 4} = \dfrac{2}{25}$

38. $62.5\% = 0.625 = \dfrac{625}{1000} = \dfrac{625 \div 125}{1000 \div 125} = \dfrac{5}{8}$

39. $175\% = 1.75 = 1\dfrac{75}{100} = 1\dfrac{75 \div 25}{100 \div 25} = 1\dfrac{3}{4}$

40. $\dfrac{7}{8} = \dfrac{x}{100}$

$8 \cdot x = 7 \cdot 100$

$\dfrac{8 \cdot x}{8} = \dfrac{700}{8}$

$x = 87.5$

Thus, $\dfrac{7}{8} = 87.5\%$ or $87\frac{1}{2}\%$.

41. $4\dfrac{1}{5} = 4\dfrac{1 \cdot 2}{5 \cdot 2} = 4\dfrac{2}{10} = 4.2 = 420\%$

42. 65% of 3280 DVDs

$$\text{part} = (0.65)(3280)$$
$$x = 2132 \text{ DVDs}$$

43. part is 76.5; percent is $4\frac{1}{2}$; whole is unknown

$$\dfrac{76.5}{x} = \dfrac{4\frac{1}{2}}{100}$$
$$(4.5)(x) = 76.5 \cdot 100$$
$$\dfrac{(4.5)(x)}{4.5} = \dfrac{7650}{4.5}$$
$$x = 1700$$

$4\frac{1}{2}\%$ of 1700 miles is 76.5 miles.

44. part is 252; whole is 180; percent is unknown

$$\dfrac{252}{180} = \dfrac{x}{100} \quad \textbf{OR} \quad \dfrac{7}{5} = \dfrac{x}{100}$$
$$5 \cdot x = 7 \cdot 100$$
$$\dfrac{5 \cdot x}{5} = \dfrac{700}{5}$$
$$x = 140$$

252 hours is 140% of 180 hours.

45. *sales tax = rate of tax · cost of item*

$$\$29.90 = r \cdot \$460$$
$$\dfrac{29.90}{460} = \dfrac{r \cdot 460}{460}$$
$$0.065 = r$$

0.065 is 6.5% or $6\frac{1}{2}\%$. The tax rate was 6.5% or $6\frac{1}{2}\%$.

46. *commission = rate of commission · sales*

$$= (14\%)(\$837)$$
$$= (0.14)(\$837)$$
$$= \$117.18$$

The amount of her tips is $117.18.

Copyright © 2018 Pearson Education, Inc.

47. $discount = rate\ of\ discount \cdot original\ price$
$$= (45\%)(\$456)$$
$$= (0.45)(\$456)$$
$$= \$205.20$$

The amount of discount is $205.20 and the sale price is $456 − $205.20 = $250.80.

48. $46,300 at 3% for 9 months
$$I = p \cdot r \cdot t$$
$$= (46{,}300)(3\%)\left(\tfrac{9}{12}\right)$$
$$= (46{,}300)(0.03)(0.75)$$
$$= 1041.75$$

The total amount to be repaid is
$$\$46{,}300 + \$1041.75 = \$47{,}341.75.$$

49. Set up a proportion.
$$\frac{9\ \text{children tested}}{4\ \text{hours}} = \frac{x}{20\ \text{hours}}$$
$$4 \cdot x = 9 \cdot 20$$
$$\frac{4 \cdot x}{4} = \frac{9 \cdot \overset{5}{\cancel{20}}}{\underset{1}{\cancel{4}}}$$
$$x = 45$$

45 children can be tested in 20 hours.

50. Set up a proportion.
$$\frac{12.5\ \text{ounces}}{5\ \text{gallons}} = \frac{x}{102\ \text{gallons}}$$
$$5 \cdot x = (12.5)(102)$$
$$\frac{5 \cdot x}{5} = \frac{1275}{5}$$
$$x = 255$$

255 ounces of Roundup is needed.

51. The increase in sales is
$$36{,}000 - 32{,}000 = 4000.$$

Use the percent proportion.
$$\frac{\text{increase}}{\text{original}} = \frac{x}{100}$$
$$\frac{4000}{32{,}000} = \frac{x}{100} \quad \textbf{OR} \quad \frac{1}{8} = \frac{x}{100}$$
$$8 \cdot x = 1 \cdot 100$$
$$\frac{8 \cdot x}{8} = \frac{100}{8}$$
$$x = 12.5$$

The percent of increase in sales in the northeastern region is 12.5%.

52. The increase in sales is
$$66{,}300 - 65{,}000 = 1300.$$
$$\frac{\text{increase}}{\text{original}} = \frac{x}{100}$$
$$\frac{1300}{65{,}000} = \frac{x}{100} \quad \textbf{OR} \quad \frac{1}{50} = \frac{x}{100}$$
$$50 \cdot x = 1 \cdot 100$$
$$\frac{50 \cdot x}{50} = \frac{100}{50}$$
$$x = 2$$

The percent of increase in sales in the midwestern region is 2%.

53. The decrease in sales is
$$82{,}000 - 77{,}500 = 4500.$$
$$\frac{\text{decrease}}{\text{original}} = \frac{x}{100}$$
$$\frac{4500}{82{,}000} = \frac{x}{100} \quad \textbf{OR} \quad \frac{9}{164} = \frac{x}{100}$$
$$164 \cdot x = 9 \cdot 100$$
$$\frac{164 \cdot x}{164} = \frac{900}{164}$$
$$x \approx 5.5$$

The percent of decrease in sales in the southern region is 5.5% (rounded).

54. The decrease in sales is
$$54{,}000 - 49{,}600 = 4400.$$
$$\frac{\text{decrease}}{\text{original}} = \frac{x}{100}$$
$$\frac{4400}{54{,}000} = \frac{x}{100} \quad \textbf{OR} \quad \frac{11}{135} = \frac{x}{100}$$
$$135 \cdot x = 11 \cdot 100$$
$$\frac{135 \cdot x}{135} = \frac{1100}{135}$$
$$x \approx 8.1$$

The percent of decrease in sales in the western region is 8.1% (rounded).

Copyright © 2018 Pearson Education, Inc.

CHAPTER 7 MEASUREMENT

7.1 Problem Solving with U.S. Measurement Units

7.1 Margin Exercises

1. (a) 1 c = $\underline{8}$ fl oz

(b) $\underline{4}$ qt = 1 gal

(c) 1 wk = $\underline{7}$ days

(d) $\underline{3}$ ft = 1 yd

(e) 1 ft = $\underline{12}$ in.

(f) $\underline{16}$ oz = 1 lb

(g) 1 ton = $\underline{2000}$ lb

(h) $\underline{60}$ min = 1 hr

(i) 1 pt = $\underline{2}$ c

(j) $\underline{24}$ hr = 1 day

(k) 1 min = $\underline{60}$ sec

(l) 1 qt = $\underline{2}$ pt

(m) $\underline{5280}$ ft = 1 mi

2. (a) $5\frac{1}{2}$ ft to inches

You are converting from a *larger* unit to a *smaller* unit, so $\underline{\text{multiply}}$.

Because 1 ft = 12 in., multiply by 12.

$$5\frac{1}{2} \text{ ft} = 5\frac{1}{2} \cdot 12 = \frac{11}{\overset{}{\underset{1}{\cancel{2}}}} \cdot \frac{\overset{6}{\cancel{12}}}{1} = \frac{66}{1} = 66 \text{ in.}$$

(b) 64 oz to pounds

You are converting from a *smaller* unit to a *larger* unit, so $\underline{\text{divide}}$.

Because 16 oz = 1 lb, divide by 16.

$$64 \text{ oz} = \frac{64}{16} = 4 \text{ lb}$$

(c) 6 yd to feet

You are converting from a *larger* unit to a *smaller* unit, so multiply.

Because 1 yd = 3 ft, multiply by 3.

$$6 \text{ yd} = 6 \cdot 3 = 18 \text{ ft}$$

(d) 2 tons to pounds

You are converting from a *larger* unit to a *smaller* unit, so multiply.

Because 1 ton = 2000 lb, multiply by 2000.

$$2 \text{ tons} = 2 \cdot 2000 = 4000 \text{ lb}$$

(e) 35 pt to quarts

You are converting from a *smaller* unit to a *larger* unit, so divide.

Because 1 qt = 2 pt, divide by 2.

$$35 \text{ pt} = \frac{35}{2} = 17\frac{1}{2} \text{ qt}$$

(f) 20 min to hours

You are converting from a *smaller* unit to a *larger* unit, so divide.

Because 60 min = 1 hr, divide by 60.

$$20 \text{ min} = \frac{20}{60} = \frac{20 \div 20}{60 \div 20} = \frac{1}{3} \text{ hr}$$

3. (a) 36 in. to feet

$\text{unit fraction} \left\} \dfrac{1 \text{ ft}}{12 \text{ in.}} \right.$

$$36 \text{ in.} = \frac{\overset{3}{\cancel{36 \text{ in.}}}}{1} \cdot \frac{1 \text{ ft}}{\underset{1}{\cancel{12 \text{ in.}}}} = \frac{3 \cdot 1 \text{ ft}}{1} = 3 \text{ ft}$$

(b) 14 ft to inches

$\text{unit fraction} \left\} \dfrac{12 \text{ in.}}{1 \text{ ft}} \right.$

$$14 \text{ ft} = \frac{14 \cancel{\text{ft}}}{1} \cdot \frac{12 \text{ in.}}{1 \cancel{\text{ft}}} = \frac{14 \cdot 12 \text{ in.}}{1} = 168 \text{ in.}$$

(c) 60 in. to feet

$\text{unit fraction} \left\} \dfrac{1 \text{ ft}}{12 \text{ in.}} \right.$

$$60 \text{ in.} = \frac{\overset{5}{\cancel{60 \text{ in.}}}}{1} \cdot \frac{1 \text{ ft}}{\underset{1}{\cancel{12 \text{ in.}}}} = \frac{5 \cdot 1 \text{ ft}}{1} = 5 \text{ ft}$$

(d) 4 yd to feet

$\text{unit fraction} \left\} \dfrac{3 \text{ ft}}{1 \text{ yd}} \right.$

$$4 \text{ yd} = \frac{4 \cancel{\text{yd}}}{1} \cdot \frac{3 \text{ ft}}{1 \cancel{\text{yd}}} = \frac{4 \cdot 3 \text{ ft}}{1} = 12 \text{ ft}$$

Copyright © 2018 Pearson Education, Inc.

(e) 39 ft to yards

unit fraction $\}$ $\dfrac{1 \text{ yd}}{3 \text{ ft}}$

$$39 \text{ ft} = \dfrac{\overset{13}{\cancel{39} \text{ ft}}}{1} \cdot \dfrac{1 \text{ yd}}{\underset{1}{\cancel{3} \text{ ft}}} = \dfrac{13 \cdot 1 \text{ yd}}{1} = 13 \text{ yd}$$

(f) 2 mi to feet

unit fraction $\}$ $\dfrac{5280 \text{ ft}}{1 \text{ mi}}$

$$2 \text{ mi} = \dfrac{2 \text{ mi}}{1} \cdot \dfrac{5280 \text{ ft}}{1 \text{ mi}} = \dfrac{2 \cdot 5280 \text{ ft}}{1} = 10{,}560 \text{ ft}$$

4.   (a) 16 qt to gallons

unit fraction $\}$ $\dfrac{1 \text{ gallon}}{4 \text{ quarts}}$

$$16 \text{ qt} = \dfrac{\overset{4}{\cancel{16} \text{ qt}}}{1} \cdot \dfrac{1 \text{ gal}}{\underset{1}{\cancel{4} \text{ qt}}} = \dfrac{4 \cdot 1 \text{ gal}}{1} = 4 \text{ gal}$$

(b) 3 c to pints

unit fraction $\}$ $\dfrac{1 \text{ pt}}{2 \text{ c}}$

$$3 \text{ c} = \dfrac{3 \text{ c}}{1} \cdot \dfrac{1 \text{ pt}}{2 \text{ c}} = \dfrac{3 \cdot 1 \text{ pt}}{2} = 1\tfrac{1}{2} \text{ pt or } 1.5 \text{ pt}$$

(c) $3\tfrac{1}{2}$ tons to pounds

unit fraction $\}$ $\dfrac{2000 \text{ lb}}{1 \text{ ton}}$

$$3\tfrac{1}{2} \text{ tons} = \dfrac{3\tfrac{1}{2} \text{ tons}}{1} \cdot \dfrac{2000 \text{ lb}}{1 \text{ ton}} = \dfrac{7}{\underset{1}{\cancel{2}}} \cdot \dfrac{\overset{1000}{\cancel{2000}}}{1}$$
$$= 7000 \text{ lb}$$

(d) $1\tfrac{3}{4}$ lb to ounces

unit fraction $\}$ $\dfrac{16 \text{ oz}}{1 \text{ lb}}$

$$1\tfrac{3}{4} \text{ lb} = \dfrac{1\tfrac{3}{4} \text{ lb}}{1} \cdot \dfrac{16 \text{ oz}}{1 \text{ lb}} = \dfrac{7}{\underset{1}{\cancel{4}}} \cdot \dfrac{\overset{4}{\cancel{16}}}{1} = 28 \text{ oz}$$

(e) 4 oz to pounds

unit fraction $\}$ $\dfrac{1 \text{ lb}}{16 \text{ oz}}$

$$4 \text{ oz} = \dfrac{\overset{1}{\cancel{4} \text{ oz}}}{1} \cdot \dfrac{1 \text{ lb}}{\underset{4}{\cancel{16} \text{ oz}}} = \dfrac{1}{4} \text{ lb or } 0.25 \text{ lb}$$

5.   (a) 4 tons to ounces

unit fractions $\}$ $\dfrac{2000 \text{ lb}}{1 \text{ ton}}$, $\dfrac{16 \text{ oz}}{1 \text{ lb}}$

$$\dfrac{4 \text{ tons}}{1} \cdot \dfrac{2000 \text{ lb}}{1 \text{ ton}} \cdot \dfrac{16 \text{ oz}}{1 \text{ lb}} = 4 \cdot 2000 \cdot 16 \text{ oz}$$
$$= 128{,}000 \text{ oz}$$

(b) 3 mi to inches

unit fractions $\}$ $\dfrac{5280 \text{ ft}}{1 \text{ mi}}$, $\dfrac{12 \text{ in.}}{1 \text{ ft}}$

$$\dfrac{3 \text{ mi}}{1} \cdot \dfrac{5280 \text{ ft}}{1 \text{ mi}} \cdot \dfrac{12 \text{ in.}}{1 \text{ ft}} = 3 \cdot 5280 \cdot 12 \text{ in.}$$
$$= 190{,}080 \text{ in.}$$

(c) 36 pt to gallons

unit fractions $\}$ $\dfrac{1 \text{ qt}}{2 \text{ pt}}$, $\dfrac{1 \text{ gal}}{4 \text{ qt}}$

$$\dfrac{36 \text{ pt}}{1} \cdot \dfrac{1 \text{ qt}}{2 \text{ pt}} \cdot \dfrac{1 \text{ gal}}{4 \text{ qt}} = \dfrac{\overset{9}{\cancel{36}}}{1} \cdot \dfrac{1}{2} \cdot \dfrac{1}{\underset{1}{\cancel{4}}} \text{ gal} = \dfrac{9}{2} \text{ gal}$$
$$= 4\tfrac{1}{2} \text{ gal or } 4.5 \text{ gal}$$

(d) 2 wk to minutes

unit fractions $\}$ $\dfrac{7 \text{ days}}{1 \text{ wk}}$, $\dfrac{24 \text{ hr}}{1 \text{ day}}$, $\dfrac{60 \text{ min}}{1 \text{ hr}}$

$$\dfrac{2 \text{ wk}}{1} \cdot \dfrac{7 \text{ days}}{1 \text{ wk}} \cdot \dfrac{24 \text{ hr}}{1 \text{ day}} \cdot \dfrac{60 \text{ min}}{1 \text{ hr}} = 2 \cdot 7 \cdot 24 \cdot 60 \text{ min}$$
$$= 20{,}160 \text{ min}$$

6.   (a) *Step 1*
The problem asks for the price per <u>pound</u>.

Step 2
Convert ounces to <u>pounds</u>. Then divide the cost by the pounds.

Step 3
To estimate, round $3.29 to $3. There are 16 oz in a pound, so 12 oz is about 1 lb.
Thus, $3 ÷ 1 pound = $3 per pound is our estimate.

Step 4
$$\dfrac{\overset{3}{\cancel{12} \text{ oz}}}{1} \cdot \dfrac{1 \text{ lb}}{\underset{4}{\cancel{16} \text{ oz}}} = \dfrac{3}{4} \text{ lb} = 0.75 \text{ lb}$$

Then divide: $\dfrac{\$3.29}{0.75 \text{ lb}} = \$4.38\overline{6} \approx \$4.39$ per lb

(continued)

Copyright © 2018 Pearson Education, Inc.

Step 5
The cheese costs $4.39 per pound (to the nearest cent).

Step 6
The answer, $4.39, is close to our estimate of $3.

(b) *Step 1*
The problem asks for the number of tons of furnishings they moved.

Step 2
First convert pounds to tons. Then multiply to find the number of tons of furnishings for 5 houses.

Step 3
To estimate, round 11,000 lb to 10,000 lb. There are 2000 lb in a ton, so 10,000 ÷ 2000 = 5 tons. Multiply by 5 to get 25 tons as our estimate.

Step 4

$$\frac{\overset{11}{\cancel{11{,}000}\ \cancel{lb}}}{1} \cdot \frac{1\ ton}{\underset{2}{\cancel{2000}\ \cancel{lb}}} = \frac{11}{2}\ tons = 5.5\ tons$$

$$5(5.5\ tons) = 27.5\ tons$$

Step 5
The company moved 27.5 tons or $27\frac{1}{2}$ tons of furnishings.

Step 6
The answer, $27\frac{1}{2}$ tons, is close to our estimate of 25 tons.

7.1 Section Exercises

1. **(a)** foot, mile, inch, yard arranged from smallest to largest is inch, foot, yard, mile

    (b) pint, fluid ounce, gallon, quart, cup arranged from smallest to largest is fluid ounce, cup, pint, quart, gallon

3. **(a)** 1 yd = $\underline{3}$ ft; **(b)** $\underline{12}$ in. = 1 ft

5. **(a)** $\underline{8}$ fl oz = 1 c; **(b)** 1 qt = $\underline{2}$ pt

7. **(a)** $\underline{2000}$ lb = 1 ton; **(b)** 1 lb = $\underline{16}$ oz

9. **(a)** 1 min = $\underline{60}$ sec; **(b)** $\underline{60}$ min = 1 hr

11. **(a)** 120 sec to minutes

    You are converting from a *smaller* unit to a *larger* unit, so divide. Because 60 sec = 1 min, divide by 60.

    $$120\ sec = \frac{120}{60} = \underline{2}\ min$$

(b) 4 hr to minutes

You are converting from a *larger* unit to a *smaller* unit, so multiply. Because 1 hr = 60 min, multiply by 60.

$$4\ hr = 4 \cdot 60 = \underline{240}\ min$$

13. **(a)** 2 qt to gallons

    You are converting from a *smaller* unit to a *larger* unit, so divide. Because 4 qt = 1 gal, divide by 4.

    $$2\ qt = \frac{2}{4} = \frac{2 \div 2}{4 \div 2} = \frac{1}{2}\ or\ \underline{0.5}\ gal$$

(b) $6\frac{1}{2}$ ft to inches

You are converting from a *larger* unit to a *smaller* unit, so multiply. Because 1 ft = 12 inches, multiply by 12.

$$6\frac{1}{2}\ ft = 6\frac{1}{2} \cdot 12 = \frac{13}{\underset{1}{\cancel{2}}} \cdot \frac{\overset{6}{\cancel{12}}}{1} = \frac{78}{1} = \underline{78}\ in.$$

15. 7 to 8 tons to pounds

    You are converting from a *larger* unit to a *smaller* unit, so multiply. Because 1 ton = 2000 pounds, multiply by 2000.

    $$7\ tons = 7 \cdot 2000 = 14{,}000\ lb$$
    $$8\ tons = 8 \cdot 2000 = 16{,}000\ lb$$

    An adult African elephant may weigh 14,000 to 16,000 lb.

17. 9 yd to feet; $\left.\begin{array}{l}\text{unit}\\\text{fraction}\end{array}\right\}\ \frac{3\ ft}{1\ yd}$

    $$\frac{9\ \cancel{yd}}{1} \cdot \frac{3\ ft}{1\ \cancel{yd}} = 9 \cdot 3\ ft = 27\ ft$$

19. 7 lb to ounces

    $$\frac{7\ \cancel{lb}}{1} \cdot \frac{16\ oz}{1\ \cancel{lb}} = 7 \cdot 16\ oz = 112\ oz$$

21. 5 qt to pints

    $$\frac{5\ \cancel{qt}}{1} \cdot \frac{2\ pt}{1\ \cancel{qt}} = 5 \cdot 2\ pt = 10\ pt$$

23. 90 min to hours

    $$\frac{90\ \cancel{min}}{1} \cdot \frac{1\ hr}{60\ \cancel{min}} = \frac{90}{60}\ hr = 1\frac{1}{2}\ or\ 1.5\ hr$$

25. 3 in. to feet

    $$\frac{\overset{1}{\cancel{3}}\ \cancel{in.}}{1} \cdot \frac{1\ ft}{\underset{4}{\cancel{12}}\ \cancel{in.}} = \frac{1}{4}\ or\ 0.25\ ft$$

Copyright © 2018 Pearson Education, Inc.

27. 24 oz to pounds

$$\frac{24 \text{ oz}}{1} \cdot \frac{1 \text{ lb}}{16 \text{ oz}} = \frac{24}{16} \text{ lb} = 1\frac{1}{2} \text{ or } 1.5 \text{ lb}$$

29. 5 c to pints

$$\frac{5 \text{ c}}{1} \cdot \frac{1 \text{ pt}}{2 \text{ c}} = \frac{5}{2} \text{ pt} = 2\frac{1}{2} \text{ or } 2.5 \text{ pt}$$

31. $\frac{1}{2} \text{ ft} = \frac{\frac{1}{2} \text{ ft}}{1} \cdot \frac{12 \text{ in.}}{1 \text{ ft}} = \frac{1}{2} \cdot 12 \text{ in.} = 6 \text{ in.}$

The ice will safely support a snowmobile or ATV or a person walking.

33. $2\frac{1}{2}$ tons to lb

$$\frac{2\frac{1}{2} \text{ tons}}{1} \cdot \frac{2000 \text{ lb}}{1 \text{ ton}} = \frac{5}{2} \cdot \frac{1000 \atop 2000}{1} \text{ lb} = 5000 \text{ lb}$$

35. $4\frac{1}{4}$ gal to quarts

$$\frac{4\frac{1}{4} \text{ gal}}{1} \cdot \frac{4 \text{ qt}}{1 \text{ gal}} = \frac{17}{4} \cdot \frac{\overset{1}{4}}{1} \text{ qt} = 17 \text{ qt}$$

37. $\frac{1}{3} \text{ ft} = \frac{\frac{1}{3} \text{ ft}}{1} \cdot \frac{12 \text{ in.}}{1 \text{ ft}} = \frac{1}{3} \cdot \frac{\overset{4}{12}}{1} \text{ in.} = 4 \text{ in.}$

Two-thirds of a foot would be twice as high; that is, $2(4 \text{ in.}) = 8 \text{ in.}$ The cactus could be 4 to 8 in. tall.

39. 6 yd to inches

$$\frac{6 \text{ yd}}{1} \cdot \frac{3 \text{ ft}}{1 \text{ yd}} \cdot \frac{12 \text{ in.}}{1 \text{ ft}} = 6 \cdot 3 \cdot 12 \text{ in.} = 216 \text{ in.}$$

41. 112 c to quarts

$$\frac{\overset{28}{\overset{56}{112}} \text{ c}}{1} \cdot \frac{1 \text{ pt}}{\underset{1}{2} \text{ c}} \cdot \frac{1 \text{ qt}}{\underset{1}{2} \text{ pt}} = 28 \text{ qt}$$

43. 6 days to seconds

$$\frac{6 \text{ days}}{1} \cdot \frac{24 \text{ hr}}{1 \text{ day}} \cdot \frac{60 \text{ min}}{1 \text{ hr}} \cdot \frac{60 \text{ sec}}{1 \text{ min}}$$
$$= 6 \cdot 24 \cdot 60 \cdot 60 \text{ sec}$$
$$= 518,400 \text{ sec}$$

45. $1\frac{1}{2}$ tons to ounces

$$\frac{1\frac{1}{2} \text{ tons}}{1} \cdot \frac{2000 \text{ lb}}{1 \text{ ton}} \cdot \frac{16 \text{ oz}}{1 \text{ lb}} = \frac{3}{2} \cdot \frac{\overset{1000}{2000}}{1} \cdot \frac{16}{1} \text{ oz}$$
$$= 48,000 \text{ oz}$$

47. **(a)** There is only one relationship that uses 1 to 16: 1 pound = 16 ounces

(b) 10 to 20 is the same as 1 to 2. There are 2 relationships like this: 1 pint = 2 cups and 1 quart = 2 pints.
10 quarts = 20 pints or 10 pints = 20 cups

(c) 120 to 2 is the same as 60 to 1. There are 2 relationships like this: 60 minutes = 1 hour and 60 seconds = 1 minute.
120 minutes = 2 hours or
120 seconds = 2 minutes

(d) 2 to 24 is the same as 1 to 12. Use 1 foot = 12 inches.
2 feet = 24 inches

(e) 6000 to 3 is the same as 2000 to 1. Use 2000 pounds = 1 ton.
6000 pounds = 3 tons

(f) 35 to 5 is the same as 7 to 1. Use 7 days = 1 week.
35 days = 5 weeks

49. $2\frac{3}{4}$ miles to inches

$$\frac{2\frac{3}{4} \text{ mi}}{1} \cdot \frac{5280 \text{ ft}}{1 \text{ mi}} \cdot \frac{12 \text{ in.}}{1 \text{ ft}} = \frac{11}{4} \cdot \frac{5280}{1} \cdot \frac{\overset{3}{12}}{1} \text{ in.}$$
$$= 174,240 \text{ in.}$$

51. $6\frac{1}{4}$ gal to fluid ounces

$$\frac{6\frac{1}{4} \text{ gal}}{1} \cdot \frac{4 \text{ qt}}{1 \text{ gal}} \cdot \frac{32 \text{ fl oz}}{1 \text{ qt}} = \frac{25}{4} \cdot \overset{1}{4} \cdot 32 \text{ fl oz}$$
$$= 800 \text{ fl oz}$$

53. 24,000 oz to tons

$$\frac{24,000 \text{ oz}}{1} \cdot \frac{1 \text{ lb}}{16 \text{ oz}} \cdot \frac{1 \text{ ton}}{2000 \text{ lb}}$$
$$= \frac{\overset{3}{\overset{12}{24,000}}}{1} \cdot \frac{1}{\underset{4}{16}} \cdot \frac{1}{\underset{1}{2000}} \text{ ton}$$
$$= \frac{3}{4} \text{ or } 0.75 \text{ ton}$$

Copyright © 2018 Pearson Education, Inc.

For Exercises 55–62, the six problem-solving steps should be used, but are only shown for Exercises 55 and 56.

55. *Step 1*
The problem asks for the price per pound of strawberries.

Step 2
Convert ounces to pounds. Then divide the cost by the pounds.

Step 3
To estimate, round $2.79 to $3. Then, there are 16 oz in a pound, so 20 oz is a little more than 1 lb. Thus, $3 \div 1 = $3 per pound is our estimate.

Step 4
$$\frac{20 \ \text{oz}}{1} \cdot \frac{1 \ \text{lb}}{16 \ \text{oz}} = \frac{5}{4} \ \text{lb} = 1.25 \ \text{lb}$$

$$\frac{\$2.79}{1.25 \ \text{lb}} = 2.232 \approx \$2.23/\text{lb} \ (\text{rounded})$$

Step 5
The strawberries are $2.23 per pound (to the nearest cent).

Step 6
The answer, $2.23, is close to our estimate of $3.

57. Find the total number of feet needed.
$$24 \cdot 2 = 48 \ \text{ft}$$

Then convert feet to yards.
$$\frac{48 \ \text{ft}}{1} \cdot \frac{1 \ \text{yd}}{3 \ \text{ft}} = 16 \ \text{yd}$$

Then multiply to find the cost.
$$\frac{16 \ \text{yd}}{1} \cdot \frac{\$8.75}{\text{yd}} = \$140$$

It will cost $140 to equip all the stations.

59. (a) Convert seconds per foot to seconds per mile.
$$\frac{1 \ \text{sec}}{8 \ \text{ft}} \cdot \frac{5280 \ \text{ft}}{1 \ \text{mi}} = \frac{5280}{8} \ \text{sec/mi} = 660 \ \text{sec/mi}$$

It would take the cockroach 660 seconds to travel 1 mile.

(b) $$\frac{660 \ \text{sec}}{1 \ \text{mi}} \cdot \frac{1 \ \text{min}}{60 \ \text{sec}} = \frac{660}{60} \ \text{min/mi}$$
$$= 11 \ \text{min/mi}$$

It would take the cockroach 11 minutes to travel 1 mile.

61. (a) Find the total number of cups per week. Then convert cups to quarts.
$$\frac{2}{3} \cdot 15 \cdot 5 = \frac{2}{3} \cdot \frac{15}{1} \cdot \frac{5}{1} = 50 \ \text{cups}$$

$$\frac{50 \ \text{cups}}{1} \cdot \frac{1 \ \text{pt}}{2 \ \text{cups}} \cdot \frac{1 \ \text{qt}}{2 \ \text{pt}} = \frac{25}{2} \ \text{qt} = 12\frac{1}{2} \ \text{qt}$$

The center needs $12\frac{1}{2}$ qt of milk per week.

(b) Convert $12\frac{1}{2}$ quarts to gallons.
$$\frac{12\frac{1}{2} \ \text{qt}}{1} \cdot \frac{1 \ \text{gal}}{4 \ \text{qt}} = \frac{25}{2} \cdot \frac{1}{4} \ \text{gal} = 3.125 \ \text{gal}$$

The center should order 4 containers, because you can't buy part of a container.

63. (a) Convert 380 feet to inches.
$$\frac{380 \ \text{ft}}{1} \cdot \frac{12 \ \text{in.}}{1 \ \text{ft}} = 380 \cdot 12 \ \text{in.} = 4560 \ \text{in.}$$

(b) Convert 550 inches to yards.
$$\frac{4560 \ \text{in.}}{1} \cdot \frac{1 \ \text{ft}}{12 \ \text{in.}} \cdot \frac{1 \ \text{yd}}{3 \ \text{ft}} = \frac{380}{3} \ \text{yd} \approx 126.6 \ \text{yd}$$

The tree is about 127 (rounded) yards tall.

(c) Find the height of each story.
$$\frac{380 \ \text{ft}}{40 \ \text{stories}} = 9.5 \ \text{feet/story}$$

64. (a) Convert 24 feet to yards.
$$\frac{24 \ \text{ft}}{1} \cdot \frac{1 \ \text{yd}}{3 \ \text{ft}} = 8 \ \text{yd}$$

(b) $3.14 \cdot 24 \ \text{ft} \approx 75.4 \ \text{ft}$

The distance around the tree is about 75.4 ft.

(c) Convert 75.4 feet to yards.
$$\frac{75.4 \ \text{ft}}{1} \cdot \frac{1 \ \text{yd}}{3 \ \text{ft}} \approx 25.1 \ \text{yd}$$

The distance around the tree is about 25 yd.

65. (a) Convert 30 inches to feet.
$$\frac{30 \ \text{in.}}{1} \cdot \frac{1 \ \text{ft}}{12 \ \text{in.}} = \frac{5}{2} \ \text{or} \ 2\frac{1}{2} \ \text{ft}$$

The young tree is $2\frac{1}{2}$ ft, or 2.5 ft, tall.

Copyright © 2018 Pearson Education, Inc.

(b) Set up and solve a proportion with age (in years) and tree height (in feet).

$$\frac{700 \text{ years}}{380 \text{ feet}} = \frac{x \text{ years}}{100 \text{ feet}}$$
$$380 \cdot x = 700 \cdot 100$$
$$\frac{380 \cdot x}{380} = \frac{700 \cdot 100}{380}$$
$$x = \frac{700 \cdot 100}{380} \approx 184.2 \text{ years}$$

The tree will be about 184 (rounded) years old when it reaches 100 feet.

66.   (a) Multiply the distance across the widest tree, 38.1 feet, by 3.14 to get 119.6, or about 120 feet (rounded).

(b) Multiply the distance across the heaviest tree, 36.5 feet, by 3.14 to get 114.6, or about 115 feet (rounded).

7.2 The Metric System—Length

7.2 Margin Exercises

1.   The length of a baseball bat, the height of a doorknob from the floor, and a basketball player's arm length are each about 1 meter in length.

2.   (a) The woman's height is 168 <u>cm</u>.

(b) The man's waist is 90 <u>cm</u> around.

(c) Louise ran the 100 <u>m</u> dash in the track meet.

(d) A postage stamp is 22 <u>mm</u> wide.

(e) Michael paddled his canoe 2 <u>km</u> down the river.

(f) The pencil lead is 1 <u>mm</u> thick.

(g) A stick of gum is 7 <u>cm</u> long.

(h) The highway speed limit is 90 <u>km</u> per hour.

(i) The classroom was 12 <u>m</u> long.

(j) A penny is about 18 <u>mm</u> across.

3.   (a) 3.67 m to cm

unit fraction $\Big\}$ $\dfrac{100 \text{ cm}}{1 \text{ m}}$

$$\frac{3.67 \text{ m}}{1} \cdot \frac{100 \text{ cm}}{1 \text{ m}} = \frac{3.67 \cdot 100 \text{ cm}}{1} = 367 \text{ cm}$$

(b) 92 cm to m

unit fraction $\Big\}$ $\dfrac{1 \text{ m}}{100 \text{ cm}}$

$$\frac{92 \text{ cm}}{1} \cdot \frac{1 \text{ m}}{100 \text{ cm}} = \frac{92}{100} \text{ m} = 0.92 \text{ m}$$

(c) 432.7 cm to m

unit fraction $\Big\}$ $\dfrac{1 \text{ m}}{100 \text{ cm}}$

$$\frac{432.7 \text{ cm}}{1} \cdot \frac{1 \text{ m}}{100 \text{ cm}} = \frac{432.7}{100} \text{ m} = 4.327 \text{ m}$$

(d) 65 mm to cm

unit fraction $\Big\}$ $\dfrac{1 \text{ cm}}{10 \text{ mm}}$

$$\frac{65 \text{ mm}}{1} \cdot \frac{1 \text{ cm}}{10 \text{ mm}} = \frac{65}{10} \text{ cm} = 6.5 \text{ cm}$$

(e) 0.9 m to mm

unit fraction $\Big\}$ $\dfrac{1000 \text{ mm}}{1 \text{ m}}$

$$\frac{0.9 \text{ m}}{1} \cdot \frac{1000 \text{ mm}}{1 \text{ m}} = (0.9)(1000) \text{ mm} = 900 \text{ mm}$$

(f) 2.5 cm to mm

unit fraction $\Big\}$ $\dfrac{10 \text{ mm}}{1 \text{ cm}}$

$$\frac{2.5 \text{ cm}}{1} \cdot \frac{10 \text{ mm}}{1 \text{ cm}} = (2.5)(10) \text{ mm} = 25 \text{ mm}$$

4.   (a) $(43.5)(10) = \underline{435}$
$43.5_\wedge$ gives 435.

(b) $43.5 \div 10 = \underline{4.35}$
$4_\wedge 3.5$ gives $\underline{4.35}$.

(c) $(28)(100) = \underline{2800}$
$28.00_\wedge$ gives $\underline{2800}$.

(d) $28 \div 100 = \underline{0.28}$
$_\wedge 28.$ gives $\underline{0.28}$.

(e) $(0.7)(1000) = \underline{700}$
$0.700_\wedge$ gives $\underline{700}$.

(f) $0.7 \div 1000 = \underline{0.0007}$
$_\wedge 000.7$ gives $\underline{0.0007}$.

Copyright © 2018 Pearson Education, Inc.

5. (a) 12.008 km to m
Count <u>three</u> places to the <u>right</u> on the conversion line.
12.008∧ km = <u>12,008</u> m
⌣⌣⌣

(b) 561.4 m to km
Count <u>three</u> places to the <u>left</u> on the conversion line.
∧561.4 m = <u>0.5614</u> km
⌣⌣⌣

(c) 20.7 cm to m
Count 2 places to the *left* on the conversion line.
∧20.7 cm = 0.207 m
⌣⌣

(d) 20.7 cm to mm
Count 1 place to the *right* on the conversion line.
20.7∧ cm = 207 mm
⌣

(e) 4.66 m to cm
Count 2 places to the *right* on the conversion line.
4.66∧ m = 466 cm
⌣⌣

6. (a) 9 m to mm
Count <u>three</u> places to the <u>right</u> on the conversion line. Three zeros are written in as placeholders.
9.000∧ m = <u>9000</u> mm
⌣⌣⌣

(b) 3 cm to m
Count <u>two</u> places to the <u>left</u> on the conversion line. One zero is written in as a placeholder.
∧03. cm = <u>0.03</u> m
⌣⌣

(c) 5 mm to cm
Count 1 place to the *left* on the conversion line.
∧5. mm = 0.5 cm
⌣

(d) 70 m to km
Count 3 places to the *left* on the conversion line. One zero is written in as a placeholder.
∧070. m = 0.07 km
⌣⌣⌣

(e) 0.8 m to cm
Count 2 places to the *right* on the conversion line. One zero is written in as a placeholder.
0.80∧ m = 80 cm
⌣⌣

7.2 Section Exercises

1. *Kilo* means <u>1000</u>, so 1 km = <u>1000</u> m.

3. *Milli* means $\frac{1}{1000}$ or 0.001, so
1 mm = $\frac{1}{1000}$ or 0.001 m.

5. *Centi* means $\frac{1}{100}$ or 0.01, so
1 cm = $\frac{1}{100}$ or 0.01 m.

7. (a) <u>cm</u> is shorter than m

(b) <u>mm</u> is shorter than m

(c) <u>cm</u> is shorter than km

9. The width of your hand in centimeters
Answers will vary; about 7 to 12 cm.

11. The width of your thumb in millimeters
Answers will vary; about 15 to 25 mm.

13. The child was 91 <u>cm</u> tall.

15. Ming-Na swam in the 200 <u>m</u> backstroke race.

17. Adriana drove 400 <u>km</u> on her vacation.

19. An aspirin tablet is 10 <u>mm</u> across.

21. A paper clip is about 3 <u>cm</u> long. Both meters and kilometers would be much too long, and 3 mm is a very tiny length.

23. Dave's truck is 5 <u>m</u> long.

25. 7 m to cm
$$\frac{7 \not{m}}{1} \cdot \frac{100 \text{ cm}}{1 \not{m}} = \frac{7 \cdot 100 \text{ cm}}{1} = 700 \text{ cm}$$

27. 40 mm to m
$$\frac{40 \not{mm}}{1} \cdot \frac{1 \text{ m}}{1000 \not{mm}} = \frac{40}{1000} \text{ m} = 0.040 \text{ m or } 0.04 \text{ m}$$

29. 9.4 km to m
$$\frac{9.4 \not{km}}{1} \cdot \frac{1000 \text{ m}}{1 \not{km}} = (9.4)(1000) \text{ m} = 9400 \text{ m}$$

31. 509 cm to m
$$\frac{509 \not{cm}}{1} \cdot \frac{1 \text{ m}}{100 \not{cm}} = \frac{509}{100} \text{ m} = 5.09 \text{ m}$$

33. 400 mm to cm
From mm to cm is *one* place to the *left* on the conversion line, so move the decimal point one place to the *left* also.
40∧0. mm = 40.0 cm = 40 cm
⌣

Copyright © 2018 Pearson Education, Inc.

35. 0.91 m to mm
Count 3 places to the *right* on the conversion line. One zero is written in as a placeholder.
0.910∧ m = 910 mm

37. 82 cm to m
From cm to m is *two* places to the left on the conversion line, so move the decimal point two places to the *left* also.

∧82. cm = 0.82 m

0.82 m is less than 1 m, so 82 cm is **less than** 1 m. The difference in length is
1 m − 0.82 m = 0.18 m or
100 cm − 82 cm = 18 cm.

39.
┌──── **5 mm = 0.5 cm**
↓
▭ ←── **1 mm = 0.1 cm**

5 mm to centimeters
Count one place to the *left* on the conversion line.
∧5. mm = 0.5 cm

Similarly, 1 mm = 0.1 cm.

41. 61 m to km
Count 3 places to the *left* on the conversion line. One zero is written in as a placeholder.
∧061. m = 0.061 km

The north fork of the Roe River is just under 0.061 km long.

43. 1.63 m to centimeters and millimeters
Count two (three) places to the *right* on the conversion line.
1.63∧ m = 163 cm;

1.630∧ m = 1630 mm

The median height for U.S. females who are 20 to 29 years old is about 163 cm, or equivalently, 1630 mm.

45. 5.6 mm to km

$$\frac{5.6 \text{ mm}}{1} \cdot \frac{1 \text{ m}}{1000 \text{ mm}} \cdot \frac{1 \text{ km}}{1000 \text{ m}} = \frac{5.6}{1,000,000} \text{ km}$$
$$= 0.0000056 \text{ km}$$

7.3 The Metric System—Capacity and Weight (Mass)

7.3 Margin Exercises

1. The liter would be used to measure the amount of water in the bathtub, the amount of gasoline you buy for your car, and the amount of water in a pail.

2. **(a)** I bought 8 <u>L</u> of soda at the store.

(b) The nurse gave me 10 <u>mL</u> of cough syrup.

(c) This is a 100 <u>L</u> garbage can.

(d) It took 10 <u>L</u> of paint to cover the bedroom walls.

(e) My car's gas tank holds 50 <u>L</u>.

(f) I added 15 <u>mL</u> of oil to the pancake mix.

(g) The can of orange soda holds 350 <u>mL</u>.

(h) My friend gave me a 30 <u>mL</u> bottle of expensive perfume.

3. **(a)** 9 L to mL
On the (metric capacity) conversion line, L to mL is <u>three</u> places to the <u>right</u>, which gives us 9000 mL. Alternatively, we could use a unit fraction as follows:

$$\frac{9 \text{ L}}{1} \cdot \frac{1000 \text{ mL}}{1 \text{ L}} = \underline{9000} \text{ mL}$$

(b) 0.75 L to mL
Count 3 places to the *right* on the (metric capacity) conversion line.
0.750∧ L = 750 mL

(c) 500 mL to L

$$\frac{500 \text{ mL}}{1} \cdot \frac{1 \text{ L}}{1000 \text{ mL}} = \frac{500}{1000} \text{ L} = 0.5 \text{ L}$$

(d) 5 mL to L
Count 3 places to the *left* on the (metric capacity) conversion line.
∧005. mL = 0.005 L

(e) 2.07 L to mL

$$\frac{2.07 \text{ L}}{1} \cdot \frac{1000 \text{ mL}}{1 \text{ L}} = 2070 \text{ mL}$$

Copyright © 2018 Pearson Education, Inc.

(f) 3275 mL to L
Count 3 places to the *left* on the (metric capacity) conversion line.

$3{\scriptstyle\wedge}275.$ mL = 3.275 L

4. A small paper clip, one playing card, and a five-dollar bill each weigh about 1 gram.

5. **(a)** A thumbtack weights 800 mg.

(b) A teenager weighs 50 kg.

(c) This large cast-iron frying pan weighs 1 kg.

(d) Jerry's basketball weighed 600 g.

(e) Tamlyn takes a 500 mg calcium tablet every morning.

(f) On his diet, Greg can eat 90 g of meat for lunch.

(g) One strand of hair weighs 2 mg.

(h) One banana might weigh 150 g.

6. **(a)** 10 kg to g
On the [metric weight (mass)] conversion line, kg to g is <u>three</u> places to the <u>right</u>, which gives us 10,000 g. Alternatively, we could use a unit fraction as follows:

$$\frac{10 \ \cancel{kg}}{1} \cdot \frac{1000 \ g}{1 \ \cancel{kg}} = \underline{10,000} \ g$$

(b) 45 mg to g

$$\frac{45 \ \cancel{mg}}{1} \cdot \frac{1 \ g}{1000 \ \cancel{mg}} = \frac{45}{1000} \ g = 0.045 \ g$$

(c) 6.3 kg to g
Count 3 places to the *right* on the [metric weight (mass)] conversion line.

$6.300{\scriptstyle\wedge}$ kg = 6300 g

(d) 0.077 g to mg
Count 3 places to the *right* on the [metric weight (mass)] conversion line.

$0.077{\scriptstyle\wedge}$ g = 77 mg

(e) 5630 g to kg
Count 3 places to the *left* on the [metric weight (mass)] conversion line.

$5{\scriptstyle\wedge}630.$ g = 5.63 kg

(f) 90 g to kg

$$\frac{90 \ \cancel{g}}{1} \cdot \frac{1 \ kg}{1000 \ \cancel{g}} = \frac{9}{100} \ kg = 0.09 \ kg$$

7. **(a)** Gail bought a 4 <u>L</u> can of paint.
Use <u>capacity</u> units.

(b) The bag of chips weighed 450 <u>g</u>.
Use <u>weight</u> units.

(c) Give the child 5 <u>mL</u> of cough syrup.
Use <u>capacity</u> units.

(d) The width of the window is 55 <u>cm</u>.
Use <u>length</u> units.

(e) Akbar drives 18 <u>km</u> to work.
Use <u>length</u> units.

(f) The laptop computer weighs 2 <u>kg</u>.
Use <u>weight</u> units.

(g) A credit card is 55 <u>mm</u> wide.
Use <u>length</u> units.

7.3   Section Exercises

1. **(a)** For a dose of cough syrup, use <u>milliliters</u>.

(b) For a large carton of milk, use <u>liters</u>.

(c) For a vitamin pill, use <u>milligrams</u>.

(d) For a heavyweight wrestler, use <u>kilograms</u>.

3. The glass held 250 <u>mL</u> of water. (Liquids are measured in mL or L.)

5. Dolores can make 10 <u>L</u> of soup in that pot. (Liquids are measured in mL or L.)

7. Our yellow Labrador dog grew up to weigh 40 <u>kg</u>. (Weight is measured in mg, g, or kg.)

9. Lori caught a small sunfish weighing 150 <u>g</u>. Weight is measured in mg, g, or kg; 150 mg is too light; 150 is too heavy.

11. Andre donated 500 <u>mL</u> of blood today. (Blood is a liquid, and liquids are measured in mL or L.)

13. The patient received a 250 <u>mg</u> tablet of medication each hour. (Weight is measured in mg, g, or kg.)

15. The gas can for the lawnmower holds 4 <u>L</u>. (Gasoline is a liquid, and liquids are measured in mL or L.)

17. Kingston's backpack weighs 5 <u>kg</u> when it is full of books. (Weight is measured in mg, g, or kg.)

19. This is unreasonable (too much) since 4.1 liters would be about 4 quarts.

21. This is unreasonable (too much) since 5 kilograms of Epsom salts would be about 11 pounds with approximately a quart of water.

Copyright © 2018 Pearson Education, Inc.

23. This is reasonable because 15 milliliters would be 3 teaspoons.

25. This is reasonable because 350 milligrams would be a little more than 1 tablet, which is about 325 milligrams.

27. Marta multiplied by the wrong unit fraction. The correct unit fraction is $\frac{1000\,g}{1\,kg}$, so that Marta can divide out kg in the numerator with kg in the denominator. The correct solution is

$$\frac{6.5\ \cancel{kg}}{1}\cdot\frac{1000\ g}{1\ \cancel{kg}}=\underline{6500}\ g$$

29. 15 L to mL

$$\left.\begin{array}{c}\text{unit}\\\text{fraction}\end{array}\right\}\ \frac{1000\ mL}{1\ L}$$

$$\frac{15\ \cancel{L}}{1}\cdot\frac{1000\ mL}{1\ \cancel{L}}=\underline{15{,}000}\ mL$$

OR
Count <u>three</u> places to the <u>right</u> on the conversion line.

$$15.000_\wedge\ L=\underline{15{,}000}\ mL$$

In exercises 31–45, we use a unit fraction or a conversion line, but not both.

31. 3000 mL to L
Count 3 places to the *left* on the conversion line.

$$3_\wedge000.\ mL=3\ L$$

33. 925 mL to L

$$\frac{925\ \cancel{mL}}{1}\cdot\frac{1\ L}{1000\ \cancel{mL}}=\frac{925}{1000}\ L=0.925\ L$$

35. 8 mL to L

$$\frac{8\ \cancel{mL}}{1}\cdot\frac{1\ L}{1000\ \cancel{mL}}=\frac{8}{1000}\ L=0.008\ L$$

37. 4.15 L to mL
Count 3 places to the *right* on the conversion line.

$$4.150_\wedge\ L=4150\ mL$$

39. 8000 g to kg
Count 3 places to the *left* on the conversion line.

$$8_\wedge000.\ g=8\ kg$$

41. 5.2 kg to g
From kg to g is *three* places to the *right* on the metric conversion line.

$$5.200_\wedge\ kg=5200\ g$$

43. 0.85 g to mg
Count 3 places to the *right* on the conversion line.

$$0.850_\wedge\ g=850\ mg$$

45. 30,000 mg to g
Count 3 places to the *left* on the conversion line.

$$30_\wedge000.\ mg=30\ g$$

47. 598 mg to g
Count 3 places to the *left* on the conversion line.

$$_\wedge598.\ mg=0.598\ g$$

49. 60 mL to L

$$\frac{60\ \cancel{mL}}{1}\cdot\frac{1\ L}{1000\ \cancel{mL}}=\frac{60}{1000}\ L=0.06\ L$$

51. 3 g to kg

$$\frac{3\ \cancel{g}}{1}\cdot\frac{1\ kg}{1000\ g}=\frac{3}{1000}\ kg=0.003\ kg$$

53. 0.99 L to mL

$$\frac{0.99\ \cancel{L}}{1}\cdot\frac{1000\ mL}{1\ \cancel{L}}=990\ mL$$

55. The masking tape is 19 <u>mm</u> wide. (Length is measured in mm, cm, m, or km.)

57. Buy a 60 <u>mL</u> jar of acrylic paint for art class. (Paint is a liquid and liquids are measured in mL or L.)

59. My waist measurement is 65 <u>cm</u>. (Length is measured in mm, cm, m, or km.)

61. A single postage stamp weighs 90 <u>mg</u>. (Weight is measured in mg, g, or kg.)

63. Convert 300 mL to L.
Count 3 places to the *left* on the conversion line.

$$_\wedge300.\ mL=0.3\ L$$

Each day 0.3 L of sweat is released.

Copyright © 2018 Pearson Education, Inc.

65. Convert 1.34 kg to g.
Count 3 places to the *right* on the conversion line.

$$1.340_\wedge \text{ kg} = 1340 \text{ g}$$

The average weight of a human brain is 1340 g.

Convert 9 kg to g.
Count 3 places to the *right* on the conversion line.

$$9.000_\wedge \text{ kg} = 9000 \text{ g}$$

An adult human's skin weighs about 5000 g.

67. Convert 500 mL to L.

$$\frac{500 \text{ mL}}{1} \cdot \frac{1 \text{ L}}{1000 \text{ mL}} = \frac{500}{1000} \text{ L} = 0.5 \text{ L}$$

On average, we breathe in and out roughly 0.5 L of air every 4 seconds.

69. Convert 3000 g to kg and 4000 g to kg.

$$\frac{3000 \text{ g}}{1} \cdot \frac{1 \text{ kg}}{1000 \text{ g}} = \frac{3000}{1000} \text{ kg} = 3 \text{ kg}$$

$$\frac{4000 \text{ g}}{1} \cdot \frac{1 \text{ kg}}{1000 \text{ g}} = \frac{4000}{1000} \text{ kg} = 4 \text{ kg}$$

A small adult cat weighs from 3 kg to 4 kg.

71. There are 1000 milligrams in a gram so 1005 mg is *greater* than 1 g. The difference in weight is
1005 mg − 1000 mg = 5 mg
or 1.005 g − 1 g = 0.005 g.

73. Convert 1 kg to g. Then divide the result by the weight of 1 nickel.

$$\frac{1 \text{ kg}}{1} \cdot \frac{1000 \text{ g}}{1 \text{ kg}} = 1000 \text{ g}$$

Divide the 1000 g by the weight of 1 nickel.

$$\frac{1000 \text{ g}}{5 \text{ g}} = 200$$

There are 200 nickels in 1 kg of nickels.

75. **(a)** 1 Mm = 1,000,000 m

(b) 3.5 Mm to m

$$\frac{3.5 \text{ Mm}}{1} \cdot \frac{1,000,000 \text{ m}}{1 \text{ Mm}} = 3,500,000 \text{ m}$$

76. **(a)** 1 Gm = 1,000,000,000 m

(b) 2500 m to Gm

$$\frac{2500 \text{ m}}{1} \cdot \frac{1 \text{ Gm}}{1,000,000,000 \text{ m}} = 0.0000025 \text{ Gm}$$

77. **(a)** 1 Tm = 1,000,000,000,000 m

(b) $\dfrac{1 \text{ Tm}}{1} \cdot \dfrac{1,000,000,000,000 \text{ m}}{1 \text{ Tm}} \cdot \dfrac{1 \text{ Gm}}{1,000,000,000 \text{ m}}$

$$= \frac{1,000,000,000,000}{1,000,000,000} \text{ Gm} = 1000 \text{ Gm}$$

So 1 Tm = 1000 Gm.

$\dfrac{1 \text{ Tm}}{1} \cdot \dfrac{1,000,000,000,000 \text{ m}}{1 \text{ Tm}} \cdot \dfrac{1 \text{ Mm}}{1,000,000 \text{ m}}$

$$= \frac{1,000,000,000,000}{1,000,000} \text{ Mm} = 1,000,000 \text{ Mm}$$

So 1 Tm = 1,000,000 Mm.

78. 1 MB = 1,000,000 bytes
1 GB = 1,000,000,000 bytes
1 TB = 1,000,000,000,000 bytes
$2^{20} = 1,048,576$
$2^{30} = 1,073,741,824$
$2^{40} = 1,099,511,627,776$
If your calculator does not show enough digits, ask your instructor for help.

Summary Exercises *U.S. and Metric Measurement Units*

1. **U.S. Measurement Units and Abbreviations**

Length	Weight	Capacity
inch (in.)	ounce (oz)	fluid ounce (fl oz)
foot (ft)	pound (lb)	cup (c)
yard (yd)	ton (ton)	pint (pt)
mile (mi)		quart (qt)
		gallon (gal)

3. **(a)** 1 ft = 12 in.

(b) 3 ft = 1 yd

(c) 1 mi = 5280 ft

5. **(a)** 1 cup = 8 fl oz

(b) 4 qt = 1 gal

(c) 2 pt = 1 qt

7. My water bottle holds 450 mL.

9. The child weighed 23 kg.

11. The red pen is 14 cm long.

13. Merlene made 12 L of fruit punch for her daughter's birthday party.

Copyright © 2018 Pearson Education, Inc.

15. 45 cm to meters

$$\frac{45 \text{ cm}}{1} \cdot \frac{1 \text{ m}}{100 \text{ cm}} = \frac{45}{100} \text{ m} = 0.45 \text{ m}$$

17. 0.6 L to milliliters

$$\frac{0.6 \text{ L}}{1} \cdot \frac{1000 \text{ mL}}{1 \text{ L}} = 600 \text{ mL}$$

19. 300 mm to centimeters

$$\frac{300 \text{ mm}}{1} \cdot \frac{1 \text{ cm}}{10 \text{ mm}} = \frac{300}{10} \text{ cm} = 30 \text{ cm}$$

21. 50 mL to liters
Count 3 places to the *left* on the conversion line.

$$\wedge 050. \text{ mL} = 0.050 \text{ or } 0.05 \text{ L}$$

Note: You could also use unit fractions.

23. 7.28 kg to grams
Count 3 places to the *right* on the conversion line.

$$7.280 \wedge \text{ kg} = 7280 \text{ g}$$

25. 9 g to kilograms

$$\frac{9 \text{ g}}{1} \cdot \frac{1 \text{ kg}}{1000 \text{ g}} = \frac{9}{1000} \text{ kg} = 0.009 \text{ kg}$$

27. 2.72 m to centimeters
Count two places to the *right* on the conversion line.

$$2.72 \wedge \text{ m} = 272 \text{ cm};$$

Now convert 2.72 m to millimeters. Count three places to the *right* on the conversion line.

$$2.720 \wedge \text{ m} = 2720 \text{ mm}$$

29. The tallest person is 2.72 m tall and the second tallest person is 268 cm tall. Since

$$2 \wedge 68. \text{ cm} = 2.68 \text{ m},$$

the difference is $2.72 - 2.68 = 0.04$ m.

Now convert 0.04 m to cm. Count two places to the *right* on the conversion line.

$$0.04 \wedge \text{ m} = 4 \text{ cm};$$

Finally, convert 0.04 m to mm. Count three places to the *right* on the conversion line.

$$0.040 \wedge \text{ m} = 40 \text{ mm}$$

31. We are given dollars and ounces, so we need to use a unit fraction involving ounces and pounds to get an answer in dollars and pounds.

$$\frac{\$3.89}{\underset{3}{\cancel{12} \text{ oz}}} \cdot \frac{\overset{4}{\cancel{16} \text{ oz}}}{1 \text{ lb}} = \frac{\$15.56}{3 \text{ lb}} \approx \$5.19/\text{lb (rounded)}$$

The price per pound to the nearest cent was $5.19 (rounded).

7.4 Problem Solving with Metric Measurement

7.4 Margin Exercises

1. *Step 1*
The problem asks for the cost of 75 <u>cm</u> of ribbon.

Step 2
The price is $0.89 per <u>meter</u>, but the length wanted is cm. Convert cm to <u>m</u>.

Step 3
There are 100 cm in a meter, so 75 cm is $\frac{3}{4}$ of a meter. Round the cost of 1 m from $0.89 to $0.88 so that it is divisible by 4.

An estimate is: $\frac{3}{4} \cdot \$0.88 = \$0.66$

Step 4
Convert 75 cm to m: 75 cm = 0.75 m
Find the cost.

$$\frac{0.75 \text{ m}}{1} \cdot \frac{\$0.89}{1 \text{ m}} = \$0.6675 \approx \$0.67 \text{ (rounded)}$$

Step 5
The cost for 75 cm of ribbon is $0.67 (rounded).

Step 6
The rounded answer, $0.67, is close to our estimate of $0.66.

2. Convert 1.2 grams to milligrams.

$$1.200 \wedge \text{ g} = 1200 \text{ mg}$$

Divide by the number of doses.

$$\frac{1200 \text{ mg}}{3 \text{ doses}} = 400 \text{ mg/dose}$$

Each dose should be 400 mg.

3. Convert centimeters to meters.
2 m 35 cm = 2.35 m
1 m 85 cm = 1.85 m

Add the measurements of the fabric.

one piece	2.35	m
other piece	+ 1.85	m
	4.20	m

Andrea has 4.2 m of fabric.

Copyright © 2018 Pearson Education, Inc.

7.4   Section Exercises

1.   An important thing to include in the answer is the units, choice **(c)**.

3.   The price is \$0.98 per <u>kg</u>, so convert 850 g to <u>kg</u>.

$$\frac{850 \text{ g}}{1} \cdot \frac{1 \text{ kg}}{1000 \text{ g}} = 0.85 \text{ kg}$$

$$\frac{\$0.98}{1 \text{ kg}} \cdot \frac{0.85 \text{ kg}}{1} = \$0.833 \approx \$0.83$$

She will pay \$0.83 (rounded) for the rice.

5.   Convert 500 g to kg.

$$\frac{500 \text{ g}}{1} \cdot \frac{1 \text{ kg}}{1000 \text{ g}} = 0.5 \text{ kg}$$

Find the difference.
90 kg − 0.5 kg = 89.5 kg

The difference in the weights of the two dogs is 89.5 kg.

7.   Write 5 L in terms of mL.

$$5 \text{ L} = \frac{5 \text{ L}}{1} \cdot \frac{1000 \text{ mL}}{1 \text{ L}} = 5000 \text{ mL}$$

$$\frac{5000 \text{ mL}}{1} \cdot \frac{1 \text{ beat}}{70 \text{ mL}} \approx 71.42857 \text{ beats} \approx 71 \text{ beats}$$

It takes 71 beats (rounded) to pass all the blood through the heart.

9.   Multiply to find the total length in mm, then convert mm to cm.
60 mm · 30 = 1800 mm

$$\frac{\overset{180}{1800} \text{ mm}}{1} \cdot \frac{1 \text{ cm}}{\underset{1}{10} \text{ mm}} = 180 \text{ cm}$$

The total length is 180 cm.
Divide to find the cost per cm.

$$\frac{\$3.29}{180 \text{ cm}} \approx \$0.0183/\text{cm} \approx \$0.02/\text{cm}$$

The cost is \$0.02/cm (rounded).

11.   Convert centimeters to meters, then add the lengths of the two pieces.

$$\begin{array}{rcl} 2 \text{ m } 8 \text{ cm} & = & 2.08 \text{ m} \\ 2 \text{ m } 95 \text{ cm} & = & + 2.95 \text{ m} \\ \hline & & 5.03 \text{ m} \end{array}$$

The total length of the two boards is 5.03 m.

13.   Convert milliliters to liters.
85 mL = 0.085 L
Multiply by the number of students.
(0.085 L)(45) = 3.825 L
Four one-liter bottles should be ordered.

15.   Three cups a day for one week is 3 · 7 = 21 cups in one week.

$$\frac{21 \text{ cups}}{1} \cdot \frac{90 \text{ mg}}{1 \text{ cup}} = 1890 \text{ mg}$$

Convert mg to g.

$$\frac{1890 \text{ mg}}{1} \cdot \frac{1 \text{ g}}{1000 \text{ mg}} = 1.89 \text{ g}$$

Agnete will consume 1.89 grams of caffeine in one week.

17.   Boston: 111 cm = 1.11 m (222 cm in 2 years)
Lander: 2.60 m = 260 cm (520 cm in 2 years)

(a)   In one year, the difference in meters is

$$2.60 - 1.11 = 1.49.$$

(b)   In two years, the difference in centimeters is

$$520 - 222 = 298.$$

19.   A box of 50 envelopes weighs 255 g. Subtract the packaging weight to find the net weight.

$$\begin{array}{r} 255 \text{ g} \\ - 40 \text{ g} \\ \hline 215 \text{ g} \end{array}$$

There are 50 envelopes. Divide to find the weight of one envelope.

$$\frac{215 \text{ g}}{50 \text{ envelopes}} = \frac{4.3 \text{ g}}{\text{envelope}}$$

One envelope weighs 4.3 grams.
Convert grams to milligrams.

$$\frac{4.3 \text{ g}}{1} \cdot \frac{1000 \text{ mg}}{1 \text{ g}} = 4300 \text{ mg}$$

20.   Box of 1000 staples
Subtract the weight of the packaging to find the net weight.

$$\begin{array}{r} 350 \text{ g} \\ - 20 \text{ g} \\ \hline 330 \text{ g} \end{array}$$

There are 1000 staples, so divide the net weight by 1000 or move the decimal point 3 places to the left to find the weight of one staple.

$$330 \text{ g} \div 1000 = 0.33 \text{ g}$$

The weight of one staple is 0.33 g or 330 mg.

21.   Given that the weight of one sheet of paper is 3000 mg, find the weight of one sheet in grams.

$$3000 \text{ mg} = 3{\scriptstyle\wedge}000. \text{ g} = 3 \text{ g}$$

(continued)

Copyright © 2018 Pearson Education, Inc.

To find the net weight of the ream of paper, multiply 500 sheets by 3 g per sheet.

$$\frac{500 \text{ sheets}}{1} \cdot \frac{3 \text{ g}}{1 \text{ sheet}} = 1500 \text{ g}$$

The net weight of the paper is 1500 g.
Since the weight of the packaging is 50 g, the

22. Box of 100 small paper clips
Use the metric conversion line to find the weight of one paper clip in grams.

$$500 \text{ mg} = {}_{\wedge}500. \text{ g} = 0.5 \text{ g}$$

The net weight is $(0.5 \text{ g})(100) = 50 \text{ g}$.
The total weight is $50 \text{ g} + 5 \text{ g} = 55 \text{ g}$.

7.5 Metric–U.S. Measurement Conversions and Temperature

7.5   Margin Exercises

1. **(a)** 23 m to yards

Look at the "Metric to U.S. Units" side of the table, where you will see that

1 meter $\approx$ 1.09 yards.

$$\frac{23 \text{ m}}{1} \cdot \frac{1.09 \text{ yd}}{1 \text{ m}} \approx 25.1 \text{ yd}$$

(b) 40 cm to inches

$$\frac{40 \text{ cm}}{1} \cdot \frac{0.39 \text{ in.}}{1 \text{ cm}} \approx 15.6 \text{ in.}$$

(c) 5 mi to kilometers

Look at the "U.S. to Metric Units" side of the table, where you will see that

1 mile $\approx$ 1.61 kilometers.

$$\frac{5 \text{ mi}}{1} \cdot \frac{1.61 \text{ km}}{1 \text{ mi}} \approx 8.1 \text{ km}$$

(d) 12 in. to centimeters

$$\frac{12 \text{ in.}}{1} \cdot \frac{2.54 \text{ cm}}{1 \text{ in.}} \approx 30.5 \text{ cm}$$

2. **(a)** 17 kg to pounds
From the table, 1 kg $\approx$ 2.20 lb.

$$\frac{17 \text{ kg}}{1} \cdot \frac{2.20 \text{ lb}}{1 \text{ kg}} = \frac{(17)(2.20 \text{ lb})}{1} \approx 37.4 \text{ lb}$$

(b) 5 L to quarts

$$\frac{5 \text{ L}}{1} \cdot \frac{1.06 \text{ qt}}{1 \text{ L}} \approx 5.3 \text{ qt}$$

(c) 90 g to ounces

$$\frac{90 \text{ g}}{1} \cdot \frac{0.035 \text{ oz}}{1 \text{ g}} \approx 3.2 \text{ oz}$$

(d) 3.5 gal to liters

$$\frac{3.5 \text{ gal}}{1} \cdot \frac{3.79 \text{ L}}{1 \text{ gal}} \approx 13.3 \text{ L}$$

(e) 145 lb to kilograms

$$\frac{145 \text{ lb}}{1} \cdot \frac{0.45 \text{ kg}}{1 \text{ lb}} \approx 65.3 \text{ kg}$$

(f) 8 oz to grams

$$\frac{8 \text{ oz}}{1} \cdot \frac{28.35 \text{ g}}{1 \text{ oz}} \approx 226.8 \text{ g}$$

3. **(a)** Set the living room thermostat at: 21 °C

(b) The baby has a fever of: 39 °C

(c) Wear a sweater outside because it's: 15 °C

(d) My iced tea is: 5 °C

(e) Let's go swimming! It's: 35 °C

(f) Inside a refrigerator (not the freezer) it's: 3 °C

(g) There is a blizzard outside. It's: -20 °C

(h) I need hot water to get these clothes clean. The water should be: 55 °C

4. Use $C = \dfrac{5(F - 32)}{9}$.

(a) 59°F

$$C = \frac{5(59 - 32)}{9} = \frac{5 \cdot \overset{3}{\cancel{27}}}{\underset{1}{\cancel{9}}} = 15$$

59°F = 15°C

(b) 41°F

$$C = \frac{5(41 - 32)}{9} = \frac{5 \cdot \overset{1}{\cancel{9}}}{\underset{1}{\cancel{9}}} = 5$$

41°F = 5°C

(c) 212°F

$$C = \frac{5(212 - 32)}{9} = \frac{5 \cdot \overset{20}{\cancel{180}}}{\underset{1}{\cancel{9}}} = 100$$

212°F = 100°C

Copyright © 2018 Pearson Education, Inc.

(d) 98.6°F

$$C = \frac{5(98.6 - 32)}{9} = \frac{5 \cdot \overset{7.4}{\cancel{66.6}}}{\underset{1}{\cancel{9}}} = 37$$

98.6°F = 37°C

5. Use $F = \dfrac{9 \cdot C}{5} + 32$.

(a) 100°C

$$F = \frac{9 \cdot \overset{20}{\cancel{100}}}{\underset{1}{\cancel{5}}} + 32 = 180 + 32 = 212$$

100°C = 212°F

(b) 25°C

$$F = \frac{9 \cdot \overset{5}{\cancel{25}}}{\underset{1}{\cancel{5}}} + 32 = 45 + 32 = 77$$

25°C = 77°F

(c) 80°C

$$F = \frac{9 \cdot \overset{16}{\cancel{80}}}{\underset{1}{\cancel{5}}} + 32 = 144 + 32 = 176$$

80°C = 176°F

(d) 5°C

$$F = \frac{9 \cdot \overset{1}{\cancel{5}}}{\underset{1}{\cancel{5}}} + 32 = 9 + 32 = 41$$

5°C = 41°F

7.5 Section Exercises

1. **(a)** On the "Metric to U.S." side of the table, we see that 1 kilogram ≈ 2.20 pounds. So two unit fractions using that information are

$$\frac{1 \text{ kg}}{2.20 \text{ lb}} \quad \text{and} \quad \frac{2.20 \text{ lb}}{1 \text{ kg}}.$$

(b) When converting 5 kg to pounds, we want to eliminate <u>kg</u>, so use the unit fraction with kg in the <u>denominator</u>.

3. 20 m to yards

$$\frac{20 \text{ } \cancel{m}}{1} \cdot \frac{1.09 \text{ yd}}{1 \text{ } \cancel{m}} \approx 21.8 \text{ yd}$$

5. 80 m to feet

$$\frac{80 \text{ } \cancel{m}}{1} \cdot \frac{3.28 \text{ ft}}{1 \text{ } \cancel{m}} \approx 262.4 \text{ ft}$$

7. 16 ft to meters

$$\frac{16 \text{ } \cancel{ft}}{1} \cdot \frac{0.30 \text{ m}}{1 \text{ } \cancel{ft}} \approx 4.8 \text{ m}$$

9. 150 g to ounces

$$\frac{150 \text{ } \cancel{g}}{1} \cdot \frac{0.035 \text{ oz}}{1 \text{ } \cancel{g}} \approx 5.3 \text{ oz}$$

11. 248 lb to kilograms

$$\frac{248 \text{ } \cancel{lb}}{1} \cdot \frac{0.45 \text{ kg}}{1 \text{ } \cancel{lb}} \approx 111.6 \text{ kg}$$

13. 28.6 L to quarts

$$\frac{28.6 \text{ } \cancel{L}}{1} \cdot \frac{1.06 \text{ qt}}{1 \text{ } \cancel{L}} \approx 30.3 \text{ qt}$$

15. **(a)** Convert 5 g to ounces.

$$\frac{5 \text{ } \cancel{g}}{1} \cdot \frac{0.035 \text{ oz}}{1 \text{ } \cancel{g}} \approx 0.2 \text{ oz}$$

About 0.2 oz of gold was used.

(b) The extra weight probably did not slow him down.

17. Convert 3.9 gal to liters.

$$\frac{3.9 \text{ } \cancel{gal}}{1} \cdot \frac{3.79 \text{ L}}{1 \text{ } \cancel{gal}} \approx 14.8 \text{ L}$$

The dishwasher uses about 14.8 L.

19. Write $\frac{1}{2}$ in. as 0.5 in. Then convert 0.5 in. to centimeters.

$$\frac{0.5 \text{ } \cancel{in.}}{1} \cdot \frac{2.54 \text{ cm}}{1 \text{ } \cancel{in.}} \approx 1.3$$

The dwarf gobie is about 1.3 cm long.

21. A snowy day
−8°C is the most reasonable temperature. 12°C and 28°C are temperatures well above freezing.

23. A high fever
40°C is the most reasonable temperature because normal body temperature is about 37°C.

25. Oven temperature
150°C is the most reasonable temperature. 50°C is the temperature of hot bath water, which is clearly not hot enough for an oven temperature.

For exercises 27–30, use the formula $C = \dfrac{5(F - 32)}{9}$.

27. Change 60°F to Celsius.

$$C = \frac{5(60 - 32)}{9} = \frac{5 \cdot 28}{9} = \frac{140}{9} \approx 15.6 \approx 16$$

60°F ≈ 16°C (rounded)

Copyright © 2018 Pearson Education, Inc.

29. 104°F

$$C = \frac{5(104 - 32)}{9} = \frac{5 \cdot 72}{9} = \frac{5 \cdot \overset{8}{\cancel{72}}}{\underset{1}{\cancel{9}}} = 40$$

$$104°F = 40°C$$

For exercises 31–34, use the formula $F = \frac{9 \cdot C}{5} + 32$.

31. 8°C

$$F = \frac{9 \cdot 8}{5} + 32 = \frac{72}{5} + 32 = 14.4 + 32 = 46.4$$

$$8°C \approx 46°F \text{ (rounded)}$$

33. 35°C

$$F = \frac{9 \cdot 35}{5} + 32 = \frac{9 \cdot \overset{7}{\cancel{35}}}{\underset{1}{\cancel{5}}} + 32 = 63 + 32 = 95$$

$$35°C = 95°F$$

35. The first mistake was multiplying 5(42) first instead of doing the subtraction inside the parenthese first. The second mistake was rounding the answer incorrectly. The correct solution is:

$$C = \frac{5(42 - 32)}{9} = \frac{5 \cdot 10}{9} = \frac{50}{9} \approx 5.56 \approx 6° \, C$$

37. **(a)** Since the comfort range of the boots is from 24°C to 4°C, you would wear these boots in pleasant weather—above freezing, but not hot.

(b) Change 24°C to Fahrenheit.

$$F = \frac{9 \cdot C}{5} + 32 = \frac{9 \cdot 24}{5} + 32 = \frac{216}{5} + 32$$
$$F = 43.2 + 32 = 75.2$$

Thus, 24°C ≈ 75°F (rounded).

Change 4°C to Fahrenheit.

$$F = \frac{9 \cdot C}{5} + 32 = \frac{9 \cdot 4}{5} + 32 = \frac{36}{5} + 32$$
$$F = 7.2 + 32 = 39.2$$

Thus, 4°C ≈ 39°F (rounded).

The boots are designed for Fahrenheit temperatures of about 75°F to about 39°F.

(c) The range of metric temperatures in January would depend on where you live. In Minnesota, it's 0°C to −40°C, and in California it's 24°C to 0°C.

39. Length of model plane

$$\frac{6 \cancel{\text{ft}}}{1} \cdot \frac{0.30 \text{ m}}{1 \cancel{\text{ft}}} \approx 6(0.30 \text{ m}) = 1.8 \text{ m}$$

40. Weight of plane

$$\frac{11 \cancel{\text{lb}}}{1} \cdot \frac{0.45 \text{ kg}}{1 \cancel{\text{lb}}} \approx 11(0.45 \text{ kg}) = 4.95 \text{ kg}$$
$$\approx 5.0 \text{ kg}$$

41. Length of flight path

$$\frac{1888.3 \cancel{\text{mi}}}{1} \cdot \frac{1.61 \text{ km}}{1 \cancel{\text{mi}}} \approx 1888.3(1.61 \text{ km})$$
$$\approx 3040.2 \text{ km}$$

42. Time of flight = 38 hr 23 min

43. Cruising altitude

$$\frac{1000 \cancel{\text{ft}}}{1} \cdot \frac{0.30 \text{ m}}{1 \cancel{\text{ft}}} \approx 1000(0.30 \text{ m}) = 300 \text{ m}$$

44. Fuel at the start

$$\frac{1 \cancel{\text{gal}}}{1} \cdot \frac{3.79 \cancel{\text{L}}}{1 \cancel{\text{gal}}} \cdot \frac{1000 \text{ mL}}{1 \cancel{\text{L}}} \approx (3.79)(1000 \text{ mL})$$
$$= 3790 \text{ mL}$$

Since the plane carried "less than a gallon of fuel," there would be fewer than 3790 mL.

45. Fuel left after landing

$$\frac{2 \cancel{\text{fl oz}}}{1} \cdot \frac{1 \cancel{\text{qt}}}{32 \cancel{\text{fl oz}}} \cdot \frac{0.95 \cancel{\text{L}}}{1 \cancel{\text{qt}}} \cdot \frac{1000 \text{ mL}}{1 \cancel{\text{L}}}$$
$$= \frac{950}{16} = 59.375 \approx 59.4 \text{ mL}$$

46.
$$\frac{1 \cancel{\text{gal}}}{1} \cdot \frac{4 \cancel{\text{qt}}}{1 \cancel{\text{gal}}} \cdot \frac{32 \text{ fl oz}}{1 \cancel{\text{qt}}} = 128 \text{ fl oz}$$
$$\frac{2 \text{ fl oz}}{128 \text{ fl oz}} = \frac{1}{64} = 0.015625 \approx 1.6\%$$

Chapter 7 Review Exercises

1. 1 lb = <u>16</u> oz

2. <u>3</u> ft = 1 yd

3. 1 ton = <u>2000</u> lb

4. <u>4</u> qt = 1 gal

5. 1 hr = <u>60</u> min

6. 1 c = <u>8</u> fl oz

7. <u>60</u> sec = 1 min

8. <u>5280</u> ft = 1 mi

9. <u>12</u> in. = 1 ft

Copyright © 2018 Pearson Education, Inc.

10. 4 ft to inches

$$\frac{4\,\cancel{ft}}{1}\cdot\frac{12\text{ in.}}{1\,\cancel{ft}}=4\cdot12\text{ in.}=48\text{ in.}$$

11. 6000 lb to tons

$$\frac{\overset{3}{6000}\,\cancel{lb}}{1}\cdot\frac{1\text{ ton}}{\underset{1}{2000}\,\cancel{lb}}=3\text{ tons}$$

12. 64 oz to pounds

$$\frac{\overset{4}{64}\,\cancel{oz}}{1}\cdot\frac{1\text{ lb}}{\underset{1}{16}\,\cancel{oz}}=4\text{ lb}$$

13. 18 hr to days

$$\frac{\overset{3}{18}\,\cancel{hr}}{1}\cdot\frac{1\text{ day}}{\underset{4}{24}\,\cancel{hr}}=\frac{3}{4}\text{ day or }0.75\text{ day}$$

14. 150 min to hours

$$\frac{\overset{5}{150}\,\cancel{min}}{1}\cdot\frac{1\text{ hr}}{\underset{2}{60}\,\cancel{min}}=\frac{5}{2}\text{ hr}=2\frac{1}{2}\text{ hr or }2.5\text{ hr}$$

15. $1\frac{3}{4}$ lb to ounces

$$\frac{1\frac{3}{4}\,\cancel{lb}}{1}\cdot\frac{16\text{ oz}}{1\,\cancel{lb}}=\frac{7}{\cancel{4}}\cdot\frac{\overset{4}{\cancel{16}}}{1}\text{ oz}=28\text{ oz}$$

16. $6\frac{1}{2}$ ft to inches

$$\frac{6\frac{1}{2}\,\cancel{ft}}{1}\cdot\frac{12\text{ in.}}{1\,\cancel{ft}}=\frac{13}{\cancel{2}}\cdot\frac{\overset{6}{\cancel{12}}}{1}\text{ in.}=78\text{ in.}$$

17. 7 gal to cups

$$\frac{7\text{ gal}}{1}\cdot\frac{4\text{ qt}}{1\text{ gal}}\cdot\frac{2\text{ pt}}{1\text{ qt}}\cdot\frac{2\text{ c}}{1\text{ pt}}=7\cdot4\cdot2\cdot2\text{ c}=112\text{ c}$$

18. 4 days to seconds

$$\frac{4\text{ days}}{1}\cdot\frac{24\text{ hr}}{1\text{ day}}\cdot\frac{60\text{ min}}{1\text{ hr}}\cdot\frac{60\text{ sec}}{1\text{ min}}$$
$$=4\cdot24\cdot60\cdot60\text{ sec}$$
$$=345,600\text{ sec}$$

19. (a) 14,000 ft to yards

$$\frac{14,000\,\cancel{ft}}{1}\cdot\frac{1\text{ yd}}{3\,\cancel{ft}}=\frac{14,000}{3}\text{ yd}$$
$$=4666\frac{2}{3}\text{ yd (exact)}$$
$$\text{or }4666.7\text{ yd (rounded).}$$

The average depth is $4666\frac{2}{3}$ yd (exact) or 4666.7 yd (rounded).

(b) 14,000 ft to miles

$$\frac{14,000\,\cancel{ft}}{1}\cdot\frac{1\text{ mi}}{5280\,\cancel{ft}}\approx2.65\text{ mi}\approx2.7\text{ mi}$$

The average depth is 2.7 mi (rounded).

20. *Step 1*
The problem asks for the amount of money the company made.

Step 2
Convert pounds to tons. Then multiply to find the total amount.

Step 3
To estimate, round 123,260 lb to 100,000 lb. Then, there are 2000 lb in a ton, so 100,000 lb is 50 tons. So $50\cdot\$40=\$2000$ is our estimate.

Step 4
$$\frac{123,260\,\cancel{lb}}{1}\cdot\frac{1\text{ ton}}{2000\,\cancel{lb}}=61.63\text{ tons}$$
$$61.63\cdot\$40=\$2465.20$$

Step 5
The company made $2465.20.

Step 6
The answer, $2465.20, is close to our estimate of $2000.

21. My thumb is 20 <u>mm</u> wide.

22. Her waist measurement is 66 <u>cm</u>.

23. The two towns are 40 <u>km</u> apart.

24. A basketball court is 30 <u>m</u> long.

25. The height of the picnic bench is 45 <u>cm</u>.

26. The eraser on the end of my pencil is 5 <u>mm</u> long.

27. 5 m to cm
Count 2 places to the *right* on the metric conversion line.
$$5.00_\wedge\text{ m}=500\text{ cm}$$

28. 8.5 km to m

$$\frac{8.5\,\cancel{km}}{1}\cdot\frac{1000\text{ m}}{1\,\cancel{km}}=8500\text{ m}$$

29. 85 mm to cm
Count 1 place to the *left* on the metric conversion line.
$$8_\wedge5.\text{ mm}=8.5\text{ cm}$$

30. 370 cm to m

$$\frac{370\,\cancel{cm}}{1}\cdot\frac{1\text{ m}}{100\,\cancel{cm}}=3.7\text{ m}$$

Copyright © 2018 Pearson Education, Inc.

31. 70 m to km
Count 3 places to the *left* on the metric conversion line.

$$\wedge070. \text{ m} = 0.07 \text{ km}$$

32. 0.93 m to mm

$$\frac{0.93 \text{ m}}{1} \cdot \frac{1000 \text{ mm}}{1 \text{ m}} = 930 \text{ mm}$$

33. The eye dropper holds 1 mL.

34. I can heat 3 L of water in this saucepan.

35. Loretta's hammer weighed 650 g.

36. Yongshu's suitcase weighed 20 kg when it was packed.

37. My fish tank holds 80 L of water.

38. I'll buy the 500 mL bottle of mouthwash.

39. Mara took a 200 mg antibiotic pill.

40. This piece of chicken weighs 100 g.

41. 5000 mL to L
Count 3 places to the *left* on the metric conversion line.

$$5\wedge000. \text{ mL} = 5 \text{ L}$$

42. 8 L to mL
Count 3 places to the *right* on the metric conversion line.

$$8.000\wedge \text{ L} = 8000 \text{ mL}$$

43. 4.58 g to mg

$$\frac{4.58 \text{ g}}{1} \cdot \frac{1000 \text{ mg}}{1 \text{ g}} = 4580 \text{ mg}$$

44. 0.7 kg to g

$$\frac{0.7 \text{ kg}}{1} \cdot \frac{1000 \text{ g}}{1 \text{ kg}} = 700 \text{ g}$$

45. 6 mg to g
Count 3 places to the *left* on the metric conversion line.

$$\wedge006. \text{ mg} = 0.006 \text{ g}$$

46. 35 mL to L
Count 3 places to the *left* on the metric conversion line.

$$\wedge035. \text{ mL} = 0.035 \text{ L}$$

47. Convert milliliters to liters.

$$\frac{180 \text{ mL}}{1} \cdot \frac{1 \text{ L}}{1000 \text{ mL}} = \frac{180}{1000} \text{ L} = 0.18 \text{ L}$$

Multiply 0.18 L by the number of servings.
$(0.18 \text{ L})(175) = 31.5 \text{ L}$

For 175 servings, 31.5 L of punch are needed.

48. Convert kilograms to grams
10 kg = 10,000 g
Divide by the number of people.

$$\frac{10,000 \text{ g}}{28 \text{ people}} \approx 357$$

Jason is allowing 357 g (rounded) of turkey for each person.

49. Convert grams to kilograms.
Using the metric conversion line,
4 kg 750 g = 4.75 kg.
Subtract the weight loss from his original weight.

$$\begin{array}{rl} 57.00 & \text{kg} \\ - \ 4.75 & \text{kg} \\ \hline 52.25 & \text{kg} \end{array}$$

Chandni weighs 52.25 kg.

50. Convert 950 g to kilograms.
Using the metric conversion line,

$$950 \text{ g} = 0.95 \text{ kg}.$$

Multiply by the price per kilogram.

$$\begin{array}{rl} \$1.49 & \leftarrow \ 2 \ \textit{decimal places} \\ \times \ 0.95 & \leftarrow \ 2 \ \textit{decimal places} \\ \hline 745 & \\ 1\ 341 \ \ & \\ \hline \$1.4155 & \leftarrow \ 4 \ \textit{decimal places} \end{array}$$

Young-Mi paid $1.42 (rounded) for the onions.

51. 6 m to yards

$$\frac{6 \text{ m}}{1} \cdot \frac{1.09 \text{ yd}}{1 \text{ m}} \approx 6.5 \text{ yd (rounded)}$$

52. 30 cm to inches

$$\frac{30 \text{ cm}}{1} \cdot \frac{0.39 \text{ in.}}{1 \text{ cm}} \approx 11.7 \text{ in. (rounded)}$$

53. 108 km to miles

$$\frac{108 \text{ km}}{1} \cdot \frac{0.62 \text{ mi}}{1 \text{ km}} \approx 67.0 \text{ mi (rounded)}$$

54. 800 mi to kilometers

$$\frac{800 \text{ mi}}{1} \cdot \frac{1.61 \text{ km}}{1 \text{ mi}} \approx 1288 \text{ km}$$

Copyright © 2018 Pearson Education, Inc.

55. 23 qt to liters

$$\frac{23 \; \cancel{qt}}{1} \cdot \frac{0.95 \; L}{1 \; \cancel{qt}} \approx 21.9 \; L \text{ (rounded)}$$

56. 41.5 L to quarts

$$\frac{41.5 \; \cancel{L}}{1} \cdot \frac{1.06 \; qt}{1 \; \cancel{L}} \approx 44.0 \; qt \text{ (rounded)}$$

57. Water freezes at 0°C.

58. Water boils at 100°C.

59. Normal body temperature is about 37°C.

60. Comfortable room temperature is about 20°C.

61. 77°F

$$C = \frac{5(77-32)}{9} = \frac{5(\overset{5}{\cancel{45}})}{\underset{1}{\cancel{9}}} = 25$$

77°F = 25°C

62. 92°F

$$C = \frac{5(92-32)}{9} = \frac{5(\overset{20}{\cancel{60}})}{\underset{3}{\cancel{9}}} \approx 33.3$$

92°F ≈ 33°C (rounded)

63. 6 °C

$$F = \frac{9 \cdot 6}{5} + 32 = \frac{54}{5} + 32 = 10.8 + 32 = 42.8$$

6 °C ≈ 43 °F (rounded)

64. 49°C

$$F = \frac{9 \cdot 49}{5} + 32 = 88.2 + 32 = 120.2 \approx 120$$

49°C ≈ 120°F (rounded)

Chapter 7 Mixed Review Exercises

1. I added 1 L of oil to my car.

2. The box of books weighed 15 kg.

3. Larry's shoe is 30 cm long.

4. Jan used 15 mL of shampoo on her hair.

5. My fingernail is 10 mm wide.

6. I walked 2 km to school.

7. The tiny bird weighed 15 g.

8. The new library building is 18 m wide.

9. The cookie recipe uses 250 mL of milk.

10. Renee's pet mouse weighs 30 g.

11. One postage stamp weighs 90 mg.

12. I bought 30 L of gas for my car.

13. 10.5 cm to millimeters
Count 1 place to the *right* on the metric conversion line.

$$10.5_{\wedge} \text{ cm} = 105 \text{ mm}$$

14. 45 min to hours

$$\frac{\overset{3}{\cancel{45}} \; \cancel{min}}{1} \cdot \frac{1 \; hr}{\underset{4}{\cancel{60}} \; \cancel{min}} = \frac{3}{4} \text{ hr or } 0.75 \text{ hr}$$

15. 90 in. to feet

$$\frac{\overset{15}{\cancel{90}} \; \cancel{in.}}{1} \cdot \frac{1 \; ft}{\underset{2}{\cancel{12}} \; \cancel{in.}} = \frac{15}{2} \text{ ft} = 7\frac{1}{2} \text{ ft or } 7.5 \text{ ft}$$

16. 1.3 m to centimeters
Count 2 places to the *right* on the metric conversion line.

$$1.30_{\wedge} \text{ m} = 130 \text{ cm}$$

17. 25 °C to Fahrenheit

$$F = \frac{9 \cdot \overset{5}{\cancel{25}}}{\underset{1}{\cancel{5}}} + 32 = 45 + 32 = 77$$

25 °C = 77 °F

18. $3\frac{1}{2}$ gal to quarts

$$\frac{3\frac{1}{2} \; \cancel{gal}}{1} \cdot \frac{4 \; qt}{1 \; \cancel{gal}} = \frac{7}{\underset{1}{\cancel{2}}} \cdot \frac{\overset{2}{\cancel{4}}}{1} \; qt = 14 \; qt$$

19. 700 mg to grams
Count 3 places to the *left* on the metric conversion line.

$$_{\wedge}700. \text{ mg} = 0.7 \text{ g}$$

20. 0.81 L to milliliters
Count 3 places to the *right* on the metric conversion line.

$$0.810_{\wedge} \text{ L} = 810 \text{ mL}$$

21. 5 lb to ounces

$$\frac{5 \; \cancel{lb}}{1} \cdot \frac{16 \; oz}{1 \; \cancel{lb}} = 5 \cdot 16 \; oz = 80 \; oz$$

Copyright © 2018 Pearson Education, Inc.

22. 60 kg to g

$$\frac{60 \text{ kg}}{1} \cdot \frac{1000 \text{ g}}{1 \text{ kg}} = 60{,}000 \text{ g}$$

23. 1.8 L to milliliters
Count 3 places to the *right* on the metric conversion line.

$$1.800_\wedge \text{ L} = 1800 \text{ mL}$$

24. 86°F to Celsius

$$C = \frac{5(86 - 32)}{9} = \frac{5(\overset{6}{\cancel{54}})}{\underset{1}{\cancel{9}}} = 30$$

86°F = 30°C

25. 0.36 m to centimeters
Count 2 places to the *right* on the metric conversion line.

$$0.36_\wedge \text{ m} = 36 \text{ cm}$$

26. 55 mL to liters
Count 3 places to the *left* on the metric conversion line.

$$_\wedge 055. \text{ mL} = 0.055 \text{ L}$$

27. Convert centimeters to meters.

$$\begin{array}{ll} 2 \text{ m } 4 \text{ cm} = 2 + 0.04 \text{ m} = & 2.04 \text{ m} \\ 78 \text{ cm} \qquad\qquad\quad = & -\,0.78 \text{ m} \\ \hline & 1.26 \text{ m} \end{array}$$

The board is 1.26 m long.

28. 3000 lb to tons

$$\frac{3000 \text{ lb}}{1} \cdot \frac{1 \text{ ton}}{2000 \text{ lb}} = \frac{3}{2} \text{ T} = 1.5 \text{ tons}$$

Multiply by 12 days.
$(1.5 \text{ tons})(12) = 18 \text{ tons}$

18 tons of cookies are sold in all.

29. Convert ounces to grams.

$$\frac{4 \text{ oz}}{1} \cdot \frac{28.35 \text{ g}}{1 \text{ oz}} \approx 113 \text{ g (rounded)}$$

Convert 350 °F to Celsius.

$$C = \frac{5(350 - 32)}{9}$$
$$= \frac{5(318)}{9} = \frac{1590}{9} \approx 177 \text{ °C (rounded)}$$

30. Convert kilograms to pounds.

$$\frac{80.9 \text{ kg}}{1} \cdot \frac{2.20 \text{ lb}}{1 \text{ kg}} \approx 178.0 \text{ lb}$$

Convert meters to feet.

$$\frac{1.83 \text{ m}}{1} \cdot \frac{3.28 \text{ ft}}{1 \text{ m}} \approx 6.0 \text{ ft}$$

Jalo weighs about 178.0 lb and is about 6.0 ft.

31. 44 ft to meters

$$\frac{44 \text{ ft}}{1} \cdot \frac{0.30 \text{ m}}{1 \text{ ft}} \approx 13.2 \text{ m}$$

32. 6.7 m to feet

$$\frac{6.7 \text{ m}}{1} \cdot \frac{3.28 \text{ ft}}{1 \text{ m}} \approx 22.0 \text{ ft}$$

33. 4 yd to meters

$$\frac{4 \text{ yd}}{1} \cdot \frac{0.91 \text{ m}}{1 \text{ yd}} \approx 3.6 \text{ m}$$

34. 102 cm to inches

$$\frac{102 \text{ cm}}{1} \cdot \frac{0.39 \text{ in.}}{1 \text{ cm}} \approx 39.8 \text{ in.}$$

35. 220,000 kg to pounds

$$\frac{220{,}000 \text{ kg}}{1} \cdot \frac{2.20 \text{ lb}}{1 \text{ kg}} \approx 484{,}000 \text{ lb}$$

36. 5 gal to liters

$$\frac{5 \text{ gal}}{1} \cdot \frac{3.79 \text{ L}}{1 \text{ gal}} = 18.95 \approx 19.0 \text{ L}$$

6 gal to liters

$$\frac{6 \text{ gal}}{1} \cdot \frac{3.79 \text{ L}}{1 \text{ gal}} \approx 22.7 \text{ L}$$

37. **(a)** From Exercise 35:

$$\frac{484{,}000 \text{ lb}}{1} \cdot \frac{1 \text{ ton}}{2000 \text{ lb}} = 242 \text{ tons}$$

   (b) From Exercise 33:

$$\frac{4 \text{ yd}}{1} \cdot \frac{3 \text{ ft}}{1 \text{ yd}} \cdot \frac{12 \text{ in.}}{1 \text{ ft}} = 144 \text{ in.}$$

38. **(a)** From Exercise 34:

$$\frac{102 \text{ cm}}{1} \cdot \frac{1 \text{ m}}{100 \text{ cm}} = 1.02 \text{ m}$$

   (b) From Exercise 32:

$$\frac{6.7 \text{ m}}{1} \cdot \frac{100 \text{ cm}}{1 \text{ m}} = 670 \text{ cm}$$

Copyright © 2018 Pearson Education, Inc.

Chapter 7 Test

1. 9 gal to quarts

$$\frac{9 \text{ gal}}{1} \cdot \frac{4 \text{ qt}}{1 \text{ gal}} = 36 \text{ qt}$$

2. 45 ft to yards

$$\frac{45 \text{ ft}}{1} \cdot \frac{1 \text{ yd}}{3 \text{ ft}} = \frac{45}{3} \text{ yd} = 15 \text{ yd}$$

3. 135 min to hours

$$\frac{\overset{27}{\cancel{135}} \text{ min}}{1} \cdot \frac{1 \text{ hr}}{\underset{12}{\cancel{60}} \text{ min}} = \frac{27}{12} \text{ hr} = 2.25 \text{ hr or } 2\frac{1}{4} \text{ hr}$$

4. 9 in. to feet

$$\frac{\overset{3}{\cancel{9}} \text{ in.}}{1} \cdot \frac{1 \text{ ft}}{\underset{4}{\cancel{12}} \text{ in.}} = \frac{3}{4} \text{ ft or } 0.75 \text{ ft}$$

5. $3\frac{1}{2}$ lb to ounces

$$\frac{3\frac{1}{2} \text{ lb}}{1} \cdot \frac{16 \text{ oz}}{1 \text{ lb}} = \frac{7}{\underset{1}{\cancel{2}}} \cdot \frac{\overset{8}{\cancel{16}}}{1} \text{ oz} = 56 \text{ oz}$$

6. 5 days to minutes

$$\frac{5 \text{ days}}{1} \cdot \frac{24 \text{ hr}}{1 \text{ day}} \cdot \frac{60 \text{ min}}{1 \text{ hr}}$$

$$= \frac{5 \cdot 24 \cdot 60}{1} \text{ min} = 7200 \text{ min}$$

7. My husband weighs 75 <u>kg</u>.

8. I hiked 5 <u>km</u> this morning.

9. She bought 125 <u>mL</u> of cough syrup.

10. This apple weighs 180 <u>g</u>.

11. The page is about 21 <u>cm</u> wide.

12. My watch band is 10 <u>mm</u> wide.

13. I bought 10 <u>L</u> of soda for the picnic.

14. The bracelet is 16 <u>cm</u> long.

15. 250 cm to meters
Count 2 places to the *left* on the metric conversion line.

$$2_\wedge 50. \text{ cm} = 2.5 \text{ m}$$

16. 4.6 km to meters

$$\frac{4.6 \text{ km}}{1} \cdot \frac{1000 \text{ m}}{1 \text{ km}} = 4600 \text{ m}$$

17. 5 mm to centimeters
Count 1 place to the *left* on the metric conversion line.

$$_\wedge 5. \text{ mm} = 0.5 \text{ cm}$$

18. 325 mg to grams

$$\frac{325 \text{ mg}}{1} \cdot \frac{1 \text{ g}}{1000 \text{ mg}} = \frac{325}{1000} \text{ g} = 0.325 \text{ g}$$

19. 16 L to milliliters
Count 3 places to the *right* on the metric conversion line.

$$16.000_\wedge \text{ L} = 16,000 \text{ mL}$$

20. 0.4 kg to grams

$$\frac{0.4 \text{ kg}}{1} \cdot \frac{1000 \text{ g}}{1 \text{ kg}} = 400 \text{ g}$$

21. 10.55 m to centimeters
Count 2 places to the *right* on the metric conversion line.

$$10.55_\wedge \text{ m} = 1055 \text{ cm}$$

22. 95 mL to liters
Count 3 places to the *left* on the metric conversion line.

$$_\wedge 095. \text{ mL} = 0.095 \text{ L}$$

23. Convert 467 in. to feet.

$$\frac{467 \text{ in.}}{1} \cdot \frac{1 \text{ ft}}{12 \text{ in.}} = \frac{467}{12} \text{ ft}$$

Divide by 12 months

$$\frac{\frac{467}{12} \text{ ft}}{12 \text{ months}} = \frac{467}{12} \cdot \frac{1}{12} \approx 3.2 \text{ ft/month}$$

It rains about 3.2 ft per month.

24. (a) Convert 1.05 g to mg.

$$\frac{1.05 \text{ g}}{1} \cdot \frac{1000 \text{ mg}}{1 \text{ g}} = 1050 \text{ mg}$$

Find the difference.
1050 mg − 280 mg = 770 mg
It has 770 mg more sodium.

Copyright © 2018 Pearson Education, Inc.

(b) 1050 mg is less than 1500 mg

1500 mg − 1050 mg = 450 mg

$$\frac{450 \text{ mg}}{1} \cdot \frac{1 \text{ g}}{1000 \text{ mg}} = 0.45 \text{ g}$$

The Chicken and Bacon Ranch Melt has 450 mg or 0.45 g less sodium than the 1500 mg recommended daily amount.

25. The water is almost boiling. On the Celsius scale, water boils at 100°, so 95 °C would be almost boiling.

26. The tomato plants may freeze tonight. Water freezes at 0 °C, so tomatoes would also likely freeze at 0 °C, so choose 0 °C.

27. 6 ft to meters

$$\frac{6 \text{ ft}}{1} \cdot \frac{0.30 \text{ m}}{1 \text{ ft}} \approx 1.8 \text{ m}$$

28. 125 lb to kilograms

$$\frac{125 \text{ lb}}{1} \cdot \frac{0.45 \text{ kg}}{1 \text{ lb}} \approx 56.3 \text{ kg (rounded)}$$

29. 50 L to gallons

$$\frac{50 \text{ L}}{1} \cdot \frac{0.26 \text{ gal}}{1 \text{ L}} \approx 13 \text{ gal}$$

30. 8.1 km to miles

$$\frac{8.1 \text{ km}}{1} \cdot \frac{0.62 \text{ mi}}{1 \text{ km}} \approx 5.0 \text{ mi (rounded)}$$

31. 74 °F to Celsius

$$C = \frac{5(74 - 32)}{9} = \frac{5(\overset{14}{\cancel{42}})}{\underset{3}{\cancel{9}}} = \frac{70}{3} \approx 23 \text{ °C}$$

74 °F ≈ 23 °C (rounded)

32. 2 °C to Fahrenheit

$$F = \frac{9 \cdot 2}{5} + 32 = \frac{18}{5} + 32 = 3.6 + 32 \approx 36°$$

2 °C ≈ 36 °F (rounded)

33. Convert 120 cm to meters.

Count 2 places to the *left* on the metric conversion line.

1∧20. cm = 1.2 m

Multiply to find the number of meters for 5 pillows.

5(1.2 m) = 6 m

Convert 6 m to yards.

$$\frac{6 \text{ m}}{1} \cdot \frac{1.09 \text{ yd}}{1 \text{ m}} \approx 6.54 \text{ yd}$$

Multiply by $3.98 to find the cost.

$$\frac{6.54 \text{ yd}}{1} \cdot \frac{\$3.98}{\text{yd}} \approx \$26.03$$

It will cost about $26.03.

34. Possible answers: Use same system as rest of the world; easier system for children to learn; less use of fractional numbers; compete internationally.

Cumulative Review Exercises (Chapters 1–7)

1.   *Estimate:*   *Exact:*

$$\begin{array}{r} 56.52 \\ \times\ 4.7 \\ \hline 39\ 564 \\ 226\ 08 \\ \hline 265.644 \end{array}$$ ← 2 decimal places
  ← 1 decimal place
  ← 3 decimal places

$$\begin{array}{r} 60 \\ \times 5 \\ \hline 300 \end{array}$$

2.   *Estimate:*   *Exact:*

$$40\overline{)20{,}000} \qquad 530$$

$$37\overline{)19{,}610}$$
$$18\ 5$$
$$1\ 11$$
$$1\ 11$$
$$0\ 0$$
$$0$$
$$0$$

Estimate: $40\overline{)20{,}000} = 500$

3.   *Estimate:*   *Exact:*

$$\begin{array}{r} 2 \\ +4 \\ \hline 6 \end{array} \qquad \begin{array}{r} 1\frac{7}{10} = 1\frac{7}{10} \\ +3\frac{4}{5} = 3\frac{8}{10} \\ \hline 4\frac{15}{10} \end{array}$$

$$4\frac{15}{10} = 4 + \frac{10}{10} + \frac{5}{10} = 5\frac{1}{2}$$

4. $3 - 2\frac{5}{16} = 2\frac{16}{16} - 2\frac{5}{16} = \frac{11}{16}$

5. $12 \cdot 2\frac{2}{9} = \frac{\overset{4}{\cancel{12}}}{1} \cdot \frac{20}{\underset{3}{\cancel{9}}} = \frac{80}{3} = 26\frac{2}{3}$

Copyright © 2018 Pearson Education, Inc.

6.

$$
0.066 \wedge \overline{\smash{)}\,0.8\,6\,0_\wedge\,0\,0} \quad \begin{array}{r} 1\,3.\,0\,3 \end{array}
$$

$$
\begin{array}{r}
6\,6 \\ \hline
2\,0\,0 \\
1\,9\,8 \\ \hline
2\ \ 0 \\
0 \\ \hline
2\ \ 0\,0 \\
1\ \ 9\,8 \\ \hline
2
\end{array}
$$

$0.86 \div 0.066 \approx 13.0$ (rounded)

7. $\dfrac{3}{8} + \dfrac{5}{6} = \dfrac{9}{24} + \dfrac{20}{24} = \dfrac{29}{24} = 1\dfrac{5}{24}$

8. $8 - 0.9207$

$$
\begin{array}{r}
8.0000 \\
-\ 0.9207 \\ \hline
7.0793
\end{array}
$$

9. $3\dfrac{3}{4} \div 6 = \dfrac{15}{4} \div \dfrac{6}{1} = \dfrac{\overset{5}{\cancel{15}}}{4} \cdot \dfrac{1}{\underset{2}{\cancel{6}}} = \dfrac{5}{8}$

10. $(2.54)(0.003)$

$$
\begin{array}{rcl}
2.54 & \leftarrow & 2 \; \textit{decimal places} \\
\times \;\; 0.003 & \leftarrow & 3 \; \textit{decimal places} \\ \hline
0.00762 & \leftarrow & 5 \; \textit{decimal places needed}
\end{array}
$$

11.
$$
\begin{array}{ll}
24 - 12 \div 6(8) + (25 - 25) & \textit{Parentheses} \\
= 24 - 12 \div 6(8) + 0 & \textit{Divide.} \\
= 24 - 2(8) + 0 & \textit{Multiply.} \\
= 24 - 16 + 0 & \textit{Subtract.} \\
= 8 + 0 = 8 & \textit{Add.}
\end{array}
$$

12.
$$
\begin{array}{ll}
3^2 + 2^5 \cdot \sqrt{64} & \textit{Exponents} \\
= 9 + 32 \cdot \sqrt{64} & \textit{Square root} \\
= 9 + 32 \cdot 8 & \textit{Multiply.} \\
= 9 + 256 = 265 & \textit{Add.}
\end{array}
$$

13.
$$
\begin{array}{l}
0.67 = 0.6700 \\
0.067 = 0.0670 \\
0.6 = 0.6000 \\
0.6007 = 0.6007
\end{array}
$$

From least to greatest:
0.067, 0.6, 0.6007, 0.67

14.

Size	Cost per Unit
package of 27	$\dfrac{\$11.71}{27} \approx \$0.434$
package of 31	$\dfrac{\$19.98}{31} \approx \$0.645$

| package of 56 | $\dfrac{\$29.99}{56} \approx \$0.536$ |

The best buy is 27 diapers for $11.71.

15.
$$
\begin{array}{c}
\dfrac{x}{16} = \dfrac{3}{4} \\
4 \cdot x = 3 \cdot 16 \\
\dfrac{4 \cdot x}{4} = \dfrac{48}{4} \\
x = 12
\end{array}
$$

16.
$$
\begin{array}{c}
\dfrac{0.9}{0.75} = \dfrac{2}{x} \\
0.9 \cdot x = 2 \cdot 0.75 \\
0.9 \cdot x = 1.5 \\
\dfrac{0.9 \cdot x}{0.9} = \dfrac{1.5}{0.9} \\
x \approx 1.67 \text{ (rounded)}
\end{array}
$$

17. Use the percent proportion.
part is 4; whole is 80; percent is unknown

$$
\begin{array}{c}
\dfrac{4}{80} = \dfrac{x}{100} \\
x \cdot 80 = 4 \cdot 100 \\
\dfrac{x \cdot 80}{80} = \dfrac{400}{80} \\
x = 5
\end{array}
$$

$4 is 5% of $80.

18. Use the percent proportion.
part is 36; percent is 120; whole is unknown

$$
\begin{array}{c}
\dfrac{36}{x} = \dfrac{120}{100} \\
120 \cdot x = 36 \cdot 100 \\
120 \cdot x = 3600 \\
\dfrac{120 \cdot x}{120} = \dfrac{3600}{120} \\
x = 30
\end{array}
$$

36 hours is 120% of 30 hours.

19. $2\dfrac{1}{2}$ ft to inches

$$
\dfrac{2\frac{1}{2}\,\cancel{\text{ft}}}{1} \cdot \dfrac{12 \text{ in.}}{1\,\cancel{\text{ft}}} = \dfrac{5}{\underset{1}{\cancel{2}}} \cdot \dfrac{\overset{6}{\cancel{12}}}{1} \text{ in.} = 30 \text{ in.}
$$

20. 105 sec to minutes

$$
\dfrac{\overset{21}{\cancel{105}} \text{ sec}}{1} \cdot \dfrac{1 \text{ min}}{\underset{12}{\cancel{60}} \text{ sec}} = \dfrac{21}{12} \text{ min}
$$

$$
= 1\dfrac{3}{4} \text{ min or } 1.75 \text{ min}
$$

Copyright © 2018 Pearson Education, Inc.

21. 2.8 m to centimeters
Count 2 places to the *right* on the metric
conversion line.
$$2.80_\wedge \text{ m} = 280 \text{ cm}$$

22. 198 km to miles
$$\frac{198 \text{ km}}{1} \cdot \frac{0.62 \text{ mi}}{1 \text{ km}} \approx 122.76 \text{ miles}$$

23. Ron bought the tube of toothpaste weighing
100 g.

24. Tia added 125 mL of milk to her cereal.

25. The hallway is 3 m wide.

26. Joe's hammer weighed 1 kg.

27. *amount of discount = rate of discount · cost of item*
$$= (0.10)(\$189.94)$$
$$\approx \$18.99$$

The amount of discount is $18.99 (rounded).
The sale price is $189.94 − $18.99 = $170.95.

28. The first 20 prints were free, the other 25 prints
cost $0.12 each. 25($0.12) = $3.00

The amount of tax is 0.065($3) ≈ $0.20.
The total cost including shipping is
$$\$3.00 + \$0.20 + \$2.95 = \$6.15.$$

The cost per print for the 45 total prints is then
$$\frac{\$6.15}{45} \approx \$0.14 \text{ (rounded)}.$$

29. Find the total ounces, then convert oz to pounds.
12 oz + 48 · 4 oz = 12 oz + 192 oz = 204 oz

$$\frac{\overset{51}{\cancel{204} \text{ oz}}}{1} \cdot \frac{1 \text{ lb}}{\underset{4}{\cancel{16} \text{ oz}}} = 12\frac{3}{4} \text{ lb or } 12.75 \text{ lb}$$

The carton would weigh $12\frac{3}{4}$ lb or 12.75 lb.

30. **(a)** Set up a proportion.
$$\frac{1 \text{ cm}}{12 \text{ km}} = \frac{7.8 \text{ cm}}{x}$$
$$x \cdot 1 = (7.8)(12)$$
$$x = 93.6$$

The distance is 93.6 kilometers.

(b) $\dfrac{93.6 \text{ km}}{1} \cdot \dfrac{0.62 \text{ mi}}{1 \text{ km}} \approx 58.0 \text{ mi}$

The distance is 58.0 mi (rounded).

31. $18,750 at 4.3% for $3\frac{1}{2}$ years
$$I = p \cdot r \cdot t$$
$$= (\$18,750)(0.043)(3.5)$$
$$= \$2821.88 \text{ (rounded)}$$

The interest is $2821.88 and the total amount due
is $18,750 + $2821.88 = $21,571.88.

32. Use the percent proportion.
part is 31; whole is 35; percent is unknown
$$\frac{31}{35} = \frac{x}{100}$$
$$35 \cdot x = 31 \cdot 100$$
$$\frac{35 \cdot x}{35} = \frac{3100}{35}$$
$$x \approx 88.57$$

The percent of the problems that were correct is
88.6% (rounded).

33. Convert 650 g to kg.
$$_\wedge 650. \text{ g} = 0.650 \text{ kg}$$

Multiply to find the cost.
$(0.650)(\$14.98) \approx \$9.74$

Mark paid $9.74 (rounded).

34. Add the amounts needed for the recipes.
$$2\frac{1}{4} = 2\frac{1}{4}$$
$$1\frac{1}{2} = 1\frac{2}{4}$$
$$+ \frac{3}{4} = \frac{3}{4}$$
$$\overline{} \quad \overline{}$$
$$3\frac{6}{4} = 4\frac{2}{4} = 4\frac{1}{2} \text{ cups}$$

Find the total amount of brown sugar.
$$2\frac{1}{3}$$
$$+ 2\frac{1}{3}$$
$$\overline{}$$
$$4\frac{2}{3} \text{ cups}$$

Subtract the amount to be used from the total
amount.
$$4\frac{2}{3} = 4\frac{4}{6}$$
$$- 4\frac{1}{2} = 4\frac{3}{6}$$
$$\overline{}$$
$$\frac{1}{6}$$

The Jackson family will have $\frac{1}{6}$ cup more than
the amount needed.

Copyright © 2018 Pearson Education, Inc.

35. Set up a proportion.

$$\frac{6 \text{ rows}}{5 \text{ cm}} = \frac{x}{100 \text{ cm}}$$
$$5 \cdot x = 6 \cdot 100$$
$$\frac{5 \cdot x}{5} = \frac{600}{5}$$
$$x = 120$$

Akuba will have to knit 120 rows to make a 100 cm scarf.

36. $\frac{3}{8}$ of 5600 students

$$\frac{3}{8} \cdot 5600 = \frac{3}{\overset{}{\underset{1}{8}}} \cdot \frac{\overset{700}{\cancel{5600}}}{1} = 2100$$

The survey showed that 2100 students work 20 hours or more per week.

37. $\dfrac{\$9.5}{50 \text{ applications}} = \$0.19/\text{application}$

The cost per application of a spray-on bandage is $0.19.

38. decrease $= 8 - 5 = 3$ days
$$\frac{3 \text{ days}}{8 \text{ days}} = \frac{x}{100}$$
$$8 \cdot x = 3 \cdot 100$$
$$\frac{8 \cdot x}{8} = \frac{300}{8}$$
$$x = 37.5$$

The length of a hospital stay has decreased by 37.5%.

Copyright © 2018 Pearson Education, Inc.

CHAPTER 8   GEOMETRY

8.1 Basic Geometric Terms

8.1   Margin Exercises

1. **(a)** The figure has two endpoints, so it is a *line segment* named $\overline{EF}$ or $\overline{FE}$.

    (b) The figure starts at point *S* and goes on forever in one direction, so it is a *ray* named $\overrightarrow{SR}$.

    (c) The figure goes on forever in both directions, so it is a *line* named $\overleftrightarrow{WX}$ or $\overleftrightarrow{XW}$.

2. **(a)** The lines cross at *E*, so they are <u>intersecting</u> lines.

    (b) The lines never intersect (cross), so they appear to be <u>parallel</u> lines.

    (c) The lines never intersect (cross), so they appear to be *parallel lines*.

3. **(a)** The angle is named $\angle 3$, $\angle CQD$, or $\angle DQC$.

    (b) Darken rays $\overrightarrow{TW}$ and $\overrightarrow{TZ}$.

    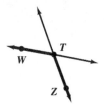

    (c) The angle can be named $\angle 1$, $\angle R$, $\angle MRN$, and $\angle NRM$.

4. **(a)** The angle measures exactly 90° (indicated by the small square at the vertex), so it is a <u>right</u> angle.

    (b) The angle measures exactly 180°, so it is a <u>straight</u> angle.

    (c) The angle measures between 90° and 180°, so it is an <u>obtuse</u> angle.

    (d) The angle measures between 0° and 90°, so it is an <u>acute</u> angle.

5. The lines in (b) show perpendicular lines, because they intersect at right angles. The lines in (a) show intersecting lines. They cross, but not at right angles.

8.1   Section Exercises

1. Answer wording may vary slightly: a **line** is a row of points continuing in both directions forever; the drawing should have arrows on both ends. A **line segment** is a piece of a line; the drawing should have an endpoint (dot) on each end. A **ray** has one endpoint and goes on forever in one direction; the drawing should have one endpoint (dot) and an arrowhead on the other end.

3. This is a *line* named $\overleftrightarrow{CD}$ or $\overleftrightarrow{DC}$. A line is a straight row of points that goes on forever in both directions.

5. The figure has two endpoints so it is a *line segment* named $\overline{GF}$ or $\overline{FG}$.

7. This is a *ray* named $\overrightarrow{PQ}$. A ray is a part of a line that has only one endpoint and goes on forever in one direction.

9. The lines are *perpendicular* because they intersect at right angles; the small red square indicates a right angle.

11. These lines appear to be *parallel* lines. Parallel lines are lines in the same plane that never intersect (cross).

12. These lines appear to be *parallel* lines. Parallel lines are lines in the same plane that never intersect (cross).

13. The lines intersect so they are *not* parallel. At their intersection they *do not* form a right angle so they are not perpendicular. The lines are *intersecting*.

15. The middle letter, *P*, identifies the vertex. The angle can be named $\angle APS$ or $\angle SPA$.

17. The middle letter, *Q*, identifies the vertex. The angle can be named $\angle AQC$ or $\angle CQA$.

19. The angle is a *right angle*, as indicated by the small square at the vertex. Right angles measure exactly 90°.

21. The measure of the angle is between 0° and 90°, so it is an *acute angle*.

23. Two rays in a straight line pointing opposite directions measure 180°. An angle that measures 180° is called a *straight angle*.

Copyright © 2018 Pearson Education, Inc.

25. **(a)** If your car "did a 360," the car turned around in a complete circle.

(b) If the governor's view on taxes "took a 180° turn," he or she took the opposite view. For example, he or she may have opposed taxes but now supports them.

27. "∠*UST* is 90°" is *true* because $\overleftrightarrow{UQ}$ is perpendicular to $\overleftrightarrow{ST}$.

28. "$\overleftrightarrow{SQ}$ and $\overleftrightarrow{PQ}$ are perpendicular" is *true* because they form a 90° angle, as indicated by the small red square.

29. "The measure of ∠*USQ* is less than the measure of ∠*PQR*" is *false*. ∠*USQ* is a straight angle and so is ∠*PQR*, therefore each measures 180°. A true statement would be: ∠*USQ* has the same measure as ∠*PQR*.

30. "$\overleftrightarrow{ST}$ and $\overleftrightarrow{PR}$ are intersecting" is *false*. $\overleftrightarrow{ST}$ and $\overleftrightarrow{PR}$ are parallel and will never intersect.

31. "$\overleftrightarrow{QU}$ and $\overleftrightarrow{TS}$ are parallel" is *false*. $\overleftrightarrow{QU}$ and $\overleftrightarrow{TS}$ are perpendicular because they intersect at right angles.

32. "∠*UST* and ∠*UQR* measure the same number of degrees" is *true* because both angles are formed by perpendicular lines, so they both measure 90°.

8.2 Angles and Their Relationships

8.2 Margin Exercises

1. ∠*COD* and ∠*DOE* are complementary angles because 70° + 20° = 90°.
∠*RST* and ∠*XPY* are complementary angles because 45° + 45° = 90°.

2. **(a)** The complement of 35° is 55°, because 90° − 35° = 55°.

(b) The complement of 80° is 10°, because 90° − 80° = 10°.

3. ∠*CRF* and ∠*BRF* are supplementary angles because 127° + 53° = 180°.
∠*CRE* and ∠*BRE* are supplementary angles because 53° + 127° = 180°.
∠*FRB* and ∠*ERB* are supplementary angles because 53° + 127° = 180°.
∠*ERC* and ∠*FRC* are supplementary angles because 53° + 127° = 180°.

4. **(a)** The supplement of 175° is 5°, because 180° − 175° = 5°.

(b) The supplement of 30° is 150°, because 180° − 30° = 150°.

5. ∠*BOC* ≅ ∠*AOD* because they each measure 150°.
∠*AOB* ≅ ∠*DOC* because they each measure 30°.

6. ∠*SPB* and ∠*MPD* are vertical angles because they do *not* share a common side and the angles are formed by intersecting lines.
Similarly, ∠*BPD* and ∠*SPM* are vertical angles.

7. **(a)** ∠*TOS*
∠*TOS* is a right angle, so its measure is 90°.

(b) ∠*QOR*
∠*QOR* and ∠*TOS* are vertical angles, so they are also congruent and have the same measure. The measure of ∠*TOS* is 90°, so the measure of ∠*QOR* is 90°.

(c) ∠*POQ*
∠*POQ* and ∠*TOV* are vertical angles, so they are congruent and have the same measure. The measure of ∠*TOV* is 52°, so the measure of ∠*POQ* is 52°.

(d) ∠*VOR*
∠*VOR* and ∠*TOV* are complementary angles, so the sum of their angle measures is 90°. Since the degree measure of ∠*TOV* is 52°, the measure of ∠*VOR* equals 90° − 52° = 38°.

(e) ∠*POS*
∠*POS* and ∠*VOR* are vertical angles, so they are congruent. From part (d), the measure of ∠*VOR* is 38°, so the measure of ∠*POS* is 38°.

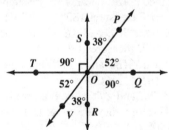

8.2 Section Exercises

1. Yes, because you can subtract 180° − 5° to get 175°. The sum of the measures of a 5° angle and a 175° angle is 180°, so they are supplements of each other.

3. The pairs of complementary angles are:
∠*EOD* and ∠*COD* because 75° + 15° = 90°;
∠*AOB* and ∠*BOC* because 25° + 65° = 90°.

Copyright © 2018 Pearson Education, Inc.

5. The pairs of supplementary angles are:
$\angle HNE$ and $\angle ENF$ because $77° + 103° = 180°$;
$\angle ACB$ and $\angle KOL$ because $120° + 60° = 180°$.

7. The complement of $40°$ is $50°$, because
$90° - 40° = \underline{50°}$.

9. Find the complement of $86°$ by subtracting:
$90° - 86° = 4°$, so the complement of $86°$ is $4°$.

11. The supplement of $130°$ is $50°$, because
$180° - 130° = \underline{50°}$.

13. Find the supplement of $90°$ by subtracting:
$180° - 90° = 90°$, so the supplement of $90°$ is
$90°$.

15. No; obtuse angles have measures greater than
$(>) 90°$, so the sum of their measures would be
greater than $(>) 180°$.

17. $\angle SON \cong \angle TOM$ because they are vertical angles.
$\angle TOS \cong \angle MON$ because they are vertical angles.

19. Because $\angle COE$ and $\angle GOH$ are vertical angles,
they are also congruent. This means they have
the same measure. $\angle COE$ measures $63°$, so
$\angle GOH$ measures $63°$.

The sum of the measures of $\angle COE$, $\angle AOC$, and
$\angle AOH$ equals $180°$ because $\angle EOH$ is a straight
angle. Therefore, $\angle AOC$ measures
$180° - (63° + 37°) = 180° - 100° = 80°$.
$\angle AOC$ and $\angle GOF$ are vertical angles, so they are
congruent and both measure $80°$. $\angle AOH$ and
$\angle EOF$ are vertical angles, so they are congruent
and both measure $37°$.

21. $\angle ABC$ and $\angle BCD$ are congruent because they
both measure $42°$, so $\angle ABC \cong \angle BCD$.
Since $\angle ABC$ and $\angle ABF$ are supplementary,
$\angle ABF$ measures $180° - 42° = 138°$.
Since $\angle BCD$ and $\angle ECD$ are supplementary,
$\angle ECD$ measures $180° - 42° = 138°$.
Since $\angle ABF$ and $\angle ECD$ each measure $138°$,
$\angle ABF \cong \angle ECD$.
$\angle FBC$ and $\angle BCE$ are both straight angles, and
therefore, they are congruent. They both measure
$180°$.

8.3 Rectangles and Squares

8.3 Margin Exercises

1. Figures (a), (b), and (e) are figures with four
sides that meet to form $90°$ angles, so they are
rectangles. Figures (c) and (d) do not have four
sides. Figures (f) and (g) have four sides, but
they do not meet to form right angles.

2. **(a)** The length of the rectangle is 17 cm and the
width is 10 cm.

$P = 17 \text{ cm} + 17 \text{ cm} + \underline{10 \text{ cm}} + \underline{10 \text{ cm}}$
$P = \underline{54 \text{ cm}}$

(b) The length is 10.5 ft and the width is 7 ft.

$P = 2 \cdot \text{length} + 2 \cdot \text{width}$
$P = 2 \cdot \underline{10.5 \text{ ft}} + 2 \cdot \underline{7 \text{ ft}}$
$P = \underline{21 \text{ ft}} + \underline{14 \text{ ft}}$
$P = \underline{35 \text{ ft}}$

(c) 6 m wide and 11 m long

$P = 2 \cdot \text{length} + 2 \cdot \text{width}$
$\quad = 2 \cdot 11 \text{ m} + 2 \cdot 6 \text{ m}$
$\quad = 22 \text{ m} + 12 \text{ m}$
$\quad = 34 \text{ m}$

(d) 0.9 km by 2.8 km

$P = 2 \cdot \text{length} + 2 \cdot \text{width}$
$\quad = 2 \cdot 2.8 \text{ km} + 2 \cdot 0.9 \text{ km}$
$\quad = 5.6 \text{ km} + 1.8 \text{ km}$
$\quad = 7.4 \text{ km}$

3. **(a)** The length is 9 ft and the width is 4 ft.

$A = \text{length} \cdot \text{width}$
$\quad = 9 \text{ ft} \cdot 4 \text{ ft}$
$\quad = \underline{36 \text{ ft}^2}$

(b) A rectangle is 6 m long and 0.5 m wide.

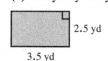
6 m
0.5 m

$A = \text{length} \cdot \text{width}$
$\quad = 6 \text{ m} \cdot 0.5 \text{ m}$
$\quad = 3 \text{ m}^2$

(c) 3.5 yd by 2.5 yd

2.5 yd
3.5 yd

$A = \text{length} \cdot \text{width}$
$\quad = 3.5 \text{ yd} \cdot 2.5 \text{ yd}$
$\quad = 8.75 \text{ yd}^2$

4. **(a)** Each side measures 3 ft.
For the perimeter of a square, use $P = 4s$.

$P = 4 \cdot s$
$\quad = 4 \cdot \underline{3 \text{ ft}}$
$\quad = \underline{12 \text{ ft}}$

(continued)

Copyright © 2018 Pearson Education, Inc.

For the area of a square, use $A = s^2$ or $A = s \cdot s$.

$$A = s \cdot s$$
$$= \underline{3 \text{ ft}} \cdot \underline{3 \text{ ft}}$$
$$= \underline{9 \text{ ft}^2}$$

(b) 10.5 cm on each side

$$P = 4 \cdot s \qquad\qquad A = s \cdot s$$
$$= 4 \cdot 10.5 \text{ cm} \qquad = 10.5 \text{ cm} \cdot 10.5 \text{ cm}$$
$$= 42 \text{ cm} \qquad\qquad = 110.25 \text{ cm}^2$$

(c) 2.1 miles on a side

$$P = 4 \cdot s \qquad\qquad A = s \cdot s$$
$$= 4 \cdot 2.1 \text{ mi} \qquad = 2.1 \text{ mi} \cdot 2.1 \text{ mi}$$
$$= 8.4 \text{ mi} \qquad\qquad = 4.41 \text{ mi}^2$$

5. The cost of the carpet is $19.95 per square yard.

(a) The length is 6.5 yd and the width is 5 yd.

$$A = \text{length} \cdot \text{width}$$
$$= 6.5 \text{ yd} \cdot 5 \text{ yd}$$
$$= \underline{32.5 \text{ yd}^2}$$

The cost of the carpet is

$$\frac{32.5 \text{ yd}^2}{1} \cdot \frac{\$19.95}{1 \text{ yd}^2} = \underline{\$648.375} \approx \$648.38.$$

(b) First change the measurements from feet to yards.

$$\frac{21 \text{ ft}}{3} = 7 \text{ yd}; \qquad \frac{9 \text{ ft}}{3} = 3 \text{ yd}; \qquad \frac{12 \text{ ft}}{3} = 4 \text{ yd}$$

Next, break up the room into two pieces.

Area of rectangle $= \text{length} \cdot \text{width}$
$$= 7 \text{ yd} \cdot 4 \text{ yd}$$
$$= 28 \text{ yd}^2$$

Area of square $= s \cdot s$
$$= 3 \text{ yd} \cdot 3 \text{ yd}$$
$$= 9 \text{ yd}^2$$

Total area $= 28 \text{ yd}^2 + 9 \text{ yd}^2$
$$= 37 \text{ yd}^2$$

The cost of the carpet is

$$\frac{37 \text{ yd}^2}{1} \cdot \frac{\$19.95}{1 \text{ yd}^2} = \$738.15.$$

(c) A rectangular classroom that is 24 ft long and 18 feet wide.

Change feet to yards.

$$\frac{24 \text{ ft}}{3} = 8 \text{ yd}; \qquad \frac{18 \text{ ft}}{3} = 6 \text{ yd}$$

$$A = \text{length} \cdot \text{width}$$
$$= 8 \text{ yd} \cdot 6 \text{ yd}$$
$$= 48 \text{ yd}^2$$

The cost of the carpet is

$$\frac{48 \text{ yd}^2}{1} \cdot \frac{\$19.95}{1 \text{ yd}^2} = \$957.60.$$

8.3 Section Exercises

1. Perimeter is the distance around the outside edges; add the lengths of all the sides to find the perimeter. The area is the surface inside the figure; multiply length times width to find the area. The units for the perimeter will be **yd**; the units for the area will be $\mathbf{yd^2}$.

3. $P = 2 \cdot \text{length} + 2 \cdot \text{width}$ (or use $2 \cdot l + 2 \cdot w$)
$$= 2 \cdot 8 \text{ yd} + 2 \cdot 6 \text{ yd}$$
$$= 16 \text{ yd} + 12 \text{ yd}$$
$$= 28 \text{ yd}$$

$$A = \text{length} \cdot \text{width}$$
$$= 8 \text{ yd} \cdot 6 \text{ yd}$$
$$= 48 \text{ yd}^2$$

5. $P = 4 \cdot s \qquad\qquad A = s \cdot s$
$P = 4 \cdot 0.9 \text{ km} \qquad A = 0.9 \text{ km} \cdot 0.9 \text{ km}$
$P = 3.6 \text{ km} \qquad\qquad A = 0.81 \text{ km}^2$

7. 10 ft by 10 ft

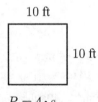

$$P = 4 \cdot s \qquad\qquad A = s \cdot s$$
$$= 4 \cdot 10 \text{ ft} \qquad = 10 \text{ ft} \cdot 10 \text{ ft}$$
$$= 40 \text{ ft} \qquad\qquad = 100 \text{ ft}^2$$

Copyright © 2018 Pearson Education, Inc.

9.   A storage building that is 76.1 ft by 22 ft

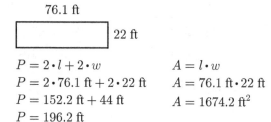

76.1 ft

22 ft

$P = 2 \cdot l + 2 \cdot w$ $A = l \cdot w$
$P = 2 \cdot 76.1 \text{ ft} + 2 \cdot 22 \text{ ft}$ $A = 76.1 \text{ ft} \cdot 22 \text{ ft}$
$P = 152.2 \text{ ft} + 44 \text{ ft}$ $A = 1674.2 \text{ ft}^2$
$P = 196.2 \text{ ft}$

11.   A square nature preserve 3 mi wide

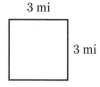

3 mi

3 mi

$P = 4 \cdot s$ $A = s \cdot s$
$\quad = 4 \cdot 3 \text{ mi}$ $\quad = 3 \text{ mi} \cdot 3 \text{ mi}$
$\quad = 12 \text{ mi}$ $\quad = 9 \text{ mi}^2$

13. $P = 12 \text{ m} + 7 \text{ m} + 3 \text{ m} + 5 \text{ m} + 9 \text{ m} + 2 \text{ m}$
$\quad = 38 \text{ m}$

Draw a horizontal line and break up the figure into two rectangles.

rectangle with   rectangle with
length: 7 m   length: 9 m
width: 3 m   width: 2 m
$A = 7 \text{ m} \cdot 3 \text{ m}$ $A = 9 \text{ m} \cdot 2 \text{ m}$
$\quad = 21 \text{ m}^2$ $\quad = 18 \text{ m}^2$

Total area $= 21 \text{ m}^2 + 18 \text{ m}^2 = 39 \text{ m}^2$

15. $P = 28 \text{ m} + 17 \text{ m} + 12 \text{ m} + 4 \text{ m} + 4 \text{ m} +$
$\quad 4 \text{ m} + 12 \text{ m} + 17 \text{ m}$
$\quad = 98 \text{ m}$

Draw a vertical line and break up the figure into a rectangle and a square.

rectangle with   square with
length: 28 m   side: 4 m
width: 17 m
$A = 28 \text{ m} \cdot 17 \text{ m}$ $A = 4 \text{ m} \cdot 4 \text{ m}$
$\quad = 476 \text{ m}^2$ $\quad = 16 \text{ m}^2$

Total area $= 476 \text{ m}^2 + 16 \text{ m}^2 = 492 \text{ m}^2$

17.   The unlabeled side is 6 in. + 6 in. or 12 in. Add the lengths of the sides to find the perimeter.

$P = 15 \text{ in.} + 6 \text{ in.} + 6 \text{ in.} + 6 \text{ in.} + 6 \text{ in.}$
$\quad + 9 \text{ in.} + 12 \text{ in.} + 6 \text{ in.} + 6 \text{ in.}$
$P = 78 \text{ in.}$

Draw two horizontal lines to break up the figure into two smaller squares and one larger rectangle.

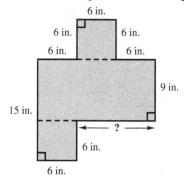

6 in.
6 in.   6 in.
6 in.   6 in.
                       9 in.
15 in.
           ?
                6 in.
           6 in.

Add the areas of these squares and rectangle to find the total area.

$A = (6 \text{ in.} \cdot 6 \text{ in.}) + (18 \text{ in.} \cdot 9 \text{ in.}) + (6 \text{ in.} \cdot 6 \text{ in.})$
$A = 36 \text{ in.}^2 + 162 \text{ in.}^2 + 36 \text{ in.}^2$
$A = 234 \text{ in.}^2$

You could also draw two vertical lines to break up the figure into three rectangles.

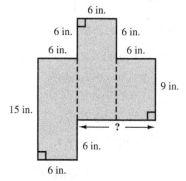

6 in.
6 in.   6 in.
6 in.   6 in.
                       9 in.
15 in.
           ?
                6 in.
           6 in.

$A = (15 \text{ in.} \cdot 6 \text{ in.}) + (15 \text{ in.} \cdot 9 \text{ in.}) + (9 \text{ in.} \cdot 6 \text{ in.})$
$A = 90 \text{ in.}^2 + 90 \text{ in.}^2 + 54 \text{ in.}^2$
$A = 234 \text{ in.}^2 \quad \leftarrow \text{Same result}$

19. $P = 4 \cdot s$ $A = s \cdot s$
$\quad = 4 \cdot 12 \text{ m}$ $\quad = 12 \text{ m} \cdot 12 \text{ m}$
$\quad = 48 \text{ m}$ $\quad = 144 \text{ m}^2$

21.   Find the area of the laptop's screen.

$A = \text{length} \cdot \text{width}$
$\quad = 12.9 \text{ in.} \cdot 9.0 \text{ in.}$
$\quad = 116.1 \text{ in.}^2 \approx 116 \text{ in.}^2 \text{ (rounded)}$

Find the area of the smartphone's screen.

$A = \text{length} \cdot \text{width}$
$\quad = 5.6 \text{ in.} \cdot 2.8 \text{ in.}$
$\quad = 15.68 \text{ in.}^2 \approx 16 \text{ in.}^2 \text{ (rounded)}$

The difference in the areas is

$116 \text{ in.}^2 - 16 \text{ in.}^2 = 100 \text{ in.}^2.$

Copyright © 2018 Pearson Education, Inc.

23. Find the perimeter of the room.

$$P = 2 \cdot 4.4 \text{ m} + 2 \cdot 5.1 \text{ m}$$
$$P = 8.8 \text{ m} + 10.2 \text{ m} = 19 \text{ m}$$

She will need 19 m of the strip. Multiply by the cost per meter to find the total cost.

$$\text{Cost} = \frac{19 \text{ m}}{1} \cdot \frac{\$4.99}{1 \text{ m}} = \$94.81$$

Tyra will have to spend $94.81 for the strip.

25. The *largest* field would be 80 yards wide by 120 yards long.

$$A = 80 \text{ yd} \cdot 120 \text{ yd}$$
$$= 9600 \text{ yd}^2$$

The *smallest* field would be 65 yards wide by 110 yards long.

$$A = 65 \text{ yd} \cdot 110 \text{ yd}$$
$$= 7150 \text{ yd}^2$$

The *difference* in the playing room between the smallest and the largest fields is

$$9600 \text{ yd}^2 - 7150 \text{ yd}^2 = 2450 \text{ yd}^2.$$

27.
$$A = \text{length} \cdot \text{width}$$
$$5300 \text{ yd}^2 = 100 \text{ yd} \cdot \text{width}$$
$$\frac{5300 \text{ yd} \cdot \text{yd}}{100 \text{ yd}} = \frac{\overset{1}{100} \text{ yd} \cdot \text{width}}{\underset{1}{100} \text{ yd}}$$
$$53 \text{ yd} = \text{width}$$

The width is 53 yd.

29. Length of inner rectangle = 172 ft − 12 ft − 12 ft
$$= 148 \text{ ft}$$
Width of inner rectangle = 124 ft − 12 ft − 12 ft
$$= 100 \text{ ft}$$

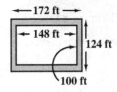

Shaded area = 172 ft · 124 ft − 148 ft · 100 ft
$$= 21{,}328 \text{ ft}^2 - 14{,}800 \text{ ft}^2$$
$$= 6528 \text{ ft}^2$$

31. Divide 12 by 2 to get 6, which is what the length plus the width must equal. The only whole number possibilities are 1 and 5, 2 and 4, and 3 and 3.

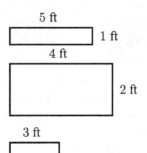

32. **(a)** 5 ft by 1 ft: $A = 5 \text{ ft} \cdot 1 \text{ ft} = 5 \text{ ft}^2$
4 ft by 2 ft: $A = 4 \text{ ft} \cdot 2 \text{ ft} = 8 \text{ ft}^2$
3 ft by 3 ft: $A = 3 \text{ ft} \cdot 3 \text{ ft} = 9 \text{ ft}^2$

(b) The square plot 3 ft by 3 ft has the greatest area.

33. The four possibilities for 16 ft of fencing are:

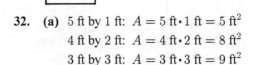

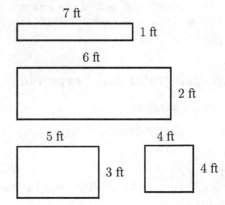

34. **(a)** 7 ft × 1 ft: $A = 7 \text{ ft} \cdot 1 \text{ ft} = 7 \text{ ft}^2$
6 ft × 2 ft: $A = 6 \text{ ft} \cdot 2 \text{ ft} = 12 \text{ ft}^2$
5 ft × 3 ft: $A = 5 \text{ ft} \cdot 3 \text{ ft} = 15 \text{ ft}^2$
4 ft × 4 ft: $A = 4 \text{ ft} \cdot 4 \text{ ft} = 16 \text{ ft}^2$

(b) The square plots have the greatest area.

35. **(a)**

3 ft

2 ft

$$P = 2 \cdot 3 \text{ ft} + 2 \cdot 2 \text{ ft} \qquad A = 3 \text{ ft} \cdot 2 \text{ ft}$$
$$= 6 \text{ ft} + 4 \text{ ft} \qquad\qquad = 6 \text{ ft}^2$$
$$= 10 \text{ ft}$$

Copyright © 2018 Pearson Education, Inc.

(b)   6 ft

4 ft

$P = 2 \cdot 6 \text{ ft} + 2 \cdot 4 \text{ ft}$ $A = 6 \text{ ft} \cdot 4 \text{ ft}$
   $= 12 \text{ ft} + 8 \text{ ft}$ $= 24 \text{ ft}^2$
   $= 20 \text{ ft}$

(c) The perimeter of the enlarged plot is twice the perimeter of the original plot. The area of the enlarged plot is four times $(2^2 = 4)$ the area of the original plot.

36.   **(a)**   9 ft

6 ft

$P = 2 \cdot 9 \text{ ft} + 2 \cdot 6 \text{ ft}$ $A = 9 \text{ ft} \cdot 6 \text{ ft}$
   $= 18 \text{ ft} + 12 \text{ ft}$ $= 54 \text{ ft}^2$
   $= 30 \text{ ft}$

(b) The perimeter is three times the original and the area is nine times $(3^2 = 9)$ the original.

(c) The perimeter will be four times the original and the area will be 16 times $(4^2 = 16)$ the original.

8.4 Parallelograms and Trapezoids

8.4 Margin Exercises

1.   **(a)** $P = 27 \text{ yd} + 27 \text{ yd} + 15 \text{ yd} + \underline{15} \text{ yd}$
    $= \underline{84 \text{ yd}}$

(b)
$P = 6.91 \text{ km} + 10.3 \text{ km} + 6.91 \text{ km} + 10.3 \text{ km}$
 $= 34.42 \text{ km}$

2.   **(a)** base: 50 ft, height: 42 ft

$A = b \cdot h$
 $= 50 \text{ ft} \cdot 42 \text{ ft}$
 $= \underline{2100 \text{ ft}^2}$

(b) base: 3.8 cm, height: 2.3 cm

$A = b \cdot h$
 $= 3.8 \text{ cm} \cdot 2.3 \text{ cm}$
 $= 8.74 \text{ cm}^2$

3.   **(a)** $P = 9.8 \text{ in.} + 6.3 \text{ in.} + 6.9 \text{ in.} + 5.6 \text{ in.}$
    $= 28.6 \text{ in.}$

(b) $P = 1.28 \text{ km} + 0.85 \text{ km} + 1.7 \text{ km} + 2 \text{ km}$
    $= 5.83 \text{ km}$

4.   **(a)** height: 30 ft, short base: 40 ft, long base: 60 ft

Area $= \frac{1}{2} \cdot$ height $\cdot$ (short base + long base)
$A = \frac{1}{2} \cdot h \cdot (b + B)$
$A = \frac{1}{2} \cdot 30 \text{ ft} \cdot (40 \text{ ft} + 60 \text{ ft})$
 $= 15 \text{ ft} \cdot (100 \text{ ft})$
 $= \underline{1500 \text{ ft}^2}$

(b) height: 12 cm, short base: 14 cm, long base: 19 cm

$A = \frac{1}{2} \cdot h \cdot (b + B)$
$A = \frac{1}{2} \cdot 12 \text{ cm} \cdot (14 \text{ cm} + 19 \text{ cm})$
 $= 6 \text{ cm} \cdot (33 \text{ cm})$
 $= 198 \text{ cm}^2$

(c) height: 4.7 m, short base: 9 m, long base: 10.5 m

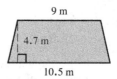

9 m

4.7 m

10.5 m

$A = 0.5 \cdot h \cdot (b + B)$
 $= 0.5 \cdot 4.7 \text{ m} \cdot (9 \text{ m} + 10.5 \text{ m})$
 $= 0.5 \cdot 4.7 \text{ m} \cdot 19.5 \text{ m}$
 $= 45.825 \text{ m}^2$

5.   **(a)** Break up the figure into two pieces, a parallelogram and a trapezoid.
Area of the parallelogram:

$A = b \cdot h = 8 \text{ m} \cdot 5 \text{ m} = 40 \text{ m}^2$

Area of the trapezoid:

$A = \frac{1}{2} \cdot h \cdot (b + B)$
 $= \frac{1}{2} \cdot 4 \text{ m} \cdot (8 \text{ m} + 14 \text{ m})$
 $= 2 \text{ m} \cdot (22 \text{ m})$
 $= 44 \text{ m}^2$

Total area $= 40 \text{ m}^2 + 44 \text{ m}^2 = 84 \text{ m}^2$

(b) Break up the figure into two pieces.
Area of the square:

$A = s \cdot s$
 $= 5 \text{ yd} \cdot 5 \text{ yd}$
 $= 25 \text{ yd}^2$

Area of the trapezoid:
$A = \frac{1}{2} \cdot h \cdot (b + B)$
 $= \frac{1}{2} \cdot 5 \text{ yd} \cdot (5 \text{ yd} + 10 \text{ yd})$
 $= 0.5 \cdot 5 \text{ yd} \cdot 15 \text{ yd}$
 $= 37.5 \text{ yd}^2$

Total area $= 25 \text{ yd}^2 + 37.5 \text{ yd}^2 = 62.5 \text{ yd}^2$

Copyright © 2018 Pearson Education, Inc.

6. **(a)** Floor (a) area is 84 m²:

$$\text{Cost} = \frac{84 \; \cancel{m^2}}{1} \cdot \frac{\$18.50}{1 \; \cancel{m^2}} = \$1554$$

(b) Floor (b) area is 62.5 yd²:

$$\text{Cost} = \frac{62.5 \; \cancel{yd^2}}{1} \cdot \frac{\$28}{1 \; \cancel{yd^2}} = \$1750$$

8.4 Section Exercises

1. In a parallelogram, opposite sides are parallel and have the same length. Drawings my vary; an example is shown.

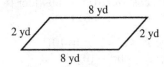

3. $P = 46 \text{ m} + \underline{46 \text{ m}} + \underline{58 \text{ m}} + \underline{58 \text{ m}}$
   $= \underline{208 \text{ m}}$

   Do not include the height (43 m).

5. $P = 4 \cdot 51.8 \text{ m}$
   $= 207.2 \text{ m}$

   Do not include the height (48.3 m).

7. $P = 3 \text{ km} + 0.8 \text{ km} + 0.95 \text{ km} + 1.31 \text{ km}$
   $= 6.06 \text{ km}$

   Do not include the height (0.4 km).

9. $A = b \cdot h$
   $= 31 \text{ mm} \cdot 25 \text{ mm}$
   $= 775 \text{ mm}^2$

11. $A = b \cdot h$
    $= 5.5 \text{ ft} \cdot 3.5 \text{ ft}$
    $= 19.25 \text{ ft}^2$

    Do not include the side that is *not* perpendicular to the height (5 ft).

13. The figure is a trapezoid. The height (*h*) is 42 cm, the short base (*b*) is 61.4 cm, and the long base (*B*) is 86.2 cm.

    $A = \frac{1}{2} \cdot h \cdot (b + B)$
    $A = 0.5 \cdot 42 \text{ cm} \cdot (61.4 \text{ cm} + 86.2 \text{ cm})$
    $A = 21 \text{ cm} \cdot (147.6 \text{ cm})$
    $A = 3099.6 \text{ cm}^2$

15. The first mistake was including height in the perimeter. The second mistake was using square units, which are for area, not perimeter. The correct answer is
    $P = 2.5 \text{ cm} + 2.5 \text{ cm} + 2.5 \text{ cm} + 2.5 \text{ cm}$
    $= 10 \text{ cm}$

17. $A = \frac{1}{2} \cdot h \cdot (b + B)$
    $= \frac{1}{2} \cdot 45 \text{ ft} \cdot (80 \text{ ft} + 110 \text{ ft})$
    $= \frac{1}{\cancel{2}} \cdot 45 \text{ ft} \cdot (\overset{95}{\cancel{190}} \text{ ft}) = 4275 \text{ ft}^2$

    $$\text{Cost} = \frac{4275 \; \cancel{ft^2}}{1} \cdot \frac{\$0.33}{1 \; \cancel{ft^2}} = \$1410.75$$

    It will cost \$1410.75 to put sod on the yard.

19. Find the area of one parallelogram piece of fabric.

    $A = b \cdot h$
    $A = 5 \text{ in.} \cdot 3.5 \text{ in.}$
    $A = 17.5 \text{ in.}^2$

    To find the total area of the 25 parallelogram pieces, multiply the area of one parallelogram (17.5 in.²) by the number of pieces (25).

    Total area $= 25 \cdot 17.5 \text{ in.}^2 = 437.5 \text{ in.}^2$

21. Break the figure into two pieces, a rectangle on the left side and a parallelogram on the right side. Find the area of each piece, and then add the areas.

    Area of rectangle:

    $A = \text{length} \cdot \text{width}$
    $= 1.3 \text{ m} \cdot 0.8 \text{ m}$
    $= 1.04 \text{ m}^2$

    Area of parallelogram:

    $A = \text{base} \cdot \text{height}$
    $= 1.8 \text{ m} \cdot 1.1 \text{ m}$
    $= 1.98 \text{ m}^2$

    Total area $= 1.04 \text{ m}^2 + 1.98 \text{ m}^2$
    $= 3.02 \text{ m}^2$

23. **(1)** Top figure (square):

    $A = s \cdot s$
    $= 96 \text{ ft} \cdot 96 \text{ ft}$
    $= 9216 \text{ ft}^2$

    (2) Middle figure (parallelogram):

    $A = b \cdot h$
    $= 96 \text{ ft} \cdot 72 \text{ ft}$
    $= 6912 \text{ ft}^2$

    (3) Bottom figure (square), same as top figure:

    $A = 9216 \text{ ft}^2$

    Total area $= 9216 \text{ ft}^2 + 6912 \text{ ft}^2 + 9216 \text{ ft}^2$
    $= 25{,}344 \text{ ft}^2$

Copyright © 2018 Pearson Education, Inc.

8.5   Triangles

8.5   Margin Exercises

1. **(a)** $P = 31 \text{ mm} + 25 \text{ mm} + \underline{16 \text{ mm}}$
    $P = \underline{72 \text{ mm}}$

    (b) $P = 25.9 \text{ m} + 16.2 \text{ m} + 11.7 \text{ m} = 53.8 \text{ m}$

    (c) $P = 6\frac{1}{2} \text{ yd} + 9\frac{3}{4} \text{ yd} + 11\frac{1}{4} \text{ yd}$
    $= 6\frac{2}{4} \text{ yd} + 9\frac{3}{4} \text{ yd} + 11\frac{1}{4} \text{ yd}$
    $= 26\frac{6}{4} \text{ yd}$
    $= 27\frac{1}{2} \text{ yd or } 27.5 \text{ yd}$

2. **(a)** base: 26 m, height: 20 m
    $A = \frac{1}{2} \cdot \text{base} \cdot \text{height} = \frac{1}{2} \cdot b \cdot h$
    $= \frac{1}{2} \cdot 26 \text{ m} \cdot 20 \text{ m}$
    $= \underline{260 \text{ m}^2}$

    (b) base: 2.1 cm, height: 1.7 cm
    $A = 0.5 \cdot b \cdot h$
    $= 0.5 \cdot 2.1 \text{ cm} \cdot 1.7 \text{ cm}$
    $= 1.785 \text{ cm}^2$

    (c) base: 6 in., height: 5.5 in.
    $A = 0.5 \cdot b \cdot h$
    $= 0.5 \cdot 6 \text{ in.} \cdot 5.5 \text{ in.}$
    $= 16.5 \text{ in.}^2$

    (d) base: $9\frac{1}{2}$ ft, height: 7 ft
    $A = 0.5 \cdot b \cdot h$
    $= 0.5 \cdot 9.5 \text{ ft} \cdot 7 \text{ ft}$
    $= 33.25 \text{ ft}^2 \text{ or } 33\frac{1}{4} \text{ ft}^2$

3. Find the area of the entire figure, which is a square.
    $A = s \cdot s$
    $= 25 \text{ m} \cdot 25 \text{ m}$
    $= \underline{625 \text{ m}^2}$

    Find the area of the unshaded parts, which are triangles.
    $A = \frac{1}{2} \cdot b \cdot h$
    $= \frac{1}{2} \cdot 25 \text{ m} \cdot 10 \text{ m}$
    $= 125 \text{ m}^2$

    There are 2 equal unshaded triangles, so the total unshaded area is
    $2 \cdot 125 \text{ m}^2 = 250 \text{ m}^2.$

    Subtract to find the area of the shaded part.

    $$\begin{array}{ccccc} entire & & unshaded & & shaded \\ area & - & triangles & = & part \\ 625 \text{ m}^2 & - & 250 \text{ m}^2 & = & 375 \text{ m}^2 \end{array}$$

    The area of the shaded part is 375 m².

4. The shaded part is 375 m². The cost of the carpet is \$27 per square meter.

    $$\frac{375 \text{ m}^2}{1} \cdot \frac{\$27}{1 \text{ m}^2} = \$10,125$$

    The cost for carpeting is \$10,125.
    The unshaded part is $2(125 \text{ m}^2) = 250 \text{ m}^2$.
    The cost for the vinyl floor covering is \$18 per square meter.

    $$\frac{250 \text{ m}^2}{1} \cdot \frac{\$18}{1 \text{ m}^2} = \$4500$$

    The cost for the vinyl floor covering is \$4500.
    The total cost of covering the floor is
    $\$10,125 + \$4500 = \$14,625.$

5. **(a)** *Step 1* $48° + 59° = \underline{107°}$
    Step 2 $180° - \underline{107°} = \underline{73°}$
    $\angle G$ measures $\underline{73°}$.

    (b) *Step 1* $90° + 55° = 145°$
    Step 2 $180° - 145° = 35°$
    $\angle C$ measures 35°.

    (c) *Step 1* $60° + 60° = 120°$
    Step 2 $180° - 120° = 60°$
    $\angle X$ measures 60°.

8.5   Section Exercises

1. Multiply $\frac{1}{2}$ times 58 m times 66 m; the units in the answer will be m².

3. $P = 72 \text{ m} + 58 \text{ m} + 72 \text{ m} = 202 \text{ m}$
    $A = \frac{1}{2} \cdot b \cdot h$
    $= \frac{1}{2} \cdot 58 \text{ m} \cdot 66 \text{ m}$
    $= 1914 \text{ m}^2$

5. $P = 15.6 \text{ cm} + 25.3 \text{ cm} + 18 \text{ cm} = 58.9 \text{ cm}$
    $A = \frac{1}{2} \cdot b \cdot h$
    $= \frac{1}{2} \cdot 25.3 \text{ cm} \cdot 11 \text{ cm}$
    $= 139.15 \text{ cm}^2$

7. $P = 9 \text{ yd} + 7 \text{ yd} + 10\frac{1}{4} \text{ yd} = 26\frac{1}{4} \text{ yd or } 26.25 \text{ yd}$
    $A = \frac{1}{2} \cdot b \cdot h$
    $= \frac{1}{2} \cdot 10\frac{1}{4} \text{ yd} \cdot 6 \text{ yd}$
    $= 30\frac{3}{4} \text{ yd}^2 \text{ or } 30.75 \text{ yd}^2$

9. To find the perimeter, add the lengths of the three sides.

    $P = 35.5 \text{ cm} + 21.3 \text{ cm} + 28.4 \text{ cm} = 85.2 \text{ cm}$
    $P = 85.2 \text{ cm}$

    When finding the area of this right triangle, the base and height are the perpendicualr sides. So the base is 28.4 cm and the height is 21.3 cm.

    (*continued*)

Copyright © 2018 Pearson Education, Inc.

$A = \frac{1}{2} \cdot b \cdot h$

$A = \frac{1}{2} \cdot 28.4 \text{ cm} \cdot 21.3 \text{ cm}$

$A = 302.46 \text{ cm}^2$

11. Find the area of the triangle.

$$A = 0.5 \cdot b \cdot h$$
$$= 0.5 \cdot 12 \text{ m} \cdot \underline{9 \text{ m}}$$
$$= \underline{54 \text{ m}^2}$$

Find the area of the square.

$$A = 12 \text{ m} \cdot \underline{12 \text{ m}}$$
$$= \underline{144 \text{ m}^2}$$

Now add to find the area of the entire figure.

Entire area $= 54 \text{ m}^2 + 144 \text{ m}^2 = 198 \text{ m}^2$

13. First, find the area of the entire figure, which is a rectangle.

$$A = l \cdot w$$
$$A = 52 \text{ m} \cdot 37 \text{ m}$$
$$A = 1924 \text{ m}^2$$

Now find the area of the unshaded triangle.

$$A = \frac{1}{2} \cdot b \cdot h$$
$$= \frac{1}{2} \cdot 52 \text{ m} \cdot 10 \text{ m}$$
$$= 260 \text{ m}^2$$

Subtract the unshaded area from the area of the entire figure to find the shaded area.

Shaded area $= 1924 \text{ m}^2 - 260 \text{ m}^2 = 1664 \text{ m}^2$

15. *Step 1* Add the two measures of the angles given: $90° + 58° = 148°$
Step 2 Subtract the sum from $180°$:
$180° - 148° = 32°$
The third angle measures $32°$.

17. *Step 1* $72° + 60° = 132°$
Step 2 $180° - 132° = 48°$
The third angle measures $48°$.

19. The first mistake was finding the perimeter instead of the area. The second mistake was forgetting to include square units for area. The correct answer is
$A = \frac{1}{2} \cdot 15 \text{ ft} \cdot 8 \text{ ft} = 60 \text{ ft}^2$

21. $A = \frac{1}{2} \cdot b \cdot h$
$$= \frac{1}{2} \cdot 3\frac{1}{2} \text{ ft} \cdot 4\frac{1}{2} \text{ ft}$$
$$= \frac{1}{2} \cdot \frac{7}{2} \text{ ft} \cdot \frac{9}{2} \text{ ft}$$
$$= \frac{63}{8} \text{ ft}^2$$
$$= 7\frac{7}{8} \text{ ft}^2 \text{ or } 7.875 \text{ ft}^2$$

To make the flap, $7\frac{7}{8}$ ft^2 or 7.875 ft^2 of canvas is needed.

23. **(a)** To find the amount of curbing needed to go around the space, find the perimeter of the triangle.

$$P = 42 \text{ m} + 32 \text{ m} + 52.8 \text{ m} = 126.8 \text{ m}$$

126.8 m of curbing will be needed.

(b) To find the amount of grass needed to cover the space, find the area. The space forms a right triangle, so the perpendicular sides are the base and height.

$$A = \frac{1}{2} \cdot 32 \text{ m} \cdot 42 \text{ m} = 672 \text{ m}^2$$

672 m^2 of grass will be needed.

25. **(a)** Area of one side of the house:

$$A = \frac{1}{2} \cdot b \cdot h$$
$$= \frac{1}{2} \cdot 8 \text{ m} \cdot 8 \text{ m}$$
$$= 32 \text{ m}^2$$

(b) Area of one roof section:

$$A = 0.5 \cdot b \cdot h$$
$$= 0.5 \cdot 3 \text{ m} \cdot 8.94 \text{ m}$$
$$= 13.41 \text{ m}^2$$

27. **(a)** The frontage of the lot is

$$100 \text{ yd} + 75 \text{ yd} = 175 \text{ yd}.$$

(b) Draw a vertical line to break the region into a triangle and a rectangle. For the triangular region, the base is 75 yd and the height is 100 yd.

$$A = 0.5 \cdot b \cdot h$$
$$= 0.5 \cdot 75 \text{ yd} \cdot 100 \text{ yd}$$
$$= 3750 \text{ yd}^2$$

Rectangular area:

$$A = l \cdot w$$
$$= 100 \text{ yd} \cdot 50 \text{ yd}$$
$$= 5000 \text{ yd}^2$$

Total area $= 3750 \text{ yd}^2 + 5000 \text{ yd}^2 = 8750 \text{ yd}^2$

8.6 Circles

8.6 Margin Exercises

1. **(a)** diameter: 40 ft

$$r = \frac{d}{2} = \frac{40 \text{ ft}}{2} = \underline{20 \text{ ft}}$$

(b) diameter: 11 cm

$$r = \frac{d}{2} = \frac{11 \text{ cm}}{2} = 5.5 \text{ cm}$$

(c) radius: 32 yd

$$d = 2 \cdot r = 2 \cdot 32 \text{ yd} = 64 \text{ yd}$$

Copyright © 2018 Pearson Education, Inc.

(d) radius: 9.5 m

$$d = 2 \cdot r = 2 \cdot 9.5 \text{ m} = 19 \text{ m}$$

2.   **(a)** diameter: 150 ft

$$C = \pi \cdot d$$
$$\approx 3.14 \cdot 150 \text{ ft}$$
$$= \underline{471 \text{ ft}}$$

(b) radius: 7 in.

$$C = 2 \cdot \pi \cdot r$$
$$\approx 2 \cdot 3.14 \cdot 7 \text{ in.}$$
$$\approx 44.0 \text{ in. (rounded)}$$

(c) diameter: 0.9 km

$$C = \pi \cdot d$$
$$\approx 3.14 \cdot 0.9 \text{ km}$$
$$\approx 2.8 \text{ km (rounded)}$$

(d) radius: 4.6 m

$$C = 2 \cdot \pi \cdot r$$
$$\approx 2 \cdot 3.14 \cdot 4.6 \text{ m}$$
$$\approx 28.9 \text{ m (rounded)}$$

3.   **(a)** radius: 4 ft

$$A = \pi \cdot r^2$$
$$= \pi \cdot r \cdot r$$
$$\approx 3.14 \cdot 4 \text{ ft} \cdot 4 \text{ ft}$$
$$\approx \underline{50.2 \text{ ft}^2} \text{ (rounded)}$$

(b) diameter: 12 in., so $r = \dfrac{12 \text{ in.}}{2} = \underline{6} \text{ in.}$

$$A = \pi \cdot r \cdot r$$
$$\approx 3.14 \cdot 6 \text{ in.} \cdot 6 \text{ in.}$$
$$\approx 113.0 \text{ in.}^2 \text{ (rounded)}$$

(c) radius: 1.8 km

$$A = \pi \cdot r \cdot r$$
$$\approx 3.14 \cdot 1.8 \text{ km} \cdot 1.8 \text{ km}$$
$$\approx 10.2 \text{ km}^2 \text{ (rounded)}$$

(d) diameter: 8.4 cm, so $r = \dfrac{8.4 \text{ cm}}{2} = 4.2 \text{ cm}$

$$A = \pi \cdot r \cdot r$$
$$\approx 3.14 \cdot 4.2 \text{ cm} \cdot 4.2 \text{ cm}$$
$$\approx 55.4 \text{ cm}^2 \text{ (rounded)}$$

4.   **(a)** radius: 24 m
Area of circle

$$A = \pi \cdot r \cdot r$$
$$\approx 3.14 \cdot 24 \text{ m} \cdot 24 \text{ m}$$
$$= 1808.64 \text{ m}^2$$

Area of semicircle

$$\frac{1808.64 \text{ m}^2}{2} \approx 904.3 \text{ m}^2$$

(b) diameter: 35.4 ft, so $r = \dfrac{35.4 \text{ ft}}{2} = 17.7 \text{ ft}$
Area of circle

$$A = \pi \cdot r \cdot r$$
$$\approx 3.14 \cdot 17.7 \text{ ft} \cdot 17.7 \text{ ft}$$
$$= 983.7306 \text{ ft}^2$$

Area of semicircle

$$\frac{983.7306 \text{ ft}^2}{2} \approx 491.9 \text{ ft}^2$$

(c) radius: 9.8 m
Area of circle

$$A = \pi \cdot r \cdot r$$
$$\approx 3.14 \cdot 9.8 \text{ m} \cdot 9.8 \text{ m}$$
$$= 301.5656 \text{ m}^2$$

Area of semicircle

$$\frac{301.5656 \text{ m}^2}{2} \approx 150.8 \text{ m}^2$$

5. $C = \pi \cdot d$
$$\approx 3.14 \cdot 3 \text{ m}$$
$$= 9.42 \text{ m}$$

$$\text{Cost of binding} = \frac{9.42 \text{ m}}{1} \cdot \frac{\$4.50}{1 \text{ m}} = \$42.39$$

7.   **(a)** Some possibilities are:

*dia*meter;	*dia*gonal
*fra*ction;	*fra*cture
*par*allel;	*par*amedic
*per*cent;	*per* capita
*peri*meter;	*peri*scope
*rad*ius;	*rad*iate
*rect*angle;	*rect*ify
*sub*tract;	*sub*marine

(b) *Peri-* in perimeter means "around," so perimeter is the distance *around* the edges of a shape.

8.6   Section Exercises

1.   Finding the circumference of a circle is like finding the <u>perimeter</u> of a triangle. If the circle's radius and diameter are measured in feet, then the units for the circumference will be <u>ft</u>.

3.   The radius is 9 mm, so the diameter is

$$d = 2 \cdot r$$
$$= 2 \cdot \underline{9} \text{ mm}$$
$$= \underline{18 \text{ mm}}.$$

Copyright © 2018 Pearson Education, Inc.

5. The diameter is 0.7 km, so the radius is

$$\frac{d}{2} = \frac{0.7 \text{ km}}{2} = 0.35 \text{ km.}$$

7. radius $r = 11$ ft

$$C = 2 \cdot \pi \cdot r$$
$$\approx 2 \cdot 3.14 \cdot 11 \text{ ft}$$
$$\approx 69.1 \text{ ft}$$
$$A = \pi \cdot r \cdot r$$
$$\approx 3.14 \cdot 11 \text{ ft} \cdot 11 \text{ ft}$$
$$\approx 379.9 \text{ ft}^2$$

9. diameter $d = 2.6$ m

$$C = \pi \cdot d$$
$$\approx 3.14 \cdot 2.6 \text{ m}$$
$$\approx 8.2 \text{ m}$$

$$\text{radius } r = \frac{d}{2} = \frac{2.6 \text{ m}}{2} = 1.3 \text{ m}$$

$$A = \pi \cdot r \cdot r$$
$$\approx 3.14 \cdot 1.3 \text{ m} \cdot 1.3 \text{ m}$$
$$\approx 5.3 \text{ m}^2$$

11. $d = 15$ cm, so $r = \frac{15 \text{ cm}}{2} = 7.5$ cm.

$$C = \pi \cdot d$$
$$\approx 3.14 \cdot 15 \text{ cm}$$
$$= \underline{47.1 \text{ cm}}$$
$$A = \pi \cdot r \cdot r$$
$$\approx 3.14 \cdot 7.5 \text{ cm} \cdot 7.5 \text{ cm}$$
$$\approx 176.6 \text{ cm}^2$$

13. $d = 7\frac{1}{2}$ ft, so $r = \frac{7\frac{1}{2} \text{ ft}}{2} = 3.75$ ft.

$$C = \pi \cdot d$$
$$C \approx 3.14 \cdot 7.5 \text{ ft}$$
$$C \approx 23.6 \text{ ft}$$
$$A = \pi \cdot r \cdot r$$
$$A \approx 3.14 \cdot 3.75 \text{ ft} \cdot 3.75 \text{ ft}$$
$$A \approx 44.2 \text{ ft}^2 \text{ (rounded to tenths)}$$

15. radius $r = 7$ in.
Area of circle:

$$A = \pi \cdot r \cdot r$$
$$\approx 3.14 \cdot 7 \text{ in.} \cdot 7 \text{ in.}$$
$$= 153.86 \text{ in.}^2$$

Area of semicircle:

$$\frac{\text{area of circle}}{2} = \frac{153.86 \text{ in.}^2}{2} \approx 76.9 \text{ in.}^2$$

17. Find the area of a whole circle with a radius of 10 cm:

$$A = \pi \cdot r \cdot r$$
$$A \approx 3.14 \cdot 10 \text{ cm} \cdot 10 \text{ cm}$$
$$A = 314 \text{ cm}^2$$

Divide the area of the whole circle by 2 to find the area of the semicircle:

$$\frac{314 \text{ cm}^2}{2} = 157 \text{ cm}^2$$

Area of the triangle:

$$A = \frac{1}{2} \cdot b \cdot h$$
$$A = \frac{1}{2} \cdot 20 \text{ cm} \cdot 10 \text{ cm}$$
$$A = 100 \text{ cm}^2$$

Subtract to find the shaded area. $157 \text{ cm}^2 - 100 \text{ cm}^2 = 57 \text{ cm}^2$.

19. The first mistake was using the diameter instead of the radius. The second mistake was forgetting to use square units for area. The correct answer is $A = \pi \cdot 3 \text{ in.} \cdot 3 \text{ in.} \approx 3.14 \cdot 3 \text{ in.} \cdot 3 \text{ in.} \approx 28.26 \text{ in.}^2$

21. $A = \pi \cdot r \cdot r$
$$\approx 3.14 \cdot 50 \text{ yd} \cdot 50 \text{ yd}$$
$$= 7850 \text{ yd}^2$$

The watered area is about 7850 yd^2.

23. A point on the tire tread moves the length of the circumference in one complete turn.

$$C = \pi \cdot d$$
$$C \approx 3.14 \cdot 29.10 \text{ in.}$$
$$C \approx 91.4 \text{ in.}$$

Thus, a point on the tire tread moves about 91.4 inches (rounded to tenths) in one complete turn.

Bonus question:

$$\frac{1 \text{ revolution}}{91.4 \text{ inches}} \cdot \frac{12 \text{ inches}}{1 \text{ foot}} \cdot \frac{5280 \text{ feet}}{1 \text{ mile}}$$
$$\approx 693 \text{ revolutions/mile}$$

The tire makes about 693 revolutions/mile.

25.

Station 150 mi

Find the area of a circle with a radius of 150 mi.

$$A = \pi \cdot r \cdot r$$
$$\approx 3.14 \cdot 150 \text{ mi} \cdot 150 \text{ mi}$$
$$= 70,650 \text{ mi}^2$$

There are about $70,650 \text{ mi}^2$ in the broadcast area.

Copyright © 2018 Pearson Education, Inc.

27.

Watch: $d = 1$ in., so $r = \frac{1}{2}$ in.

$$C = \pi \cdot 1 \text{ in.}$$
$$\approx 3.14 \cdot 1 \text{ in.}$$
$$\approx 3.1 \text{ in.}$$
$$A = \pi \cdot r \cdot r$$
$$\approx 3.14 \cdot \frac{1}{2} \text{ in.} \cdot \frac{1}{2} \text{ in.}$$
$$\approx 0.8 \text{ in.}^2$$

Wall clock: $r = 3$ in.

$$C = 2 \cdot \pi \cdot 3 \text{ in.}$$
$$\approx 2 \cdot 3.14 \cdot 3 \text{ in.}$$
$$\approx 18.8 \text{ in.}$$
$$A = \pi \cdot r \cdot r$$
$$\approx 3.14 \cdot 3 \text{ in.} \cdot 3 \text{ in.}$$
$$\approx 28.3 \text{ in.}^2$$

29.

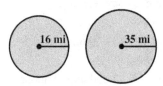

Area of larger circle:

$$A = \pi \cdot r \cdot r$$
$$\approx 3.14 \cdot 35 \text{ mi} \cdot 35 \text{ mi}$$
$$= 3846.5 \text{ mi}^2$$

Area of smaller circle:

$$A = \pi \cdot r \cdot r$$
$$\approx 3.14 \cdot 16 \text{ mi} \cdot 16 \text{ mi}$$
$$\approx 803.8 \text{ mi}^2$$

The difference in the area covered is about

$$3846.5 \text{ mi}^2 - 803.8 \text{ mi}^2 = 3042.7 \text{ mi}^2.$$

31. (a)

$$C = \pi \cdot d$$
$$144 \text{ cm} = \pi \cdot d \qquad \text{(since } C = 144 \text{ cm)}$$
$$144 \text{ cm} \approx 3.14 \cdot d$$
$$\frac{144 \text{ cm}}{3.14} \approx d$$
$$d \approx 45.9 \text{ cm}$$

The diameter is about 45.9 cm.

(b) Divide the circumference by π (3.14).

33. Find the area of the rectangle with length 29 ft and width 16 ft.

$$A = \text{length} \cdot \text{width}$$
$$= 29 \text{ ft} \cdot 16 \text{ ft}$$
$$= 464 \text{ ft}^2$$

The semicircles are of equal area, and the sum of their areas equals that of a circle. The radius is 8 ft.

$$A = \pi \cdot r \cdot r$$
$$\approx 3.14 \cdot 8 \text{ ft} \cdot 8 \text{ ft}$$
$$= 200.96 \text{ ft}^2$$

The total area is approximately

$$464 \text{ ft}^2 + 200.96 \text{ ft}^2 = 664.96 \text{ ft}^2.$$

$$\text{Cost of the sod} = \frac{664.96 \text{ ft}^2}{1} \cdot \frac{\$0.49}{1 \text{ ft}^2} \approx \$325.83$$

35. The prefix *rad*- tells you that radius is a ray from the center of the circle. The prefix *dia*- means the diameter goes through the circle, and the prefix *circum*- means the circumference is the distance around.

Refer to the following figure for exercises 37–41.

First-level floor plan	Second-level floor plan

37. The first floor consists of the semicircular entrance area and a rectangular area. Find the area of each section and then add the two areas.

The circular entrance area is a semicircle with radius is 30 ft.

$$A = \frac{1}{2} \cdot \pi \cdot r \cdot r$$
$$\approx \frac{1}{2} \cdot 3.14 \cdot 30 \text{ ft} \cdot 30 \text{ ft}$$
$$= 1413 \text{ ft}^2$$

(continued)

Copyright © 2018 Pearson Education, Inc.

The rest of the first floor is a rectangle with width 60 ft and length $= 45 + 30 + 30 + 45 = 150$ ft.

$$P = l \cdot w$$
$$= 60 \text{ ft} \cdot 150 \text{ ft}$$
$$= 9000 \text{ ft}^2$$

The total area of the first floor is $1413 \text{ ft}^2 + 9000 \text{ ft}^2 = 10{,}413 \text{ ft}^2$.

Use similar reasoning to find the perimeter of the first level. Find the lengths of each side of the rectangular region and add them. The circumference of the semicircular section is

$$C = 2 \cdot \pi \cdot r \cdot \frac{1}{2} = \pi \cdot r$$
$$C \approx 3.14 \cdot 30 \text{ ft}$$
$$C \approx 94.2 \text{ ft}$$

The length of the back wall of the first floor is $45 + 30 + 30 + 45 = 150$ ft while the length of the front wall is $45 + 94.2 + 45 = 184.2$ ft. The perimeter of the first floor is
$150 \text{ ft} + 60 \text{ ft} + 184.2 + 60 \approx 454$ ft (rounded).

38. The second floor is a rectangle with a semicircular region "cut" out of it. Find the area of each section and then subtract the area of the semicircular region from the area of the entire rectangle.

From Exercise 37, the area of the rectangular section is 9000 ft^2, and the area of the semicircular section is 1413 ft^2. The area of the second floor is $9000 \text{ ft}^2 - 1413 \text{ ft}^2 = 7587 \text{ ft}^2$.

The length of the back wall of the first floor is $45 + 30 + 30 + 45 = 150$ ft while the length of the front wall is $45 + 94.2 + 45 = 184.2$ ft. The perimeter of the first floor is
$150 \text{ ft} + 60 \text{ ft} + 184.2 + 60 \approx 454$ ft (rounded).

39. From Exercise 37, the area of the rectangular section is 9000 ft^2, and the area of the semicircular section is 1413 ft^2, so the area of the entire circular region is $2(1413) = 2826 \text{ ft}^2$. However, the region to be carpeted has the semicircular piece "cut" out of the rectangular region. From Exercise 38, the area of this region is 7587 ft^2.

The cost of the tile is $2826 \cdot \$13 = \$36{,}738$. The cost of the carpeting is $7587 \cdot \$7 = \$53{,}109$. Therefore, the total cost of the flooring is $\$36{,}738 + \$53{,}109 = \$89{,}847$.

40. Using the results of Exercises 37 and 38, the building has a total area of
$10{,}413 \text{ ft}^2 + 7587 \text{ ft}^2 = 18{,}000 \text{ ft}^2$, which is $18{,}000 \text{ ft}^2 - 16{,}500 \text{ ft}^2 = 1500 \text{ ft}^2$ more than the law firm needs.

41. $\dfrac{\$3{,}698{,}700}{18{,}000 \text{ ft}^2} \approx \$205$ per square foot (rounded)

Summary Exercises *Perimeter, Circumference, and Area*

1. **(a)**

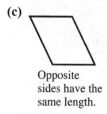

All sides have the same length.

(b)

Opposite sides have the same length.

(c)

Opposite sides have the same length.

3. Add up the lengths of all the sides to find the perimeter.

5. parallelogram: **(f)** $A = b \cdot h$

square: **(e)** $A = s^2$

trapezoid: **(d)** $A = \frac{1}{2} \cdot h \cdot (b + B)$

circle: **(b)** $A = \pi \cdot r^2$

rectangle: **(c)** $A = l \cdot w$

triangle: **(a)** $A = \frac{1}{2} \cdot b \cdot h$

7. Leftmost triangle:
$$P = 18 \text{ ft} + 21 \text{ ft} + 24 \text{ ft} = 63 \text{ ft}$$

$$A = \tfrac{1}{2} \cdot b \cdot h$$
$$= \tfrac{1}{2} \cdot 24 \text{ ft} \cdot 15 \text{ ft}$$
$$= 180 \text{ ft}^2$$

(continued)

Copyright © 2018 Pearson Education, Inc.

Rightmost triangle:
$P = 9.5 \text{ cm} + 9.5 \text{ cm} + 13.4 \text{ cm} = 32.4 \text{ cm}$

$$\begin{aligned} A &= \tfrac{1}{2} \cdot b \cdot h \\ &= \tfrac{1}{2} \cdot 9.5 \text{ cm} \cdot 9.5 \text{ cm} \\ &= 45.125 \text{ cm}^2 \approx 45.1 \text{ cm}^2 \end{aligned}$$

9. The figure is a rectangle.

$$\begin{aligned} P &= 2 \cdot l + 2 \cdot w \\ &= 2 \cdot 10\tfrac{1}{2} \text{ in.} + 2 \cdot 3 \text{ in.} \\ &= 21 \text{ in.} + 6 \text{ in.} \\ &= 27 \text{ in.} \end{aligned}$$

$$\begin{aligned} A &= l \cdot w \\ &= 10\tfrac{1}{2} \text{ in.} \cdot 3 \text{ in.} \\ &= 31\tfrac{1}{2} \text{ in.}^2 \text{ or } 31.5 \text{ in.}^2 \end{aligned}$$

11. The figure is a parallelogram.

$$\begin{aligned} P &= 1.3 \text{ m} + 1.9 \text{ m} + 1.3 \text{ m} + 1.9 \text{ m} \\ &= 6.4 \text{ m} \end{aligned}$$

$$\begin{aligned} A &= b \cdot h \\ &= 1.3 \text{ m} \cdot 1.7 \text{ m} \\ &= 2.21 \text{ m}^2 \approx 2.2 \text{ m}^2 \text{ (rounded)} \end{aligned}$$

13. (a) $r = 6 \text{ cm}; d = 2 \cdot 6 \text{ cm} = 12 \text{ cm}$

(b) $\begin{aligned} C &= 2 \cdot \pi \cdot r \\ &\approx 2 \cdot 3.14 \cdot 6 \text{ cm} \\ &\approx 37.7 \text{ cm (rounded)} \end{aligned}$

(c) $\begin{aligned} A &= \pi \cdot r \cdot r \\ &\approx 3.14 \cdot 6 \text{ cm} \cdot 6 \text{ cm} \\ &\approx 113.0 \text{ cm}^2 \text{ (rounded)} \end{aligned}$

15. (a) $d = 9 \text{ ft}; r = \dfrac{9 \text{ ft}}{2} = 4.5 \text{ ft}$

(b) $\begin{aligned} C &= \pi \cdot d \\ &\approx 3.14 \cdot 9 \text{ ft} \\ &\approx 28.3 \text{ ft (rounded)} \end{aligned}$

(c) $\begin{aligned} A &= \pi \cdot r \cdot r \\ &\approx 3.14 \cdot 4.5 \text{ ft} \cdot 4.5 \text{ ft} \\ &\approx 63.6 \text{ ft}^2 \text{ (rounded)} \end{aligned}$

17. Subtract the area of the triangle from the area of the rectangle.

Triangle: $\begin{aligned} A &= \tfrac{1}{2} \cdot b \cdot h \\ &= \tfrac{1}{2} \cdot 15 \text{ yd} \cdot 13 \text{ yd} = 97.5 \text{ yd}^2 \end{aligned}$

Rectangle: $A = l \cdot w = 21 \text{ yd} \cdot 15 \text{ yd} = 315 \text{ yd}^2$

Shaded area $= 315 \text{ yd}^2 - 97.5 \text{ yd}^2 = 217.5 \text{ yd}^2$

The shaded area is 217.5 yd^2.

19. $C = 2 \cdot \pi \cdot r \approx 2 \cdot 3.14 \cdot 2.33 \text{ ft} \approx 14.6 \text{ ft}$
The rag traveled about 14.6 ft with each revolution of the wheel.

Bonus question:

$$\frac{1 \text{ revolution}}{14.6 \text{ ft}} \cdot \frac{5280 \text{ ft}}{1 \text{ mi}} \approx 361.6 \text{ revolutions/mile}$$

The wheel makes about 361.6 revolutions in one mile. The Mormons used 360, which is the answer you get using $2\tfrac{1}{3}$ ft instead of 2.33 ft as the radius.

8.7 Volume

8.7 Margin Exercises

1. (a) $\begin{aligned} V &= \text{length} \cdot \text{width} \cdot \text{height (or } l \cdot w \cdot h) \\ &= 8 \text{ m} \cdot 3 \text{ m} \cdot \underline{3} \text{ m} \\ &= \underline{72 \text{ m}^3} \end{aligned}$

(b) $\begin{aligned} V &= \text{length} \cdot \text{width} \cdot \text{height} \\ &= 23.4 \text{ cm} \cdot 15.2 \text{ cm} \cdot 52.3 \text{ cm} \\ &= 18{,}602.064 \text{ cm}^3 \\ &\approx 18{,}602.1 \text{ cm}^3 \text{ (rounded)} \end{aligned}$

(c) $\begin{aligned} V &= \text{length} \cdot \text{width} \cdot \text{height} \\ &= 6\tfrac{1}{4} \text{ ft} \cdot 3\tfrac{1}{2} \text{ ft} \cdot 2 \text{ ft} \\ &= 6.25 \text{ ft} \cdot 3.5 \text{ ft} \cdot 2 \text{ ft} \\ &= 43\tfrac{3}{4} \text{ ft}^3 \text{ or } 43.75 \text{ ft}^3 \\ &\approx 43.8 \text{ ft}^3 \text{ (rounded)} \end{aligned}$

2. (a) $\begin{aligned} V &= \tfrac{4}{3} \cdot \pi \cdot r^3 \\ &\approx \frac{4 \cdot 3.14 \cdot 12 \text{ in.} \cdot 12 \text{ in.} \cdot 12 \text{ in.}}{3} \\ &\approx \underline{7234.6 \text{ in.}^3} \text{ (rounded)} \end{aligned}$

(b) $\begin{aligned} V &= \tfrac{4}{3} \cdot \pi \cdot r^3 \\ &\approx \frac{4 \cdot 3.14 \cdot 3.5 \text{ m} \cdot 3.5 \text{ m} \cdot 3.5 \text{ m}}{3} \\ &\approx 179.5 \text{ m}^3 \text{ (rounded)} \end{aligned}$

(c) $\begin{aligned} V &= \tfrac{4}{3} \cdot \pi \cdot r^3 \\ &\approx \frac{4 \cdot 3.14 \cdot 2.7 \text{ cm} \cdot 2.7 \text{ cm} \cdot 2.7 \text{ cm}}{3} \\ &\approx 82.4 \text{ cm}^3 \text{ (rounded)} \end{aligned}$

3. (a) $\begin{aligned} V &= \tfrac{2}{3} \cdot \pi \cdot r^3 \\ &\approx \frac{2 \cdot 3.14 \cdot 15 \text{ ft} \cdot 15 \text{ ft} \cdot 15 \text{ ft}}{3} \\ &\approx 7065 \text{ ft}^3 \end{aligned}$

Copyright © 2018 Pearson Education, Inc.

(b) $V = \frac{2}{3} \cdot \pi \cdot r^3$

$\approx \dfrac{2 \cdot 3.14 \cdot 6 \text{ cm} \cdot 6 \text{ cm} \cdot 6 \text{ cm}}{3}$

$\approx 452.2 \text{ cm}^3$ (rounded)

4. (a) $V = \pi \cdot r^2 \cdot h$

$\approx 3.14 \cdot 4 \text{ ft} \cdot 4 \text{ ft} \cdot 12 \text{ ft}$

$\approx \underline{602.9 \text{ ft}^3}$ (rounded)

(b) The diameter is 7 cm, so the radius r is given by $r = \frac{7 \text{ cm}}{2} = 3.5$ cm.

$V = \pi \cdot r \cdot r \cdot h$

$\approx 3.14 \cdot 3.5 \text{ cm} \cdot 3.5 \text{ cm} \cdot 6 \text{ cm}$

$\approx 230.8 \text{ cm}^3$ (rounded)

5. First find the area of the circular base.

$B = \pi \cdot r \cdot r$

$\approx 3.14 \cdot 2 \text{ ft} \cdot 2 \text{ ft}$

$= \underline{12.56 \text{ ft}^2}$

Now find the volume of the cone.

$V = \dfrac{B \cdot h}{3}$

$\approx \dfrac{12.56 \text{ ft}^2 \cdot 11 \text{ ft}}{3}$

$\approx \underline{46.1 \text{ ft}^3}$ (rounded)

6. First find the area of the square base.

$B = s \cdot s$

$= 10 \text{ m} \cdot 10 \text{ m}$

$= \underline{100 \text{ m}^2}$

Now find the volume of the pyramid.

$V = \dfrac{B \cdot h}{3}$

$= \dfrac{100 \text{ m}^2 \cdot 8 \text{ m}}{3}$

$\approx \underline{266.7 \text{ m}^3}$ (rounded)

8.7 Section Exercises

1. It is not a cube because the edges have different measurements; in a cube, all edges measure the same. The units will be cm³.

3. The figure is a rectangular solid.

$V = l \cdot w \cdot h$

$= 12.5 \text{ cm} \cdot 4 \text{ cm} \cdot 11 \text{ cm}$

$= 550 \text{ cm}^3$

5. The figure is a sphere.

$V = \dfrac{4}{3} \cdot \pi \cdot r^3$

$\approx \dfrac{4 \cdot 3.14 \cdot 22 \text{ m} \cdot 22 \text{ m} \cdot 22 \text{ m}}{3}$

$\approx 44{,}579.6 \text{ m}^3$ (rounded)

7. The figure is a hemisphere.

$V = \dfrac{2}{3} \cdot \pi \cdot r^3$

$V \approx \dfrac{2 \cdot 3.14 \cdot 12 \text{ in.} \cdot 12 \text{ in.} \cdot 12 \text{ in.}}{3}$

$V \approx 3617.3 \text{ in.}^3$ (rounded to tenths)

9. The figure is a cylinder.

$V = \pi \cdot r^2 \cdot h$

$\approx 3.14 \cdot 5 \text{ ft} \cdot 5 \text{ ft} \cdot 6 \text{ ft}$

$= 471 \text{ ft}^3$

11. The figure is a cone.
First find B, the area of the circular base.

$B = \pi \cdot r \cdot r$

$B \approx 3.14 \cdot 5 \text{ m} \cdot 5 \text{ m}$

$B = 78.5 \text{ m}^2$

Now find the volume of the cone.

$V = \dfrac{B \cdot h}{3}$

$V \approx \dfrac{78.5 \text{ m}^2 \cdot 16 \text{ m}}{3}$

$V \approx 418.7 \text{ m}^3$ (rounded to tenths)

13. The figure is a pyramid.

First find the area of the rectangular base.

$B = l \cdot w$

$= 8 \text{ cm} \cdot 15 \text{ cm}$

$= 120 \text{ cm}^2$

Now find the volume of the pyramid.

$V = \dfrac{B \cdot h}{3}$

$= \dfrac{120 \text{ cm}^2 \cdot 20 \text{ cm}}{3}$

$= 800 \text{ cm}^3$

Copyright © 2018 Pearson Education, Inc.

15. Use the formula for the volume of a rectangular solid.

$$V = l \cdot w \cdot h$$
$$= 3 \text{ in.} \cdot 8 \text{ in.} \cdot \frac{3}{4} \text{ in.}$$
$$= \frac{\overset{6}{\cancel{24}}}{1} \text{ in.}^2 \cdot \frac{3}{\underset{1}{\cancel{4}}} \text{ in.}$$
$$= 18 \text{ in.}^3$$

The volume of the pencil box is 18 in.^{3}.

17. Use the formula for the volume of a sphere. The radius is $\frac{16.8 \text{ cm}}{2} = 8.4$ cm.

$$V = \frac{4}{3} \cdot \pi \cdot r^3$$
$$\approx \frac{4 \cdot 3.14 \cdot 8.4 \text{ cm} \cdot 8.4 \text{ cm} \cdot 8.4 \text{ cm}}{3}$$
$$\approx 2481.5 \text{ cm}^3$$

The volume of the globe is about 2481.5 cm^3.

19. Use the formula for the volume of a pyramid. First find B, the area of the square base.

$$B = s \cdot s$$
$$= 230 \text{ m} \cdot 230 \text{ m}$$
$$= 52,900 \text{ m}^2$$

Now find the volume of the pyramid.

$$V = \frac{B \cdot h}{3}$$
$$= \frac{52,900 \text{ m}^2 \cdot 138 \text{ m}}{3}$$
$$= 2,433,400 \text{ m}^3$$

The volume of the ancient stone pyramid is 2,433,400 m^3.

21. Use the formula for the volume of a cylinder. First find the radius of the pipe.

$$\text{radius} = \frac{5 \text{ ft}}{2} = 2.5 \text{ ft.}$$

Now find the volume.

$$V = \pi \cdot r^2 \cdot h$$
$$V \approx 3.14 \cdot 2.5 \text{ ft} \cdot 2.5 \text{ ft} \cdot 200 \text{ ft}$$
$$V = 3925 \text{ ft}^3$$

The volume of the city sewer pipe is about 3925 ft^3.

23. The first mistake was using the diameter instead of the radius. The second mistake was using square units instead of cubic units. The correct answer is $V = 3.14 \cdot 3.5 \cdot 3.5 \cdot 5 \approx 192.3$ cm^3.

8.8 Pythagorean Theorem

8.8 Margin Exercises

1. **(a)** $\sqrt{36} = 6$ because $6 \cdot 6 = 36$.

   (b) $\sqrt{25} = 5$ because $5 \cdot 5 = 25$.

   (c) $\sqrt{9} = 3$ because $3 \cdot 3 = 9$.

   (d) $\sqrt{100} = 10$ because $10 \cdot 10 = 100$.

   (e) $\sqrt{121} = 11$ because $11 \cdot 11 = 121$.

2. **(a)** $\sqrt{11}$
   Calculator shows 3.31662479; round to 3.317.

   (b) $\sqrt{40}$
   Calculator shows 6.32455532; round to 6.325.

   (c) $\sqrt{56}$
   Calculator shows 7.48331477; round to 7.483.

   (d) $\sqrt{196}$
   Calculator shows 14;
   $\sqrt{196} = 14$ because $14 \cdot 14 = 196$.

   (e) $\sqrt{147}$
   Calculator shows 12.12435565; round to 12.124.

3. **(a)** legs: 5 in. and 12 in.

   $$\text{hypotenuse} = \sqrt{(\text{leg})^2 + (\text{leg})^2}$$
   $$= \sqrt{(5)^2 + (12)^2}$$
   $$= \sqrt{25 + \underline{144}}$$
   $$= \sqrt{\underline{169}}$$
   $$= \underline{13 \text{ in.}}$$

   (b) hypotenuse: 25 cm, leg: 7 cm

   $$\text{leg} = \sqrt{(\text{hypotenuse})^2 - (\text{leg})^2}$$
   $$= \sqrt{(25)^2 - (7)^2}$$
   $$= \sqrt{625 - \underline{49}}$$
   $$= \sqrt{\underline{576}}$$
   $$= \underline{24 \text{ cm}}$$

   (c) legs: 17 m and 13 m

   $$\text{hypotenuse} = \sqrt{(\text{leg})^2 + (\text{leg})^2}$$
   $$= \sqrt{(17)^2 + (13)^2}$$
   $$= \sqrt{289 + 169}$$
   $$= \sqrt{458}$$
   $$\approx 21.4 \text{ m}$$

Copyright © 2018 Pearson Education, Inc.

(d) hypotenuse: 20 ft, leg: 18 ft

$$\text{leg} = \sqrt{(\text{hypotenuse})^2 - (\text{leg})^2}$$
$$= \sqrt{(20)^2 - (18)^2}$$
$$= \sqrt{400 - 324}$$
$$= \sqrt{76}$$
$$\approx 8.7 \text{ ft}$$

4.   **(a)** hypotenuse: 25 ft, leg: 20 ft

$$\text{leg} = \sqrt{(\text{hypotenuse})^2 - (\text{leg})^2}$$
$$= \sqrt{(25)^2 - (20)^2}$$
$$= \sqrt{625 - 400}$$
$$= \sqrt{225}$$
$$= 15$$

The bottom of the ladder is 15 ft from the building.

(b) legs: 11 ft and 8 ft

$$\text{hypotenuse} = \sqrt{(\text{leg})^2 + (\text{leg})^2}$$
$$= \sqrt{(11)^2 + (8)^2}$$
$$= \sqrt{121 + 64}$$
$$= \sqrt{185}$$
$$\approx 13.6 \text{ ft}$$

The ladder is about 13.6 ft long.

(c) hypotenuse: 17 ft, leg: 10 ft

$$\text{leg} = \sqrt{(\text{hypotenuse})^2 - (\text{leg})^2}$$
$$= \sqrt{(17)^2 - (10)^2}$$
$$= \sqrt{289 - 100}$$
$$= \sqrt{189}$$
$$\approx 13.7 \text{ ft}$$

The ladder will reach about 13.7 ft high up on the building.

8.8 Section Exercises

1. $\sqrt{16} = 4$ because $4 \cdot 4 = 16$.

3. $\sqrt{64} = 8$ because $8 \cdot 8 = 64$.

5. $\sqrt{11}$
Calculator shows 3.31662479; round to 3.317.

7. $\sqrt{5}$
Calculator shows 2.236067977; rounds to 2.236.

9. $\sqrt{73}$
Calculator shows 8.544003745; rounds to 8.544.

11. $\sqrt{101}$
Calculator shows 10.04987562; rounds to 10.050.

13. $\sqrt{190}$
Calculator shows 13.78404875; rounds to 13.784.

15. $\sqrt{1000}$
Calculator shows 31.6227766; rounds to 31.623.

17.   30 is about halfway between 25 and 36, so $\sqrt{30}$ should be about halfway between 5 and 6, or about 5.5. Using a calculator, $\sqrt{30} \approx 5.477$. Similarly, $\sqrt{26}$ should be a little more than $\sqrt{25}$. By calculator, $\sqrt{26} \approx 5.099$. Also, $\sqrt{35}$ should be a little less than $\sqrt{36}$. By calculator, $\sqrt{35} \approx 5.916$.

19.   legs: 15 ft and 36 ft

$$\text{hypotenuse} = \sqrt{(\text{leg})^2 + (\text{leg})^2}$$
$$= \sqrt{(15)^2 + (36)^2}$$
$$= \sqrt{225 + 1296}$$
$$= \sqrt{1521}$$
$$= 39 \text{ ft}$$

21.   legs: 8 in. and 15 in.

$$\text{hypotenuse} = \sqrt{(\text{leg})^2 + (\text{leg})^2}$$
$$= \sqrt{(8)^2 + (15)^2}$$
$$= \sqrt{64 + 225}$$
$$= \sqrt{289}$$
$$= 17 \text{ in.}$$

23.   hypotenuse: 20 mm, leg: 16 mm

$$\text{leg} = \sqrt{(\text{hypotenuse})^2 - (\text{leg})^2}$$
$$= \sqrt{(20)^2 - (16)^2}$$
$$= \sqrt{400 - 256}$$
$$= \sqrt{144}$$
$$= 12 \text{ mm}$$

Copyright © 2018 Pearson Education, Inc.

25. The unknown length is the side opposite the right angle, which is the hypotenuse. The lengths of the legs are 8 in. and 3 in.

$$\text{hypotenuse} = \sqrt{(\text{leg})^2 + (\text{leg})^2}$$
$$\text{hypotenuse} = \sqrt{(8)^2 + (3)^2}$$
$$\text{hypotenuse} = \sqrt{64 + 9}$$
$$\text{hypotenuse} = \sqrt{73}$$
$$\text{hypotenuse} \approx 8.5 \text{ in.}$$

27. legs: 7 yd and 4 yd

$$\text{hypotenuse} = \sqrt{(\text{leg})^2 + (\text{leg})^2}$$
$$= \sqrt{(7)^2 + (4)^2}$$
$$= \sqrt{49 + 16}$$
$$= \sqrt{65}$$
$$\approx 8.1 \text{ yd}$$

29. hypotenuse: 22 cm, leg: 17 cm

$$\text{leg} = \sqrt{(\text{hypotenuse})^2 - (\text{leg})^2}$$
$$= \sqrt{(22)^2 - (17)^2}$$
$$= \sqrt{484 - 289}$$
$$= \sqrt{195}$$
$$\approx 14.0 \text{ cm}$$

31. legs: 1.3 m and 2.5 m

$$\text{hypotenuse} = \sqrt{(\text{leg})^2 + (\text{leg})^2}$$
$$= \sqrt{(1.3)^2 + (2.5)^2}$$
$$= \sqrt{1.69 + 6.25}$$
$$= \sqrt{7.94}$$
$$\approx 2.8 \text{ m}$$

33. hypotenuse: 11.5 cm, leg: 8.2 cm

$$\text{leg} = \sqrt{(\text{hypotenuse})^2 - (\text{leg})^2}$$
$$= \sqrt{(11.5)^2 - (8.2)^2}$$
$$= \sqrt{132.25 - 67.24}$$
$$= \sqrt{65.01}$$
$$\approx 8.1 \text{ cm}$$

35. The length of the hypotenuse is 21.6 km. The length of one of the legs is 13.2 km.

$$\text{leg} = \sqrt{(\text{hypotenuse})^2 - (\text{leg})^2}$$
$$\text{leg} = \sqrt{(21.6)^2 - (13.2)^2}$$
$$\text{leg} = \sqrt{466.56 - 174.24}$$
$$\text{leg} = \sqrt{292.32}$$
$$\text{leg} \approx 17.1 \text{ km}$$

37. The first mistake was squaring the numbers incorrectly: 9^2 is 81 and 7^2 is 49. The second mistake was rounding the final answer to the nearest thousandth instead of the nearest tenth. The correct answer is:

$$\sqrt{(9)^2 + (7)^2} = \sqrt{81 + 49}$$
$$= \sqrt{130}$$
$$\approx 11.4 \text{ in. (rounded)}$$

39. legs: 4 ft and 7 ft

$$\text{hypotenuse} = \sqrt{(\text{leg})^2 + (\text{leg})^2}$$
$$= \sqrt{(4)^2 + (7)^2}$$
$$= \sqrt{16 + 49}$$
$$= \sqrt{65}$$
$$\approx 8.1 \text{ ft}$$

The length of the loading ramp is about 8.1 ft.

41. hypotenuse: 1000 m, leg: 800 m

$$\text{leg} = \sqrt{(\text{hypotenuse})^2 - (\text{leg})^2}$$
$$= \sqrt{(1000)^2 - (800)^2}$$
$$= \sqrt{1,000,000 - 640,000}$$
$$= \sqrt{360,000}$$
$$= 600 \text{ m}$$

The airplane is 600 m above the ground.

43. The diagonal brace is the hypotenuse. The lengths of the legs are 4.5 ft and 3.5 ft.

$$\text{hypotenuse} = \sqrt{(\text{leg})^2 + (\text{leg})^2}$$
$$\text{hypotenuse} = \sqrt{(4.5)^2 + (3.5)^2}$$
$$\text{hypotenuse} = \sqrt{20.25 + 12.25}$$
$$\text{hypotenuse} = \sqrt{32.5}$$
$$\text{hypotenuse} \approx 5.7 \text{ ft}$$

The diagonal brace is about 5.7 ft long.

45.

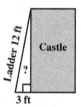

The ladder is opposite the right angle so it is the hypotenuse. Thus, the length of the hypotenuse is 12 ft and the length of one of the legs is 3 ft.

$$\text{leg} = \sqrt{(\text{hypotenuse})^2 - (\text{leg})^2}$$
$$\text{leg} = \sqrt{(12)^2 - (3)^2}$$
$$\text{leg} = \sqrt{144 - 9}$$
$$\text{leg} = \sqrt{135}$$
$$\text{leg} \approx 11.6 \text{ ft}$$

The ladder will reach about 11.6 ft high on the castle.

Copyright © 2018 Pearson Education, Inc.

47. legs: 90 ft and 90 ft

$$\text{hypotenuse} = \sqrt{(\text{leg})^2 + (\text{leg})^2}$$
$$= \sqrt{(90)^2 + (90)^2}$$
$$= \sqrt{8100 + 8100}$$
$$= \sqrt{16,200}$$
$$\approx 127.3$$

The distance from home plate to second base is about 127.3 ft.

48. (a)

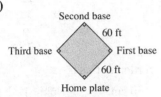

Second base

60 ft

Third base   First base

60 ft

Home plate

(b) legs: 60 ft and 60 ft

$$\text{hypotenuse} = \sqrt{(\text{leg})^2 + (\text{leg})^2}$$
$$= \sqrt{(60)^2 + (60)^2}$$
$$= \sqrt{3600 + 3600}$$
$$= \sqrt{7200}$$
$$\approx 84.9$$

The distance from home plate to second base is about 84.9 ft.

49. The distance from third to first is the same as the distance from home to second because the baseball diamond is a square.

50. One possibility is:

$$\left. \begin{array}{c} \text{major} \\ \text{league} \end{array} \right\{ \ \frac{90 \text{ ft}}{127.3 \text{ ft}} = \frac{60 \text{ ft}}{x} \left. \right\} \text{softball}$$

$$90 \cdot x = 127.3 \cdot 60$$
$$\frac{90 \cdot x}{90} = \frac{7638}{90}$$
$$x \approx 84.9 \text{ ft}$$

8.9 Similar Triangles

8.9 Margin Exercises

1. (a) Corresponding angles have the same measure. The corresponding angles are 1 and $\underline{4}$, 2 and $\underline{5}$, 3 and $\underline{6}$.
$\overline{PN}$ and $\overline{ZX}$ are opposite corresponding angles 3 and 6.
$\overline{PM}$ and $\overline{ZY}$ are opposite corresponding angles 1 and 4.
$\overline{NM}$ and $\overline{XY}$ are opposite corresponding angles 2 and 5.

Thus, the corresponding sides are $\overline{PN}$ and $\overline{ZX}$, $\overline{PM}$ and $\overline{ZY}$, $\overline{NM}$ and $\overline{XY}$.

(b) The corresponding angles are 1 and $\underline{6}$, 2 and $\underline{4}$, 3 and $\underline{5}$.
The corresponding sides are $\overline{AB}$ and $\overline{EF}$, $\overline{BC}$ and $\overline{FG}$, $\overline{AC}$ and $\overline{EG}$.

2. Find the length of $\overline{EF}$.

$$\frac{EF}{CB} = \frac{ED}{CA} \qquad \textit{Corresponding sides}$$

$$\frac{EF}{CB} = \frac{5}{15} \qquad \begin{array}{l} \textit{Replace ED with 5} \\ \textit{and CA with 15.} \end{array}$$

$$\frac{x}{33} = \frac{1}{3} \qquad \begin{array}{l} \textit{Replace EF with x} \\ \textit{and CB with 33.} \end{array}$$

$$3 \cdot x = 33 \cdot 1$$
$$\frac{3 \cdot x}{3} = \frac{33}{3}$$
$$x = 11$$

$\overline{EF}$ has a length of 11 m.

3. (a) Find the length of $\overline{AB}$.
From Example 2 in the text,

$$\frac{PR}{AC} = \frac{7}{14} = \frac{1}{2}, \text{ so } \frac{PQ}{AB} = \frac{1}{2}.$$

Replace PQ with 3 and AB with y.

$$\frac{3}{y} = \frac{1}{2}$$
$$y \cdot 1 = 3 \cdot 2$$
$$y = 6$$

$\overline{AB}$ is 6 ft.
Find the perimeter.
Perimeter $= 14 \text{ ft} + 10 \text{ ft} + 6 \text{ ft} = 30 \text{ ft}$

(b) Set up ratios.

$$\frac{PQ}{AB} = \frac{10 \text{ m}}{30 \text{ m}} = \frac{1}{3}, \text{ so } \frac{QR}{BC} = \frac{1}{3} \text{ and } \frac{PR}{AC} = \frac{1}{3}.$$

Replace QR with x and BC with 18.

$$\frac{x}{18} = \frac{1}{3}$$
$$3 \cdot x = 18 \cdot 1$$
$$\frac{3 \cdot x}{3} = \frac{18}{3}$$
$$x = 6$$

$\overline{QR}$ is 6 m.
The perimeter of triangle PQR is

$$10 \text{ m} + 8 \text{ m} + 6 \text{ m} = 24 \text{ m}.$$

(continued)

Copyright © 2018 Pearson Education, Inc.

Next replace PR with 8 and AC with y.

$$\frac{8}{y} = \frac{1}{3}$$
$$y \cdot 1 = 8 \cdot 3$$
$$y = 24$$

$\overline{AC}$ is 24 m.
The perimeter of triangle ABC is

$$30\text{ m} + 18\text{ m} + 24\text{ m} = 72\text{ m}.$$

4. **(a)** Write a proportion.

$$\frac{\text{longer height}}{\text{shorter height}} = \frac{\text{longer shadow}}{\text{shorter shadow}}$$
$$\frac{h}{5} = \frac{48}{12}$$
$$\frac{h}{5} = \frac{4}{1} \quad \textit{Reduce.}$$
$$1 \cdot h = 5 \cdot 4$$
$$h = 20$$

The flagpole is 20 ft high.

(b) Write a proportion.

$$\frac{\text{longer height}}{\text{shorter height}} = \frac{\text{longer base}}{\text{shorter base}}$$
$$\frac{h}{7.2} = \frac{12.5}{5}$$
$$5 \cdot h = 7.2 \cdot 12.5$$
$$\frac{5 \cdot h}{5} = \frac{90}{5}$$
$$h = 18$$

The flagpole is 18 m high.

8.9 Section Exercises

1. The triangles have the same shape, so they are *similar*.

3. The triangles do not have the same shape, so they are *not similar*.

5. **(a)** In two triangles that are similar, corresponding angles have the same measure.

(b) In two triangles that are similar, the ratios of the lengths of corresponding sides are equal.

7. The corresponding angles are
$\angle 1$ and $\angle 4$, $\angle 2$ and $\angle 5$, $\angle 3$ and $\angle 6$.
The corresponding sides are
$\overline{AB}$ and $\overline{PQ}$, $\overline{BC}$ and $\overline{QR}$, $\overline{AC}$ and $\overline{PR}$.

9. The corresponding angles are
$\angle 1$ and $\angle 6$, $\angle 2$ and $\angle 5$, $\angle 3$ and $\angle 4$.
The corresponding sides are
$\overline{MP}$ and $\overline{QS}$, $\overline{MN}$ and $\overline{QR}$, $\overline{NP}$ and $\overline{RS}$.

11. $\dfrac{AB}{PQ} = \dfrac{9\text{ m}}{6\text{ m}} = \dfrac{9}{6} = \dfrac{3}{2}$
$\dfrac{AC}{PR} = \dfrac{12\text{ m}}{8\text{ m}} = \dfrac{12}{8} = \dfrac{3}{2}$
$\dfrac{BC}{QR} = \dfrac{15\text{ m}}{10\text{ m}} = \dfrac{15}{10} = \dfrac{3}{2}$

13. Write a proportion to find a.

$$\frac{a}{12\text{ cm}} = \frac{6\text{ cm}}{12\text{ cm}} \quad \textbf{OR} \quad \frac{a}{12} = \frac{1}{2}$$
$$2 \cdot a = 12 \cdot 1$$
$$\frac{2 \cdot a}{2} = \frac{12}{2}$$
$$a = 6\text{ cm}$$

Write a proportion to find b.

$$\frac{7.5\text{ cm}}{b} = \frac{6\text{ cm}}{12\text{ cm}} \quad \textbf{OR} \quad \frac{7.5}{b} = \frac{1}{2}$$
$$1 \cdot b = 2 \cdot 7.5$$
$$b = 15\text{ cm}$$

15. Set up a ratio of corresponding sides, using the longest side in each triangle.

$$\frac{6\text{ mm}}{12\text{ mm}} = \frac{6}{12} = \frac{1}{2}$$

Write a proportion to find a.

$$\frac{a}{10} = \frac{1}{2}$$
$$a \cdot 2 = 10 \cdot 1$$
$$\frac{a \cdot 2}{2} = \frac{10}{2}$$
$$a = 5\text{ mm}$$

Write a proportion to find b.

$$\frac{b}{6} = \frac{1}{2}$$
$$b \cdot 2 = 6 \cdot 1$$
$$\frac{b \cdot 2}{2} = \frac{6}{2}$$
$$b = 3\text{ mm}$$

17. Write a proportion to find x.

$$\frac{x}{18.6} = \frac{28}{21} \quad \textbf{OR} \quad \frac{x}{18.6} = \frac{4}{3}$$
$$3 \cdot x = 4 \cdot 18.6$$
$$\frac{3 \cdot x}{3} = \frac{74.4}{3}$$
$$x = 24.8\text{ m}$$

$$P = 24.8\text{ m} + 28\text{ m} + 20\text{ m} = 72.8\text{ m}$$

(continued)

Copyright © 2018 Pearson Education, Inc.

Write a proportion to find y.

$$\frac{y}{20} = \frac{21}{28} \quad \textbf{OR} \quad \frac{y}{20} = \frac{3}{4}$$
$$4 \cdot y = 20 \cdot 3$$
$$\frac{4 \cdot y}{4} = \frac{60}{4}$$
$$y = 15 \text{ m}$$

$$P = 15 \text{ m} + 21 \text{ m} + 18.6 \text{ m} = 54.6 \text{ m}$$

19. First, find the lengths of $\overline{FH}$ and $\overline{FG}$ in triangle FGH. Set up a ratio of corresponding sides where x is the length of side FH.

$$\frac{DE}{GH} = \frac{CE}{FH}$$
$$\frac{12}{8} = \frac{12}{x}$$
$$12 \cdot x = 8 \cdot 12$$
$$\frac{12 \cdot x}{12} = \frac{96}{12}$$
$$x = 8$$

Note that when you set up a proportion to find the length of $\overline{FG}$, it is the same as that used to find the length of $\overline{FH}$. Therefore the length of $\overline{FG}$ is also 8 cm.

Perimeter of triangle FGH
$$P = 8 \text{ cm} + 8 \text{ cm} + 8 \text{ cm}$$
$$P = 24 \text{ cm}$$

Set up a ratio of corresponding sides to find the height h of triangle FGH.

$$\frac{10.4}{12} = \frac{h}{8}$$
$$12 \cdot h = 8 \cdot 10.4$$
$$\frac{12 \cdot h}{12} = \frac{83.2}{12}$$
$$h \approx 6.9 \text{ cm}$$

Area of triangle FGH

$$A = 0.5 \cdot b \cdot h$$
$$A \approx 0.5 \cdot 8 \text{ cm} \cdot 6.9 \text{ cm}$$
$$A = 27.6 \text{ cm}^2$$

21. Write a proportion to find h.

$$\frac{2}{16} = \frac{3}{h}$$
$$2 \cdot h = 3 \cdot 16$$
$$\frac{2 \cdot h}{2} = \frac{48}{2}$$
$$h = 24 \text{ ft}$$

The height of the house is 24 ft.

23. Let s be the length of the stick's shadow and h is the height of the house (24 ft) from Exercise 21. Write a proportion to find s.

$$\frac{h}{3} = \frac{6}{s}$$
$$\frac{24}{3} = \frac{6}{s} \quad (h = 24 \text{ from Exercise 21})$$
$$\frac{8}{1} = \frac{6}{s} \quad \text{Write } \frac{24}{3} \text{ in lowest terms as } \frac{8}{1}.$$
$$8 \cdot s = 1 \cdot 6$$
$$\frac{8 \cdot s}{8} = \frac{6}{8}$$
$$s = \frac{3}{4} \text{ ft} \quad \text{or} \quad 0.75 \text{ ft}$$

The stick's shadow would be $\frac{3}{4}$ foot at that time.

25. Let x be Lu's height.

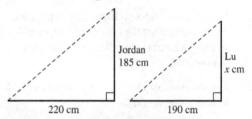

Write a proportion to find x.

$$\frac{x}{185} = \frac{190}{220}$$
$$220 \cdot x = 185 \cdot 190$$
$$\frac{220 \cdot x}{220} = \frac{35{,}150}{220}$$
$$x \approx 160 \text{ cm}$$

Lu is about 160 cm (rounded) tall.

27. Using the hint, we can write a proportion to find x.

$$\frac{x}{120} = \frac{100}{100 + 140} \quad \textbf{OR} \quad \frac{x}{120} = \frac{5}{12}$$
$$x \cdot 12 = 120 \cdot 5$$
$$\frac{x \cdot 12}{12} = \frac{600}{12}$$
$$x = 50 \text{ m}$$

29. Write a proportion to find n.

$$\frac{50}{n} = \frac{100}{100 + 120} \quad \textbf{OR} \quad \frac{50}{n} = \frac{5}{11}$$
$$5 \cdot n = 50 \cdot 11$$
$$\frac{5 \cdot n}{5} = \frac{550}{5}$$
$$n = 110 \text{ m}$$

The length of the lake is 110 m

Copyright © 2018 Pearson Education, Inc.

Chapter 8 Review Exercises

1. The figure has two endpoints so it is a *line segment* named $\overline{AB}$ or $\overline{BA}$.

2. This is a *line* named $\overleftrightarrow{CD}$ or $\overleftrightarrow{DC}$. A line is a straight row of points that goes on forever in both directions.

3. This is a *ray* named $\overrightarrow{OP}$. A ray is a part of a line that has only one endpoint and goes on forever in one direction.

4. These lines appear to be *parallel* lines. Parallel lines are lines in the same plane that never intersect (cross).

5. The lines are *perpendicular* because they intersect at right angles.

6. The lines intersect so they are *not* parallel. At their intersection they *do not* form a right angle so they are not perpendicular. The lines are *intersecting*.

7. The measure of the angle is between $0°$ and $90°$, so it is an *acute angle.*

8. The measure of the angle is between $90°$ and $180°$, so it is an *obtuse angle.*

9. Two rays in a straight line pointing opposite directions measure $180°$. An angle that measures $180°$ is called a *straight angle.*

10. The angle is a *right angle*, as indicated by the small square at the vertex. Right angles measure exactly $90°$.

11. $\angle 4$ measures $60°$ since $\angle 4$ and the $60°$ angle are vertical angles, which are congruent.
$\angle 4 + \angle 3 = 90°$, so $\angle 3 = 90° - 60° = 30°$.
$\angle 1$ also measures $30°$ since $\angle 1$ and $\angle 3$ are vertical angles.
$\angle 2$ measures $90°$ since $\angle 2$ and the $90°$ angle marked with the small square are vertical angles.

12. $\angle 1$ measures $100°$ since it is a vertical angle with the $100°$ angle and vertical angles are congruent.
$\angle 4 + 35° + 100° = 180°$ since they form a straight angle. Therefore, $\angle 4$ measures $180° - 35° - 100° = 45°$.
$\angle 2$ also measures $45°$ since $\angle 2$ and $\angle 4$ are vertical angles and vertical angles are congruent.
$\angle 3$ measures $35°$ since it is a vertical angle with the $35°$ angle.

13. $\angle AOB$ and $\angle BOC$,
$\angle BOC$ and $\angle COD$,
$\angle COD$ and $\angle DOA$, and

$\angle DOA$ and $\angle AOB$
are supplementary because the sum of each pair of angles is $180°$.

14. $\angle ERH$ and $\angle HRG$,
$\angle HRG$ and $\angle GRF$,
$\angle GRF$ and $\angle FRE$, and
$\angle FRE$ and $\angle ERH$
are supplementary because the sum of each pair of angles is $180°$.

15. **(a)** The complement of $80°$ is $10°$, because $90° - 80° = 10°$.

   (b) The complement of $45°$ is $45°$, because $90° - 45° = 45°$.

   (c) The complement of $7°$ is $83°$, because $90° - 7° = 83°$.

16. **(a)** The supplement of $155°$ is $25°$, because $180° - 155° = 25°$.

   (b) The supplement of $90°$ is $90°$, because $180° - 90° = 90°$.

   (c) The supplement of $33°$ is $147°$, because $180° - 33° = 147°$.

17. The figure is a rectangle.

$$P = 1.5\,\text{m} + 0.92\,\text{m} + 1.5\,\text{m} + 0.92\,\text{m}$$
$$= 4.84\,\text{m}$$

18. The figure is a square. So all 4 sides have the same measurement.

$$P = 32\,\text{in.} + 32\,\text{in.} + 32\,\text{in.} + 32\,\text{in.}$$
$$= 128\,\text{in.}$$

19. $P = 4 \cdot s$
$= 4 \cdot 38\,\text{cm}$
$= 152\,\text{cm}$

To trim all the edges, 152 cm of trim is needed.

20. $P = 2 \cdot \text{length} + 2 \cdot \text{width}$
$= 2 \cdot 12\,\text{ft} + 2 \cdot 8\frac{1}{2}\,\text{ft}$
$= 24\,\text{ft} + 17\,\text{ft}$
$= 41\,\text{ft}$

41 feet of fencing is needed to surround the garden.

21. $A = \text{length} \cdot \text{width}$
$= 27\,\text{mm} \cdot 18\,\text{mm}$
$= 486\,\text{mm}^2$

Copyright © 2018 Pearson Education, Inc.

22. $A = \text{length} \cdot \text{width}$
$= 5\frac{1}{2}$ ft $\cdot$ 3 ft
$= 5.5$ ft $\cdot$ 3 ft
$= 16.5$ ft^2 or $16\frac{1}{2}$ ft^2

23. $A = s^2$
$= 6.3$ m $\cdot$ 6.3 m
$= 39.69$ m^2
≈ 39.7 m^2 (rounded)

24. The figure is a parallelogram.
$P = 11$ cm $+ 14$ cm $+ 11$ cm $+ 14$ cm
$= 50$ cm
$A = b \cdot h$
$= 14$ cm $\cdot$ 10 cm
$= 140$ cm^2

25. The figure is a trapezoid.
$P = 26$ ft $+ 21.1$ ft $+ 37$ ft $+ 18$ ft
$= 102.1$ ft
$A = \frac{1}{2} \cdot h \cdot (b + B)$
$= \frac{1}{2} \cdot 18$ ft $\cdot (26$ ft $+ 37$ ft$)$
$= 9$ ft $\cdot 63$ ft
$= 567$ ft^2

26. The figure is a trapezoid.
$P = 33.9$ m $+ 59.7$ m $+ 34.2$ m $+ 72.4$ m
$= 200.2$ m
$A = 0.5 \cdot h \cdot (b + B)$
$= 0.5 \cdot 33.4$ m $\cdot (59.7$ m $+ 72.4$ m$)$
$= 0.5 \cdot 33.4$ m $\cdot 132.1$ m
$= 2206.07$ m$^2 \approx 2206.1$ m^2 (rounded)

27. $P = 153$ cm $+ 153$ cm $+ 212$ cm
$= 518$ cm
$A = \frac{1}{2} \cdot b \cdot h$
$= \frac{1}{2} \cdot 212$ cm $\cdot 110$ cm
$= 11,660$ cm^2

28. $P = 5$ m $+ 12.3$ m $+ 9.8$ m
$= 27.1$ m
$A = \frac{1}{2} \cdot b \cdot h$
$= \frac{1}{2} \cdot 9.8$ m $\cdot 4.8$ m
$= 23.52$ m^2

29. $P = 8$ ft $+ 3\frac{1}{2}$ ft $+ 8\frac{3}{4}$ ft
$= 8$ ft $+ 3.5$ ft $+ 8.75$ ft
$= 20.25$ ft or $20\frac{1}{4}$ ft
$A = \frac{1}{2} \cdot b \cdot h$
$= \frac{1}{2} \cdot 3.5$ ft $\cdot 8$ ft
$= 14$ ft^2

30. *Step 1* $70° + 40° = 110°$
Step 2 $180° - 110° = 70°$
The unlabeled angle measures $70°$.

31. *Step 1* $90° + 66° = 156°$
Step 2 $180° - 156° = 24°$
The unlabeled angle measures $24°$.

32. $d = 2 \cdot r = 2 \cdot 68.9$ m $= 137.8$ m
The diameter of the field is 137.8 m.

33. $r = \dfrac{d}{2} = \dfrac{3 \text{ in.}}{2} = 1\frac{1}{2}$ in. or 1.5 in.

The radius of the juice can is $1\frac{1}{2}$ in. or 1.5 in.

34. radius $= 1$ cm, so $d = 2 \cdot 1$ cm $= 2$ cm.
$C = \pi \cdot d$
$\approx 3.14 \cdot 2$ cm
≈ 6.3 cm (rounded)
$A = \pi \cdot r \cdot r$
$\approx 3.14 \cdot 1$ cm $\cdot 1$ cm
≈ 3.1 cm^2 (rounded)

35. radius $= 17.4$ m, so $d = 2 \cdot 17.4$ m $= 34.8$ m.
$C = \pi \cdot d$
$\approx 3.14 \cdot 34.8$ m
≈ 109.3 m (rounded)
$A = \pi \cdot r \cdot r$
$\approx 3.14 \cdot 17.4$ m $\cdot 17.4$ m
≈ 950.7 m^2 (rounded)

36. diameter $= 12$ in., so $r = \frac{1}{2} \cdot 12$ in. $= 6$ in.
$C = \pi \cdot d$
$\approx 3.14 \cdot 12$ in.
≈ 37.7 in. (rounded)
$A = \pi \cdot r \cdot r$
$\approx 3.14 \cdot 6$ in. $\cdot 6$ in.
≈ 113.0 in.^{2} (rounded)

37. radius $= 3.6$ m
Area of circle:
$A = \pi \cdot r \cdot r$
$\approx 3.14 \cdot 3.6$ m $\cdot 3.6$ m
$= 40.6944$ m^2

(*continued*)

Copyright © 2018 Pearson Education, Inc.

Area of semicircle:

$$\frac{40.6944 \text{ m}^2}{2} \approx 20.3 \text{ m}^2 \text{ (rounded)}$$

38. Draw a horizontal line to break up the shaded area into 2 rectangles that have a length of 8 inches and a width of 4 inches.

$$\begin{aligned} A &= \text{length} \cdot \text{width} \\ &= 8 \text{ in.} \cdot 4 \text{ in.} \\ &= 32 \text{ in.}^2 \end{aligned}$$

Total area $= 2 \times 32 \text{ in.}^2 = 64 \text{ in.}^2$

39. Draw a vertical line to break up the figure into 3 parts.

(1) a rectangle with length 24 km and width 12 km

$$\begin{aligned} A &= 24 \text{ km} \cdot 12 \text{ km} \\ &= 288 \text{ km}^2 \end{aligned}$$

(2) a parallelogram with base 24 km and height 11 km

$$\begin{aligned} A &= b \cdot h \\ &= 24 \text{ km} \cdot 11 \text{ km} \\ &= 264 \text{ km}^2 \end{aligned}$$

(3) a square with side 11 km

$$\begin{aligned} A &= s \cdot s \\ &= 11 \text{ km} \cdot 11 \text{ km} \\ &= 121 \text{ km}^2 \end{aligned}$$

Total area $= 288 \text{ km}^2 + 264 \text{ km}^2 + 121 \text{ km}^2$
$$= 673 \text{ km}^2$$

40. Draw a vertical line to break up the figure into 2 parts.

(1) a rectangle with length 45 m and width 21 m

$$\begin{aligned} A &= 45 \text{ m} \cdot 21 \text{ m} \\ &= 945 \text{ m}^2 \end{aligned}$$

(2) a triangle with base 15 m and height 10 m

$$\begin{aligned} A &= \tfrac{1}{2} \cdot b \cdot h \\ &= \tfrac{1}{2} \cdot 15 \text{ m} \cdot 10 \text{ m} \\ &= 75 \text{ m}^2 \end{aligned}$$

Total area $= 945 \text{ m}^2 + 75 \text{ m}^2$
$$= 1020 \text{ m}^2$$

41. Draw 2 horizontal lines to break up the figure into 3 parts.

(1) 2 rectangles with length 15 ft and width 6 ft

$$\begin{aligned} A &= \text{length} \cdot \text{width} \\ &= 15 \text{ ft} \cdot 6 \text{ ft} \\ &= 90 \text{ ft}^2 \end{aligned}$$

(2) a square 7 ft by 7 ft

$$\begin{aligned} A &= s \cdot s \\ &= 7 \text{ ft} \cdot 7 \text{ ft} \\ &= 49 \text{ ft}^2 \end{aligned}$$

Total area $= 2 \times 90 \text{ ft}^2 + 49 \text{ ft}^2$
$$= 229 \text{ ft}^2$$

42. Find the area of the trapezoid and subtract the area of the triangle.

Area of trapezoid:

$$\begin{aligned} A &= \tfrac{1}{2} \cdot h \cdot (b + B) \\ &= \tfrac{1}{2} \cdot 12 \text{ ft} \cdot (11 \text{ ft} + 18 \text{ ft}) \\ &= \tfrac{1}{2} \cdot 12 \text{ ft} \cdot (29 \text{ ft}) \\ &= 174 \text{ ft}^2 \end{aligned}$$

Area of triangle:

$$\begin{aligned} A &= \tfrac{1}{2} \cdot b \cdot h \\ &= \tfrac{1}{2} \cdot 12 \text{ ft} \cdot 7 \text{ ft} \\ &= 42 \text{ ft}^2 \end{aligned}$$

Total shaded area $= 174 \text{ ft}^2 - 42 \text{ ft}^2$
$$= 132 \text{ ft}^2$$

43. Find the area of the rectangle and subtract the area of the two unshaded triangles.

Area of rectangle:

$$\begin{aligned} A &= \text{length} \cdot \text{width} \\ &= (48 \text{ cm} + 48 \text{ cm}) \cdot 74 \text{ cm} \\ &= 96 \text{ cm} \cdot 74 \text{ cm} \\ &= 7104 \text{ cm}^2 \end{aligned}$$

Area of triangle:

$$\begin{aligned} A &= \tfrac{1}{2} \cdot b \cdot h \\ &= \tfrac{1}{2} \cdot 48 \text{ cm} \cdot 36 \text{ cm} \\ &= 864 \text{ cm}^2 \end{aligned}$$

Total shaded area
$$= 7104 \text{ cm}^2 - \left(864 \text{ cm}^2 + 864 \text{ cm}^2 \right)$$
$$= 7104 \text{ cm}^2 - 1728 \text{ cm}^2$$
$$= 5376 \text{ cm}^2$$

Copyright © 2018 Pearson Education, Inc.

44. Find the area of the rectangle and subtract the area of the semicircle.

Area of rectangle:

$$A = \text{length} \cdot \text{width}$$
$$= 32 \text{ ft} \cdot 21 \text{ ft}$$
$$= 672 \text{ ft}^2$$

Area of circle $\left(\text{radius} = \frac{d}{2} = \frac{21 \text{ ft}}{2} = 10.5 \text{ ft}\right)$:

$$A = \pi \cdot r \cdot r$$
$$\approx 3.14 \cdot 10.5 \text{ ft} \cdot 10.5 \text{ ft}$$
$$= 346.185 \text{ ft}^2$$

Area of semicircle:

$$\frac{346.185 \text{ ft}^2}{2} \approx 173.1 \text{ ft}^2 \text{ (rounded)}$$

Total shaded area is approximately

$$672 \text{ ft}^2 - 173.1 \text{ ft}^2 = 498.9 \text{ ft}^2.$$

45. Find the area of the rectangle.

$$A = \text{length} \cdot \text{width}$$
$$= 21 \text{ yd} \cdot 14 \text{ yd}$$
$$= 294 \text{ yd}^2$$

The two semicircles at the ends make one circle.

$$A = \pi \cdot r \cdot r$$
$$\approx 3.14 \cdot 7 \text{ yd} \cdot 7 \text{ yd}$$
$$\approx 153.9 \text{ yd}^2$$

The total area is approximately

$$294 \text{ yd}^2 + 153.9 \text{ yd}^2 = 447.9 \text{ yd}^2 \text{ (rounded)}.$$

46. The figure is a rectangular solid.

$$V = \text{length} \cdot \text{width} \cdot \text{height}$$
$$= 4 \text{ in.} \cdot 3 \text{ in.} \cdot 2\frac{1}{2} \text{ in.}$$
$$= 30 \text{ in.}^3$$

47. The figure is a rectangular solid.

$$V = \text{length} \cdot \text{width} \cdot \text{height}$$
$$= 6 \text{ cm} \cdot 4 \text{ cm} \cdot 4 \text{ cm}$$
$$= 96 \text{ cm}^3$$

48. The figure is a rectangular solid.

$$V = \text{length} \cdot \text{width} \cdot \text{height}$$
$$= 30 \text{ mm} \cdot 20 \text{ mm} \cdot 75 \text{ mm}$$
$$= 45{,}000 \text{ mm}^3$$

49. The figure is a sphere.

$$V = \frac{4}{3} \cdot \pi \cdot r^3$$
$$\approx \frac{4}{3} \cdot 3.14 \cdot (4 \text{ m})^3$$
$$= \frac{4}{3} \cdot 3.14 \cdot 64 \text{ m}^3$$
$$\approx 267.9 \text{ m}^3 \text{ (rounded)}$$

50. The figure is a hemisphere.

$$V = \frac{2}{3} \cdot \pi \cdot r^3$$
$$\approx \frac{2}{3} \cdot 3.14 \cdot (6 \text{ ft})^3$$
$$= \frac{2}{3} \cdot 3.14 \cdot 216 \text{ ft}^3$$
$$\approx 452.2 \text{ ft}^3 \text{ (rounded)}$$

51. The figure is a cylinder.

$$V = \pi \cdot r^2 \cdot h$$
$$\approx 3.14 \cdot 5 \text{ cm} \cdot 5 \text{ cm} \cdot 7 \text{ cm}$$
$$= 549.5 \text{ cm}^3$$

52. The figure is a cylinder.

$$V = \pi \cdot r^2 \cdot h \quad (d = 24 \text{ m, so } r = 12 \text{ m})$$
$$\approx 3.14 \cdot 12 \text{ m} \cdot 12 \text{ m} \cdot 4 \text{ m}$$
$$\approx 1808.6 \text{ m}^3 \text{ (rounded)}$$

53. The figure is a cone.

$$V = \frac{1}{3} \cdot \pi \cdot r^2 \cdot h$$
$$\approx \frac{1}{3} \cdot 3.14 \cdot 7 \text{ m} \cdot 7 \text{ m} \cdot 10 \text{ m}$$
$$\approx 512.9 \text{ m}^3 \text{ (rounded)}$$

54. The figure is a pyramid.

$$V = \frac{1}{3} \cdot B \cdot h$$
$$= \frac{1}{3}(4 \text{ yd} \cdot 3 \text{ yd}) \cdot 4 \text{ yd}$$
$$= \frac{1}{3} \cdot 12 \text{ yd}^2 \cdot 4 \text{ yd}$$
$$= 16 \text{ yd}^3$$

55. $\sqrt{49} = 7$ because $7 \cdot 7 = 49$.

56. $\sqrt{8}$
Calculator shows 2.828427125; round to 2.828.

57. $\sqrt{3000}$
Calculator shows 54.77225575; round to 54.772.

58. $\sqrt{144} = 12$ because $12 \cdot 12 = 144$.

59. $\sqrt{58}$
Calculator shows 7.615773106; round to 7.616.

60. $\sqrt{625} = 25$ because $25 \cdot 25 = 625$.

61. $\sqrt{105}$
Calculator shows 10.24695077; round to 10.247.

62. $\sqrt{80}$
Calculator shows 8.94427191; round to 8.944.

Copyright © 2018 Pearson Education, Inc.

63. $\text{hypotenuse} = \sqrt{(\text{leg})^2 + (\text{leg})^2}$
$= \sqrt{(15)^2 + (8)^2}$
$= \sqrt{225 + 64}$
$= \sqrt{289}$
$= 17 \text{ in.}$

64. $\text{leg} = \sqrt{(\text{hypotenuse})^2 - (\text{leg})^2}$
$= \sqrt{(25)^2 - (24)^2}$
$= \sqrt{625 - 576}$
$= \sqrt{49}$
$= 7 \text{ cm}$

65. $\text{leg} = \sqrt{(\text{hypotenuse})^2 - (\text{leg})^2}$
$= \sqrt{(15)^2 - (11)^2}$
$= \sqrt{225 - 121}$
$= \sqrt{104}$
$\approx 10.2 \text{ cm}$

66. $\text{hypotenuse} = \sqrt{(\text{leg})^2 + (\text{leg})^2}$
$= \sqrt{(6)^2 + (4)^2}$
$= \sqrt{36 + 16}$
$= \sqrt{52}$
$\approx 7.2 \text{ in.}$

67. $\text{hypotenuse} = \sqrt{(\text{leg})^2 + (\text{leg})^2}$
$= \sqrt{(2.2)^2 + (1.3)^2}$
$= \sqrt{4.84 + 1.69}$
$= \sqrt{6.53}$
$\approx 2.6 \text{ m}$

68. $\text{leg} = \sqrt{(\text{hypotenuse})^2 - (\text{leg})^2}$
$= \sqrt{(12)^2 - (8.5)^2}$
$= \sqrt{144 - 72.25}$
$= \sqrt{71.75}$
$\approx 8.5 \text{ km}$

69. Write a proportion to find y.
$$\frac{y}{15 \text{ ft}} = \frac{40 \text{ ft}}{20 \text{ ft}} \quad \textbf{OR} \quad \frac{y}{15} = \frac{2}{1}$$
$$y \cdot 1 = 2 \cdot 15$$
$$y = 30 \text{ ft}$$

Write a proportion to find x.
$$\frac{x}{17 \text{ ft}} = \frac{40 \text{ ft}}{20 \text{ ft}} \quad \textbf{OR} \quad \frac{x}{17} = \frac{2}{1}$$
$$x \cdot 1 = 17 \cdot 2$$
$$x = 34 \text{ ft}$$

$P = 34 \text{ ft} + 30 \text{ ft} + 40 \text{ ft}$
$= 104 \text{ ft}$

70. Write a proportion to find x.
$$\frac{6 \text{ m}}{x} = \frac{4 \text{ m}}{6 \text{ m}} \quad \textbf{OR} \quad \frac{6}{x} = \frac{2}{3}$$
$$2 \cdot x = 3 \cdot 6$$
$$\frac{2 \cdot x}{2} = \frac{18}{2}$$
$$x = 9 \text{ m}$$

Write a proportion to find y.
$$\frac{5 \text{ m}}{y} = \frac{4 \text{ m}}{6 \text{ m}} \quad \textbf{OR} \quad \frac{5}{y} = \frac{2}{3}$$
$$2 \cdot y = 5 \cdot 3$$
$$\frac{2 \cdot y}{2} = \frac{15}{2}$$
$$y = 7.5 \text{ m}$$

$P = 9 \text{ m} + 7.5 \text{ m} + 6 \text{ m}$
$= 22.5 \text{ m}$

71. Write a proportion to find x.
$$\frac{x}{9 \text{ mm}} = \frac{16 \text{ mm}}{12 \text{ mm}} \quad \textbf{OR} \quad \frac{x}{9} = \frac{4}{3}$$
$$3 \cdot x = 9 \cdot 4$$
$$\frac{3 \cdot x}{3} = \frac{36}{3}$$
$$x = 12 \text{ mm}$$

Write a proportion to find y.
$$\frac{10 \text{ mm}}{y} = \frac{16 \text{ mm}}{12 \text{ mm}} \quad \textbf{OR} \quad \frac{10}{y} = \frac{4}{3}$$
$$y \cdot 4 = 10 \cdot 3$$
$$\frac{y \cdot 4}{4} = \frac{30}{4}$$
$$y = 7.5 \text{ mm}$$

$P = 12 \text{ mm} + 10 \text{ mm} + 16 \text{ mm}$
$= 38 \text{ mm}$

Chapter 8 Mixed Review Exercises

1. The figure is a square.
$P = 4 \cdot s$
$= 4 \cdot 4\frac{1}{2} \text{ in.}$
$= 18 \text{ in.}$
$A = s \cdot s$
$= 4\frac{1}{2} \text{ in.} \cdot 4\frac{1}{2} \text{ in.}$
$= \frac{9}{2} \text{ in.} \cdot \frac{9}{2} \text{ in.}$
$= \frac{81}{4} \text{ in.}^2$
$= 20\frac{1}{4} \text{ in.}^2 \text{ or } 20.3 \text{ in.}^2 \text{ (rounded)}$

Copyright © 2018 Pearson Education, Inc.

2. The figure is a trapezoid.

$$P = 2.3 \text{ cm} + 2.5 \text{ cm} + 2.1 \text{ cm} + 3.4 \text{ cm}$$
$$= 10.3 \text{ cm}$$
$$A = 0.5 \cdot h \cdot (b + B)$$
$$= 0.5 \cdot 2.1 \text{ cm} \cdot (2.5 \text{ cm} + 3.4 \text{ cm})$$
$$= 0.5 \cdot 2.1 \text{ cm} \cdot 5.9 \text{ cm}$$
$$\approx 6.2 \text{ cm}^2 \text{ (rounded)}$$

3. The figure is a circle.

$$C = \pi \cdot d$$
$$\approx 3.14 \cdot 13 \text{ m}$$
$$\approx 40.8 \text{ m (rounded)}$$
$$A = \pi \cdot r^2 \left(r = \frac{13 \text{ m}}{2} = 6.5 \text{ m} \right)$$
$$\approx 3.14 \cdot 6.5 \text{ m} \cdot 6.5 \text{ m}$$
$$\approx 132.7 \text{ m}^2 \text{ (rounded)}$$

4. The figure is a parallelogram.

$$P = 17 \text{ ft} + 10 \text{ ft} + 17 \text{ ft} + 10 \text{ ft}$$
$$= 54 \text{ ft}$$
$$A = b \cdot h$$
$$= 10 \text{ ft} \cdot 14 \text{ ft}$$
$$= 140 \text{ ft}^2$$

5. The figure is a triangle.

$$P = 6\frac{1}{4} \text{ yd} + 7\frac{1}{2} \text{ yd} + 6\frac{1}{4} \text{ yd}$$
$$= 20 \text{ yd}$$
$$A = \frac{1}{2} \cdot b \cdot h = \frac{1}{2} \cdot 7\frac{1}{2} \text{ yd} \cdot 5 \text{ yd}$$
$$= 18\frac{3}{4} \text{ yd}^2 \text{ or } 18.8 \text{ yd}^2 \text{ (rounded)}$$

6. The figure is a rectangle.

$$P = 0.7 \text{ km} + 2.8 \text{ km} + 0.7 \text{ km} + 2.8 \text{ km}$$
$$= 7 \text{ km}$$
$$A = \text{length} \cdot \text{width}$$
$$= 2.8 \text{ km} \cdot 0.7 \text{ km}$$
$$= 1.96 \text{ km}^2$$
$$\approx 2.0 \text{ km}^2 \text{ (rounded)}$$

7. The figure is a circle.

$$C = 2 \cdot \pi \cdot r$$
$$\approx 2 \cdot 3.14 \cdot 8.5 \text{ m}$$
$$\approx 53.4 \text{ m (rounded)}$$
$$A = \pi \cdot r^2$$
$$\approx 3.14 \cdot 8.5 \text{ m} \cdot 8.5 \text{ m}$$
$$\approx 226.9 \text{ m}^2 \text{ (rounded)}$$

8. The figure is a parallelogram.

$$P = 24 \text{ mm} + 15 \text{ mm} + 24 \text{ mm} + 15 \text{ mm}$$
$$= 78 \text{ mm}$$

$$A = b \cdot h$$
$$= 24 \text{ mm} \cdot 12 \text{ mm}$$
$$= 288 \text{ mm}^2$$

9. The figure is a triangle.

$$P = 13 \text{ mi} + 9 \text{ mi} + 15.8 \text{ mi}$$
$$= 37.8 \text{ mi}$$
$$A = 0.5 \cdot b \cdot h$$
$$= 0.5 \cdot 9 \text{ mi} \cdot 13 \text{ mi}$$
$$= 58.5 \text{ mi}^2$$

10. $\overleftrightarrow{WX}$ and $\overleftrightarrow{YZ}$ are parallel lines.

11. $\overline{QR}$ is a line segment.

12. $\angle CQD$ is an acute angle.

13. $\overleftrightarrow{PQ}$ and $\overleftrightarrow{NO}$ are intersecting lines.

14. $\angle APB$ is a right angle measuring 90°.

15. $\overrightarrow{AB}$ is a ray.

16. $\overleftrightarrow{T}$ is a straight angle measuring 180°.

17. $\angle FEG$ is an obtuse angle.

18. $\overleftrightarrow{LM}$ and $\overleftrightarrow{JK}$ are perpendicular lines.

19. The complement of an angle measuring 9°
$$90° - 9° = 81°.$$

20. The supplement of an angle measuring 42° is
$$180° - 42° = 138°.$$

21. Find the unknown length of the horizontal line.

$$17 \text{ m} - 1 \text{ m} - 1 \text{ m} = 15 \text{ m}$$

$$P = 17 \text{ m} + 16 \text{ m} + 1 \text{ m} + 12 \text{ m}$$
$$+ 15 \text{ m} + 12 \text{ m} + 1 \text{ m} + 16 \text{ m}$$
$$= 90 \text{ m}$$

Draw two short horizontal lines to break up the figure into 3 rectangles to find the area.
Two rectangles with length 12 m and width 1 m:

$$A = 12 \text{ m} \cdot 1 \text{ m}$$
$$= 12 \text{ m}^2$$

A rectangle with length 17 m and width 16 m − 12 m = 4 m:

$$A = 17 \text{ m} \cdot 4 \text{ m}$$
$$= 68 \text{ m}^2$$

Total area $= 12 \text{ m}^2 + 12 \text{ m}^2 + 68 \text{ m}^2$
$$= 92 \text{ m}^2$$

Copyright © 2018 Pearson Education, Inc.

22. $P = 72 \text{ cm} + 19 \text{ cm} + 50 \text{ cm}$
$+ 19 \text{ cm} + 72 \text{ cm} + 50 \text{ cm}$
$= 282 \text{ cm}$

Break up the figure into a parallelogram and a rectangle.
A parallelogram with base 72 cm and height 45 cm:

$$A = 72 \text{ cm} \cdot 45 \text{ cm}$$
$$= 3240 \text{ cm}^2$$

A rectangle with base 19 cm and height 50 cm:

$$A = 19 \text{ cm} \cdot 50 \text{ cm}$$
$$= 950 \text{ cm}^2$$

Total area $= 3240 \text{ cm}^2 + 950 \text{ cm}^2$
$$= 4190 \text{ cm}^2$$

23. The figure is a cylinder.

$$V = \pi \cdot r^2 \cdot h$$
$$\approx 3.14 \cdot 2 \text{ ft} \cdot 2 \text{ ft} \cdot 8 \text{ ft}$$
$$\approx 100.5 \text{ ft}^3 \text{ (rounded)}$$

24. The figure is a rectangular solid or cube.

$$V = \text{length} \cdot \text{width} \cdot \text{height}$$
$$= 1\tfrac{1}{2} \text{ in.} \cdot 1\tfrac{1}{2} \text{ in.} \cdot 1\tfrac{1}{2} \text{ in.}$$
$$= \tfrac{3}{2} \text{ in.} \cdot \tfrac{3}{2} \text{ in.} \cdot \tfrac{3}{2} \text{ in.}$$
$$= \tfrac{27}{8} \text{ in.}^3 \left(\text{or } 3\tfrac{3}{8} \text{ in.}^3\right)$$
$$\approx 3.4 \text{ in.}^3 \text{ (rounded)}$$

25. The figure is a rectangular solid.

$$V = \text{length} \cdot \text{width} \cdot \text{height}$$
$$= 3.5 \text{ m} \cdot 3 \text{ m} \cdot 0.7 \text{ m}$$
$$= 7.35 \text{ m}^3$$
$$\approx 7.4 \text{ m}^3 \text{ (rounded)}$$

26. The figure is a pyramid.

$$V = \tfrac{1}{3} \cdot B \cdot h$$
$$= \tfrac{1}{3} \cdot (11 \text{ cm} \cdot 9 \text{ cm}) \cdot 17 \text{ cm}$$
$$= \tfrac{1}{3} \cdot 99 \text{ cm}^2 \cdot 17 \text{ cm}$$
$$= 561 \text{ cm}^3$$

27. The figure is a cone.

$$V = \tfrac{1}{3} \cdot \pi \cdot r^2 \cdot h$$
$$\approx \tfrac{1}{3} \cdot 3.14 \cdot 9 \text{ cm} \cdot 9 \text{ cm} \cdot 15 \text{ cm}$$
$$\approx 1271.7 \text{ cm}^3$$

28. The figure is a sphere.

$$V = \tfrac{4}{3} \cdot \pi \cdot r^3$$
$$\approx \tfrac{4}{3} \cdot 3.14 \cdot 7 \text{ m} \cdot 7 \text{ m} \cdot 7 \text{ m}$$
$$\approx 1436.0 \text{ m}^3 \text{ (rounded)}$$

29. $\text{leg} = \sqrt{(\text{hypotenuse})^2 - (\text{leg})^2}$
$$= \sqrt{(14)^2 - (6)^2}$$
$$= \sqrt{196 - 36}$$
$$= \sqrt{160}$$
$$\approx 12.6 \text{ km (rounded)}$$

30. *Step 1* $60° + 48° = 108°$
Step 2 $180° - 108° = 72°$
$\angle D$ measures $72°$.

31. Write a proportion to find x.

$$\frac{x}{10 \text{ m}} = \frac{18 \text{ m}}{15 \text{ m}} \quad \textbf{OR} \quad \frac{x}{10} = \frac{6}{5}$$
$$5 \cdot x = 10 \cdot 6$$
$$\frac{5 \cdot x}{5} = \frac{60}{5}$$
$$x = 12 \text{ mm}$$

Write a proportion to find y.

$$\frac{16.8 \text{ m}}{y} = \frac{18 \text{ m}}{15 \text{ m}} \quad \textbf{OR} \quad \frac{16.8}{y} = \frac{6}{5}$$
$$y \cdot 6 = 16.8 \cdot 5$$
$$\frac{y \cdot 6}{6} = \frac{84}{6}$$
$$y = 14 \text{ mm}$$

32. The prefix *dec-* in decade means 10 and the prefix *cent-* in century means 100, so divide 200 (two centuries) by 10. The answer is 20 decades.

Chapter 8 Test

1. $\angle LOM$ is an acute angle, so the answer is (e).

2. $\angle YOX$ is a right angle, so the answer is (a). Its measure is $90°$.

3. $\overrightarrow{GH}$ is a ray, so the answer is (d).

4. $\overleftrightarrow{W}$ is a straight angle, so the answer is (g). Its measure is $180°$.

5. **Parallel lines** are lines in the same plane that never intersect. **Perpendicular lines** intersect to form a right angle.

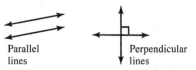

Parallel lines   Perpendicular lines

Copyright © 2018 Pearson Education, Inc.

6. The complement of an angle measuring $81°$ is $90° - 81° = 9°$.

7. The supplement of an angle measuring $20°$ is $180° - 20° = 160°$.

8. $\angle 1$ measures $50°$ since $\angle 1$ and the $50°$ angle are vertical angles, so they are congruent.
$\angle 3$ measures $90°$ since it is a vertical angle with the $90°$ angle (the one marked with the square).
$\angle 4 + 50° + \angle 3 = 180°$
$\angle 4 + 50° + 90° = 180°$
$\angle 4$ measures $40°$ ($180° - 50° - 90°$).
$\angle 2$ also measures $40°$ since $\angle 2$ and $\angle 4$ are vertical angles.

9. The figure is a rectangle.
$$P = 4\text{ ft} + 7\tfrac{1}{2}\text{ ft} + 4\text{ ft} + 7\tfrac{1}{2}\text{ ft}$$
$$= 23\text{ ft}$$
$$A = \text{length} \cdot \text{width}$$
$$= 7\tfrac{1}{2}\text{ ft} \cdot 4\text{ ft}$$
$$= 30\text{ ft}^2$$

10. The figure is a square.
$$P = 4 \cdot s$$
$$= 4 \cdot 18\text{ mm}$$
$$= 72\text{ mm}$$
$$A = s \cdot s$$
$$= 18\text{ mm} \cdot 18\text{ mm}$$
$$= 324\text{ mm}^2$$

11. The figure is a parallelogram.
$$P = 5.9\text{ m} + 7.2\text{ m} + 5.9\text{ m} + 7.2\text{ m}$$
$$= 26.2\text{ m}$$
$$A = b \cdot h$$
$$= 7.2\text{ m} \cdot 4.6\text{ m}$$
$$= 33.12\text{ m}^2$$

12. The figure is a trapezoid.
$$P = 29\text{ cm} + 46.4\text{ cm} + 57\text{ cm} + 37\text{ cm}$$
$$= 169.4\text{ cm}$$
$$A = \tfrac{1}{2} \cdot h \cdot (b + B)$$
$$= \tfrac{1}{2} \cdot 37\text{ cm} \cdot (29\text{ cm} + 57\text{ cm})$$
$$= \tfrac{1}{2} \cdot 37\text{ cm} \cdot 86\text{ cm}$$
$$= 1591\text{ cm}^2$$

13. $P = 11.8\text{ m} + 8.65\text{ m} + 12\text{ m}$
$$= 32.45\text{ m}$$
$$A = 0.5 \cdot b \cdot h$$
$$= 0.5 \cdot 12\text{ m} \cdot 8\text{ m}$$
$$= 48\text{ m}^2$$

14. $P = 9\text{ yd} + 15\tfrac{4}{5}\text{ yd} + 13\text{ yd}$
$$= 37\tfrac{4}{5}\text{ yd or } 37.8\text{ yd}$$
$$A = \tfrac{1}{2} \cdot b \cdot h$$
$$= \tfrac{1}{2} \cdot 13\text{ yd} \cdot 9\text{ yd}$$
$$= 58.5\text{ yd}^2 \text{ or } 58\tfrac{1}{2}\text{ yd}^2$$

15. *Step 1* $90° + 35° = 125°$
Step 2 $180° - 125° = 55°$
The third angle measures $55°$.

16. $\text{radius} = \dfrac{d}{2} = \dfrac{25\text{ in.}}{2} = 12.5\text{ in. or } 12\dfrac{1}{2}\text{ in.}$

17. $C = 2 \cdot \pi \cdot r$
$$\approx 2 \cdot 3.14 \cdot 0.9\text{ km}$$
$$\approx 5.7\text{ km (rounded)}$$

18. $r = \dfrac{d}{2} = \dfrac{16.2\text{ cm}}{2} = 8.1\text{ cm}$
$$A = \pi \cdot r^2$$
$$\approx 3.14 \cdot 8.1\text{ cm} \cdot 8.1\text{ cm}$$
$$\approx 206.0\text{ cm}^2 \text{ (rounded)}$$

19. $A = \dfrac{\pi \cdot r^2}{2}$
$$\approx \dfrac{3.14 \cdot 5\text{ m} \cdot 5\text{ m}}{2}$$
$$= \dfrac{78.5}{2}\text{ m}^2$$
$$= 39.25\text{ m}^2$$
$$\approx 39.3\text{ m}^2 \text{ (rounded)}$$

20. The figure is a rectangular solid.
$$V = \text{length} \cdot \text{width} \cdot \text{height}$$
$$= 30\text{ m} \cdot 18\text{ m} \cdot 12\text{ m}$$
$$= 6480\text{ m}^3$$

21. The figure is a sphere.
$$V = \tfrac{4}{3} \cdot \pi \cdot r^3$$
$$\approx \tfrac{4}{3} \cdot 3.14 \cdot 2\text{ ft} \cdot 2\text{ ft} \cdot 2\text{ ft}$$
$$= \dfrac{100.48}{3}\text{ ft}^3$$
$$\approx 33.5\text{ ft}^3 \text{ (rounded)}$$

22. The figure is a cylinder.
$$V = \pi \cdot r^2 \cdot h$$
$$\approx 3.14 \cdot 18\text{ ft} \cdot 18\text{ ft} \cdot 5\text{ ft}$$
$$= 5086.8\text{ ft}^3$$

Copyright © 2018 Pearson Education, Inc.

23. $\text{hypotenuse} = \sqrt{(\text{leg})^2 + (\text{leg})^2}$
$= \sqrt{(7)^2 + (6)^2}$
$= \sqrt{49 + 36}$
$= \sqrt{85}$
$\approx 9.2 \text{ cm (rounded)}$

24. Write a proportion to find y.

$$\frac{y}{18 \text{ cm}} = \frac{10 \text{ cm}}{15 \text{ cm}} \quad \textbf{OR} \quad \frac{y}{18} = \frac{2}{3}$$
$$y \cdot 3 = 18 \cdot 2$$
$$\frac{y \cdot 3}{3} = \frac{36}{3}$$
$$y = 12 \text{ cm}$$

Write a proportion to find z.

$$\frac{z}{9 \text{ cm}} = \frac{10 \text{ cm}}{15 \text{ cm}} \quad \textbf{OR} \quad \frac{z}{9} = \frac{2}{3}$$
$$z \cdot 3 = 9 \cdot 2$$
$$\frac{z \cdot 3}{3} = \frac{18}{3}$$
$$z = 6 \text{ cm}$$

25. Linear units like cm are used to measure perimeter, radius, diameter, and circumference. Area is measured in square units like cm^2 (squares that measure 1 cm on each side). Volume is measured in cubic units like cm^3 (cubes that measure 1 cm on each side). Examples will vary.

Cumulative Review Exercises (Chapters 1–8)

1. $3\dfrac{3}{5} \div 8 = \dfrac{18}{5} \div \dfrac{8}{1} = \dfrac{\overset{9}{\cancel{18}}}{5} \cdot \dfrac{1}{\underset{4}{\cancel{8}}} = \dfrac{9}{20}$

2. $1 - 0.0868$

$$\begin{array}{r} 1.0000 \\ - \, 0.0868 \\ \hline 0.9132 \end{array}$$

3. $(0.006)(0.013)$

$$\begin{array}{r} 0.006 \leftarrow 3 \textit{ decimal places} \\ \times \, 0.013 \leftarrow 3 \textit{ decimal places} \\ \hline 18 \\ 6 \\ \hline 0.000078 \leftarrow 6 \textit{ decimal places needed} \end{array}$$

4. $0.7 \div 0.036 \approx 19.44 \text{ (rounded)}$
Move the decimal point three places to the right in the dividend and divisor.

$$\begin{array}{r} 1\,9.4\,4\,4 \\ 36\overline{\smash)7\,0\,0.0\,0\,0} \\ \underline{3\,6} \\ 3\,4\,0 \\ \underline{3\,2\,4} \\ 1\,6\,0 \\ \underline{1\,4\,4} \\ 1\,6\,0 \\ \underline{1\,4\,4} \\ 1\,6\,0 \\ \underline{1\,4\,4} \\ 1\,6 \end{array}$$

5. $6\frac{1}{6} - 1\frac{3}{4}$

$$6\frac{1}{6} = 6\frac{2}{12} = 5\frac{14}{12}$$
$$-1\frac{3}{4} = 1\frac{9}{12} = 1\frac{9}{12}$$
$$\overline{\phantom{-1\frac{3}{4} = 1\frac{9}{12} =}\, 4\frac{5}{12}}$$

6. $16 - (10 - 2) \div 2(3) + 5 = 16 - 8 \div 2(3) + 5$
$= 16 - 4(3) + 5$
$= 16 - 12 + 5$
$= 4 + 5 = 9$

7. 0.0208 is two hundred eight ten-thousandths.

8. From least to greatest:

$$2.55 = 2.5500 \quad \textit{greatest}$$
$$2.505 = 2.5050$$
$$2.055 = 2.0550 \quad \textit{least}$$
$$2.5005 = 2.5005$$

Answer: 2.055; 2.5005; 2.505; 2.55

9. $$\frac{5}{13} = \frac{x}{91}$$
$$13 \cdot x = 5 \cdot 91$$
$$\frac{13 \cdot x}{13} = \frac{455}{13}$$
$$x = 35$$

10. $$\frac{4.5}{x} = \frac{6.7}{3}$$
$$6.7 \cdot x = 3 \cdot 4.5$$
$$\frac{6.7 \cdot x}{6.7} = \frac{13.5}{6.7}$$
$$x \approx 2.01 \text{ (rounded)}$$

Copyright © 2018 Pearson Education, Inc.

11. part is 72; whole is 45; percent is unknown.
Use the percent proportion.

$$\frac{72}{45} = \frac{x}{100}$$
$$45 \cdot x = 72 \cdot 100$$
$$\frac{45 \cdot x}{45} = \frac{7200}{45}$$
$$x = 160$$

72 patients is 160% of 45 patients.

12. part is 18; percent is 3; whole is unknown.
Use the percent proportion.

$$\frac{18}{x} = \frac{3}{100}$$
$$3 \cdot x = 18 \cdot 100$$
$$\frac{3 \cdot x}{3} = \frac{1800}{3}$$
$$x = 600$$

$18 is 3% of $600.

13. $2\frac{1}{4}$ hours to minutes

$$\frac{2\frac{1}{4} \text{ hours}}{1} \cdot \frac{60 \text{ minutes}}{1 \text{ hour}} = \frac{9}{4} \cdot \frac{\overset{15}{\cancel{60}}}{1} \text{ minutes}$$
$$= 135 \text{ minutes}$$

14. 40 oz to pounds

$$\frac{40 \text{ ounces}}{1} \cdot \frac{1 \text{ pound}}{16 \text{ ounces}} = \frac{40}{16} \text{ pounds}$$
$$= 2\frac{1}{2} \text{ lb or } 2.5 \text{ lb}$$

15. 8 cm to meters
Count 2 places to the *left* on the metric conversion line.
8 cm = 0.08 m

16. 1.8 L to mL
Count 3 places to the *right* on the metric conversion line.
1.8 L = 1800 mL

17. Her wristwatch strap is 15 <u>mm</u> wide.

18. Jon added 2 <u>L</u> of oil to his car.

19. The child weighs 15 <u>kg</u>.

20. The bookcase is 90 <u>cm</u> high.

21. The figure is a rectangle.

$$P = 3\frac{1}{2} \text{ in.} + 2 \text{ in.} + 3\frac{1}{2} \text{ in.} + 2 \text{ in.}$$
$$= 11 \text{ in.}$$
$$A = \text{length} \cdot \text{width}$$
$$= 3\frac{1}{2} \text{ in.} \cdot 2 \text{ in.}$$
$$= 7 \text{ in.}^2$$

22. The figure is a triangle.

$$P = 2.1 \text{ m} + 2 \text{ m} + 1.7 \text{ m}$$
$$= 5.8 \text{ m}$$
$$A = 0.5 \cdot b \cdot h$$
$$= 0.5 \cdot 2.1 \text{ m} \cdot 1.5 \text{ m}$$
$$= 1.575 \text{ m}^2$$

23. The figure is a circle.

$$C = 2 \cdot \pi \cdot r$$
$$\approx 2 \cdot 3.14 \cdot 5 \text{ ft}$$
$$= 31.4 \text{ ft}$$
$$A = \pi \cdot r \cdot r$$
$$\approx 3.14 \cdot 5 \text{ ft} \cdot 5 \text{ ft}$$
$$= 78.5 \text{ ft}^2$$

24. The figure is a parallelogram.

$$P = 24 \text{ cm} + 14 \text{ cm} + 24 \text{ cm} + 14 \text{ cm}$$
$$= 76 \text{ cm}$$
$$A = \text{base} \cdot \text{height}$$
$$= 24 \text{ cm} \cdot 11 \text{ cm}$$
$$= 264 \text{ cm}^2$$

25.

$$\text{hypotenuse} = \sqrt{(\text{leg})^2 + (\text{leg})^2}$$
$$= \sqrt{(19)^2 + (15)^2}$$
$$= \sqrt{361 + 225}$$
$$= \sqrt{586}$$
$$\approx 24.2$$

$y \approx 24.2$ mm (rounded)

26. part is 53; whole is 90; percent is unknown.

$$\frac{53}{90} = \frac{x}{100}$$
$$90 \cdot x = 53 \cdot 100$$
$$\frac{90 \cdot x}{90} = \frac{5300}{90}$$
$$x \approx 58.89 \approx 59$$

Mei Ling has 59% (rounded) of the necessary credits.

27. Use the formula for the volume of a cylinder.

$$V = \pi \cdot r^2 \cdot h \left(r = \frac{d}{2} = \frac{15.6 \text{ cm}}{2} = 7.8 \text{ cm} \right)$$
$$\approx 3.14 \cdot 7.8 \text{ cm} \cdot 7.8 \text{ cm} \cdot 16 \text{ cm}$$
$$= 3056.6016 \text{ cm}^3$$
$$\approx 3056.6 \text{ cm}^3 \text{ (rounded)}$$

The volume of the can is about 3056.6 cm³.

Copyright © 2018 Pearson Education, Inc.

28. $8\frac{1}{2}$ ounces of Brand T for
$3.09 − $0.75 (coupon) = $2.34

$$\frac{\$2.34}{8\frac{1}{2} \text{ ounces}} \approx \$0.275 \text{ per ounce } (*)$$

$10\frac{1}{2}$ ounces of Brand F for $3.00

$$\frac{\$3.00}{10\frac{1}{2} \text{ ounces}} \approx \$0.286 \text{ per ounce}$$

$9\frac{1}{2}$ ounces of Brand H for
$3.69 − $0.50 (coupon) = $3.19

$$\frac{\$3.19}{9\frac{1}{2} \text{ ounces}} \approx \$0.336 \text{ per ounce}$$

The best buy is Brand T at $8\frac{1}{2}$ ounces for $3.09 with a $0.75 coupon.

29. First add the amount of canvas material that was used.

$$1\frac{2}{3} = 1\frac{8}{12}$$
$$+ 1\frac{3}{4} = 1\frac{9}{12}$$
$$\overline{\phantom{+ 1\frac{3}{4}} = 2\frac{17}{12} = 3\frac{5}{12} \text{ yd}}$$

Subtract $3\frac{5}{12}$ from the amount of canvas material that Steven bought.

$$4\frac{1}{2} = 4\frac{6}{12}$$
$$- 3\frac{5}{12} = 3\frac{5}{12}$$
$$\overline{\phantom{- 3\frac{5}{12}} = 1\frac{1}{12} \text{ yd}}$$

There will be $1\frac{1}{12}$ yd left.

30.
$$\frac{25 \text{ pounds}}{140 \text{ pounds}} = \frac{x}{200 \text{ people}}$$
$$140 \cdot x = 25 \cdot 200$$
$$\frac{140 \cdot x}{140} = \frac{5000}{140}$$
$$x \approx 35.7$$

To feed 200 people, the cooks will need about 35.7 pounds of meat.

Similarly, for potatoes:
$$\frac{140 \cdot x}{140} = \frac{35 \cdot 200}{140}$$
$$x = 50 \text{ pounds}$$

And for carrots:
$$\frac{140 \cdot x}{140} = \frac{15 \cdot 200}{140}$$
$$x \approx 21.4 \text{ pounds}$$

To feed 200 people, the cooks will need 50 pounds of potatoes and about 21.4 pounds of carrots.

31. Convert 85 cm to m.
85 cm = 0.85 m
She will need $4 \cdot 0.85$ m = 3.4 m.

32. Find the discount amount.

$$\begin{aligned}
\text{discount amount} &= \text{discount rate} \cdot \text{cost} \\
&= 65\% \cdot \$64 \\
&= 0.65 \cdot \$64 \\
&= \$41.6
\end{aligned}$$

Subtract the discount amount from the regular price.

$$\$64 − \$41.6 = \$22.4$$

Lanece will pay $22.40.

Copyright © 2018 Pearson Education, Inc.

CHAPTER 9   BASIC ALGEBRA

9.1 Signed Numbers

9.1 Margin Exercises

1. **(a)** A temperature at the North Pole of 70 degrees below 0 is written $-70°$.

    (b) Your bank account is overdrawn by 15 dollars is written as $-\$15$.

    (c) The altitude of a place 284 feet below sea level is written -284 ft.

2. **(a)** -8 is negative.

    (b) $-\frac{3}{4}$ is negative.

    (c) 1 is positive.

    (d) 0 is neither positive nor negative.

3. **(a)** $-1, 1, -3, 3$

    To graph -1, put a dot at -1 on the number line. Now put dots in the correct places for $1, -3,$ and 3.

    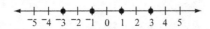

    (b) $-2, 4, 0, -1, -4$

4. **(a)** $4 \underline{>} 0$ because 4 is to the *right* of 0 on the number line.

    (b) $-1 \underline{<} 0$ because -1 is to the *left* of 0 on the number line.

    (c) $-3 \underline{<} -1$ because -3 is to the *left* of -1 on the number line.

    (d) $-8 \underline{>} -9$ because -8 is to the *right* of -9 on the number line.

    (e) $0 \underline{>} -3$ because 0 is to the *right* of -3 on the number line.

5. **(a)** The *distance* from 0 to 5 is 5, so $|5| = 5$.

    (b) The *distance* from 0 to -5 is also 5, so $|-5| = 5$.

    (c) $-|-17|$
    First, $|-17| = 17$. But there is a negative sign outside the absolute value bars. So, $-\underline{17}$ is the simplified expression.

    (d) $-|-9|$
    First, $|-9| = 9$. But there is a negative sign outside the absolute value bars. So, -9 is the simplified expression.

    (e) $-|2| = -(2) = -2$
    Because the negative sign is outside the absolute value bars, the answer is negative.

6. **(a)** The opposite of 4 is $-(4) = -4$.

    (b) The opposite of 10 is $-(10) = -10$.

    (c) The opposite of 49 is $-(49) = -49$.

    (d) The opposite of $\frac{2}{5}$ is $-\left(\frac{2}{5}\right) = -\frac{2}{5}$.

    (e) The opposite of 0 is 0. Zero is neither positive nor negative.

7. **(a)** The opposite of -4 is $-(-4) = 4$.

    (b) The opposite of -10 is $-(-10) = 10$.

    (c) The opposite of -25 is $-(-25) = 25$.

    (d) The opposite of -1.9 is $-(-1.9) = 1.9$.

    (e) The opposite of -0.85 is $-(-0.85) = 0.85$.

    (f) The opposite of $-\frac{3}{4}$ is $-\left(-\frac{3}{4}\right) = \frac{3}{4}$.

9.1 Section Exercises

1. Water freezes at 32 degrees above 0 on the Fahrenheit temperature scale.

    $+32$ degrees or 32 degrees

3. The price of the stock fell $12.

    $-\$12$

5. Keith lost $6\frac{1}{2}$ lb while he was sick with the flu.

    $-6\frac{1}{2}$ lb

7. Mount McKinley, the tallest mountain in the United States, rises 20,237 ft above sea level.

    $+20{,}237$ or 20,237 ft

9. $4, -1, 2, 0, -5$

    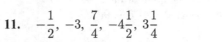

11. $-\dfrac{1}{2}, -3, \dfrac{7}{4}, -4\dfrac{1}{2}, 3\dfrac{1}{4}$

    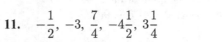

13. **(a)** The statement $0 < 5$ in words is zero is less than five (or positive five).

Copyright © 2018 Pearson Education, Inc.

(b) The statement $-10 > -17$ in words is negative ten is greater than negative seventeen.

15. $9 < 14$ because 9 is to the *left* of 14 on a number line.

17. $0 > -2$ because 0 is to the *right* of -2 on a number line.

19. $-6 < 3$ because -6 is to the *left* of 3 on a number line.

21. Compare 1 and -1. Because 1 is to the *right* of -1 on a number line, $1 > -1$.

23. Compare -11 and -2. Because -11 is to the *left* of -2 on a number line, $-11 < -2$.

25. $-72 > -75$ because -72 is to the *right* of -75 on a number line.

27. The *distance* from 0 to 3 is 3, so $|3| = 3$.

29. The *distance* from 0 to -10 is 10, so $|-10| = 10$.

31. The *distance* from 0 to 0 is 0, so $|0| = 0$.

33. $-|-18|$
First, $|-18| = 18$. But there is also a negative sign *outside* the absolute value bars. So, -18 is the simplified expression.

35. $-|32| = -(32) = -32$
Because the negative sign is outside the absolute value bars, the answer is negative.

37. The opposite of 7 is -7.

39. The opposite of -14 is $-(-14) = 14$.

41. The opposite of $\frac{2}{3}$ is $-\frac{2}{3}$.

43. The opposite of -8.3 is $-(-8.3) = 8.3$ because the opposite of any negative number is positive.

45. The opposite of $-\frac{1}{6}$ is $-\left(-\frac{1}{6}\right) = \frac{1}{6}$.

47. $|-5| = 5$

5 is greater than 0, so $|-5| > 0$ is true.

49. $-(-6) = 6$

0 is less than 6, so $0 < -(-6)$ is true.

51. Because $|-4| = 4$, then $-|-4| = -4$. Because $|-7| = 7$, then $-|-7| = -7$. Since -4 is to the right of -7 on the number line, $-4 > -7$; therefore, $-|-4| < -|-7|$ is false.

53. $-2.8,\ 0,\ -1.5,\ -1$

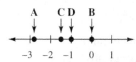

54. From lowest to highest, the patients' scores are:
$$-2.8,\ -1.5,\ -1,\ 0$$

55. Since $-2.8 < -2.5$, Patient A has osteoporosis.

Since $0 > -1$, Patient B is normal.

Since $-2.5 \le -1.5 \le -1$, Patient C is at risk.

Since $-2.5 \le -1 \le -1$, Patient D is at risk.

56. **(a)** The patient would think the interpretation was "above normal" and wouldn't get treatment.

(b) The sign of the score makes no difference in Patient B's score of 0 because 0 is neither positive nor negative.

9.2 Adding and Subtracting Signed Numbers

9.2 Margin Exercises

1. **(a)**

(b)

(c)

(d)

2. **(a)** $3 + (-2) = \underline{1}$

(b) $-4 + 1 = \underline{-3}$

Copyright © 2018 Pearson Education, Inc.

(c) $-3 + 7 = \underline{4}$

(d) $-1 + (-4) = \underline{-5}$

3. **(a)** $-4 + (-4)$

Add the absolute values.

$$|-4| = 4 \quad \text{and} \quad |-4| = 4$$
$$4 + 4 = 8$$

Write a negative sign in front of the sum.

$$-4 + (-4) = -8$$

(b) $-3 + (-20)$

Add the absolute values.

$$|-3| = 3 \quad \text{and} \quad |-20| = 20$$
$$3 + 20 = 23$$

Write a negative sign in front of the sum.

$$-3 + (-20) = -23$$

(c) $-31 + (-5)$

Add the absolute values.

$$|-31| = 31 \quad \text{and} \quad |-5| = 5$$
$$31 + 5 = 36$$

Write a negative sign in front of the sum.

$$-31 + (-5) = -36$$

(d) $-10 + (-8)$

Add the absolute values.

$$|-10| = 10 \quad \text{and} \quad |-8| = 8$$
$$10 + 8 = 18$$

Write a negative sign in front of the sum.

$$-10 + (-8) = -18$$

(e) $-\dfrac{9}{10} + \left(-\dfrac{3}{5}\right)$

Add the absolute values.

$$\left|-\dfrac{9}{10}\right| = \dfrac{9}{10} \quad \text{and} \quad \left|-\dfrac{3}{5}\right| = \dfrac{3}{5}$$

$$\dfrac{9}{10} + \dfrac{3}{5} = \dfrac{9}{10} + \dfrac{6}{10} = \dfrac{15}{10}$$

Write $\dfrac{15}{10}$ in lowest terms:

$$\dfrac{15}{10} = \dfrac{15 \div \underline{5}}{10 \div \underline{5}} = \dfrac{3}{2}$$

Write a negative sign in front of the sum.

$$-\dfrac{9}{10} + \left(-\dfrac{3}{5}\right) = -\dfrac{3}{2}$$

4. **(a)** $10 + (-2)$

The signs are different, so subtract the absolute values.

$$|10| = 10 \quad \text{and} \quad |-2| = 2$$
$$10 - 2 = 8$$

The positive number, 10, has the greater absolute value, so the answer is positive.

$$10 + (-2) = \underline{8}$$

(b) $-7 + 8$

The signs are different, so subtract the absolute values.

$$|-7| = 7 \quad \text{and} \quad |8| = 8$$
$$8 - 7 = 1$$

The positive number, 8, has the greater absolute value, so the answer is positive.

$$-7 + 8 = 1$$

(c) $-11 + 11$

The signs are different, so subtract the absolute values.

$$|-11| = 11 \quad \text{and} \quad |11| = 11$$
$$11 - 11 = 0$$

The numbers have equal absolute values, so the answer is neither positive nor negative.

(d) $23 + (-32)$

The signs are different, so subtract the absolute values.

$$|23| = 23 \quad \text{and} \quad |-32| = 32$$
$$32 - 23 = 9$$

The negative number, -32, has the greater absolute value, so the answer is negative.

$$23 + (-32) = -9$$

(e) $-\dfrac{7}{8} + \dfrac{1}{4}$

The signs are different, so subtract the absolute values.

$$\left|-\dfrac{7}{8}\right| = \dfrac{7}{8} \quad \text{and} \quad \left|\dfrac{1}{4}\right| = \dfrac{1}{4}$$

$$\dfrac{7}{8} - \dfrac{1}{4} = \dfrac{7}{8} - \dfrac{2}{8} = \dfrac{5}{8}$$

(continued)

Copyright © 2018 Pearson Education, Inc.

The negative number, $-\frac{7}{8}$, has the greater absolute value, so the answer is negative.

$$-\frac{7}{8} + \frac{1}{4} = -\frac{5}{8}$$

5. **(a)** The additive inverse of 12 is -12. The sum of the number and its inverse is $12 + (-12) = \underline{0}$.

(b) The additive inverse of -9 is $-(-9)$ or 9. The sum of the number and its inverse is $(-9) + 9 = 0$.

(c) The additive inverse of 3.5 is -3.5. The sum of the number and its inverse is $3.5 + (-3.5) = 0$.

(d) The additive inverse of $-\frac{7}{10}$ is $-\left(-\frac{7}{10}\right)$ or $\frac{7}{10}$. The sum of the number and its inverse is $\left(-\frac{7}{10}\right) + \frac{7}{10} = 0$.

(e) The additive inverse of 0 is 0. The sum of the number and its inverse is $0 + 0 = 0$.

6. **(a)** $-7 - 2$ in words is negative seven minus positive two.

(b) -10 in words is negative ten.

(c) $3 - (-5)$ in words is positive three minus negative five.

(d) 4 in words is positive four.

(e) $-8 - (-6)$ in words is negative eight minus negative six.

(f) $2 - 9$ in words is positive two minus positive nine.

7. **(a)** $-6 - 5$
Change positive 5 to its opposite (-5) and add.
$$-6 + (-5) = \underline{-11}$$

(b) $3 - (-10)$
Change negative 10 to its opposite $(+10)$ and add.
$$3 - (-10) = 3 + (+10) = 13$$

(c) $-8 - (-2)$
Change negative 2 to its opposite $(+2)$ and add.
$$-8 - (-2) = -8 + (+2) = -6$$

(d) $4 - 9$
Change positive 9 to its opposite (-9) and add.
$$4 - 9 = 4 + (-9) = -5$$

(e) $-7 - (-15)$
Change negative 15 to its opposite $(+15)$ and add.
$$-7 - (-15) = -7 + (+15) = 8$$

(f) $100 - 300$
Change positive 300 to its opposite (-300) and add.
$$100 - 300 = 100 + (-300) = -200$$

8. **(a)** $-0.8 - 0.5$
Change positive 0.5 to its opposite (-0.5) and add.
$$-0.8 + (-0.5) = \underline{-1.3}$$

(b) $4 - 12.03$
Change positive 12.03 to its opposite (-12.03) and add.
$$4 - 12.03 = 4 + (-12.03) = -8.03$$

(c) $-1.65 - (-20.4)$
Change negative 20.4 to its opposite $(+20.4)$ and add.
$$-1.65 - (-20.4) = -1.65 + (+20.4) = 18.75$$

(d) $\dfrac{1}{2} - \left(-\dfrac{2}{3}\right)$
Change negative $\frac{2}{3}$ to its opposite $\left(+\frac{2}{3}\right)$ and add.
$$\frac{1}{2} + \left(+\frac{2}{3}\right) = \frac{3}{6} + \left(+\frac{4}{6}\right) = \frac{7}{6}$$

(e) $\dfrac{1}{10} - \dfrac{4}{5}$
Change positive $\frac{4}{5}$ to its opposite $\left(-\frac{4}{5}\right)$ and add.
$$\frac{1}{10} - \frac{4}{5} = \frac{1}{10} + \left(-\frac{4}{5}\right)$$
$$= \frac{1}{10} + \left(-\frac{8}{10}\right) = -\frac{7}{10}$$

(f) $-\dfrac{7}{9} - \left(-\dfrac{1}{6}\right)$
Change negative $\frac{1}{6}$ to its opposite $\left(+\frac{1}{6}\right)$ and add.
$$-\frac{7}{9} + \left(+\frac{1}{6}\right) = -\frac{14}{18} + \left(+\frac{3}{18}\right)$$
$$= -\frac{11}{18}$$

9. **(a)** $6 - 7 + (-3)$
$$= 6 + (-7) + (-3)$$
$$= -1 + (-3)$$
$$= \underline{-4}$$

(b) $-2 + (-3) - (-5)$
$$= -5 - (-5)$$
$$= -5 + (+5)$$
$$= \underline{0}$$

Copyright © 2018 Pearson Education, Inc.

(c) $-3 - (-9) - (-5)$

$\qquad = -3 + (+9) - (-5)$

$\qquad = 6 + (+5) = 11$

(d) $8 - (-2) + (-6)$

$\qquad = 8 + (+2) + (-6)$

$\qquad = 10 + (-6) = 4$

(e) $-1 - 2 + 3 - 4$

$\qquad = -1 + (-2) + 3 - 4$

$\qquad = -3 + 3 + (-4)$

$\qquad = 0 + (-4) = -4$

(f) $7 - 6 - 5 + (-4)$

$\qquad = 7 + (-6) - 5 + (-4)$

$\qquad = 1 + (-5) + (-4)$

$\qquad = -4 + (-4) = -8$

(g) $-6 + (-15) - (-19)$

$\qquad = -21 + (+19) = -2$

(h)

-19.2			
-6.7	$\rightarrow$	-25.9	
15.8	15.8	$\rightarrow$	-10.1
17.1	17.1	$17.1 \rightarrow$	7
-5.4	-5.4	-5.4	-5.4
			1.6

9.2 Section Exercises

1. $-5 + (-2) = -7$

3. $3 + (-4) = -1$

5. Subtract the lesser absolute value from the greater absolute value. Write the sign of the number with the greater absolute value in front of the answer. Examples will vary. One possibility is shown.

$12 + (-15)$

The signs are different, so subtract the absolute values.

$$|12| = 12 \quad \text{and} \quad |-15| = 15$$

$$15 - 12 = 3$$

The negative number, -15, has the greater absolute value, so the answer is negative.

$$12 + (-15) = -3$$

7. $-8 + 5$

The signs are different, so subtract the absolute values.

$$|-8| = 8 \quad \text{and} \quad |5| = 5$$

$$8 - 5 = 3$$

The negative number, -8, has the greater absolute value, so the answer is negative.

$$-8 + 5 = -3$$

9. $-1 + 8$

The signs are different, so subtract the absolute values.

$$|-1| = 1 \quad \text{and} \quad |8| = 8$$

$$8 - 1 = 7$$

The positive number, 8, has the greater absolute value, so the answer is positive.

$$-1 + 8 = 7$$

11. $-2 + (-5)$

Add the absolute values.

$$|-2| = 2 \quad \text{and} \quad |-5| = 5$$

$$2 + 5 = 7$$

Write a negative sign in front of the sum, because both numbers are negative.

$$-2 + (-5) = -7$$

13. $6 + (-5)$

The signs are different, so subtract the absolute values.

$$|6| = 6 \quad \text{and} \quad |-5| = 5$$

$$6 - 5 = 1$$

The positive number, 6, has the greater absolute value, so the answer is positive.

$$6 + (-5) = 1$$

15. $4 + (-12)$

The signs are different, so subtract the absolute values.

$$|4| = 4 \quad \text{and} \quad |-12| = 12$$

$$12 - 4 = 8$$

The negative number, -12, has the greater absolute value, so the answer is negative.

$$4 + (-12) = -8$$

Copyright © 2018 Pearson Education, Inc.

17. $-10 + (-10)$
Add the absolute values.

$$|-10| = 10 \quad \text{and} \quad |-10| = 10$$

$$10 + 10 = 20$$

Write a negative sign in front of the sum because both numbers are negative.

$$-10 + (-10) = -20$$

19. 13 yards gained (positive) and
17 yards lost (negative)

$$13 + (-17) = -4 \text{ yd}$$

21. The overdrawn amount is negative $(-\$52.50)$ and the deposit is positive.

$$-\$52.50 + \$50 = -\$2.50$$

23. Lost (negative); Won (positive)
Jeff: $-20 + 75 + (-55) = 0$ pts
Terry: $42 + (-15) + 20 = 47$ pts

25. $7.8 + (-14.6)$
The signs are different, so subtract the absolute values. First find the absolute values.

$$|7.8| = 7.8 \quad \text{and} \quad |-14.6| = 14.6$$

Subtract the lesser absolute value from the greater absolute value.

$$14.6 - 7.8 = 6.8$$

The negative number, -14.6, has the greater absolute value, so the answer is negative.

$$7.8 + (-14.6) = -6.8$$

27. $-\dfrac{1}{2} + \dfrac{3}{4}$
The signs are different, so subtract the absolute values.

$$\left|-\frac{1}{2}\right| = \frac{1}{2} \quad \text{and} \quad \left|\frac{3}{4}\right| = \frac{3}{4}$$

$$\frac{3}{4} - \frac{1}{2} = \frac{3}{4} - \frac{2}{4} = \frac{1}{4}$$

The positive number, $\frac{3}{4}$, has the greater absolute value, so the answer is positive.

$$-\frac{1}{2} + \frac{3}{4} = \frac{1}{4}$$

29. $-\dfrac{7}{10} + \dfrac{2}{5}$
The signs are different, so subtract the absolute values. First, find the absolute values.

$$\left|-\frac{7}{10}\right| = \frac{7}{10} \quad \text{and} \quad \left|\frac{2}{5}\right| = \frac{2}{5}$$

$$\frac{7}{10} - \frac{2}{5} = \frac{7}{10} - \frac{4}{10} = \frac{3}{10}$$

Subtract the lesser absolute value from the greater absolute value.

$$\frac{7}{10} - \frac{2}{5} = \frac{7}{10} - \frac{4}{10} = \frac{3}{10}$$

The negative number, $-\frac{7}{10}$, has the *greater* absolute value, so the answer is negative.

$$-\frac{7}{10} + \frac{2}{5} = -\frac{3}{10}$$

31. $-\dfrac{7}{3} + \left(-\dfrac{5}{9}\right)$
Add the absolute values.

$$\left|-\frac{7}{3}\right| = \frac{7}{3} \quad \text{and} \quad \left|-\frac{5}{9}\right| = \frac{5}{9}$$

$$\frac{7}{3} + \frac{5}{9} = \frac{21}{9} + \frac{5}{9} = \frac{26}{9}$$

Write a negative sign in front of the sum because both numbers are negative.

$$-\frac{7}{3} + \left(-\frac{5}{9}\right) = -\frac{26}{9}$$

33. The additive inverse of 3 is -3.

35. The additive inverse of -9 is $-(-9)$, or 9.

37. The additive inverse of $\frac{1}{2}$ is $-\frac{1}{2}$.

39. The additive inverse of -6.2 is $-(-6.2)$, or 6.2.

41. $-7 - 3$ in words is negative seven minus positive three.

43. $-20 + (-3)$ in words is negative twenty plus negative three.

45. $19 - 5 = 19 + (-5)$
$ = \underline{14}$

47. $10 - 12 = 10 + (-12)$
$ = -2$

49. The first number, 7, stays the same. Change subtraction to addition. Change the sign of the second number to its opposite.

$$c = 7 + (-19)$$

Now add.

$$7 + (-19) = -12$$

So, $7 - 19 = -12$ also.

51. $-15 - 10 = -15 + (-10)$
$ = -25$

Copyright © 2018 Pearson Education, Inc.

53. $-14 - (-9) = -14 + 9$
$$= -5$$

55. The first number, -3, stays the same. Change subtraction to addition. Change the sign of the second number to its opposite.
$$-3 - (-8) = -3 + (+8)$$
Now add.
$$-3 + 8 = 5$$
So, $-3 - (-8) = 5$ also.

57. $6 - (-14) = 6 + (+14)$
$$= 20$$

59. $1 - (-10) = 1 + (+10)$
$$= 11$$

61. The first number, -30, stays the same. Change subtraction to addition. Change the sign of the second number to its opposite.
$$-30 + (-30) = -60$$
So, $-30 - 30 = -60$ also.

63. $-16 - (-16) = -16 + (+16)$
$$= 0$$

65. $-\dfrac{4}{5} - \left(-\dfrac{7}{10}\right) = -\dfrac{4}{5} + \dfrac{7}{10}$
$$= -\dfrac{8}{10} + \dfrac{7}{10}$$
$$= -\dfrac{1}{10}$$

67. $\dfrac{1}{2} - \dfrac{9}{10} = \dfrac{1}{2} + \left(-\dfrac{9}{10}\right)$ *Change subtraction to addition.*
$$= \dfrac{5}{10} + \left(-\dfrac{9}{10}\right) \quad \text{\textit{Common denominator is 10.}}$$
$$= -\dfrac{4}{10} = -\dfrac{2}{5}$$

Once the subtraction is changed to addition, notice that the numbers have different signs. So, subtract the lesser absolute value from the greater absolute value. $\left|\dfrac{5}{10}\right|$ is $\dfrac{5}{10}$ and $\left|-\dfrac{9}{10}\right|$ is $\dfrac{9}{10}$.
Then $\dfrac{9}{10} - \dfrac{5}{10} = \dfrac{4}{10} = \dfrac{2}{5}$.
The *negative* number $-\dfrac{9}{10}$ has the greater absolute value, so the answer is *negative*.

69. $-8.3 - (-9) = -8.3 + (+9)$
$$= 0.7$$

71. Valerie did not change 6 to its opposite, -6. The correct answer is $-6 + (-6) = -12$.

73. **(a)** The 30 °F column and the 10 mph wind row intersect at 21 °F. The difference between the actual temperature and the wind chill temperature is $30 - 21 = 9$ degrees.

(b) The 15 °F column and the 15 mph wind row intersect at 0 °F. The difference between the actual temperature and the wind chill temperature is $15 - 0 = 15$ degrees.

75. **(a)** The 5 °F column and the 25 mph wind row intersect at -17 °F. The difference between the actual temperature and the wind chill temperature is $5 - (-17) = 5 + 17 = 22$ degrees.

(b) The -10 °F column and the 35 mph wind row intersect at -41 °F. The difference between the actual temperature and the wind chill temperature is $-10 - (-41) = -10 + 41 = 31$ degrees.

77. $-2 + (-11) - (-3) = -13 - (-3)$
$$= -13 + 3$$
$$= -10$$

79. Change subtraction to addition. Change -13 to its opposite, $(+13)$.
$$4 - (-13) + (-5) = 4 + (+13) + (-5)$$
$$= 17 + (-5)$$
$$= 12$$

81. $-12 - (-3) - (-2) = -12 + 3 - (-2)$
$$= -9 - (-2)$$
$$= -9 + 2$$
$$= -7$$

83. $4 - (-4) - 3 = 4 + (+4) - 3$
$$= 8 - 3$$
$$= 5$$

85. $\dfrac{1}{2} - \dfrac{2}{3} + \left(-\dfrac{5}{6}\right) = \dfrac{1}{2} + \left(-\dfrac{2}{3}\right) + \left(-\dfrac{5}{6}\right)$
Change subtraction to addition.
Change $\dfrac{2}{3}$ to its opposite, $\left(-\dfrac{2}{3}\right)$.
$$= \dfrac{3}{6} + \left(-\dfrac{4}{6}\right) + \left(-\dfrac{5}{6}\right)$$
$$= -\dfrac{1}{6} + \left(-\dfrac{5}{6}\right)$$
$$= -\dfrac{6}{6} = -1$$

87. $-5.7 - (-9.4) - 8.1 = -5.7 + (+9.4) - 8.1$
$$= 3.7 + (-8.1)$$
$$= -4.4$$

Copyright © 2018 Pearson Education, Inc.

89. $-2 + (-11) + |-10 + 8| = -2 + (-11) + |-2|$
$$= -13 + |-2|$$
$$= -13 + 2$$
$$= -11$$

91.   Work inside the parentheses: $(-2 + 4)$ is 2.

$-3 - (-2 + 4) + (-5) = -3 - (+2) + (-5)$
Change subtraction to addition.
Change 2 to its opposite, (-2).
$$= -3 + (-2) + (-5)$$
$$= -5 + (-5)$$
$$= -10$$

93. $\$37 - \$626 + \$908 - \$60$
$$= \$37 + (-\$626) + \$908 + (-\$60)$$
$$= -\$589 + \$908 + (-\$60)$$
$$= \$319 + (-\$60)$$
$$= \$259$$
The balance is $259.

95. $-\$89.62 - \$110.70 - \$99.68 + 3(\$100)$
$$= -\$89.62 + (-\$110.70) + (-\$99.68) + \$300$$
$$= -\$200.32 + (-\$99.68) + \$300$$
$$= -\$300 + \$300$$
$$= \$0$$
The balance is $0.

97.   **(a)** $-5 + 3 = \underline{-2}$ $\qquad$ $3 + (-5) = \underline{-2}$

(b) $-2 + (-6) = \underline{-8}$ $\qquad$ $-6 + (-2) = \underline{-8}$

(c) $17 + (-7) = \underline{10}$ $\qquad$ $-7 + 17 = \underline{10}$

Addition is commutative, so changing the order of the addends does *not* change the sum.

98.   **(a)** $-3 - 5 = \underline{-8}$
$5 - (-3) = 5 + 3 = \underline{8}$

(b) $-4 - (-6) = -4 + 6 = \underline{2}$
$-6 - (-4) = -6 + 4 = \underline{-2}$

(c) $3 - 10 = \underline{-7}$
$10 - 3 = \underline{7}$

Changing the order of the numbers in subtraction *does change* the result.

99.   **(a)** The answers have the same absolute value but opposite signs.

(b) When the order of the numbers in a subtraction problem is switched, change the sign on the answer to its opposite.

100. **(a)** $-18 + 0 = \underline{-18}$ $\qquad$ $20 + 0 = \underline{20}$
$0 + (-5) = \underline{-5}$ $\qquad$ $0 + 4 = \underline{4}$

Adding 0 to any number leaves the number unchanged.

(b) $2 - 0 = \underline{2}$ $\qquad$ $0 - 10 = \underline{-10}$
$-3 - 0 = \underline{-3}$ $\qquad$ $0 - (-7) = 0 + 7 = \underline{7}$

Subtracting 0 from a number leaves the number unchanged, but subtracting a *number from 0* changes the sign of the number.

9.3 Multiplying and Dividing Signed Numbers

9.3   Margin Exercises

1.   **(a)** $5 \cdot (-4) = -20$
The product of two numbers with *different* signs is *negative*.

(b) $(-9)(15) = -135$

(c) $12(-1) = -12$

(d) $-6(6) = -36$

(e) $\left(-\dfrac{7}{8}\right)\left(\dfrac{4}{3}\right) = \left(-\dfrac{7}{\overset{\;}{\underset{2}{8}}}\right)\left(\dfrac{\overset{1}{4}}{3}\right) = -\dfrac{7}{6}$

2.   **(a)** $(-5)(-5) = 25$
The product of two numbers with the *same* sign is *positive*.

(b) $(-14)(-1) = 14$

(c) $-7(-8) = 56$

(d) $3(12) = 36$

(e) $\left(-\dfrac{2}{3}\right)\left(-\dfrac{6}{5}\right) = \left(-\dfrac{2}{\overset{\;}{\underset{1}{3}}}\right)\left(-\dfrac{\overset{2}{6}}{5}\right) = \dfrac{4}{5}$

3.   **(a)** $\dfrac{-20}{4} = -5$
The numbers have *different* signs, so the quotient is *negative*.

(b) $\dfrac{-50}{-5} = 10$
The numbers have the *same* sign, so the quotient is *positive*.

(c) $\dfrac{44}{2} = 22$

(d) $\dfrac{6}{-6} = -1$

(e) $\dfrac{-15}{-1} = 15$

Copyright © 2018 Pearson Education, Inc.

(f) $\dfrac{-\frac{3}{5}}{\frac{9}{10}} = -\dfrac{3}{5} \div \dfrac{9}{10} = -\dfrac{\overset{1}{\cancel{3}}}{\cancel{5}} \cdot \dfrac{\overset{2}{\cancel{10}}}{\cancel{9}} = -\dfrac{2}{3}$

(g) $\dfrac{-35}{0}$ is undefined.

9.3 Section Exercises

1. The product of two numbers with the *same* sign is *positive*. Examples will vary. One possibility is shown.

$$(-2)(-4) = 8$$

3. $-5 \cdot 7 = -35$
The numbers have *different* signs, so the product is *negative*.

5. $(-5)(9) = -45$
The numbers have *different* signs, so the product is *negative*.

7. $3(-6) = -18$
The numbers have *different* signs, so the product is *negative*.

9. $10(-5) = -50$
The numbers have *different* signs, so the product is *negative*.

11. $(-1)(40) = -40$
The numbers have *different* signs, so the product is *negative*.

13. $-8(-4) = 32$
The numbers have the *same* sign, so the product is *positive*.

15. $11(7) = 77$
The numbers have the *same* sign, so the product is *positive*.

17. $-19(-7) = 133$
The numbers have the *same* sign, so the product is *positive*.

19. $-13(-1) = 13$
The numbers have the *same* sign, so the product is *positive*.

21. $(0)(-25) = 0$

23. $-\dfrac{1}{2}(-8) = -\dfrac{1}{\cancel{2}_{1}}\left(-\dfrac{\cancel{8}^{4}}{1}\right) = \dfrac{4}{1} = 4$

25. $-10\left(\dfrac{2}{5}\right) = -\dfrac{\cancel{10}^{2}}{1} \cdot \dfrac{2}{\cancel{5}_{1}} = -2 \cdot 2 = -4$

27. $\left(\dfrac{3}{5}\right)\left(-\dfrac{1}{6}\right) = \left(\dfrac{\cancel{3}^{1}}{5}\right)\left(-\dfrac{1}{\cancel{6}_{2}}\right) = -\dfrac{1}{10}$

29. $-\dfrac{7}{5}\left(-\dfrac{10}{3}\right) = -\dfrac{7}{\cancel{5}_{1}}\left(-\dfrac{\cancel{10}^{2}}{3}\right) = \dfrac{14}{3}$

31. $-\dfrac{7}{15} \cdot \dfrac{25}{14} = -\dfrac{\cancel{7}^{1}}{\cancel{15}_{3}} \cdot \dfrac{\cancel{25}^{5}}{\cancel{14}_{2}} = -\dfrac{5}{6}$
The numbers have *diffferent* signs, so the product is *negative*.

33. $-\dfrac{5}{2}\left(-\dfrac{7}{10}\right) = -\dfrac{\cancel{5}^{1}}{2}\left(-\dfrac{7}{\cancel{10}_{2}}\right) = \dfrac{7}{4}$

35. $9(-4.7) = -42.3$

37. $(-0.5)(-12) = 6$

39. $(-6.2)(5.1) = -31.62$

41. $-1.25(-3.6) = 4.5$
The numbers have the *same* sign, so the product is *positive*.

43. $(-8.23)(-1) = 8.23$

45. $0(-58.6) = 0$

47. Thong forgot that 0 *can be* divided by a number. It is dividing by 0 that is undefined, such as $12 \div 0$. The correct answer is

$$\dfrac{0}{12} = 0$$

49. $\dfrac{-14}{7} = -2$
The numbers have *different* signs, so the quotient is *negative*.

51. $\dfrac{30}{-6} = -5$
The numbers have *different* signs, so the quotient is *negative*.

53. $\dfrac{-28}{0}$ is undefined.

55. $\dfrac{14}{-1} = -14$
The numbers have *different* signs, so the quotient is *negative*.

57. $\dfrac{-20}{-2} = 10$
The numbers have the *same* sign, so the quotient is *positive*.

Copyright © 2018 Pearson Education, Inc.

59. $\dfrac{-48}{-12} = 4$

The numbers have the *same* sign, so the quotient is *positive*.

61. $\dfrac{-18}{18} = -1$

63. $\dfrac{-573}{-3} = 191$

65. $\dfrac{0}{-9} = 0$

67. $\dfrac{-30}{-30} = 1$

69. Rewrite the division as multiplication that uses the reciprocal of $-\frac{15}{14}$, which is $-\frac{14}{15}$. Divide out all the common factors. Then multiply.

$$\dfrac{-\frac{5}{7}}{-\frac{15}{14}} = -\dfrac{\overset{1}{\cancel{5}}}{\underset{1}{7}}\left(-\dfrac{\overset{2}{\cancel{14}}}{\underset{3}{\cancel{15}}}\right) = \dfrac{2}{3}$$

71. Rewrite division as multiplication using the reciprocal of -2, which is $-\frac{1}{2}$. Divide out the common factor. Then multiply.

$$-\dfrac{2}{3} \div (-2) = -\dfrac{\overset{1}{\cancel{2}}}{3}\left(-\dfrac{1}{\underset{1}{\cancel{2}}}\right) = \dfrac{1}{3}$$

The numbers have the *same* sign, so the quotient is *positive*.

73. $5 \div \left(-\dfrac{5}{8}\right) = \dfrac{\overset{1}{\cancel{5}}}{1}\left(-\dfrac{8}{\underset{1}{\cancel{5}}}\right) = -8$

75. $-\dfrac{7}{5} \div \dfrac{3}{10} = -\dfrac{7}{\underset{1}{\cancel{5}}} \cdot \dfrac{\overset{2}{\cancel{10}}}{3} = -\dfrac{14}{3}$

77. $\dfrac{-18.92}{-4} = 4.73$

The numbers have the *same* sign, so the quotient is *positive*.

$$\begin{array}{r} 4.73 \\ 4\overline{)18.92} \\ 16 \\ \hline 29 \\ 28 \\ \hline 12 \\ 12 \\ \hline 0 \end{array}$$

79. $\dfrac{-7.05}{1.5} = -4.7$

The numbers have *different* signs, so the quotient is *negative*.

$$\begin{array}{r} 4.7 \\ 1.5_\wedge\overline{)7.0_\wedge5} \\ 60 \\ \hline 105 \\ 105 \\ \hline 0 \end{array}$$

81. $\dfrac{45.58}{-8.6} = -5.3$

The numbers have *different* signs, so the quotient is *negative*.

$$\begin{array}{r} 5.3 \\ 8.6_\wedge\overline{)45.5_\wedge8} \\ 430 \\ \hline 258 \\ 258 \\ \hline 0 \end{array}$$

83. $(-4)(-6)\dfrac{1}{2} = 24 \cdot \dfrac{1}{2} = 12$

85. $(-0.6)(-0.2)(-3) = (0.12)(-3) = -0.36$

87. $\left(-\dfrac{1}{2}\right)\left(\dfrac{2}{5}\right)\left(\dfrac{7}{8}\right) = -\dfrac{1}{5}\left(\dfrac{7}{8}\right) = -\dfrac{7}{40}$

89. $-36 \div (-2) \div (-3) \div (-3) \div (-1)$
$= 18 \div (-3) \div (-3) \div (-1)$
$= -6 \div (-3) \div (-1)$
$= 2 \div (-1) = -2$

91. $|-8| \div (-4) \cdot |-5|$
$= 8 \div (-4) \cdot (5)$
$= -2 \cdot 5 = -10$

93. $12(-9950) = -119,400$
The total loss for the year was $119,400, which we can write as $\underline{-\$119,400}$.

95. $\dfrac{-36,198}{18} = -2011$
The sub dives 2011 feet in each step, which we can write as $\underline{-2011 \text{ ft}}$.

97. $13(182) = 2366$
Wei Chen will pay $\underline{\$2366}$ for 13 credits.

99. First, divide 24,000 by 1000.
$$\dfrac{24,000}{1000} = 24$$
Then, multiply 24 by the number of degrees the temperature changes per 1000 feet. The temperature will change $24(-3) = -72$ degrees in 24,000 ft.
Finally, add the ground temperature and the change in temperature. The temperature will be $50 + (-72) = \underline{-22 \text{ degrees}}$.

Copyright © 2018 Pearson Education, Inc.

101. (a) Examples will vary. Some possibilities:

$6(-1) = -6; 2(-1) = -2; 15(-1) = -15$

(b) Examples will vary. Some possibilities:

$-6(-1) = 6; -2(-1) = 2; -15(-1) = 15$

The result of multiplying any nonzero number times -1 is the number with the opposite sign.

102. (a) Examples will vary. Some possibilities:

$$\frac{-6}{-1} = 6; \frac{-2}{-1} = 2; \frac{-15}{-1} = 15$$

(b) Examples will vary. Some possibilities:

$$\frac{6}{-1} = -6; \frac{2}{-1} = -2; \frac{15}{-1} = -15$$

When dividing by -1, the sign of the number changes to its opposite.

9.4 Order of Operations

9.4 Margin Exercises

1. (a) $-9 + (-15) + (-3)$ *Add.*
$= -24 + (-3)$ *Add.*
$= -27$

(b) $-8 - (-2) + (-6)$ *Change subtraction to addition.*
$= -8 + (+2) + (-6)$ *Add.*
$= -6 + (-6)$ *Add.*
$= -12$

(c) $3(-4) \div (-6)$ *Multiply.*
$= -12 \div (-6)$ *Divide.*
$= 2$

(d) $-18 \div 9(-4)$ *Divide.*
$= -2(-4)$ *Multiply.*
$= 8$

2. (a) $10 + 8 \div 2$ *Divide.*
$= 10 + \underline{4}$ *Add.*
$= 14$

(b) $4 - 6(-2)$ *Multiply.*
$= 4 - (-12)$ *Change subt.*
$= 4 + (+12)$ *Add.*
$= 16$

(c) $-3 + (-5) \cdot 2 - 1$ *Multiply.*
$= -3 + (-10) - 1$ *Add.*
$= -13 - 1$ *Change subt.*
$= -13 + (-1)$ *Add.*
$= -14$

(d) $-6 \div 2 + 3(-2)$ *Divide.*
$= -3 + 3(-2)$ *Multiply.*
$= -3 + (-6)$ *Add.*
$= -9$

(e) $7 - 6(2) \div (-3)$ *Multiply.*
$= 7 - 12 \div (-3)$ *Divide.*
$= 7 - (-4)$ *Change subt.*
$= 7 + (+4)$ *Add.*
$= 11$

3. (a) $2 + 40 \div (-5 + 3)$ *Parentheses*
$= 2 + 40 \div (-2)$ *Divide.*
$= 2 + (\underline{-20})$ *Add.*
$= \underline{-18}$

(b) $5 - 3(2 + 4)$ *Parentheses*
$= 5 - 3(6)$ *Multiply.*
$= 5 - 18$ *Change subt.*
$= 5 + (\underline{-18})$ *Add.*
$= \underline{-13}$

(c) $(-24 \div 2) + (15 - 3)$ *Parentheses*
$= -12 + 12$ *Add.*
$= 0$

(d) $-3(2 - 8) - 5(4 - 3)$ *Parentheses*
$= -3(-6) - 5(1)$ *Multiply.*
$= 18 - 5$ *Change subt.*
$= 18 + (-5)$ *Add.*
$= 13$

(e) $3(3) - 10(3) \div 5$ *Multiply.*
$= 9 - 10(3) \div 5$ *Multiply.*
$= 9 - 30 \div 5$ *Divide.*
$= 9 - 6$ *Change subt.*
$= 9 + (-6)$ *Add.*
$= 3$

(f) $6 - (2 + 7) \div (-4 + 1)$ *Parentheses*
$= 6 - 9 \div (-3)$ *Divide.*
$= 6 - (-3)$ *Change subt.*
$= 6 + (+3)$ *Add.*
$= 9$

4. (a) $2^3 - 3^2$ *Exponents*
$= 8 - \underline{9}$ *Change subt.*
$= 8 + (-9)$ *Add.*
$= \underline{-1}$

Copyright © 2018 Pearson Education, Inc.

(b) $4^2 - 3^2(5 - 2)$ *Parentheses*
$= 4^2 - 3^2(3)$ *Exponents*
$= 16 - 9(3)$ *Multiply.*
$= 16 - 27$ *Change subt.*
$= 16 + (-27)$ *Add.*
$= -11$

(c) $-18 \div (-3)(2^3)$ *Exponent*
$= -18 \div (-3)(8)$ *Divide.*
$= 6 \cdot 8$ *Multiply.*
$= 48$

(d) $(-3)^3 + (3 - 8)^2$ *Parentheses*
$= (-3)^3 + (-5)^2$ *Exponents*
$= -27 + 25$ *Add.*
$= -2$

(e) $\dfrac{3}{8} + \left(-\dfrac{1}{2}\right)^2 \div \dfrac{1}{4}$ *Exponent*

$= \dfrac{3}{8} + \left(\dfrac{1}{4}\right) \div \dfrac{1}{4}$ *Divide.*

$= \dfrac{3}{8} + 1$ *Add.*

$= \dfrac{3}{8} + \dfrac{8}{8}$

$= \dfrac{11}{8}$

5.   **(a)** $\dfrac{-3(2^3)}{-10 - 6 + 8}$

First do the work in the numerator.

$-3(2^3)$ *Exponent*
$= -3 \cdot 8$ *Multiply.*
$= -24$ *Numerator*

Now do the work in the denominator.

$-10 - 6 + 8$ *Change subt.*
$= -10 + (-6) + 8$ *Add.*
$= -16 + 8$ *Add.*
$= -8$ *Denominator*

The last step is the division.
$$\frac{-24}{-8} = 3$$

(b) $\dfrac{(-10)(-5)}{-6 \div 3(5)}$

Numerator:
$(-10)(-5) = 50$

Denominator:
$-6 \div 3(5)$ *Divide.*
$= -2(5)$ *Multiply.*
$= -10$

Divide: $\dfrac{50}{-10} = -5$

(c) $\dfrac{6 + 18 \div (-2)}{(1 - 10) \div 3}$

Numerator:
$6 + 18 \div (-2)$ *Divide.*
$= 6 + (-9)$ *Add.*
$= -3$

Denominator:
$(1 - 10) \div 3$ *Parentheses*
$= (-9) \div 3$ *Divide.*
$= -3$

Divide: $\dfrac{-3}{-3} = 1$

(d) $\dfrac{6^2 - 3^2(4)}{5 + (3 - 7)^2}$

Numerator:
$6^2 - 3^2(4)$ *Exponents*
$= 36 - 9(4)$ *Multiply.*
$= 36 - 36$ *Subtract.*
$= 0$

Denominator:
$5 + (3 - 7)^2$ *Parentheses*
$= 5 + (-4)^2$ *Exponent*
$= 5 + 16$ *Add.*
$= 21$

Divide: $\dfrac{0}{21} = 0$

9.4 Section Exercises

1. $(-6)^2 - 8 \div 2(-3)$ *Exponent*
$= 36 - 8 \div 2(-3)$ *Divide.*
$= 36 - 4(-3)$ *Multiply.*
$= 36 - (-12)$ *Change subt.*
$= 36 + +12$ *Add.*
$= 48$

3. $6 + 3(-4)$ *Multiply.*
$= 6 + (\underline{-12})$ *Add.*
$= \underline{-6}$

5. $-1 + 15 + 2(-7)$ *Multiply.*
$= -1 + 15 + (-14)$ *Add.*
$= 14 + (-14)$ *Add.*
$= 0$

7. $6^2 + 4^2$ *Exponents*
$= 36 + \underline{16}$ *Add.*
$= \underline{52}$

Copyright © 2018 Pearson Education, Inc.

9. $10 - 7^2$ *Exponent*

$= 10 - 49$ *Change subtraction to addition.*

$= 10 + (-49)$ *Add.*

$= -39$

11. $(-2)^5 + 2$ *Exponent*

$= -32 + 2$ *Add.*

$= -30$

13. $4^2 + 3^2 + (-8)$ *Exponents*

$= 16 + 9 + (-8)$ *Add.*

$= 25 + (-8)$ *Add.*

$= 17$

15. $2 - (-5) + 3^2$ *Exponent*

$= 2 - (-5) + 9$ *Change subt.*

$= 2 + (+5) + 9$ *Add.*

$= 7 + 9$ *Add.*

$= 16$

17. $(-4)^2 + (-3)^2 + 5$ *Exponents*

$= 16 + 9 + 5$ *Add.*

$= 25 + 5$ *Add.*

$= 30$

19. $3 + 5(6 - 2)$ *Parentheses*

$= 3 + 5(4)$ *Multiply.*

$= 3 + 20$ *Add.*

$= 23$

21. $-7 + 6(8 - 14)$ *Parentheses*

$= -7 + 6(-6)$ *Multiply.*

$= -7 + (-36)$ *Add.*

$= -43$

23. $-6 + (-5)(9 - 14)$ *Parentheses*

$= -6 + (-5)(-5)$ *Multiply.*

$= -6 + 25$ *Add.*

$= 19$

25. $(-5)(7 - 13) \div (-10)$ *Parentheses*

$= (-5)(-6) \div (-10)$ *Multiply.*

$= 30 \div (-10)$ *Divide.*

$= -3$

27. $9 \div (-3)^2 + (-1)$ *Exponent*

$= 9 \div 9 + (-1)$ *Divide.*

$= 1 + (-1)$ *Add.*

$= 0$

29. $2 - (-5)(-3)^2$ *Exponent*

$= 2 - (-5)(9)$ *Multiply.*

$= 2 - (-45)$ *Change subt.*

$= 2 + (+45)$ *Add.*

$= 47$

31. $(-2)(-7) + 3(9)$ *Multiply.*

$= 14 + 27$ *Add.*

$= 41$

33. $30 \div (-5) - 36 \div (-9)$ *Divide 30 by -5.*

$= -6 - 36 \div (-9)$ *Divide 36 by -9.*

$= -6 - (-4)$ *Change subt.*

$= -6 + 4$ *Add.*

$= -2$

35. $2(5) - 3(4) + 5(3)$ *Multiply.*

$= 10 - 12 + 15$ *Change subt.*

$= 10 + (-12) + 15$ *Add.*

$= -2 + 15$ *Add.*

$= 13$

37. $4(3^2) + 7(3 + 9) - (-6)$ *Parentheses*

$= 4 \cdot 3^2 + 7(12) - (-6)$ *Exponent*

$= 4 \cdot 9 + 7 \cdot 12 - (-6)$ *Multiply.*

$= 36 + 84 - (-6)$ *Add; change subt.*

$= 120 + (+6)$ *Add.*

$= 126$

39. $\dfrac{-1 + 5^2 - (-3)}{-6 - 9 + 12}$

Numerator:

$(-1) + 5^2 - (-3)$

$= -1 + 25 - (-3)$

$= 24 - (-3)$

$= 24 + (+3) = 27$

Denominator:

$-6 - 9 + 12$

$= -6 + (-9) + 12$

$= -15 + 12 = -3$

Divide: $\dfrac{27}{-3} = -9$

41. $\dfrac{-2(4^2) - 4(6 - 2)}{-4(8 - 13) \div (-5)}$

Numerator:

$-2(4^2) - 4(6 - 2)$

$= -2(4^2) - 4(4)$

$= -2(16) - 4(4)$

$= -32 - 16$

$= -32 + (-16)$

$= -48$

(continued)

Copyright © 2018 Pearson Education, Inc.

Denominator:

$$-4(8 - 13) \div (-5)$$
$$= -4(-5) \div (-5)$$
$$= 20 \div (-5)$$
$$= -4$$

Divide: $\dfrac{-48}{-4} = 12$

43. $\dfrac{2^3(-2 - 5) + 4(-1)}{4 + 5(-6 \cdot 2) + (5 \cdot 11)}$

Numerator:

$$2^3(-2 - 5) + 4(-1) \quad \text{Parentheses}$$
$$= 2^3(-7) + 4(-1) \quad \text{Exponent}$$
$$= 8(-7) + 4(-1) \quad \text{Multiply } 8(-7).$$
$$= -56 + (-4) \quad \text{Multiply } 4(-1).$$
$$= -60 \quad \text{Add.}$$

Denominator:

$$4 + 5(-6 \cdot 2) + (5 \cdot 11) \quad \text{First set of parens}$$
$$= 4 + 5(-12) + (5 \cdot 11) \quad \text{Second set of parens}$$
$$= 4 + 5(-12) + 55 \quad \text{Multiply } 5(-12).$$
$$= 4 + (-60) + 55 \quad \text{Add } 4 + (-60).$$
$$= -56 + 55 \quad \text{Add}$$
$$= -1$$

Divide: $\dfrac{-60}{-1} = 60$

45. $(-4)^2(7 - 9)^2 \div 2^3$
$$= (-4)^2(-2)^2 \div 2^3$$
$$= 16(4) \div 8$$
$$= 64 \div 8$$
$$= 8$$

47. $(-0.3)^2 + (-0.5)^2 + 0.9$
$$= 0.09 + 0.25 + 0.9$$
$$= 0.34 + 0.9 = 1.24$$

49. $(-0.75)(3.6 - 5)^2$
$$= (-0.75)(-1.4)^2$$
$$= (-0.75)(1.96)$$
$$= -1.47$$

51. $(0.5)^2(-8) - (0.31)$
$$= 0.25 \cdot (-8) - (0.31)$$
$$= -2 + (-0.31)$$
$$= -2.31$$

53. $\dfrac{2}{3} \div \left(-\dfrac{5}{6}\right) - \dfrac{1}{2}$

$$= \dfrac{2}{\overset{1}{\cancel{3}}} \cdot \left(-\dfrac{\overset{2}{\cancel{6}}}{5}\right) - \dfrac{1}{2}$$

$$= -\dfrac{4}{5} + \left(-\dfrac{1}{2}\right)$$

$$= -\dfrac{8}{10} + \left(-\dfrac{5}{10}\right) = -\dfrac{13}{10}$$

55. $\left(-\dfrac{1}{2}\right)^2 - \left(\dfrac{3}{4} - \dfrac{7}{4}\right)$

$$= \left(-\dfrac{1}{2}\right)^2 - \left(-\dfrac{4}{4}\right)$$

$$= \dfrac{1}{4} + \left(+\dfrac{4}{4}\right)$$

$$= \dfrac{5}{4}$$

57. $\dfrac{3}{5}\left(-\dfrac{7}{6}\right) - \left(\dfrac{1}{6} - \dfrac{5}{3}\right)$

Rewrite $\dfrac{5}{3}$ *as* $\dfrac{10}{6}$.

$$= \dfrac{\overset{1}{\cancel{3}}}{5} \cdot \left(-\dfrac{7}{\underset{2}{\cancel{6}}}\right) - \left(\dfrac{1}{6} - \dfrac{10}{6}\right)$$

Parentheses

$$= \dfrac{\overset{1}{\cancel{3}}}{5} \cdot \left(-\dfrac{7}{\underset{2}{\cancel{6}}}\right) - \left(-\dfrac{9}{6}\right)$$

Divide out the common factor.

$$= -\dfrac{7}{10} - \left(-\dfrac{9}{6}\right)$$

Change subtraction to addition
and $-\dfrac{9}{6}$ *to* $\dfrac{3}{2}$.

$$= -\dfrac{7}{10} + \left(+\dfrac{3}{2}\right)$$

Rewrite $\dfrac{3}{2}$ *as* $\dfrac{15}{10}$.

$$= -\dfrac{7}{10} + \dfrac{15}{10} \quad \text{Add.}$$
$$= \dfrac{8}{10} = \dfrac{4}{5} \quad \text{Simplify.}$$

Copyright © 2018 Pearson Education, Inc.

59. $5^2(9-11)(-3)(-2)^3$
$= 5^2(-2)(-3)(-2)^3$
$= 25(-2)(-3)(-8)$
$= -50(-3)(-8)$
$= 150 \cdot (-8)$
$= -1200$

61. $1.6(-0.8) \div (-0.32) \div 2^2$
$= 1.6(-0.8) \div (-0.32) \div 4$
$= -1.28 \div (-0.32) \div 4$
$= 4 \div 4$
$= 1$

63. $\dfrac{-9 + 18 \div (-3)(-6)}{5 - 4(12) \div 3(2)}$

Numerator:
$-9 + 18 \div (-3)(-6)$
$= -9 + (-6)(-6)$
$= -9 + 36$
$= 27$

Denominator:
$5 - 4(12) \div 3(2)$
$= 5 - 48 \div 3(2)$
$= 5 - 16(2)$
$= 5 - 32$
$= 5 + (-32)$
$= -27$

Divide: $\dfrac{27}{-27} = -1$

65. $-7\left(6 - \dfrac{5}{8} \cdot 24 + 3 \cdot \dfrac{8}{3}\right)$ *Multiply* $\dfrac{5}{8} \cdot 24$.

$= -7\left(6 - 15 + 3 \cdot \dfrac{8}{3}\right)$ *Multiply* $3 \cdot \dfrac{8}{3}$.

$= -7(6 - 15 + 8)$ *Subtract* $6 - 15$.
$= -7(-9 + 8)$ *Add* $-9 + 8$.
$= -7(-1)$ *Multiply.*
$= 7$

67. $|-12| \div 4 + 2(3^2) \div 6$
$= |-12| \div 4 + 2(9) \div 6$
$= 12 \div 4 + 2(9) \div 6$
$= 3 + 2(9) \div 6$
$= 3 + 18 \div 6$
$= 3 + 3$
$= 6$

Summary Exercises *Operations with Signed Numbers*

1. $2 - 8 = 2 + (-8) = -6$

3. $-14 - (-7) = -14 + (+7) = -7$

5. $-9(-7) = 63$

7. $(1)(-56) = -56$

9. $5 - (-7) = 5 + (+7) = 12$

11. $-18 + 5 = -13$

13. $-40 - 40 = -40 + (-40) = -80$

15. $8(-6) = -48$

17. $\left(-\dfrac{5}{6}\right)\left(\dfrac{1}{5}\right) = \left(-\dfrac{\cancel{5}^{1}}{6}\right)\left(\dfrac{1}{\cancel{5}_{1}}\right) = -\dfrac{1}{6}$

19. $0 - 14.6 = 0 + (-14.6) = -14.6$

21. $-4(-6)(2) = 24(2) = 48$

23. $-60 \div 10 \div (-3) = -6 \div (-3) = 2$

25. $64(0) \div (-8) = 0 \div (-8) = 0$

27. $-9 + 8 + (-2) = -1 + (-2) = -3$

29. $8 + 6 + (-8) = 14 + (-8) = 6$

31. $-7 + 28 + (-56) + 3 = 21 + (-56) + 3$
$= -35 + 3$
$= -32$

33. $-6(-8) \div (-5 - 7) = -6(-8) \div (-12)$
$= 48 \div (-12)$
$= -4$

35. $-80 \div 4(-5) = -20(-5)$
$= 100$

37. $-7 \cdot |7| \cdot |-7| = -7 \cdot 7 \cdot 7 = -49 \cdot 7 = -343$

39. $-2(-3)(7) \div (-7) = 6(7) \div (-7)$
$= 42 \div (-7)$
$= -6$

41. $0 - |-7 + 2| = 0 - |-5|$
$= 0 - 5$
$= 0 + (-5)$
$= -5$

Copyright © 2018 Pearson Education, Inc.

43. $12 \div 4 + 2(-2)^2 \div (-4)$
$= 12 \div 4 + 2(4) \div (-4)$
$= 3 + 2(4) \div (-4)$
$= 3 + 8 \div (-4)$
$= 3 + (-2)$
$= 1$

45. $\dfrac{5 - |2 - 4 \cdot 4| + (-5)^2 \div 5^2}{-9 \div 3(2 - 2) - (-8)}$

$= \dfrac{5 - |2 - 16| + (-5)^2 \div 5^2}{-9 \div 3(0) - (-8)}$

$= \dfrac{5 - |-14| + (-5)^2 \div 5^2}{-9 \div 3(0) - (-8)}$

$= \dfrac{5 - 14 + 25 \div 25}{-3(0) + (+8)}$

$= \dfrac{5 - 14 + 1}{0 + 8}$

$= \dfrac{5 + (-14) + 1}{8}$

$= \dfrac{-9 + 1}{8}$

$= \dfrac{-8}{8}$

$= -1$

47. $-\$4176 \div 174 = -\$24$
The average loss on each item was $24.

49. $37(730) = 27{,}010$
The plane ascended 27,010 ft.

51. $-65 + 24 + (-49) = -41 + (-49) = -90$
Her final depth was -90 ft.

53. $-\$700$ is less than any other number in the "Profit or Loss" column, so January had the greatest loss.
April had the greatest profit at $900.

55. The total expenses were
$$\$3100 + \$2000 + \$1800 + \$1400$$
$$+ \$1600 + \$1200$$
$$= \$11{,}100,$$

which can be represented as $-\$11{,}100$.

Her average monthly amount of expenses was
$-\$11{,}100 \div 6 = -\$1850$.

57. Similar: If the signs match, the result is positive. If the signs are different, the result is negative.
Different: Multiplication is commutative; division is not. You can multiply by 0, but dividing by 0 cannot be done.

9.5 Evaluating Expressions and Formulas

9.5 Margin Exercises

1. $5x - 3y$

(a) x is 5, y is 2
Replace x with 5. Replace y with 2.
$$5x - 3y = 5(5) - 3(2)$$
$$= \underline{25} - \underline{6}$$
$$= \underline{19}$$

(b) x is -3, y is 4
Replace x with -3. Replace y with 4.
$$5x - 3y = 5(-3) - 3(4)$$
$$= -15 - 12$$
$$= -27$$

(c) x is 0, y is 6
Replace x with 0. Replace y with 6.
$$5x - 3y = 5(0) - 3(6)$$
$$= 0 - 18$$
$$= -18$$

2. $\dfrac{3k + r}{2s}$

(a) k is 1, r is 1, s is 2
Replace k with 1, r with 1, and s with 2.
$$\frac{3k + r}{2s} = \frac{3(1) + (1)}{2(2)}$$
$$= \frac{3 + 1}{4}$$
$$= \frac{4}{4} = \underline{1}$$

(b) k is 8, r is -2, s is -4
Replace k with 8, r with -2, and s with -4.
$$\frac{3k + r}{2s} = \frac{3(8) + (-2)}{2(-4)}$$
$$= \frac{24 + (-2)}{-8}$$
$$= \frac{22}{-8} = -\frac{11}{4}$$

3. **(a)** $-x - 6y$; x is -1, y is -4
$$-x - 6y = -(-1) - 6(-4)$$
$$= 1 - (-24)$$
$$= 1 + 24$$
$$= \underline{25}$$

Copyright © 2018 Pearson Education, Inc.

(b) $-4a - b$; a is 4, b is -2

$$-4a - b = -4(4) - (-2)$$
$$= -16 + 2$$
$$= -14$$

4. **(a)** $A = \frac{1}{2}bh$, when b is 6 yd and h is 12 yd

$$A = \frac{1}{2}(6 \text{ yd})(12 \text{ yd})$$
$$= \frac{1}{\underset{1}{2}}(\overset{3}{\cancel{6}} \text{ yd})(12 \text{ yd})$$
$$= 36 \text{ yd}^2$$

(b) $P = 2l + 2w$, when l is 10 cm, w is 8 cm

$$P = 2(10 \text{ cm}) + 2(8 \text{ cm})$$
$$= 20 \text{ cm} + 16 \text{ cm}$$
$$= 36 \text{ cm}$$

9.5 Section Exercises

1. **(a)** $6 \cdot y$ can be written as $6y$.

(b) $-2 \cdot x - 7 \cdot w$ can be written as $-2x - 7w$.

3. r is 2, s is 6
Replace r with 2. Replace s with 6.
$$2r + 4s = 2(2) + 4(6)$$
$$= 4 + 24$$
$$= 28$$

5. r is 1, s is -3
Replace r with 1. Replace s with -3.
$$2r + 4s = 2(1) + 4(-3)$$
$$= 2 + (-12)$$
$$= -10$$

7. r is -1, s is -7
Replace r with -1. Replace s with -7.
$$2r + 4s = 2(-1) + 4(-7)$$
$$= -2 + (-28)$$
$$= -30$$

9. r is 0, s is -2
Replace r with 0. Replace s with -2.
$$2r + 4s = 2(0) + 4(-2)$$
$$= 0 + (-8)$$
$$= -8$$

11. $8x - y$; x is 1, y is 8
Replace x with 1. Replace y with 8.
$$8x - y = 8(1) - 8$$
$$= 8 - 8$$
$$= 0$$

13. $6k + 2s$; k is 1, s is -2
Replace k with 1. Replace s with -2.
$$6k + 2s = 6(1) + 2(-2)$$
$$= 6 + (-4)$$
$$= 2$$

15. $\dfrac{-m + 5n}{2s + 2}$; m is 4, n is -8, s is 0
Replace m with 4, n with -8, and s with 0.
$$\frac{-m + 5n}{2s + 2} = \frac{-4 + 5(-8)}{2(0) + 2}$$
$$= \frac{-4 + (-40)}{0 + 2}$$
$$= \frac{-44}{2} = -22$$

17. $-m - 3n$; m is $\frac{1}{2}$, n is $\frac{3}{8}$
Replace m with $\frac{1}{2}$. Replace n with $\frac{3}{8}$.
$$-m - 3n = -\tfrac{1}{2} - 3\left(\tfrac{3}{8}\right) \quad \textit{Multiply } 3\left(\tfrac{3}{8}\right)$$
$$\textit{to get } \tfrac{9}{8}.$$
$$= -\tfrac{1}{2} - \tfrac{9}{8} \qquad \textit{Change subtr. to}$$
$$\textit{addition and}$$
$$\tfrac{9}{8} \textit{ to } -\tfrac{9}{8}.$$
$$= -\tfrac{4}{8} + \left(-\tfrac{9}{8}\right) \quad \textit{Rewrite } -\tfrac{1}{2} \textit{ as } -\tfrac{4}{8}.$$
$$= -\tfrac{13}{8} \qquad \textit{Add.}$$

19. Guntis made the error when replacing x with -3. The correct answer is

$$-x - 4y \qquad \begin{array}{l}\textit{Replace } x \textit{ with } (-3) \\ \textit{and } y \textit{ with } (-1).\end{array}$$
$$-(-3) - 4(-1) \quad \textit{Opposite of } (-3) \textit{ is } +3.$$
$$(+3) - (-4) \quad \textit{Change subtraction to addition.}$$
$$\underbrace{3 + (+4)}_{7} \qquad \textit{Add.}$$

21. $-c - 5b$; c is -8, b is -4
Replace c with -8. Replace b with -4.
$$-c - 5b = -(-8) - 5(-4)$$
$$\textit{-(-8) is the opposite of } -8 \textit{ or } 8.$$
$$\textit{Multiply } 5(-4) \textit{ to get } -20.$$
$$= 8 - (\underline{-20})$$

$$\textit{Change subtraction to addition}$$
$$\textit{and } -20 \textit{ to } 20.$$
$$= 8 + (+20) \quad \textit{Add.}$$
$$= \underline{28}$$

Copyright © 2018 Pearson Education, Inc.

23. $-4x - y$; x is 5, y is -15
Replace x with 5. Replace y with -15.

$$\begin{aligned}
-4x - y &= -4(5) - (-15) \\
&= -20 - (-15) \\
&= -20 + (+15) \\
&= -5
\end{aligned}$$

25. $-k - m - 8n$; k is 6, m is -9, n is 0
Replace k with 6, m with -9, and n with 0.

$$\begin{aligned}
-k - m - 8n &= -6 - (-9) - 8(0) \\
&= -6 + (+9) - 0 \\
&= 3
\end{aligned}$$

27. $\dfrac{-3s - t - 4}{-s + 6 + t}$; s is -1, t is -13
Replace s with -1. Replace t with -13.

$$\frac{-3s - t - 4}{-s + 6 + t} = \frac{-3(-1) - (-13) - 4}{-(-1) + 6 + (-13)}$$

Multiply $-3(-1)$ *in the numerator.*
$-(-1)$ *is 1 in the denominator.*

$$= \frac{3 + (+13) - 4}{1 + 6 + (-13)}$$

Add 3 and 13 in the numerator.
Add 1 and 6 in the denominator.

$$= \frac{16 - 4}{7 + (-13)}$$

Subtract in the numerator.
Add in the denominator.

$$= \frac{12}{-6} = -2 \quad \textit{Divide.}$$

29. $P = 4s$; s is 7.5 cm

$$\begin{aligned}
P &= 4(7.5 \text{ cm}) \\
&= 30 \text{ cm}
\end{aligned}$$

31. $P = 2l + 2w$; l is 9 in., w is 5 in.

$$\begin{aligned}
P &= 2(9 \text{ in.}) + 2(5 \text{ in.}) \\
&= 18 \text{ in.} + 10 \text{ in.} \\
&= 28 \text{ in.}
\end{aligned}$$

33. $A = \pi r^2$; π is approximately 3.14, r is 5 yd
Replace π with 3.14. Replace r with 5 yd.

$$\begin{aligned}
A &\approx 3.14(5 \text{ yd})^2 \\
&= 3.14 \cdot 5 \text{ yd} \cdot 5 \text{ yd} \\
&= 78.5 \text{ yd}^2
\end{aligned}$$

35. $A = \frac{1}{2}bh$; b is 15 ft, h is 3 ft

$$\begin{aligned}
A &= \tfrac{1}{2}(15 \text{ ft})(3 \text{ ft}) \\
&= \tfrac{15}{2}(3) \text{ ft}^2 \\
&= \tfrac{45}{2} \text{ or } 22\tfrac{1}{2} \text{ ft}^2
\end{aligned}$$

37. $V = \frac{1}{3}Bh$; B is 30 mm^2, h is 60 mm

$$\begin{aligned}
V &= \tfrac{1}{3} \cdot 30 \text{ mm}^2 \cdot 60 \text{ mm} \\
&= 10 \cdot 60 \text{ mm}^3 \\
&= 600 \text{ mm}^3
\end{aligned}$$

39. $V = lwh$; l is 12 in., w is 9 in., h is 10 in.

$$\begin{aligned}
V &= 12 \text{ in.} \cdot 9 \text{ in.} \cdot 10 \text{ in.} \\
&= 108 \cdot 10 \text{ in.}^3 \\
&= 1080 \text{ in.}^3
\end{aligned}$$

41. $C = 2\pi r$; π is approximately 3.14, r is 4 km

$$\begin{aligned}
C &\approx 2 \cdot (3.14) \cdot 4 \text{ km} \\
&= 6.28 \cdot 4 \text{ km} \\
&= 25.12 \text{ km}
\end{aligned}$$

43. **(a)** $P = 3s$; s is 11
$P = 3(11 \text{ in.}) = 33 \text{ in.}$
The perimeter is 33 in.

(b) $P = 3s$; s is 3
$P = 3(3 \text{ ft}) = 9 \text{ ft}$
The perimeter is 9 ft.

45. **(a)** $\dfrac{p}{t}$; p is 332, t is 4

$$\frac{p}{t} = \frac{332}{4} = 83$$

The average score is 83 points.

(b) $\dfrac{p}{t}$; p is 637, t is 7

$$\frac{p}{t} = \frac{637}{7} = 91$$

The average score is 91 points.

9.6 Solving Equations

9.6 Margin Exercises

1. **(a)** $p + 1 = 10$; 9
Replace p with 9.

$$\begin{aligned}
9 + 1 &= 10 \\
\underline{10} &= 10 \quad \textit{True}
\end{aligned}$$

Because the statement is *true*, 9 *is* a solution of the equation $p + 1 = 10$.

(b) $30 = 5r$; 6
Replace r with 6.

$$\begin{aligned}
30 &= 5(6) \\
30 &= 30 \quad \textit{True}
\end{aligned}$$

6 is a solution.

Copyright © 2018 Pearson Education, Inc.

(c) $3k - 2 = 4$; 3
Replace k with 3.

$$3(3) - 2 = 4$$
$$7 = 4 \quad \textit{False}$$

Because the statement is *false*, 3 *is not* a solution of the equation $3k - 2 = 4$.

(d) $23 = 4y + 3$; 5
Replace y with 5.

$$23 = 4(5) + 3$$
$$23 = 23 \qquad \textit{True}$$

23 is a solution.

2. **(a)** $\qquad n - 5 = 8$
$$n - \underbrace{5 + 5}_{} = 8 + \underline{5} \quad \text{Add 5 to both sides.}$$
$$n + \underline{0} = \underline{13}$$
$$n = \underline{13}$$

Check: $13 - 5 = 8$
$$8 = 8 \quad \textit{True}$$

The solution is 13.

(b) $\qquad 5 = r - 10$
$$5 + 10 = r - 10 + 10 \quad \begin{array}{l}\text{Add 10 to}\\ \text{both sides.}\end{array}$$
$$15 = r$$

Check: $5 = 15 - 10$
$$5 = 5 \qquad \textit{True}$$

The solution is 15.

(c) $\qquad 3 = z + 1$
$$3 - 1 = z + 1 - 1 \quad \begin{array}{l}\text{Subtract 1 from}\\ \text{both sides.}\end{array}$$
$$2 = z$$

Check: $3 = 2 + 1$
$$3 = 3 \qquad \textit{True}$$

The solution is 2.

(d) $\qquad k + 9 = 0$
$$k + 9 - 9 = 0 - 9 \quad \begin{array}{l}\text{Subtract 9 from}\\ \text{both sides.}\end{array}$$
$$k = -9$$

Check: $-9 + 9 = 0$
$$0 = 0 \quad \textit{True}$$

The solution is -9.

(e) $\qquad -2 = y + 9$
$$-2 - 9 = y + 9 - 9 \quad \begin{array}{l}\text{Subtract 9 from}\\ \text{both sides.}\end{array}$$
$$-11 = y$$

Check: $-2 = -11 + 9$
$$-2 = -2 \qquad \textit{True}$$

The solution is -11.

(f) $\qquad x - 2 = -6$
$$x - 2 + 2 = -6 + 2 \quad \text{Add 2 to both sides.}$$
$$x = -4$$

Check: $-4 - 2 = -6$
$$-6 = -6 \quad \textit{True}$$

The solution is -4.

3. **(a)** $\qquad 2y = 14$
$$\frac{\overset{1}{\cancel{2}} \cdot y}{\underset{1}{\cancel{2}}} = \frac{14}{2} \quad \text{Divide both sides by 2.}$$
$$y = \underline{7}$$

Check: $2 \cdot 7 = 14$
$$14 = 14 \quad \textit{True}$$

The solution is 7.

(b) $\qquad 42 = 7p$
$$\frac{\overset{6}{\cancel{42}}}{7} = \frac{7 \cdot p}{7} \quad \text{Divide both sides by 7.}$$
$$6 = p$$

Check: $42 = 7 \cdot 6$
$$42 = 42 \qquad \textit{True}$$

The solution is 6.

(c) $\qquad -8a = 32$
$$\frac{\overset{1}{\cancel{-8}}a}{\underset{1}{\cancel{-8}}} = \frac{32}{-8} \quad \text{Divide both sides by } -8.$$
$$a = -4$$

Check: $-8(-4) = 32$
$$32 = 32 \quad \textit{True}$$

The solution is -4.

(d) $\qquad -3r = -15$
$$\frac{\overset{1}{\cancel{-3}}r}{\underset{1}{\cancel{-3}}} = \frac{-15}{-3} \quad \text{Divide both sides by } -3.$$
$$r = 5$$

Check: $-3(5) = -15$
$$-15 = -15 \quad \textit{True}$$

The solution is 5.

Copyright © 2018 Pearson Education, Inc.

(e) $-60 = -6k$

$$\frac{-60}{-6} = \frac{\overset{1}{\cancel{-6}}k}{\underset{1}{\cancel{-6}}} \quad \text{Divide both sides by } -6.$$

$$10 = k$$

Check: $-60 = -6(10)$

$\qquad\quad -60 = -60 \qquad$ *True*

The solution is 10.

(f) $10x = 0$

$$\frac{10x}{10} = \frac{0}{10} \quad \text{Divide both sides by } 10.$$

$$x = 0$$

Check: $10 \cdot 0 = 0$

$\qquad\qquad 0 = 0 \quad$ *True*

The solution is 0.

4.   **(a)** $\qquad\qquad \dfrac{x}{4} = 2$

$\qquad\qquad \dfrac{1}{4}x = 2 \qquad$ Rewrite $\dfrac{x}{4}$ as $\dfrac{1}{4}x$.

$$\left(\frac{4}{1}\right) \cdot \left(\frac{1}{4}\right)x = 4 \cdot 2 \quad \begin{array}{l}\text{Multiply both}\\ \text{sides by 4/1}\\ \text{(4/1 equals 4).}\end{array}$$

$$x = \underline{8}$$

Notes: When multiplying by the reciprocal, the product is always 1, so we omit the cancellation of all numbers. In the second to last step on the right side, you may write $4 \cdot 2$ or $2 \cdot 4$.

Check: $\dfrac{8}{4} = 2$

$\qquad\qquad 2 = 2 \quad$ *True*

The solution is 8.

(b) $\qquad\qquad \dfrac{y}{7} = -3$

$\qquad\qquad \dfrac{1}{7}y = -3$

$$\left(\frac{7}{1}\right) \cdot \left(\frac{1}{7}\right)y = -3 \cdot 7$$

$$y = -21$$

Check: $\dfrac{-21}{7} = -3$

$\qquad\qquad -3 = -3 \quad$ *True*

The solution is -21.

(c) $\qquad -8 = \dfrac{k}{6}$

$$-8 = \frac{1}{6}k$$

$$6 \cdot (-8) = \left(\frac{6}{1}\right) \cdot \left(\frac{1}{6}\right)k$$

$$-48 = k$$

Check: $-8 = \dfrac{-48}{6}$

$\qquad\qquad -8 = -8 \quad$ *True*

The solution is -48.

(d) $\qquad\qquad\qquad 8 = -\dfrac{4}{5}z$

$$\left(-\frac{5}{4}\right) \cdot \left(\frac{8}{1}\right) = \left(-\frac{5}{4}\right) \cdot \left(-\frac{4}{5}\right)z$$

$$-10 = z$$

Check: $8 = \left(-\dfrac{4}{5}\right) \cdot (-10)$

$\qquad\qquad 8 = 8 \qquad\qquad$ *True*

The solution is -10.

(e) $\qquad\qquad -\dfrac{5}{8}p = -10$

$$\left(-\frac{8}{5}\right) \cdot \left(-\frac{5}{8}\right)p = (-10) \cdot \left(-\frac{8}{5}\right)$$

$$p = 16$$

Check: $\left(-\dfrac{5}{8}\right) \cdot (16) = -10$

$\qquad\qquad\qquad -10 = -10 \quad$ *True*

The solution is 16.

9.6   Section Exercises

1. $x + 7 = 11;\ 4$
Replace x with 4.

$$4 + 7 = 11$$
$$11 = 11 \quad \text{*True*}$$

Because the statement is *true*, 4 *is* a solution of the equation $x + 7 = 11$.

3. $4y = 28;\ 7$
Replace y with 7.

$$4(7) = 28$$
$$28 = 28 \quad \text{*True*}$$

7 is a solution.

Copyright © 2018 Pearson Education, Inc.

5. $2z - 1 = -15;\ -8$
Replace z with -8.

$$2(-8) - 1 = -15$$
$$-16 - 1 = -15$$
$$-17 = -15 \quad False$$

Because the statement is *false*, -8 *is not* a solution of the equation $2z - 1 = -15$.

7. **(a)** $m - 8 = 1$
Add 8 to both sides because $-8 + 8$ gives 0 on the left side. Then m will be by itself on the left side.

(b) $-7 = w + 5$
Add -5 to both sides because $5 + (-5)$ gives 0 on the right side. Then w will be by itself on the right side.

9. $p + 5 = 9$
Subtract 5 from both sides.

$$p + 5 - \underline{5} = 9 - \underline{5}$$
$$p + \underline{0} = \underline{4}$$
$$p = \underline{4}$$

Check: $4 + 5 = 9$
$\qquad\quad\ \ 9 = 9 \quad True$

The solution is 4.

11. $k + 15 = 0$
Add -15 to both sides.

$$k + 15 - 15 = 0 - 15$$
$$k + 0 = 0 - 15$$
$$k = -15$$

Check the solution by replacing k with -15 in the original equation.
$$k + 15 = 0$$
$$-15 + 15 = 0$$
$$0 = 0 \quad True$$

The solution is -15.

13. $z - 5 = 3$
Add 5 to both sides.

$$z - 5 + 5 = 3 + 5$$
$$z = 8$$

Check: $8 - 5 = 3$
$\qquad\qquad 3 = 3 \quad True$

The solution is 8.

15. $8 = r - 2$
Add 2 to both sides.

$$8 + 2 = r - 2 + 2$$
$$10 = r$$

Check: $8 = 10 - 2$
$\qquad\qquad 8 = 8 \quad True$

The solution is 10.

17. $-5 = n + 3$
Subtract 3 from both sides.

$$-5 - 3 = n + 3 - 3$$
$$-8 = n$$

Check: $-5 = -8 + 3$
$\qquad\qquad -5 = -5 \quad True$

The solution is -8.

19. $7 = r + 13$
Subtract 13 from both sides.

$$7 - 13 = r + 13 - 13$$
$$-6 = r$$

Check: $7 = -6 + 13$
$\qquad\qquad 7 = 7 \quad True$

The solution is -6.

21. $-4 + k = 14$
Add 4 to both sides.

$$-4 + 4 + k = 14 + 4$$
$$k = 18$$

Check: $-4 + 18 = 14$
$\qquad\qquad\quad 14 = 14 \quad True$

The solution is 18.

23. $-12 + x = -1$
Add 12 to both sides.

$$-12 + 12 + x = -1 + 12$$
$$x = 11$$

Check: $-12 + 11 = -1$
$\qquad\qquad\quad -1 = -1 \quad True$

The solution is 11.

25. $-5 = -2 + r$
Add 2 to both sides.

$$-5 + 2 = -2 + 2 + r$$
$$-3 = 0 + r$$
$$-3 = r$$

Check: $-5 = -2 + (-3)$
$\qquad\qquad -5 = -5 \quad True$

The solution is -3.

Copyright © 2018 Pearson Education, Inc.

27. $d + \frac{2}{3} = 3$

Subtract $\frac{2}{3}$ from both sides.

$$d + \frac{2}{3} - \frac{2}{3} = 3 - \frac{2}{3}$$
$$d = \frac{7}{3} \text{ or } 2\frac{1}{3}$$

Check: $\frac{7}{3} + \frac{2}{3} = 3$
$$\frac{9}{3} = 3$$
$$3 = 3 \quad True$$

The solution is $\frac{7}{3}$ or $2\frac{1}{3}$.

29. $z - \frac{7}{8} = 10$

Add $\frac{7}{8}$ to both sides.

$$z - \frac{7}{8} + \frac{7}{8} = 10 + \frac{7}{8}$$
$$z = \frac{80}{8} + \frac{7}{8}$$
$$= \frac{87}{8} \text{ or } 10\frac{7}{8}$$

Check: $\frac{87}{8} - \frac{7}{8} = 10$
$$\frac{80}{8} = 10$$
$$10 = 10 \quad True$$

The solution is $\frac{87}{8}$ or $10\frac{7}{8}$.

31. $\frac{1}{2} = k - 2$

Add 2 to both sides.

$$\frac{1}{2} + 2 = k - 2 + 2$$
$$\frac{1}{2} + \frac{4}{2} = k + 0$$
$$\frac{5}{2} = k$$

Check the solution by replacing k with $\frac{5}{2}$ in the original equation.

$$\frac{1}{2} = k - 2$$
$$\frac{1}{2} = \frac{5}{2} - 2$$
$$\frac{1}{2} = \frac{1}{2} \quad True$$

The solution is $\frac{5}{2}$ or $2\frac{1}{2}$.

33. $m - \frac{7}{5} = \frac{11}{4}$

Add $\frac{7}{5}$ to both sides.

$$m - \frac{7}{5} + \frac{7}{5} = \frac{11}{4} + \frac{7}{5}$$
$$m = \frac{55}{20} + \frac{28}{20}$$
$$= \frac{83}{20} \text{ or } 4\frac{3}{20}$$

Check: $\frac{83}{20} - \frac{7}{5} = \frac{11}{4}$
$$\frac{83}{20} - \frac{28}{20} = \frac{55}{20}$$
$$\frac{55}{20} = \frac{55}{20} \quad True$$

The solution is $\frac{83}{20}$ or $4\frac{3}{20}$.

35. $x - 0.8 = 5.07$

Add 0.8 to both sides.

$$x - 0.8 + 0.8 = 5.07 + 0.8$$
$$x = 5.87$$

Check: $5.87 - 0.8 = 5.07$
$$5.07 = 5.07 \quad True$$

The solution is 5.87.

37. $3.25 = 4.76 + r$

Subtract 4.76 from both sides.

$$3.25 - 4.76 = 4.76 - 4.76 + r$$
$$-1.51 = r$$

Check: $3.25 = 4.76 + (-1.51)$
$$3.25 = 3.25 \quad True$$

The solution is -1.51.

39. (a) $-10x = 40$

Divide both sides by -10. Then, on the left side, divide out -10 to get x by itself.

(b) $-25 = 5z$

Divide both sides by 5. Then, on the right side, divide out 5 to get z by itself.

41. $6z = 12$

Divide both sides by 6.

$$\frac{\overset{1}{\cancel{6}} \cdot z}{\underset{1}{\cancel{6}}} = \frac{12}{6}$$
$$z = \underline{2}$$

Check: $6(2) = 12$
$$12 = 12 \quad True$$

The solution is 2.

43. $48 = 12r$

Divide both sides by 12.

$$\frac{48}{12} = \frac{\overset{1}{\cancel{12}} \cdot r}{\underset{1}{\cancel{12}}}$$
$$4 = r$$

Check: $48 = 12(4)$
$$48 = 48 \quad True$$

The solution is 4.

Copyright © 2018 Pearson Education, Inc.

45. $3y = 0$

Divide both sides by 3.

$$\frac{\overset{1}{\cancel{3}} \cdot y}{\underset{1}{\cancel{3}}} = \frac{0}{3}$$

$$y = 0$$

Check the solution by replacing y with 0 in the original equation.

$$3y = 0$$
$$3(0) = 0$$
$$0 = 0 \quad \textit{True}$$

The solution is 0.

47. $-6k = 36$

Divide both sides by -6.

$$\frac{\overset{1}{\cancel{-6}} \cdot k}{\underset{1}{\cancel{-6}}} = \frac{36}{-6}$$

$$k = -6$$

Check: $-6(-6) = 36$
$$36 = 36 \quad \textit{True}$$

The solution is -6.

49. $-36 = -4p$

Divide both sides by -4.

$$\frac{-36}{-4} = \frac{\overset{1}{\cancel{-4}} \cdot p}{\underset{1}{\cancel{-4}}}$$

$$9 = p$$

Check: $-36 = -4(9)$
$$-36 = -36 \quad \textit{True}$$

The solution is 9.

51. $-1.2m = 8.4$

Divide both sides by -1.2.

$$\frac{\overset{1}{\cancel{-1.2}}m}{\underset{1}{\cancel{-1.2}}} = \frac{8.4}{-1.2}$$

$$m = -7$$

Check: $-1.2(-7) = 8.4$
$$8.4 = 8.4 \quad \textit{True}$$

The solution is -7.

53. $-8.4p = -9.24$

Divide both sides by -8.4.

$$\frac{\overset{1}{\cancel{-8.4}}p}{\underset{1}{\cancel{-8.4}}} = \frac{-9.24}{-8.4}$$

$$p = 1.1$$

Check: $-8.4(1.1) = -9.24$
$$-9.24 = -9.24 \quad \textit{True}$$

The solution is 1.1.

55. (a) $\frac{x}{8} = -4$

Rewrite $\frac{x}{8}$ as $\frac{1}{8}x$. Multiply both sides by the reciprocal of $\frac{1}{8}$, which is $\frac{8}{1}$, in order to get x by itself on the left side.

(b) $-6 = -\frac{2}{3}k$

Multiply both sides by the reciprocal of $-\frac{2}{3}$, which is $-\frac{3}{2}$, in order to get k by itself on the right side.

57. $\dfrac{k}{2} = 17$ or $\dfrac{1}{2}k = 17$

Multiply both sides by 2.

$$\frac{\overset{1}{\cancel{2}}}{1} \cdot \frac{1}{\underset{1}{\cancel{2}}}k = 17 \cdot 2$$

$$k = 34$$

Check: $\dfrac{34}{2} = 17$
$$17 = 17 \quad \textit{True}$$

The solution is 34.

59. $11 = \dfrac{a}{6}$ or $11 = \dfrac{1}{6}a$

Multiply both sides by 6.

$$6 \cdot 11 = \frac{\overset{1}{\cancel{6}}}{1} \cdot \frac{1}{\underset{1}{\cancel{6}}}a$$

$$66 = a$$

Check: $11 = \dfrac{66}{6}$
$$11 = 11 \quad \textit{True}$$

The solution is 66.

61. $\frac{r}{3} = -12$ can be written as $\frac{1}{3}r = -12$.

Multiply both sides by $\frac{3}{1}$, the reciprocal of $\frac{1}{3}$.

$$\frac{\overset{1}{\cancel{3}}}{1} \cdot \frac{1}{\underset{1}{\cancel{3}}}r = -12 \cdot 3$$

$$r = -36$$

Check the solution by replacing r with -36 in the original equation.

$$\frac{r}{3} = -12$$

$$\frac{-36}{3} = -12$$

$$-12 = -12 \quad \textit{True}$$

The solution is -36.

Copyright © 2018 Pearson Education, Inc.

63. $-\dfrac{2}{5}p = 8$

Multiply both sides by $-\dfrac{5}{2}$.

$$-\dfrac{\overset{1}{\cancel{5}}}{\underset{1}{\cancel{2}}} \cdot \left(-\dfrac{\overset{1}{\cancel{2}}}{\underset{1}{\cancel{5}}}p\right) = -\dfrac{5}{\underset{1}{\cancel{2}}} \cdot \overset{4}{\cancel{8}}$$

$$p = -20$$

Check: $-\dfrac{2}{5}(-20) = 8$

$$8 = 8 \quad \textit{True}$$

The solution is -20.

65. $-\dfrac{3}{4}m = -3$

Multiply both sides by $-\dfrac{4}{3}$, the reciprocal of $-\dfrac{3}{4}$.

$$-\dfrac{\overset{1}{\cancel{4}}}{\underset{1}{\cancel{3}}} \cdot \left(-\dfrac{\overset{1}{\cancel{3}}}{\underset{1}{\cancel{4}}}m\right) = -\cancel{3} \cdot \left(-\dfrac{4}{\cancel{3}}\right)$$

$$m = 4$$

Check: $-\dfrac{3}{4}(4) = -3$

$$-3 = -3 \quad \textit{True}$$

The solution is 4.

67. $6 = \dfrac{3}{8}x$

Multiply both sides by $\dfrac{8}{3}$.

$$\dfrac{8}{\underset{1}{\cancel{3}}} \cdot \overset{2}{\cancel{6}} = \dfrac{\overset{1}{\cancel{8}}}{\underset{1}{\cancel{3}}} \cdot \dfrac{\overset{1}{\cancel{3}}}{\underset{1}{\cancel{8}}}x$$

$$16 = x$$

Check: $6 = \dfrac{3}{8}(16)$

$$6 = 6 \qquad \textit{True}$$

The solution is 16.

69. $\dfrac{y}{2.6} = 0.5$ or $\dfrac{1}{2.6}y = 0.5$

Multiply both sides by 2.6.

$$2.6 \cdot \dfrac{y}{2.6} = 0.5(2.6)$$

$$y = 1.3$$

Check: $\dfrac{1.3}{2.6} = 0.5$

$$0.5 = 0.5 \quad \textit{True}$$

The solution is 1.3.

71. $\dfrac{z}{-3.8} = 1.3$ or $\dfrac{1}{-3.8}z = 1.3$

Multiply both sides by -3.8.

$$\dfrac{\cancel{-3.8}}{1} \cdot \dfrac{1}{\cancel{-3.8}}z = (-3.8)(1.3)$$

$$z = -4.94$$

Check: $\dfrac{-4.94}{-3.8} = 1.3$

$$1.3 = 1.3 \quad \textit{True}$$

The solution is -4.94.

73. $x - 17 = 5 - 3$

$$x - 17 = 2$$

$$x - 17 + 17 = 2 + 17$$

$$x = 19$$

The solution is 19.

75. $3 = x + 9 - 15$

$$3 = x - 6$$

$$3 + 6 = x - 6 + 6$$

$$9 = x$$

The solution is 9.

77. $\dfrac{7}{2}x = \dfrac{4}{3}$

$$\dfrac{\overset{1}{\cancel{2}}}{\underset{1}{\cancel{7}}} \cdot \dfrac{\overset{1}{\cancel{7}}}{\underset{1}{\cancel{2}}}x = \dfrac{2}{7} \cdot \dfrac{4}{3}$$

$$x = \dfrac{8}{21}$$

The solution is $\dfrac{8}{21}$.

79. $\dfrac{1}{2} - \dfrac{3}{4} = \dfrac{a}{5}$ *Rewrite $\frac{1}{2}$ as $\frac{3}{4}$.*

$$\dfrac{2}{4} - \dfrac{3}{4} = \dfrac{a}{5} \qquad \textit{Subtract } \tfrac{2}{4} - \tfrac{3}{4}.$$

$$-\dfrac{1}{4} = \dfrac{a}{5} \qquad \textit{Multiply both sides by } \tfrac{5}{1},$$

the reciprocal of $\frac{1}{5}$.

$$\dfrac{5}{1} \cdot \left(-\dfrac{1}{4}\right) = \dfrac{a}{5} \cdot \dfrac{5}{1}$$

$$-\dfrac{5}{4} = a$$

The solution is $-\dfrac{5}{4}$.

Copyright © 2018 Pearson Education, Inc.

81.
$$m - 2 + 18 = |-3 - 4| + 5$$
$$m + 16 = |-7| + 5$$
$$m + 16 = 7 + 5$$
$$m + 16 = 12$$
$$m + 16 - 16 = 12 - 16$$
$$m = -4$$

The solution is -4.

9.7 Solving Equations with Several Steps

9.7 Margin Exercises

1. **(a)**
$$2r + 7 = 13$$
$$2r + 7 - 7 = 13 - 7$$

$$2r = 6 \qquad \text{Divide both sides by } \underline{2}.$$

$$\frac{\overset{1}{\cancel{2}}r}{\underset{1}{\cancel{2}}} = \frac{6}{2}$$

$$r = 3$$

Check: Replace r with 3.
$$2(3) + 7 = 13$$
$$13 = 13 \qquad \textit{True}$$

The solution is 3.

(b)
$$20 = 6y - 4$$
$$20 + 4 = 6y - 4 + 4$$

$$24 = 6y \qquad \text{Divide both sides by } 6.$$

$$\frac{24}{6} = \frac{\overset{1}{\cancel{6}}y}{\underset{1}{\cancel{6}}}$$

$$4 = y$$

Check: Replace y with 4.
$$20 = 6(4) - 4$$
$$20 = 20 \qquad \textit{True}$$

The solution is 4.

(c)
$$7m + 9 = 9$$
$$7m + 9 - 9 = 9 - 9$$
$$7m = 0$$

$$\frac{\overset{1}{\cancel{7}}m}{\underset{1}{\cancel{7}}} = \frac{0}{7}$$

$$m = 0$$

Check: Replace m with 0.
$$7(0) + 9 = 9$$
$$9 = 9 \qquad \textit{True}$$

The solution is 0.

(d)
$$-2 = 4p + 10$$
$$-2 - 10 = 4p + 10 - 10$$
$$-12 = 4p$$

$$\frac{-12}{4} = \frac{\overset{1}{\cancel{4}}p}{\underset{1}{\cancel{4}}}$$

$$-3 = p$$

Check: Replace p with -3.
$$-2 = 4(-3) + 10$$
$$-2 = -12 + 10$$
$$-2 = -2 \qquad \textit{True}$$

The solution is -3.

2. **(a)** $3(2 + 6) = 3 \cdot 2 + 3 \cdot 6$ Do multiplications.
$$= 6 + 18 \qquad \text{Now add.}$$
$$= 24$$

(b) $8(k - 3) = 8 \cdot k - 8 \cdot 3$
$$= 8k - 24$$

(c) $-6(r + 5) = -6 \cdot r + (-6) \cdot 5$
$$= -6r + (-30)$$
$$= -6r - 30$$

(d) $-9(s - 8) = -9 \cdot s - (-9) \cdot 8$
$$= -9s - (-72)$$
$$= -9s + 72$$

3. **(a)** $5y + 3y = (5 + 3)y = \underline{8}y$

(b) $10a - 28a = (10 - 28)a = -18a$

(c) $3x + 3x - 9x = (3 + 3 - 9)x = -3x$

(d) $k + k = 1k + 1k = (1 + 1)k = 2k$

(e) $6b - b - 7b = (6 - 1 - 7)b = -2b$

4. **(a)**
$$3y - 1 = 2y + 7$$
$$3y - 1 - 2y = 2y + 7 - 2y$$
$$3y - 2y - 1 = 2y - 2y + 7$$
$$1y - 1 = 7$$
$$y - 1 + 1 = 7 + \underline{1}$$
$$y = \underline{8}$$

Check: Replace y with 8.
$$3(8) - 1 = 2(8) + 7$$
$$24 - 1 = 16 + 7$$
$$23 = 23 \qquad \textit{True}$$

The solution is 8.

Copyright © 2018 Pearson Education, Inc.

(b)
$$5a + 7 = 3a - 9$$
$$5a - 3a + 7 = 3a - 3a - 9$$
$$2a + 7 = -9$$
$$2a + 7 - 7 = -9 - 7$$
$$2a = -16$$
$$\frac{2a}{2} = \frac{-16}{2}$$
$$a = -8$$

Check: Replace a with -8.
$$5(-8) + 7 = 3(-8) - 9$$
$$-40 + 7 = -24 + (-9)$$
$$-33 = -33 \qquad True$$

The solution is -8.

(c)
$$3p - 2 = p - 6$$
$$3p - p - 2 = p - p - 6$$
$$2p - 2 = -6$$
$$2p - 2 + 2 = -6 + 2$$
$$2p = -4$$
$$\frac{2p}{2} = \frac{-4}{2}$$
$$p = -2$$

Check: Replace p with -2.
$$3(-2) - 2 = -2 - 6$$
$$-6 - 2 = -8$$
$$-8 = -8 \qquad True$$

The solution is -2.

5.   (a)
$$-12 = 4(y - 1)$$
$$-12 = 4 \cdot y - 4 \cdot 1$$
$$-12 = \underline{4y} - \underline{4}$$
$$-12 + 4 = 4y - 4 + 4$$
$$-8 = 4y$$
$$\frac{-8}{4} = \frac{4y}{4}$$
$$-2 = y \quad \text{or} \quad y = -2$$

Check: Replace y with -2.
$$-12 = 4(-2 - 1)$$
$$-12 = 4(-3)$$
$$-12 = -12 \qquad True$$

The solution is -2.

(b)
$$5(m + 4) = 20$$
$$5m + 20 = 20$$
$$5m + 20 - 20 = 20 - 20$$
$$5m = 0$$
$$\frac{5m}{5} = \frac{0}{5}$$
$$m = 0$$

Check: Replace m with 0.
$$5(0 + 4) = 20$$
$$20 = 20 \qquad True$$

The solution is 0.

(c)
$$6(t - 2) = 18$$
$$6t - 12 = 18$$
$$6t - 12 + 12 = 18 + 12$$
$$6t = 30$$
$$\frac{6t}{6} = \frac{30}{6}$$
$$t = 5$$

Check: Replace t with 5.
$$6(5 - 2) = 18$$
$$6(3) = 18$$
$$18 = 18 \qquad True$$

The solution is 5.

9.7   Section Exercises

1. $-4y + 10 = 2$
Subtract 10 from both sides of the equation.
$$-4y + 10 - 10 = 2 - 10$$
$$-4y = -8$$

Note: You could also add (-10) to both sides as the first step. The result is the same.

3.
$$7p + 5 = 12$$
$$7p + 5 - 5 = 12 - 5$$
$$7p = \underline{7}$$
$$\frac{7p}{7} = \frac{7}{7}$$
$$p = 1$$

Check: Replace p with 1.
$$7(1) + 5 = 12$$
$$7 + 5 = 12$$
$$12 = 12 \qquad True$$

The solution is 1.

5. $2 = 8y - 6$
Add 6 to both sides.
$$2 + 6 = 8y - 6 + 6$$
$$8 = 8y$$
Divide both sides by 8.
$$\frac{8}{8} = \frac{8y}{8}$$
$$1 = y$$

(continued)

Copyright © 2018 Pearson Education, Inc.

Check: Replace y with 1 in the original equation.

$$2 = 8y - 6$$
$$2 = 8(1) - 6$$
$$2 = 2 \qquad \textit{True}$$

The solution is 1.

7.
$$-3m + 1 = 1$$
$$-3m + 1 - 1 = 1 - 1$$
$$-3m = 0$$
$$\frac{-3m}{-3} = \frac{0}{-3}$$
$$m = 0$$

Check: Replace m with 0.

$$-3(0) + 1 = 1$$
$$1 = 1 \qquad \textit{True}$$

The solution is 0.

9.
$$28 = -9a + 10$$
$$28 - 10 = -9a + 10 - 10$$
$$18 = -9a$$
$$\frac{18}{-9} = \frac{-9a}{-9}$$
$$-2 = a$$

Check: Replace a with -2.

$$28 = -9(-2) + 10$$
$$28 = 18 + 10$$
$$28 = 28 \qquad \textit{True}$$

The solution is -2.

11.
$$-5x - 4 = 16$$
$$-5x - 4 + 4 = 16 + 4$$
$$-5x = 20$$
$$\frac{-5x}{-5} = \frac{20}{-5}$$
$$x = -4$$

Check: Replace x with -4.

$$-5(-4) - 4 = 16$$
$$20 - 4 = 16$$
$$16 = 16 \qquad \textit{True}$$

The solution is -4.

13.
$$-\frac{1}{2}z + 2 = -1$$

Subtract 2 from both sides.

$$-\frac{1}{2}z + 2 - 2 = -1 - 2$$
$$-\frac{1}{2}z = -3$$

Multiply both sides by $-\frac{2}{1}$*, the reciprocal of* $-\frac{1}{2}$*.*

$$-\frac{2}{1} \cdot \left(-\frac{1}{2}z\right) = -3 \cdot \left(-\frac{2}{1}\right)$$
$$z = 6$$

Check: Replace z with 6 in the original equation.

$$-\frac{1}{2}z + 2 = -1$$
$$-\frac{1}{2}(6) + 2 = -1$$
$$-3 + 2 = -1$$
$$-1 = -1 \quad \textit{True}$$

The solution is 6.

15. $6(x + 4) = 6 \cdot x + 6 \cdot 4$
$$= \underline{6x} + \underline{24}$$

17. $7(p - 8) = 7 \cdot p - 7 \cdot 8$
$$= 7p - 56$$

19. $-3(m + 6) = -3 \cdot m + (-3) \cdot 6$
$$= -3m - 18$$

21. $-2(y - 3) = -2 \cdot y - (-2) \cdot 3$
$$= -2y - (-6)$$
$$= -2y + 6$$

23. $-8(c + 8) = -8 \cdot c + (-8) \cdot 8$
$$= -8c - 64$$

25. $-10(w - 9) = -10 \cdot w - (-10) \cdot 9$
$$= -10w - (-90)$$
$$= -10w + 90$$

27. In the first expression, Jamaica did not multiply $5 \cdot 4$. In the second expression, Jamaica has the wrong sign for 16. The correct answers are $5(x - 4) = 5x - 20$ and $-8(y - 2) = -8y + 16$.

29. $11r + 6r = (11 + 6)r = \underline{17r}$

31. $8z - 7z = (8 - 7)z = 1z$ or z

33. $y - 3y = (1 - 3)y = -2y$

35. $-4t + t - 4t = (-4 + 1 - 4)t = -7t$

37. $7p - 9p + 2p = (7 - 9 + 2)p = 0p = 0$

39. $\dfrac{5}{2}b - \dfrac{11}{2}b = \left(\dfrac{5}{2} - \dfrac{11}{2}\right)b = -\dfrac{6}{2}b = -3b$

Copyright © 2018 Pearson Education, Inc.

41. $3p - 5p = 32$

$(3 - 5)p = 32$ Combine like terms
            on the left side.

$-2p = 32$ $3 - 5$ is $3 + (-5)$, or -2.

$\dfrac{\overset{1}{-\cancel{2}} \cdot p}{\underset{1}{-\cancel{2}}} = \dfrac{32}{-2}$ Divide both sides by -2 to
             get p by itself on the left side.

$p = -16$ 32 divided by -2 is -16.

43. $4k + 6k = 50$

$(4 + 6)k = 50$

$10k = 50$

$\dfrac{10k}{10} = \dfrac{50}{10}$

$k = \underline{5}$

Check: Replace k with 5.

$$4(5) + 6(5) = 50$$
$$20 + 30 = 50$$
$$50 = 50 \quad True$$

The solution is 5.

45. $54 = 10m - m$

$54 = (10 - 1)m$

$54 = 9m$

$\dfrac{54}{9} = \dfrac{9m}{9}$

$6 = m$

Check: Replace m with 6.

$$54 = 10(6) - 6$$
$$54 = 60 - 6$$
$$54 = 54 \quad True$$

The solution is 6.

47. $2b - 6b = 24$

$(2 - 6)b = 24$

$-4b = 24$

$\dfrac{-4b}{-4} = \dfrac{24}{-4}$

$b = -6$

Check: Replace b with -6.

$$2(-6) - 6(-6) = 24$$
$$-12 - (-36) = 24$$
$$-12 + 36 = 24$$
$$24 = 24 \quad True$$

The solution is -6.

49. $-12 = 6y - 18y$

Use the distributive property to combine
$6y$ and $-18y$.

$-12 = (6 - 18)y$

$-12 = -12y$

Divide both sides by -12.

$\dfrac{-12}{-12} = \dfrac{-12y}{-12}$

$1 = y$

Check: Replace y with 1 in the original
equation.

$$-12 = 6y - 18y$$
$$-12 = 6(1) - 18(1)$$
$$-12 = 6 - 18$$
$$-12 = -12 \quad True$$

The solution is 1.

51. $6p - 2 = 4p + 6$

$6p - 2 + 2 = 4p + 6 + 2$

$6p = 4p + 8$

$6p - 4p = 4p - 4p + 8$

$2p = 8$

$\dfrac{2p}{2} = \dfrac{8}{2}$

$p = 4$

Check: Replace p with 4.

$$6(4) - 2 = 4(4) + 6$$
$$24 - 2 = 16 + 6$$
$$22 = 22 \quad True$$

The solution is 4.

53. $9 + 7z = 9z + 13$

$9 + 7z - 9z = 9z + 13 - 9z$

$9 - 2z = 13$

$9 - 2z - 9 = 13 - 9$

$-2z = 4$

$\dfrac{-2z}{-2} = \dfrac{4}{-2}$

$z = -2$

Check: Replace z with -2.

$$9 + 7(-2) = 9(-2) + 13$$
$$9 + (-14) = -18 + 13$$
$$-5 = -5 \quad True$$

The solution is -2.

Copyright © 2018 Pearson Education, Inc.

55.
$$-2y + 6 = 6y - 10$$
Add 10 to both sides.
$$-2y + 6 + 10 = 6y - 10 + 10$$
$$-2y + 16 = 6y$$
Add 2y to both sides.
$$-2y + 2y + 16 = 6y + 2y$$
Combine 6y and 2y because they are like terms.
$$16 = 8y$$
Divide both sides by 8.
$$\frac{16}{8} = \frac{8y}{8}$$
$$2 = y$$

Check: Replace y with 2.
$$-2(2) + 6 = 6(2) - 10$$
$$(-4) + 6 = 12 - 10$$
$$2 = 2 \qquad \textit{True}$$

The solution is 2.

57.
$$b + 3.05 = 2$$
$$b + 3.05 - 3.05 = 2 - 3.05$$
$$b = -1.05$$

Check: Replace b with -1.05.
$$-1.05 + 3.05 = 2$$
$$2 = 2 \quad \textit{True}$$

The solution is -1.05.

59.
$$2.5r + 9 = -1$$
$$2.5r + 9 - 9 = -1 - 9$$
$$2.5r = -10$$
$$\frac{2.5r}{2.5} = \frac{-10}{2.5}$$
$$r = -4$$

Check: Replace r with -4.
$$2.5(-4) + 9 = -1$$
$$-10 + 9 = -1$$
$$-1 = -1 \quad \textit{True}$$

The solution is -4.

61.
$$-10 = 2(y + 4)$$
$$-10 = 2y + 8$$
$$-10 - 8 = 2y + 8 - 8$$
$$-18 = 2y$$
$$\frac{-18}{2} = \frac{2y}{2}$$
$$-9 = y$$

Check: Replace y with -9.
$$-10 = 2(-9 + 4)$$
$$-10 = 2(-5)$$
$$-10 = -10 \qquad \textit{True}$$

The solution is -9.

63.
$$-4(t + 2) = 12$$
Use the distributive property on the left side of the equation.
$$-4t + (-4) \cdot 2 = 12$$
$$-4t + (-8) = 12$$
Add 8 to both sides.
$$-4t - 8 + 8 = 12 + 8$$
$$-4t = 20$$
Divide both sides by -4.
$$\frac{-4t}{-4} = \frac{20}{-4}$$
$$t = -5$$

Check: Replace t with -5 in the original equation.
$$-4(t + 2) = 12$$
$$-4(-5 + 2) = 12$$
$$-4(-3) = 12$$
$$12 = 12 \quad \textit{True}$$

The solution is -5.

65.
$$6(x - 5) = -30$$
$$6x - 30 = -30$$
$$6x - 30 + 30 = -30 + 30$$
$$6x = 0$$
$$\frac{6x}{6} = \frac{0}{6}$$
$$x = 0$$

Check: Replace x with 0.
$$6(0 - 5) = -30$$
$$6(-5) = -30$$
$$-30 = -30 \quad \textit{True}$$

The solution is 0.

67.
$$-2t - 10 = 3t + 5$$

$$-2t - 3t - 10 = 3t - 3t + 5 \qquad \text{Subtract } 3t \text{ from both sides (addition property).}$$

$$-5t - 10 = 5$$

$$-5t - 10 + 10 = 5 + 10 \qquad \text{Add 10 to both sides (addition property).}$$

$$-5t = 15$$

$$\frac{-\cancel{5} \cdot t}{-\cancel{5}} = \frac{15}{-5} \qquad \text{Divide both sides by } -5 \text{ (multiplication property).}$$

$$t = -3$$

Copyright © 2018 Pearson Education, Inc.

69. $30 - 40 = -2x + 7x - 4x$
$$-10 = (-2 + 7 - 4)x$$
$$-10 = 1x$$
$$-10 = x \quad \text{or} \quad x = \underline{-10}$$

The solution is -10.

71. $0 = -2(y - 2)$

Use the distributive property on the right side of the equation.
$$0 = -2y - (-2) \cdot 2$$
$$0 = -2y + 4$$

Add $2y$ to both sides.
$$0 + 2y = -2y + 2y + 4$$
$$2y = 4$$

Divide both sides by 2.
$$\frac{2y}{2} = \frac{4}{2}$$
$$y = 2$$

The solution is 2.

73. $\dfrac{y}{2} - 2 = \dfrac{y}{4} + 3$

Add 2 to both sides.
$$\frac{y}{2} - 2 + 2 = \frac{y}{4} + 3 + 2$$
$$\frac{y}{2} = \frac{y}{4} + 5$$
$$\frac{1}{2}y = \frac{1}{4}y + 5$$

Subtract $\frac{1}{4}y$ from both sides.
$$\frac{1}{2}y - \frac{1}{4}y = \frac{1}{4}y - \frac{1}{4}y + 5$$
$$\frac{1}{4}y = 5$$

Multiply both sides by $\frac{4}{1}$, the reciprocal of $\frac{1}{4}$.
$$\frac{4}{1} \cdot \frac{1}{4}y = \frac{4}{1} \cdot 5$$
$$y = 20$$

The solution is 20.

75. $-3(w - 2) = |0 - 13| + 4w$
$$-3 \cdot w - (-3) \cdot 2 = |-13| + 4w$$
$$-3w + 6 = 13 + 4w$$
$$-3w + 6 - 6 = 13 + 4w - 6$$
$$-3w = 7 + 4w$$
$$-3w - 4w = 7 + 4w - 4w$$
$$-7w = 7$$
$$\frac{-7w}{-7} = \frac{7}{-7}$$
$$w = -1$$

The solution is -1.

77. $2(a + 0.3) = 1.2(a - 4)$

Use the distributive property on both sides.
$$2 \cdot a + 2(0.3) = 1.2 \cdot a - 1.2 \cdot 4$$
$$2a + 0.6 = 1.2a - 4.8$$

Subtract 0.6 from both sides.
$$2a + 0.6 - 0.6 = 1.2a - 4.8 - 0.6$$
$$2a = 1.2a - 5.4$$

Subtract $1.2a$ from both sides.
$$2a - 1.2a = 1.2a - 5.4 - 1.2a$$

Combine like terms.
$$0.8a = -5.4$$

Divide both sides by 0.8.
$$\frac{0.8a}{0.8} = \frac{-5.4}{0.8}$$
$$a = -6.75$$

The solution is -6.75.

9.8   Using Equations to Solve Application Problems

9.8   Margin Exercises

1.   **(a)** 15 less than a number: $x - 15$

     (b) 12 more than a number:
        $x + 12 \quad \text{or} \quad 12 + x$

     (c) A number increased by 13:
        $x + 13 \quad \text{or} \quad 13 + x$

     (d) A number minus 8: $x - 8$

     (e) -10 plus a number:
        $-10 + x \quad \text{or} \quad x + (-10)$

     (f) A number subtracted from 6: $6 - x$

2.   **(a)** Double a number: $2x$

     (b) The product of -8 and a number: $-8x$

     (c) The quotient of 15 and a number: $\dfrac{15}{x}$

     (d) One-half of a number: $\dfrac{1}{2}x \quad \text{or} \quad \dfrac{x}{2}$

3.   **(a)** Let x represent the unknown number.

3 times a number	added to	4	is	19.
↓	↓	↓	↓	↓
$3x$	$+$	4	$=$	19

(continued)

Copyright © 2018 Pearson Education, Inc.

$$3x + 4 = 19$$
$$3x + 4 - 4 = 19 - 4$$
$$3x = 15$$
$$\frac{3x}{3} = \frac{15}{3}$$
$$x = 5$$

The number is 5.

Check: 3 times 5 [= 15] is added to 4 [= 19]. *True*

(b) Let x represent the unknown number.

$$\begin{array}{ccccc} \text{-6 times} & \text{added} & & & \\ \text{a number} & \text{to} & 5 & \text{is} & -13. \\ \downarrow & \downarrow & \downarrow & \downarrow & \downarrow \\ -6x & + & 5 & = & -13 \end{array}$$

$$-6x + 5 = -13$$
$$-6x + 5 - 5 = -13 - 5$$
$$-6x = -18$$
$$\frac{-6x}{-6} = \frac{-18}{-6}$$
$$x = 3$$

The number is 3.

Check: −6 times 3 [= −18] is added to 5 [= −13]. *True*

(c) Let x represent the unknown number.

$$65 - 2x = -21$$
$$65 - 2x - 65 = -21 - 65$$
$$-2x = -86$$
$$\frac{-2x}{-2} = \frac{-86}{-2}$$
$$x = 43$$

The number is 43.

Check: Twice 43 [= 86] is subtracted from 65 [= −21]. *True*

4. *Step 1*
Asks for the number of red pens ordered.

Step 2
Let x be the number of red pens ordered.

Step 3 $\quad x - 12 - 24 = 48$

Step 4 $\qquad x - 36 = 48$
$$x - 36 + 36 = 48 + 36$$
$$x = 84$$

Step 5
The bookstore ordered 84 red pens.

Step 6
$84 - 12 - 24$ is 48, which matches the number of pens left.

5. *Step 1*
Asks for amount of LuAnn's donation.

Step 2
Let d be LuAnn's donation.

Step 3 $\quad 2d + 10 = 22$ (or, $10 + 2d = 22$)

Step 4 $\quad 2d + 10 - 10 = 22 - 10$
$$2d = 12$$
$$\frac{2d}{2} = \frac{12}{2}$$
$$d = 6$$

Step 5
LuAnn donated $6.

Step 6
Susan donated $10 more than twice $6, which is $10 + 2 \cdot \$6 = \$10 + \$12 = \$22$. That matches the amount given in the problem for Susan.

6. **(a)** *Step 2*
Let x represent the amount made by the daughter and $x + 12$ the amount made by Keonda.

Step 3 $\quad x + (x + 12) = 182$

Step 4 $\qquad 2x + 12 = 182$
$$2x + 12 - 12 = 182 - 12$$
$$2x = 170$$
$$\frac{2x}{2} = \frac{170}{2}$$
$$x = 85$$

Step 5
The daughter made $85.
Keonda made $85 + \$12 = \$97$.

Step 6
Check: $97 is $12 more than $85, and the sum of $85 and $97 is $182.

(b) *Step 2* Let x represent the length of the shorter piece and $x + 3$ the length of the longer piece.

Step 3 $\quad x + (x + 3) = 21$

Step 4 $\qquad 2x + 3 = 21$
$$2x + 3 - 3 = 21 - 3$$
$$2x = 18$$
$$\frac{2x}{2} = \frac{18}{2}$$
$$x = 9$$

Step 5
The length of the shorter piece is 9 yd. The length of the longer piece is 9 yd + 3 yd = 12 yd.

Step 6
Check: 12 yards is 3 yards longer than 9 yards and the sum of 9 yards and 12 yards is 21 yards.

Copyright © 2018 Pearson Education, Inc.

7.   *Step 2*
Let x represents the width and $x + 3$ the length.

$$x + 3$$

$$x \qquad\qquad x$$

$$x + 3$$

Step 3
Use the formula for the perimeter of a rectangle.

$$P = 2 \cdot l + 2 \cdot w$$
$$22 = 2(x + 3) + 2 \cdot x$$

Step 4
$$22 = 2x + 6 + 2x$$
$$22 = 4x + 6$$
$$22 - 6 = 4x + 6 - 6$$
$$16 = 4x$$
$$\frac{16}{4} = \frac{4x}{4}$$
$$4 = x$$

Step 5
The width is 4 yd and the length is $4 + 3 = 7$ yd.

Step 6
Check: 7 yd is 3 more than 4 yd.

$$P = 2 \cdot 7 \text{ yd} + 2 \cdot 4 \text{ yd}$$
$$P = 14 \text{ yd} + 8 \text{ yd}$$
$$P = 22 \text{ yd}$$

9.8   Section Exercises

1.   14 plus a number: $14 + x$ or $x + 14$

3. -5 added to a number: $-5 + x$ or $x + (-5)$

5.   20 minus a number: $20 - x$

7.   9 less than a number: $x - 9$

9.   Subtract 4 from a number: $x - 4$

11.   Six times a number: $6x$

13.   Double a number: $2x$

15.   A number divided by 2: $\dfrac{x}{2}$

17.   Twice a number added to 8:
$8 + 2x$ or $2x + 8$

19.   10 fewer than seven times a number: $7x - 10$

21.   The sum of twice a number and the number:
$2x + x$ or $x + 2x$

23.   **(a)** A variable is a letter that represents an unknown quantity. Examples: x, w, p

(b) An expression is a combination of operations on variables and numbers. Examples: $6x$, $w - 5, 2p + 3x$

(c) An equation has an $=$ sign and shows that two expressions are equal. Examples: $2y = 14$; $x + 5 = 2x$; $8p - 10 = 54$

25.   Let n represent the unknown number.

four times a number	decreased by	2	result is	26
↓	↓	↓	↓	↓
$4n$	$-$	2	$=$	26

$$4n - 2 = 26$$
$$4n - 2 + 2 = 26 + 2$$
$$4n = 28$$
$$\frac{4n}{4} = \frac{28}{4}$$
$$n = 7$$

The number is 7.

Check: Four times 7 [$= 28$] is decreased by 2 [$= 26$]. *True*

27.   Let n represent the unknown number.

Twice a number	added to	the number	is	-15.
↓	↓	↓	↓	↓
$2n$	$+$	n	$=$	-15

$$2n + n = -15$$
$$3n = -15$$
$$\frac{3n}{3} = \frac{-15}{3}$$
$$n = -5$$

The number is -5.

Check: Twice -5 [$= -10$] is added to -5 [$= -15$]. *True*

29.   Let n represent the unknown number.

Product of a number and 5	increased by	12	the result is	7 times the number
↓	↓	↓	↓	↓
$5n$	$+$	12	$=$	$7n$

$$5n + 12 = 7n$$
$$5n + 12 - 5n = 7n - 5n$$
$$12 = 2n$$
$$\frac{12}{2} = \frac{2n}{2}$$
$$6 = n$$

The number is 6.

(*continued*)

Copyright © 2018 Pearson Education, Inc.

Check: The product of 6 and 5 [$= 30$] is increased by 12 [$= 42$] is seven times 6 [$= 42$]. *True*

31. Let n represent the unknown number.

30	subtract	3 times a number	is	2	plus	the number
↓	↓	↓	↓	↓	↓	↓
30	−	$3n$	$=$	2	+	n

$$30 - 3n = 2 + n$$

Add $3n$ to both sides.

$$30 - 3n + 3n = 2 + n + 3n$$

Combine like terms on each side.

$$30 = 2 + 4n$$

Subtract 2 from both sides.

$$30 - 2 = 2 + 4n - 2$$
$$28 = 4n$$

Divide both sides by 4.

$$\frac{28}{4} = \frac{4n}{4}$$
$$7 = n$$

The number is 7.

Check: Three times 7 [$= 21$] is subtracted from 30 [$= 9$] is 2 plus 7 [$= 9$]. *True*

33. Let n represent the unknown number.

half a number	is added	to twice the number	answer is	50
↓	↓	↓	↓	↓
$\frac{1}{2}n$	+	$2n$	$=$	50

$$\frac{1}{2}n + 2n = 50$$
$$\left(\frac{1}{2} + 2\right)n = 50$$
$$\left(\frac{1}{2} + \frac{4}{2}\right)n = 50$$
$$\frac{5}{2}n = 50$$
$$\frac{2}{5} \cdot \frac{5}{2}n = \frac{2}{5} \cdot 50$$
$$n = 20$$

The number is 20.

Check: If half of 20 [$= 10$] is added to twice 20 [$= 40$], the answer is 50. *True*

35. *Step 1*
The problem is about Ricardo's weight.
Unknown: his original weight
Known: He gained 15 pounds, lost 28, and then regained 5 pounds to weigh 177.

Step 2
Let x represent Ricardo's original weight.

Step 3

weight at beginning	+	pounds gained	−	pounds lost	+	pounds regained	=	final weight
↓		↓		↓		↓		↓
x	+	15	−	28	+	5	=	177

Step 4
$$x + 15 - 28 + 5 = 177 \quad \text{Combine like terms.}$$
$$x - 8 = 177$$
$$x = 185$$

Step 5
He weighed 185 pounds originally.

Step 6
$$185 + 15 - 28 + 5 = 177$$

The correct answer is 185 pounds.

37. *Step 2*
Let x be the amount Brenda spent.

Step 3

Twice what Brenda spent	less	$3	is	the amount Consuelo spent
↓	↓	↓	↓	↓
$2x$	−	3	=	81

Step 4 $2x - 3 = 81$

Add 3 to both sides.

$$2x - 3 + 3 = 81 + 3$$
$$2x = 84$$

Divide both sides by 2.

$$\frac{2x}{2} = \frac{84}{2}$$
$$x = 42$$

Step 5
Brenda spent $42.

Step 6
Check: Consuelo spent $3 less than twice $42 [$= \$84$], which is $81. *True*

39. *Step 2*
You know the least about my age, so let x be my age. Then $x - 9$ is my sister's age.

Step 3   The sum of the ages   is   51.

	↓		↓	↓
	$x + (x - 9)$		$=$	51

(continued)

Copyright © 2018 Pearson Education, Inc.

Step 4
$$x + x - 9 = 51$$
$$2x - 9 = 51$$
Add 9 to both sides.
$$2x - 9 + 9 = 51 + 9$$
$$2x = 60$$
Divide both sides by 2.
$$\frac{2x}{2} = \frac{60}{2}$$
$$x = 30$$

Step 5
x is my age, so I am 30; my sister is $x - 9$ or $30 - 9$, so she is 21.

Step 6
Check: 21 is 9 less than 30 and the sum of 21 and 30 is 51. *True*

41. *Step 2*
Let x be husband's earnings; $x + 1500$ is Lien's earnings.

Step 3   Together (add)   they earned   37,500.
$$\downarrow \qquad\qquad \downarrow \qquad\qquad \downarrow$$
$$x + (x + 1500) \qquad = \qquad 37{,}500$$

Step 4
$$x + x + 1500 = 37{,}500$$
$$2x + 1500 = 37{,}500$$
$$2x + 1500 - 1500 = 37{,}500 - 1500$$
$$2x = 36{,}000$$
$$\frac{2x}{2} = \frac{36{,}000}{2}$$
$$x = 18{,}000$$
$$x + 1500 = 19{,}500$$

Step 5
Lien earned $19,500 and her husband earned $18,000.

Step 6
Check: $19,500 is $1500 more than $18,000 and the sum of $18,000 and $19,500 is $37,500. *True*

43. *Step 2*
You know the least about the cost of the printer, so let x be the cost of the printer. Then $8 \cdot x$ or $8x$ is the cost of the laptop.

Step 3   The total cost (add)   is $396.
$$\downarrow \qquad\qquad \downarrow \quad \downarrow$$
$$x + 8x \qquad = \quad 396$$

Step 4 $x + 8x = 396$
$$9x = 396$$
Divide both sides by 9.
$$\frac{9x}{9} = \frac{396}{9}$$
$$x = 44$$

Step 5
The cost of the printer is x, so the printer cost $44. The laptop cost 8 times as much, so $8(\$44) = \$352$ for the laptop.

Step 6
Check: Eight times $44 is $352 and the sum of $44 and $352 is $396. *True*

45. *Step 2*
You know the least about the shorter piece, so let x be the length of the shorter piece. Then $x + 10$ is the length of the longer piece.

Step 3

The length of the shorter piece	and	the length of the longer piece	is	78 cm.
$\downarrow$	$\downarrow$	$\downarrow$	$\downarrow$	$\downarrow$
x	$+$	$x + 10$	$=$	78

Step 4 $x + x + 10 = 78$
Combine like terms.
$$2x + 10 = 78$$
Subtract 10 from both sides.
$$2x + 10 - 10 = 78 - 10$$
$$2x = 68$$
Divide both sides by 2.
$$\frac{2x}{2} = \frac{68}{2}$$
$$x = 34$$

Step 5
The length of the shorter piece is x, so the shorter piece is 34 cm. The length of the longer piece is $x + 10$, so $34 + 10 = 44$ cm.

Step 6
Check: 44 cm is 10 cm longer than 34 cm and the sum of 34 cm and 44 cm is 78 cm. *True*

47. *Step 2*
Let x be the length of the longer piece, and $x - 7$ the length of the shorter piece.

Step 3

The length of the longer piece	and	the length of the shorter piece	are	31 ft.
$\downarrow$	$\downarrow$	$\downarrow$	$\downarrow$	$\downarrow$
x	$+$	$x - 7$	$=$	31

(continued)

Copyright © 2018 Pearson Education, Inc.

Step 4 $x + x - 7 = 31$

$$2x - 7 = 31$$
$$2x - 7 + 7 = 31 + 7$$
$$2x = 38$$
$$\frac{2x}{2} = \frac{38}{2}$$
$$x = 19$$
$$x - 7 = 12$$

Step 5

The pieces are 19 ft and 12 ft.

Step 6

Check: 12 ft is 7 ft shorter than 19 ft and the sum of 12 ft and 19 ft is 31 ft. *True*

49. Let x be the length of the rectangle.

$$P = 2 \cdot l + 2 \cdot w$$
$$48 = 2x + 2 \cdot 5$$
$$48 - 10 = 2x + 10 - 10$$
$$38 = 2x$$
$$\frac{38}{2} = \frac{2x}{2}$$
$$19 = x$$

The length is 19 yd.

Check: $48 = 2 \cdot 19 + 2 \cdot 5$
$$48 = 38 + 10$$
$$48 = 48 \qquad \textit{True}$$

51. Let w be the width of the rectangle, and $2w$ the length.

$$P = 2 \cdot l + 2 \cdot w$$
$$36 = 2 \cdot 2w + 2 \cdot w$$
$$36 = 4w + 2w$$
$$36 = 6w$$
$$\frac{36}{6} = \frac{6w}{6}$$
$$6 = w$$
$$12 = 2w$$

The length is 12 ft and the width is 6 ft.

Check: The length, 12, is twice the width, 6. Now check the perimeter.

$$36 = 2(12) + 2(6)$$
$$36 = 24 + 12$$
$$36 = 36 \qquad \textit{True}$$

53.   *Step 2*: You know the least about the widht, so set x be the width. Then $2x + 3$ the length.

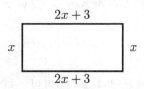

Step 3: $P = 2 \cdot l + 2 \cdot w$
$$36 = 2 \cdot (2x + 3) + 2 \cdot x$$

Step 4: $36 = 2 \cdot (2x + 3) + 2 \cdot x$

Apply the distributive property.
$$36 = 4x + 6 + 2x$$

Combine like terms.
$$36 = 6x + 6$$

Subtract 6 *from both sides.*
$$36 - 6 = 6x + 6 - 6$$
$$30 = 6x$$

Divide both sides by 6.
$$\frac{30}{6} = \frac{6x}{6}$$
$$5 = x$$

Step 5: The width is x, so the width is 5 in. The length is $2x + 3$, so $2(5) + 3$ or 13 in. is the length.

Check: The length, 13, is 3 inches more than twice the width, 5. Now check the perimeter.

$$36 = 2(13) + 2(5)$$
$$36 = 26 + 10$$
$$36 = 36 \qquad \textit{True}$$

Chapter 9 Review Exercises

1. $2, -3, 4, 1, 0, -5$

To graph 2, put a dot at 2 on the number line. Now put dots in the correct places for $-3, 4, 1, 0,$ and -5.

$$\begin{array}{c}\xleftarrow{\hspace{1em}} \; | \; | \; \bullet \; | \; \bullet \; | \; | \; \bullet \; \bullet \; \bullet \; | \; \xrightarrow{\hspace{1em}} \\ \scriptstyle -7 \; -6 \; -5 \; -4 \; -3 \; -2 \; -1 \;\; 0 \;\; 1 \;\; 2 \;\; 3 \;\; 4\end{array}$$

2. $-2, 5, -4, -1, 3, -6$

$$\begin{array}{c}\xleftarrow{\hspace{1em}} \; \bullet \; | \; \bullet \; | \; \bullet \; \bullet \; | \; | \; | \; \bullet \; | \; \bullet \; \xrightarrow{\hspace{1em}} \\ \scriptstyle -6 \; -5 \; -4 \; -3 \; -2 \; -1 \;\; 0 \;\; 1 \;\; 2 \;\; 3 \;\; 4 \;\; 5\end{array}$$

Copyright © 2018 Pearson Education, Inc.

3. $-1\frac{1}{4}, -\frac{5}{8}, -3\frac{3}{4}, 2\frac{1}{8}, \frac{3}{2}, -2\frac{1}{8}$

4. $0, -\frac{3}{4}, \frac{5}{4}, -4\frac{1}{2}, \frac{7}{8}, -7\frac{2}{3}$

5. $0 > -2$ because 0 is to the *right* of -2 on a number line.

6. $-5 < 0$ because -5 is to the *left* of 0 on a number line.

7. $-1 > -4$ because -1 is to the *right* of -4 on a number line.

8. $-9 < -6$ because -9 is to the *left* of -6 on a number line.

9. The *distance* from 0 to 8 is 8, so $|8| = 8$.

10. The *distance* from 0 to -19 is 19, so $|-19| = 19$.

11. $-|-7|$
First, $|-7| = 7$. But there is a negative sign outside the absolute value bars. So, -7 is the simplified expression.

12. $-|15| = -(15) = -15$
Because the negative sign is outside the absolute value bars, the answer is negative.

13. $-4 + 6$
The signs are different, so subtract the absolute values.

$$|-4| = 4 \quad \text{and} \quad |6| = 6$$
$$6 - 4 = 2$$

The positive number, 6, has the greater absolute value, so the answer is positive.

$$-4 + 6 = 2$$

14. $-10 + 3$
The signs are different, so subtract the absolute values.

$$|-10| = 10 \quad \text{and} \quad |3| = 3$$
$$10 - 3 = 7$$

The negative number, 10, has the greater absolute value, so the answer is negative.

$$-10 + 3 = -7$$

15. $-11 + (-8)$
Add the absolute values.

$$|-11| = 11 \quad \text{and} \quad |-8| = 8$$
$$11 + 8 = 19$$

Write a negative sign in front of the sum because both numbers are negative.

$$-11 + (-8) = -19$$

16. $-9 + (-24)$
Add the absolute values.

$$|-9| = 9 \quad \text{and} \quad |-24| = 24$$
$$9 + 24 = 33$$

Write a negative sign in front of the sum because both numbers are negative.

$$-9 + (-24) = -33$$

17. $12 + (-11)$
The signs are different, so subtract the absolute values.

$$|12| = 12 \quad \text{and} \quad |-11| = 11$$
$$12 - 11 = 1$$

The positive number, 12, has the greater absolute value, so the answer is positive.

$$12 + (-11) = 1$$

18. $1 + (-20)$
The signs are different, so subtract the absolute values.

$$|1| = 1 \quad \text{and} \quad |-20| = 20$$
$$20 - 1 = 19$$

The negative number, -20, has the greater absolute value, so the answer is negative.

$$1 + (-20) = -19$$

19. $\dfrac{9}{10} + \left(-\dfrac{3}{5}\right) = \dfrac{9}{10} + \left(-\dfrac{6}{10}\right) = \dfrac{3}{10}$

20. $-\dfrac{7}{8} + \dfrac{1}{2} = -\dfrac{7}{8} + \dfrac{4}{8} = -\dfrac{3}{8}$

21. $-6.7 + 1.5$
The signs are different, so subtract the absolute values.

$$|-6.7| = 6.7 \quad \text{and} \quad |1.5| = 1.5$$
$$6.7 - 1.5 = 5.2$$

The negative number, -6.7, has the greater absolute value, so the answer is negative.

$$-6.7 + 1.5 = -5.2$$

22. $-0.8 + (-0.7)$
Add the absolute values.

$$|-0.8| = 0.8 \quad \text{and} \quad |-0.7| = 0.7$$
$$0.8 + 0.7 = 1.5$$

Write a negative sign in front of the sum because both numbers are negative.

$$-0.8 + (-0.7) = -1.5$$

Copyright © 2018 Pearson Education, Inc.

23. The additive inverse of 6 is -6.

24. The additive inverse of -14 is $-(-14)$ or 14.

25. The additive inverse of $-\frac{5}{8}$ is $-\left(-\frac{5}{8}\right)$ or $\frac{5}{8}$.

26. The additive inverse of 3.75 is -3.75.

27. $4 - 10 = 4 + (-10) = -6$

28. $7 - 15 = 7 + (-15) = -8$

29. $-6 - 1 = -6 + (-1) = -7$

30. $-12 - 5 = -12 + (-5) = -17$

31. $8 - (-3) = 8 + (+3) = 11$

32. $2 - (-9) = 2 + (+9) = 11$

33. $-1 - (-14) = -1 + (+14) = 13$

34. $-10 - (-4) = -10 + (+4) = -6$

35. $-40 - 40 = -40 + (-40) = -80$

36. $-15 - (-15) = -15 + (+15) = 0$

37. $\dfrac{1}{3} - \dfrac{5}{6} = \dfrac{2}{6} + \left(-\dfrac{5}{6}\right) = -\dfrac{3}{6} = -\dfrac{1}{2}$

38. $2.8 - (-6.2) = 2.8 + (+6.2) = 9$

39. $-4(6) = -24$
The numbers have *different* signs, so the product is *negative*.

40. $5 \cdot (-4) = -20$
The numbers have *different* signs, so the product is *negative*.

41. $-3(-5) = 15$
The numbers have the *same* sign, so the product is *positive*.

42. $-8(-8) = 64$
The numbers have the *same* sign, so the product is *positive*.

43. $\dfrac{80}{-10} = -8$
The numbers have *different* signs, so the quotient is *negative*.

44. $\dfrac{-9}{3} = -3$
The numbers have *different* signs, so the quotient is *negative*.

45. $\dfrac{-25}{-5} = 5$
The numbers have *same* sign, so the quotient is *positive*.

46. $\dfrac{-120}{-6} = 20$
The numbers have *same* sign, so the quotient is *positive*.

47. $(-37)(0) = 0$

48. $(-1)(81) = -81$

49. $\dfrac{0}{-10} = 0$

50. $\dfrac{-20}{0}$ is undefined.

51. $\left(\dfrac{2}{3}\right) \cdot \left(-\dfrac{6}{7}\right) = \dfrac{2}{\underset{1}{\cancel{3}}} \cdot \left(-\dfrac{\overset{2}{\cancel{6}}}{7}\right) = -\dfrac{4}{7}$

52. $-\dfrac{4}{5} \div \left(-\dfrac{2}{15}\right) = -\dfrac{4}{5} \cdot \left(-\dfrac{15}{2}\right) = \dfrac{\overset{2}{\cancel{4}}}{\underset{1}{\cancel{5}}} \cdot \dfrac{\overset{3}{\cancel{15}}}{\underset{1}{\cancel{2}}} = 6$

53. $(-0.5)(-2.8) = 1.4$

The numbers have *same* sign, so the product is *positive*.

54. $\dfrac{-5.28}{0.8} = -6.6$
Move the decimal place to the right in both the divisor and the dividend.

$$
\begin{array}{r}
6.\,6 \\
8\overline{\smash{\big)}\,5\,2.\,8} \\
\underline{4\,8} \\
4\,8 \\
\underline{4\,8} \\
0
\end{array}
$$

The numbers have *different* signs, so the quotient is *negative*.

55. $2 - 11(-5)$
$= 2 - (-55)$
$= 2 + (+55)$
$= 57$

56. $(-4)(-8) - 9$
$= 32 - 9$
$= 23$

57. $48 \div (-2)^3 - (-5)$
$= 48 \div (-8) - (-5)$
$= -6 - (-5)$
$= -6 + (+5)$
$= -1$

Copyright © 2018 Pearson Education, Inc.

58. $-36 \div (-3)^2 - (-2)$
$= -36 \div 9 - (-2)$
$= -4 - (-2)$
$= -4 + (+2)$
$= -2$

59. $5(4) - 7(6) + 3(-4)$
$= 20 - 42 + (-12)$
$= 20 + (-42) + (-12)$
$= -22 + (-12)$
$= -34$

60. $2(8) - 4(9) + 2(-6)$
$= 16 - 36 + (-12)$
$= 16 + (-36) + (-12)$
$= -20 + (-12)$
$= -32$

61. $3^3(-4) - 2(5 - 9)$
$= 27(-4) - 2 \cdot (-4)$
$= -108 - (-8)$
$= -108 + (+8)$
$= -100$

62. $6(-4)^2 - 3(7 - 14)$
$= 6(-4)^2 - 3 \cdot (-7)$
$= 6(16) - 3 \cdot (-7)$
$= 96 - (-21)$
$= 96 + (+21)$
$= 117$

63. $\dfrac{3 - (5^2 - 4^2)}{14 + 24 \div (-3)}$

Numerator:
$3 - \left(5^2 - 4^2\right) = 3 - (25 - 16)$
$= 3 - 9$
$= 3 + (-9)$
$= -6$

Denominator:
$14 + 24 \div (-3) = 14 + (-8)$
$= 6$

Divide: $\dfrac{-6}{6} = -1$

64. $(-0.8)^2(0.2) - (-1.2)$
$= (0.64)(0.2) - (-1.2)$
$= 0.128 + (+1.2)$
$= 1.328$

65. $\left(-\dfrac{1}{3}\right)^2 + \dfrac{1}{4}\left(-\dfrac{4}{9}\right)$

$= \dfrac{1}{9} + \left(\dfrac{1}{\underset{1}{\cancel{4}}}\right)\left(-\dfrac{\overset{1}{\cancel{4}}}{9}\right)$

$= \dfrac{1}{9} + \left(-\dfrac{1}{9}\right)$

$= 0$

66. $\dfrac{12 \div (2 - 5) + 12(-1)}{2^3 - (-4)^2}$

Numerator:

$12 \div (2 - 5) + 12(-1) = 12 \div (-3) + 12(-1)$
$= -4 + 12(-1)$
$= -4 + (-12)$
$= -16$

Denominator:
$2^3 - (-4)^2 = 8 - 16$
$= 8 + (-16)$
$= -8$

Divide. $\dfrac{-16}{-8} = 2$

67. $3k + 5m$; k is 4, m is 3
Replace k with 4. Replace m with 3.

$3k + 5m = 3(4) + 5(3)$
$= 12 + 15$
$= 27$

68. $3k + 5m$; k is -6, m is 2
Replace k with -6. Replace m with 2.

$3k + 5m = 3(-6) + 5(2)$
$= -18 + 10$
$= -8$

69. $2f - g$; f is -5, g is -10
Replace f with -5. Replace g with -10.

$2f - g = 2(-5) - (-10)$
$= -10 + (+10)$
$= 0$

70. $2f - g$; f is 6, g is -7
Replace f with 6. Replace g with -7.

$2f - g = 2(6) - (-7)$
$= 12 + (+7)$
$= 19$

Copyright © 2018 Pearson Education, Inc.

71. $\dfrac{5a - 7y}{2 + m}$; a is 1, y is 4, m is -3

Replace a with 1, y with 4, and m with -3.

$$\begin{aligned}\frac{5a - 7y}{2 + m} &= \frac{5(1) - 7(4)}{2 + (-3)} \\ &= \frac{5 - 28}{-1} \\ &= \frac{5 + (-28)}{-1} \\ &= \frac{-23}{-1} \\ &= 23\end{aligned}$$

72. $\dfrac{5a - 7y}{2 + m}$; a is 2, y is -2, m is -26

Replace a with 2, y with -2, and m with -26.

$$\begin{aligned}\frac{5a - 7y}{2 + m} &= \frac{5(2) - 7(-2)}{2 + (-26)} \\ &= \frac{10 - (-14)}{-24} \\ &= \frac{10 + (+14)}{-24} \\ &= \frac{24}{-24} \\ &= -1\end{aligned}$$

73. $P = a + b + c$; a is 9 cm, b is 12 cm, c is 14 cm

$$\begin{aligned}P &= 9 \text{ cm} + 12 \text{ cm} + 14 \text{ cm} \\ &= 21 \text{ cm} + 14 \text{ cm} \\ &= 35 \text{ cm}\end{aligned}$$

74. $A = \frac{1}{2}bh$; b is 6 ft, h is 9 ft

$$\begin{aligned}A &= \tfrac{1}{2}(6 \text{ ft})(9 \text{ ft}) \\ &= 3(9) \text{ ft}^2 \\ &= 27 \text{ ft}^2\end{aligned}$$

75. $y + 3 = 0$

Subtract 3 from both sides.

$$\begin{aligned}y + 3 - 3 &= 0 - 3 \\ y + 0 &= -3 \\ y &= -3\end{aligned}$$

Check: $-3 + 3 = 0$

$0 = 0$ *True*

The solution is -3.

76. $a - 8 = 8$

$$\begin{aligned}a - 8 + 8 &= 8 + 8 \\ a &= 16\end{aligned}$$

Check: $16 - 8 = 8$

$8 = 8$ *True*

The solution is 16.

77. $-5 = z - 6$

$$\begin{aligned}-5 + 6 &= z - 6 + 6 \\ 1 &= z\end{aligned}$$

Check: $-5 = 1 - 6$

$-5 = -5$ *True*

The solution is 1.

78. $-8 = -9 + r$

$$\begin{aligned}-8 + 9 &= -9 + 9 + r \\ 1 &= r\end{aligned}$$

Check: $-8 = -9 + 1$

$-8 = -8$ *True*

The solution is 1.

79. $-\dfrac{3}{4} + x = -2$

$$\begin{aligned}-\frac{3}{4} + x + \frac{3}{4} &= -2 + \frac{3}{4} \\ x &= -\frac{8}{4} + \frac{3}{4} \\ &= -\frac{5}{4} \text{ or } -1\frac{1}{4}\end{aligned}$$

Check: $-\dfrac{3}{4} + \left(-\dfrac{5}{4}\right) = -2$

$$-\frac{8}{4} = -2$$

$-2 = -2$ *True*

The solution is $-\frac{5}{4}$ or $-1\frac{1}{4}$.

80. $12.92 + k = 4.87$

$$\begin{aligned}12.92 - 12.92 + k &= 4.87 + (-12.92) \\ k &= -8.05\end{aligned}$$

Check: $12.92 + (-8.05) = 4.87$

$4.87 = 4.87$ *True*

The solution is -8.05.

81. $-8r = 56$

$$\frac{\overset{1}{\cancel{-8}}r}{\underset{1}{\cancel{-8}}} = \frac{56}{-8}$$

$$r = -7$$

Check: $(-8)(-7) = 56$

$56 = 56$ *True*

The solution is -7.

82. $3p = 24$ **Check:** $3(8) = 24$

$$\frac{3p}{3} = \frac{24}{3} \qquad 24 = 24 \quad True$$

$$p = 8$$

The solution is 8.

Copyright © 2018 Pearson Education, Inc.

83. $\dfrac{z}{4} = 5$

$\dfrac{4}{1} \cdot \dfrac{z}{4} = 5 \cdot 4$

$z = 20$

Check: $\dfrac{20}{4} = 5$

$5 = 5$ *True*

The solution is 20.

84. $\dfrac{a}{5} = -11$

$\dfrac{5}{1} \cdot \dfrac{a}{5} = -11 \cdot 5$

$a = -55$

Check: $\dfrac{-55}{5} = -11$

$-11 = -11$ *True*

The solution is -55.

85. $20 = 3y - 7$

$20 + 7 = 3y - 7 + 7$

$27 = 3y$

$\dfrac{27}{3} = \dfrac{3y}{3}$

$9 = y$

Check: $20 = 3(9) - 7$

$20 = 27 - 7$

$20 = 20$ *True*

The solution is 9.

86. $-5 = 2b + 3$

$-5 - 3 = 2b + 3 - 3$

$-8 = 2b$

$\dfrac{-8}{2} = \dfrac{2b}{2}$

$-4 = b$

Check: $-5 = 2(-4) + 3$

$-5 = -8 + 3$

$-5 = -5$ *True*

The solution is -4.

87. $6(r - 5) = 6 \cdot r - 6 \cdot 5$

$= 6r - 30$

88. $11(p + 7) = 11 \cdot p + 11 \cdot 7$

$= 11p + 77$

89. $-9(z - 3) = -9 \cdot z - (-9)(3)$

$= -9z - (-27)$

$= -9z + 27$

90. $-8(x + 4) = -8 \cdot x + (-8) \cdot 4$

$= -8x - 32$

91. $3r + 8r = (3 + 8)r$

$= 11r$

92. $10z - 15z = (10 - 15)z$

$= -5z$

93. $3p - 12p + p = (3 - 12 + 1)p$

$= -8p$

94. $-6x - x + 9x = (-6 - 1 + 9)x$

$= 2x$

95. $-4z + 2z = 18$

$(-4 + 2)z = 18$

$-2z = 18$

$\dfrac{-2z}{-2} = \dfrac{18}{-2}$

$z = -9$

Check: $-4(-9) + 2(-9) = 18$

$36 + (-18) = 18$

$18 = 18$ *True*

The solution is -9.

96. $-35 = 9k - 2k$

$-35 = (9 - 2)k$

$-35 = 7k$

$\dfrac{-35}{7} = \dfrac{7k}{7}$

$-5 = k$

Check: $-35 = 9(-5) - 2(-5)$

$-35 = -45 - (-10)$

$-35 = -45 + (+10)$

$-35 = -35$ *True*

The solution is -5.

97. $4y - 3 = 7y + 12$

$4y - 3 - 4y = 7y + 12 - 4y$

$-3 = 3y + 12$

$-3 + (-12) = 3y + 12 + (-12)$

$-15 = 3y$

$\dfrac{-15}{3} = \dfrac{3y}{3}$

$-5 = y$

Check: $4(-5) + (-3) = 7(-5) + 12$

$-20 + (-3) = -35 + 12$

$-23 = -23$ *True*

The solution is -5.

Copyright © 2018 Pearson Education, Inc.

98.
$$b + 6 = 3b - 8$$
$$b + 6 - b = 3b - 8 - b$$
$$6 = 2b - 8$$
$$6 + 8 = 2b - 8 + 8$$
$$14 = 2b$$
$$\frac{14}{2} = \frac{2b}{2}$$
$$7 = b$$

Check: $7 + 6 = 3 \cdot 7 - 8$
$$13 = 21 - 8$$
$$13 = 13 \qquad \textit{True}$$

The solution is 7.

99.
$$-14 = 2(a - 3)$$
$$-14 = 2 \cdot a - 2 \cdot 3$$
$$-14 = 2a - 6$$
$$-14 + 6 = 2a - 6 + 6$$
$$-8 = 2a$$
$$\frac{-8}{2} = \frac{2a}{2}$$
$$-4 = a$$

Check: $-14 = 2(-4 - 3)$
$$-14 = 2(-7)$$
$$-14 = -14 \qquad \textit{True}$$

The solution is -4.

100.
$$42 = 7(t + 6)$$
$$42 = 7t + 7 \cdot 6$$
$$42 = 7t + 42$$
$$42 - 42 = 7t + 42 - 42$$
$$0 = 7t$$
$$0 = t$$

Check: $42 = 7(0 + 6)$
$$42 = 7(6)$$
$$42 = 42 \qquad \textit{True}$$

The solution is 0.

101. 18 plus a number: $18 + x$ or $x + 18$

102. Half a number: $\frac{1}{2}x$ or $\frac{x}{2}$

103. -5 times a number: $-5x$

104. A number subtracted from 20: $20 - x$

105. Let x represent the number.

four times a number	the sum of	6	is	-14
↓	↓	↓	↓	↓
$4x$	$+$	6	$=$	-14

$$4x + 6 = -14$$
$$4x + 6 - 6 = -14 - 6$$
$$4x = -20$$
$$\frac{4x}{4} = \frac{-20}{4}$$
$$x = -5$$

The number is -5.

Check: The sum of four times -5 [$= -20$] and 6 is -14. *True*

106. Let x represent the number.

five times a number	subtract	the number	is	100
↓	↓	↓	↓	↓
$5x$	$-$	x	$=$	100

$$5x - x = 100$$
$$4x = 100$$
$$\frac{4x}{4} = \frac{100}{4}$$
$$x = 25$$

The number is 25.

Check: If 25 is subtracted from five times 25 [$= 125$], the result is 100. *True*

107. *Step 2*
Let x be the amount received by one student, and $3x$ the amount received by the other student. The sum of the amounts is $30,000.

Step 3 $\quad x + 3x = 30{,}000$

Step 4 $\quad 4x = 30{,}000$
$$\frac{4x}{4} = \frac{30{,}000}{4}$$
$$x = 7500$$
$$3x = 22{,}500$$

Step 5
The students will receive $7500 and $22,500.

Step 6
Check: $22,500 is three times $7500 and the sum of $22,500 and $7500 is $30,000. *True*

Copyright © 2018 Pearson Education, Inc.

108. *Step 2*

Let w represent the width, and $w + 4$ the length.

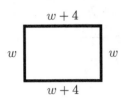

Step 3 $P = 2 \cdot l + 2 \cdot w$

$48 = 2 \cdot (w + 4) + 2 \cdot w$

Step 4 $48 = 2w + 8 + 2w$

$48 = 4w + 8$

$40 = 4w$

$\dfrac{40}{4} = \dfrac{4w}{4}$

$10 = w$

$14 = w + 4$

Step 5

The width is 10 in. and the length is 14 in.

Step 6

Check: The length, 14, is 4 inches more than the width, 10.   Now check the perimeter.

$48 = 2(14) + 2(10)$

$48 = 28 + 20$

$48 = 48$ *True*

109. *Step 2*

Let x represent a zebra's running speed and $x + 25$ a cheetah's sprinting speed.

Step 3 $x + (x + 25) = 111$

Step 4 $2x + 25 = 111$

$2x + 25 - 25 = 111 - 25$

$2x = 86$

$\dfrac{2x}{2} = \dfrac{86}{2}$

$x = 43$

Step 5

A zebra runs 43 miles per hour, and a cheetah sprints $43 + 25 =$ 68 miles per hour.

Step 6

Check: 68 mph is 25 mph faster than 43 mph and the sum of 68 mph and 43 mph is 111 mph.
True

110. *Step 2*

Let $w =$ the width, and $5w =$ the length.

Step 3 $P = 2 \cdot l + 2 \cdot w$

$1320 = 2 \cdot 5w + 2 \cdot w$

Step 4 $1320 = 10w + 2w$

$1320 = 12w$

$\dfrac{1320}{12} = \dfrac{12w}{12}$

$110 = w$

$550 = 5w$

Step 5

The length is 550 yd and the width is 110 yd.

Step 6

Check: The length, 550, is five times the width, 110.   Now check the perimeter.

$1320 = 2(550) + 2(110)$

$1320 = 1100 + 220$

$1320 = 1320$ *True*

Chapter 9 Mixed Review Exercises

1. $-6 - (-9) = -6 + (+9) = 3$

2. $(-8)(-5) = 40$

3. $-12 + 11 = -1$

4. $\dfrac{-70}{10} = -7$

5. $-4(4) = -16$

6. $5 - 14 = 5 + (-14) = -9$

7. $\dfrac{-42}{-7} = 6$

8. $16 + (-11) = 5$

9. $-10 - 10 = -10 + (-10) = -20$

10. $\dfrac{-5}{0}$ is undefined.

11. $-\dfrac{2}{3} + \dfrac{1}{9} = -\dfrac{6}{9} + \dfrac{1}{9} = -\dfrac{5}{9}$

12. $0.7(-0.5) = -0.35$

13. $|-6| + 2 - 3(-8) - 5^2$

$= |-6| + 2 - 3(-8) - 25$

$= 6 + 2 - 3(-8) - 25$

$= 6 + 2 - (-24) - 25$

$= 6 + 2 + (+24) - 25$

$= 8 + 24 - 25$

$= 32 - 25$

$= 7$

Copyright © 2018 Pearson Education, Inc.

14. $9 \div |-3| + 6(-5) + 2^3$

$= 9 \div |-3| + 6(-5) + 8$

$= 9 \div 3 + 6(-5) + 8$

$= 3 + 6(-5) + 8$

$= 3 + (-30) + 8$

$= -27 + 8$

$= -19$

15. $-45 = -5y$

$\dfrac{-45}{-5} = \dfrac{-5y}{-5}$

$9 = y$

The solution is 9.

16. $b - 8 = -12$

$b - 8 + 8 = -12 + 8$

$b = -4$

The solution is -4.

17. $6z - 3 = 3z + 9$

$6z - 3 + 3 = 3z + 9 + 3$

$6z = 3z + 12$

$6z - 3z = 3z + 12 - 3z$

$3z = 12$

$\dfrac{3z}{3} = \dfrac{12}{3}$

$z = 4$

The solution is 4.

18. $-5 = r + 5$

$-5 + (-5) = r + 5 + (-5)$

$-10 = r$

The solution is -10.

19. $-3x = 33$

$\dfrac{-3x}{-3} = \dfrac{33}{-3}$

$x = -11$

The solution is -11.

20. $2z - 7z = -15$

$-5z = -15$

$\dfrac{-5z}{-5} = \dfrac{-15}{-5}$

$z = 3$

The solution is 3.

21. $3(k - 6) = 6 - 12$

$3k - 18 = -6$

$3k - 18 + 18 = -6 + 18$

$3k = 12$

$\dfrac{3k}{3} = \dfrac{12}{3}$

$k = 4$

The solution is 4.

22. $6(t + 3) = -2 + 20$

$6t + 18 = 18$

$6t + 18 - 18 = 18 - 18$

$6t = 0$

$t = 0$

The solution is 0.

23. $-10 = \dfrac{a}{5} - 2$

$-10 = \dfrac{1}{5}a - 2$

$-10 + 2 = \dfrac{1}{5}a - 2 + 2$

$-8 = \dfrac{1}{5}a$

$5 \cdot (-8) = \left(\dfrac{5}{1}\right) \cdot \left(\dfrac{1}{5}\right)a$

$-40 = a$

The solution is -40.

24. $4 + 8p = 4p + 16$

$4 + 8p - 4p = 4p + 16 - 4p$

$4 + 4p = 16$

$4 + 4p - 4 = 16 - 4$

$4p = 12$

$\dfrac{4p}{4} = \dfrac{12}{4}$

$p = 3$

The solution is 3.

25. *Step 2*

Let x represent the recommended daily intake of vitamin C for a young child, and $4x + 15$ the recommended amount for an adult female.

Step 3 $4x + 15 = 75$

Step 4 $4x + 15 - 15 = 75 - 15$

$4x = 60$

$\dfrac{4x}{4} = \dfrac{60}{4}$

$x = 15$

Step 5

A young child should receive 15 mg of vitamin C.

Step 6

Check: 8 mg more than four times 15 mg $[= 60]$ is 60 mg. *True*

Copyright © 2018 Pearson Education, Inc.

26. *Step 2*
Let x represent the number of Senators, and $5x - 65$ the number of Representatives. There are a total (add) of 535 members.

Step 3 $x + (5x - 65) = 535$

Step 4
$$6x - 65 = 535$$
$$6x - 65 + 65 = 535 + 65$$
$$6x = 600$$
$$\frac{6x}{6} = \frac{600}{6}$$
$$x = 100$$
$$5x - 65 = 435$$

Step 5
There are 100 Senators and 435 Representatives.

Step 6
Check: 65 less than 5 times 100 [$= 500$] is 435, and the sum of 100 and 435 is 535. *True*

Chapter 9 Test

1. $-4, -1, 1\frac{1}{2}, 3, 0$

To graph -4, put a dot at -4 on the number line. Now put dots in the correct places for -1, $1\frac{1}{2}$, 3, and 0.

$\begin{array}{c}\xleftarrow{\hspace{0.3em}}\!\!\!+\!\!\bullet\!\!+\!\!+\!\!+\!\!\bullet\!\!+\!\!\bullet\!\!+\!\!\bullet\!\!+\!\!\bullet\!\!+\!\!+\!\!\xrightarrow{\hspace{0.3em}}\\ -5\,-4\,-3\,-2\,-1\ \ \mathbf{0}\ \ 1\ \ 2\ \ 3\ \ 4\ \ 5\end{array}$

2. $-3 < 0$ because -3 is to the *left* of 0 on a number line.
$-4 > -8$ because -4 is to the *right* of -8 on a number line.

3. $|-7| = 7$ and $|15| = 15$ because the distances from -7 to 0 and 15 to 0 are 7 and 15, respectively.

4. $-8 + 7 = -1$

5. $-11 + (-2) = -13$

6. $6.7 + (-1.4) = 5.3$

7. $8 - 15 = 8 + (-15) = -7$

8. $4 - (-12) = 4 + (+12) = 16$

9. $-\frac{1}{2} - \left(-\frac{3}{4}\right) = -\frac{1}{2} + \left(+\frac{3}{4}\right)$
$$= -\frac{2}{4} + \frac{3}{4} = \frac{1}{4}$$

10. $8(-4) = -32$

11. $-7(-12) = 84$

12. $(-16)(0) = 0$

13. $\frac{-100}{4} = -25$

14. $\frac{-24}{-3} = 8$

15. $-\frac{1}{4} \div \frac{5}{12} = -\frac{1}{\cancel{4}_1} \cdot \frac{\cancel{12}^3}{5} = -\frac{3}{5}$

16. $-5 + 3(-2) - (-12)$
$$= -5 + (-6) - (-12)$$
$$= -11 - (-12)$$
$$= -11 + (+12)$$
$$= 1$$

17. $2 - (6 - 8) - (-5)^2$
$$= 2 - (-2) - (-5)^2$$
$$= 2 - (-2) - (25)$$
$$= 2 + (+2) - 25$$
$$= 4 + (-25)$$
$$= -21$$

18. $8k - 3m$; k is -4, m is 2
Replace k with -4. Replace m with 2.
$$8k - 3m = 8(-4) - 3(2)$$
$$= -32 - 6$$
$$= -32 + (-6)$$
$$= -38$$

19. $8k - 3m$; k is 3, m is -1
Replace k with 3. Replace m with -1.
$$8k - 3m = 8(3) - 3(-1)$$
$$= 24 - (-3)$$
$$= 24 + (+3)$$
$$= 27$$

20. When evaluating, you are given specific values to replace each variable. When solving an equation, you are not given the value of the variable. You must find a value that "works"; that is, when your solution is substituted for the variable, the two sides of the equation are equal.

21. $A = \frac{1}{2}bh$; b is 20 ft, h is 11 ft
$$A = \frac{1}{2}(20\text{ ft})(11\text{ ft})$$
$$= 10(11)\text{ ft}^2$$
$$= 110\text{ ft}^2$$

22. $$x - 9 = -4$$
$$x - 9 + 9 = -4 + 9$$
$$x = 5$$

The solution is 5.

Copyright © 2018 Pearson Education, Inc.

23.
$$30 = -1 + r$$
$$30 + 1 = -1 + r + 1$$
$$31 = r$$
The solution is 31.

24.
$$3t - 8t = -25$$
$$(3 - 8)t = -25$$
$$-5t = -25$$
$$\frac{-5t}{-5} = \frac{-25}{-5}$$
$$t = 5$$
The solution is 5.

25.
$$\frac{p}{5} = -3$$
$$\frac{1}{5}p = -3$$
$$\left(\frac{5}{1}\right) \cdot \left(\frac{1}{5}\right)p = -3 \cdot 5$$
$$p = -15$$
The solution is -15.

26.
$$-15 = 3(a - 2)$$
$$-15 = 3 \cdot a - 3 \cdot 2$$
$$-15 = 3a - 6$$
$$-15 + 6 = 3a - 6 + 6$$
$$-9 = 3a$$
$$\frac{-9}{3} = \frac{3a}{3}$$
$$-3 = a$$
The solution is -3.

27.
$$3m - 5 = 7m - 13$$
$$3m - 7m - 5 = 7m - 7m - 13$$
$$(3 - 7)m - 5 = -13$$
$$-4m - 5 = -13$$
$$-4m - 5 + 5 = -13 + 5$$
$$-4m = -8$$
$$\frac{-4m}{-4} = \frac{-8}{-4}$$
$$m = 2$$
The solution is 2.

28. *Step 2*
Let x represent the shorter piece and $x + 4$ the longer piece.

Step 3 $x + (x + 4) = 118$

Step 4
$$2x + 4 = 118$$
$$2x + 4 - 4 = 118 - 4$$
$$2x = 114$$
$$\frac{2x}{2} = \frac{114}{2}$$
$$x = 57$$

Step 5
The length of the shorter piece is 57 cm, and the length of the longer piece is $57 + 4 = 61$ cm.

Step 6
Check: 61 cm is 4 cm longer than 57 cm and the sum of 57 cm and 61 cm is 118 cm. *True*

29. *Step 2*
Let w represent the width and $4w$ the length.

$$P = 2 \cdot l + 2 \cdot w$$

Step 3 $420 = 2(4w) + 2w$

Step 4
$$420 = 8w + 2w$$
$$420 = 10w$$
$$\frac{420}{10} = \frac{10w}{10}$$
$$42 = w$$

Step 5
The width is 42 ft. The length is $4(42 \text{ ft}) = 168$ ft.

Step 6
Check: The length, 168, is four times the width, 42. Now check the perimeter.
$$420 = 2(4w) + 2w$$
$$420 = 2(4 \cdot 42) + 2(42)$$
$$420 = 2(168) + 84$$
$$420 = 336 + 84$$
$$420 = 420 \qquad \textit{True}$$

30. *Step 2*
Let x represent Marcella's time and $x - 3$ Tim's time.

Hours Marcella spent	+	hours Tim spent	=	19 hours
↓	↓	↓	↓	↓
x	+	$x - 3$	=	19

Step 3 $x + (x - 3) = 19$

(continued)

Copyright © 2018 Pearson Education, Inc.

Step 4
$$2x - 3 = 19$$
$$2x - 3 + 3 = 19 + 3$$
$$2x = 22$$
$$\frac{2x}{2} = \frac{22}{2}$$
$$x = 11$$

Step 5
Marcella spent 11 hours redecorating their living room and Tim spent $11 - 3 = 8$ hours.

Step 6
Check: Tim's time is 3 hours less than Marcella's time and the sum of their times is 19 hours. *True*

Cumulative Review Exercises (Chapters 1–9)

1. *Estimate:*

$$\begin{array}{r} 5\,0\,0 \\ 80\overline{)4\,0,0\,0\,0} \end{array}$$

Exact:

$$\begin{array}{r} 5\,0\,3 \\ 78\overline{)3\,9,2\,3\,4} \\ 3\,9\,0 \\ \hline 2\,3 \\ 0 \\ \hline 2\,3\,4 \\ 2\,3\,4 \\ \hline 0 \end{array}$$

2. *Exact:*

$$5\frac{5}{6} \cdot \frac{9}{10} = \frac{\overset{7}{\cancel{35}}}{\cancel{6}} \cdot \frac{\overset{3}{\cancel{9}}}{\cancel{10}} = \frac{21}{4} = 5\frac{1}{4}$$

Estimate: $\underline{6} \cdot \underline{1} = \underline{6}$

3. *Exact:*

$$4\frac{1}{6} \div 1\frac{2}{3} = \frac{25}{6} \div \frac{5}{3} = \frac{\overset{5}{\cancel{25}}}{\cancel{6}} \cdot \frac{\overset{1}{\cancel{3}}}{\cancel{5}} = \frac{5}{2} = 2\frac{1}{2}$$

Estimate: $\underline{4} \div \underline{2} = \underline{2}$

4. $17 - 8.094$

$$\begin{array}{r} {}^{0\;16\;9\;9\;10} \\ \cancel{1\,7}.\cancel{0}\,\cancel{0}\,\cancel{0} \\ -\;\;8.0\,9\,4 \\ \hline 8.9\,0\,6 \end{array}$$

Check:
$$\begin{array}{r} 8.094 \\ +\,8.906 \\ \hline 17.000 \end{array}$$

5. $4.06 \div 0.072$

$$\begin{array}{r} 5\,6.\,3\,8 \\ 0.072_\wedge\overline{)4.\,0\,6\,0_\wedge0\,0} \\ 3\,6\,0 \\ \hline 4\,6\,0 \\ 4\,3\,2 \\ \hline 2\,8\;\;0 \\ 2\,1\;\;6 \\ \hline 6\;\;4\;0 \\ 5\;\;7\;6 \\ \hline 6\;\;4 \end{array}$$

56.38 rounded to the nearest tenth is 56.4.

6. $\dfrac{30}{-6} = -5$

7. $-3 - (-7) = -3 + (+7) = 4$

8. $3.2 + (-4.5) = -1.3$

9. $\dfrac{1}{4} - \dfrac{3}{4} = \dfrac{1}{4} + \left(-\dfrac{3}{4}\right) = -\dfrac{2}{4} = -\dfrac{1}{2}$

10. $45 \div \sqrt{25} - 2(3) + (10 \div 5)$
$$= 45 \div \sqrt{25} - 2(3) + 2$$
$$= 45 \div 5 - 2(3) + 2$$
$$= 9 - 2(3) + 2$$
$$= 9 - 6 + 2$$
$$= 3 + 2$$
$$= 5$$

11. $-6 - (4 - 5) + (-3)^2$
$$= -6 - (-1) + (-3)^2$$
$$= -6 - (-1) + 9$$
$$= -6 + (+1) + 9$$
$$= -5 + 9$$
$$= 4$$

12. $$\dfrac{x}{12} = \dfrac{1.5}{45}$$
$$45 \cdot x = 12 \cdot 1.5$$
$$\dfrac{45 \cdot x}{45} = \dfrac{18}{45}$$
$$x = \dfrac{18}{45} = \dfrac{18 \div 9}{45 \div 9} = \dfrac{2}{5} = 0.4$$

13. part is 90; percent is 180; whole is unknown. Use the percent proportion.

$$\dfrac{90}{x} = \dfrac{180}{100} \quad \textbf{OR} \quad \dfrac{90}{x} = \dfrac{9}{5}$$
$$9 \cdot x = 90 \cdot 5$$
$$\dfrac{9 \cdot x}{9} = \dfrac{450}{9}$$
$$x = 50$$

90 cars is 180% of 50 cars.

Copyright © 2018 Pearson Education, Inc.

14. part is 5.8; whole is 145; percent is unknown. Use the percent proportion.

$$\frac{5.8}{145} = \frac{x}{100}$$
$$145 \cdot x = 5.8 \cdot 100$$
$$\frac{145 \cdot x}{145} = \frac{580}{145}$$
$$x = 4$$

$5.80 is 4% of $145.

15. $3\frac{1}{2}$ gal to quarts

$$\frac{3\frac{1}{2} \text{ gallons}}{1} \cdot \frac{4 \text{ quarts}}{1 \text{ gallon}} = \frac{7}{2} \cdot \frac{\overset{2}{\cancel{4}}}{1} \text{ quarts}$$
$$\underset{1}{} = 14 \text{ quarts}$$

16. 72 hr to days

$$\frac{72 \text{ hours}}{1} \cdot \frac{1 \text{ day}}{24 \text{ hours}} = \frac{72 \cdot 1 \text{ day}}{24} = 3 \text{ days}$$

17. 3.75 kg to grams

$$\frac{3.75 \text{ kg}}{1} \cdot \frac{1000 \text{ g}}{1 \text{ kg}} = 3.75 \cdot 1000 \text{ g}$$
$$= 3750 \text{ g}$$

18. 40 cm to meters

$$\frac{40 \text{ cm}}{1} \cdot \frac{1 \text{ m}}{100 \text{ cm}} = \frac{40 \text{ m}}{100} = 0.4 \text{ m}$$

19. The shape is a square.

$$P = 4 \cdot s$$
$$= 4 \cdot 2\frac{1}{4} \text{ ft}$$
$$= \frac{\overset{1}{\cancel{4}}}{1} \cdot \frac{9}{\underset{1}{\cancel{4}}} \text{ ft} = 9 \text{ ft}$$

$$A = s \cdot s$$
$$= 2\frac{1}{4} \text{ ft} \cdot 2\frac{1}{4} \text{ ft}$$
$$= \frac{9}{4} \text{ ft} \cdot \frac{9}{4} \text{ ft}$$
$$= \frac{81}{16} \text{ ft}^2$$
$$= 5\frac{1}{16} \text{ ft}^2 \text{ or } 5.1 \text{ ft}^2 \text{ (rounded)}$$

20. The shape is a circle.

$$C = \pi \cdot d$$
$$\approx 3.14 \cdot 9 \text{ mm}$$
$$= 28.26 \text{ mm } (\approx 28.3)$$

$$A = \pi r^2$$
$$\approx 3.14 \cdot 4.5 \text{ mm} \cdot 4.5 \text{ mm}$$
$$= 63.585 \text{ mm}^2 \ (\approx 63.6)$$

21. The shape is a right triangle.

$$P = a + b + c$$
$$= 24 \text{ mi} + 7 \text{ mi} + 25 \text{ mi}$$
$$= 56 \text{ mi}$$

$$A = \frac{1}{2}bh$$
$$= \frac{1}{2} \cdot 24 \text{ mi} \cdot 7 \text{ mi}$$
$$= 12 \text{ mi} \cdot 7 \text{ mi}$$
$$= 84 \text{ mi}^2$$

22. The shape is a parallelogram.

$$P = 2 \cdot l + 2 \cdot w$$
$$= 2 \cdot 10 \text{ ft} + 2 \cdot 8 \text{ ft}$$
$$= 20 \text{ ft} + 16 \text{ ft}$$
$$= 36 \text{ ft}$$

$$A = bh$$
$$= 10 \text{ ft} \cdot 7 \text{ ft}$$
$$= 70 \text{ ft}^2$$

23.
$$-2t - 6t = 40$$
$$-2t + (-6t) = 40$$
$$-8t = 40$$
$$\frac{-8t}{-8} = \frac{40}{-8}$$
$$t = -5$$

The solution is -5.

24.
$$3x + 5 = 5x - 11$$
$$3x + 5 + 11 = 5x - 11 + 11$$
$$3x + 16 = 5x$$
$$3x + 16 - 3x = 5x - 3x$$
$$16 = 2x$$
$$\frac{16}{2} = \frac{2x}{2}$$
$$8 = x$$

The solution is 8.

25.
$$6(p + 3) = -6$$
$$6(p) + 6(3) = -6$$
$$6p + 18 = -6$$
$$6p + 18 - 18 = -6 - 18$$
$$6p = -24$$
$$\frac{6p}{6} = \frac{-24}{6}$$
$$p = -4$$

The solution is -4.

Copyright © 2018 Pearson Education, Inc.

26.   Let x represent the number.

40	added to	4 times a number	is	0.
↓	↓	↓	↓	↓
40	+	$4x$	=	0

$$40 + 4x = 0$$
$$40 - 40 + 4x = 0 - 40$$
$$4x = -40$$
$$\frac{4x}{4} = \frac{-40}{4}$$
$$x = -10$$

The number is -10.

Check: If 40 is added to four times $-10[= -40]$, the result is zero. *True*

27.   *Step 2*
Let m represent the money Reggie will receive and $m + 300$ the money Donald will receive.

Step 3 $m + (m + 300) = 1000$

Step 4
$$2m + 300 = 1000$$
$$2m + 300 - 300 = 1000 - 300$$
$$2m = 700$$
$$\frac{2m}{2} = \frac{700}{2}$$
$$m = 350$$

Step 5
Reggie will receive \$350 and Donald will receive $\$350 + \$300 = \$650$.

Step 6
Check: \$650 is \$300 more than \$350 and the sum of \$650 and \$350 is \$1000. *True*

28.   *Step 2*
Make a drawing.

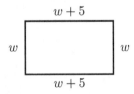

Let w represent the width and $w + 5$ the length. Use $P = 2 \cdot l + 2 \cdot w$.

Step 3 $82 = 2 \cdot (w + 5) + 2 \cdot w$

Step 4
$$82 = 2w + 10 + 2w$$
$$82 = 4w + 10$$
$$82 - 10 = 4w + 10 - 10$$
$$72 = 4w$$
$$\frac{72}{4} = \frac{4w}{4}$$
$$18 = w$$

Step 5
The width is 18 cm and the length is $18 + 5 = 23$ cm.

Step 6
Check: $82 = 2 \cdot 23 + 2 \cdot 18$
$$82 = 46 + 36$$
$$82 = 82 \qquad\qquad \textit{True}$$

29.   Hypotenuse: 20 yd, leg: 15 yd

$$y = \sqrt{(\text{hypotenuse})^2 - (\text{leg})^2}$$
$$= \sqrt{(20)^2 - (15)^2}$$
$$= \sqrt{400 - 225}$$
$$= \sqrt{175}$$
$$\approx 13.2 \text{ yd (rounded)}$$

30.   Find the cost for two Blu-rays.
$$\$14.98 + \$14.98 = \$29.96$$

$$\text{Sales Tax} = \$29.96 \cdot 6\tfrac{1}{2}\%$$
$$= \$29.96(0.065)$$
$$\approx \$1.95$$

$$\text{Total cost} = \$29.96 + \$1.95 = \$31.91 \text{ (rounded)}$$

31.   Write a proportion.

$$\frac{25}{14} = \frac{x}{30}$$
$$14 \cdot x = 25 \cdot 30$$
$$14x = 750$$
$$\frac{14x}{14} = \frac{750}{14}$$
$$x \approx 53.571$$

It will take Rich approximately 54 minutes to read 30 pages.

32.   Find the volume of the crate.

$$V = l \cdot w \cdot h$$
$$= 2.4 \text{ m} \cdot 1.2 \text{ m} \cdot 1.2 \text{ m}$$
$$= 3.456 \text{ m}^3$$

The crate's volume is
$4 - 3.456 \text{ km} = 0.544 \text{ m}^3$ less than 4 m^3.

Copyright © 2018 Pearson Education, Inc.

33. Jackie: $\dfrac{364 \text{ mi}}{14.5 \text{ gal}} \approx 25.1 \text{ mpg}$

Maya: $\dfrac{406 \text{ mi}}{16.3 \text{ gal}} \approx 24.9 \text{ mpg}$

Naomi: $\dfrac{300 \text{ mi}}{11.9 \text{ gal}} \approx 25.2 \text{ mpg} \; (*)$

Naomi's car had the highest number of miles per gallon with 25.2 miles per gallon (rounded).

34. Part is 2480; whole is 2000; percent is unknown. Use the percent proportion.

$$\frac{2480}{2000} = \frac{x}{100} \quad \textbf{OR} \quad \frac{31}{25} = \frac{x}{100}$$
$$25 \cdot x = 31 \cdot 100$$
$$\frac{25 \cdot x}{25} = \frac{3100}{25}$$
$$x = 124$$

2480 pounds is 124% of 2000 pounds.

35. Multiply $3\frac{1}{3}$ cups times $2\frac{1}{2}$.

$$3\frac{1}{3} \cdot 2\frac{1}{2} = \frac{\overset{5}{\cancel{10}}}{3} \cdot \frac{5}{\underset{1}{\cancel{2}}} = \frac{25}{3} = 8\frac{1}{3}$$

He needs $8\frac{1}{3}$ cups of tomato sauce.

Copyright © 2018 Pearson Education, Inc.

CHAPTER 10   STATISTICS

10.1   Circle Graphs

10.1   Margin Exercises

1. **(a)** The circle graph shows that the greatest number of hours is spent sleeping.

    (b)
    $$\begin{array}{r} 6 \text{ hours working} \\ -\,4 \text{ hours studying} \\ \hline 2 \text{ hours} \end{array}$$

    Two more hours are spent working than studying.

    (c)
    $$\begin{array}{r} 4 \text{ hours studying} \\ 6 \text{ hours working} \\ +\,3 \text{ hours attending class} \\ \hline 13 \text{ hours} \end{array}$$

    Thirteen hours are spent studying, working, and attending classes.

2. **(a)** Hours spent driving to whole day:
    $$\frac{2 \text{ hours (driving)}}{24 \text{ hours (whole day)}} = \frac{2 \div 2}{24 \div 2} = \frac{1}{12}$$

    (b) Hours spent sleeping to whole day:
    $$\frac{7 \text{ hours (sleeping)}}{24 \text{ hours (whole day)}} = \frac{7}{24}$$

    (c) Hours spent attending class and studying to whole day:
    $$\begin{array}{r} 3 \text{ hours attending class} \\ +\,4 \text{ hours studying} \\ \hline 7 \text{ hours} \end{array}$$
    $$\frac{7 \text{ hours}}{24 \text{ hours}} = \frac{7}{24}$$

    (d) Hours spent driving and working to whole day:
    $$\begin{array}{r} 2 \text{ hours driving} \\ +\,6 \text{ hours working} \\ \hline 8 \text{ hours} \end{array}$$
    $$\frac{8 \text{ hours}}{24 \text{ hours}} = \frac{8 \div 8}{24 \div 8} = \frac{1}{3}$$

3. **(a)** Hours spent in class to hours spent studying:
    $$\frac{3 \text{ hours (class)}}{4 \text{ hours (studying)}} = \frac{3}{4}$$

    (b) Hours spent working to hours spent sleeping:
    $$\frac{6 \text{ hours (working)}}{7 \text{ hours (sleeping)}} = \frac{6}{7}$$

    (c) Hours spent driving to hours spent working:
    $$\frac{2 \text{ hours (driving)}}{6 \text{ hours (working)}} = \frac{2 \div 2}{6 \div 2} = \frac{1}{3}$$

    (d) Hours spent in class to hours spent for "Other:"
    $$\frac{3 \text{ hours (class)}}{2 \text{ hours (other)}} = \frac{3}{2}$$

4. Use the percent equation:
    $$\text{part} = \text{percent} \cdot \text{whole}$$

    (a) whole = 36 billion;
    percent = 5.1% = 0.051; find the part.
    $$x = (0.051)(36 \text{ billion}) = \underline{1.836 \text{ billion}}$$
    The amount spent on corn chips is $1.836 billion or $1,836,000,000.

    (b) whole = 36 billion;
    percent = 14.4% = 0.144; find the part.
    $$x = (0.144)(36 \text{ billion}) = 5.184 \text{ billion}$$
    The amount spent on miscellaneous is $5.184 billion or $5,184,000,000.

    (c) whole = 36 billion;
    percent = 5.3% = 0.053; find the part.
    $$x = (0.053)(36 \text{ billion}) = 1.908 \text{ billion}$$
    The amount spent on cakes/pies is $1.908 billion or $1,908,000,000.

    (d) whole = 36 billion;
    percent = 23.8% = 0.238; find the part.
    $$x = (0.238)(36 \text{ billion}) = 8.568 \text{ billion}$$
    The amount spent on cookies/crackers is $8.568 billion or $8,568,000,000.

5. **(a)** Working poor:
    $$(360°)(10\%) = (360°)\underline{(0.10)} = \underline{36°}$$

    (b) Ex-offenders:
    $$(360°)(10\%) = (360°)(0.10) = 36°$$

    (c) At-risk youth:
    $$(360°)(5\%) = (360°)(0.05) = 18°$$

Copyright © 2018 Pearson Education, Inc.

(d) Unemployed:

$$(360°)(35\%) = (360°)(0.35) = 126°$$

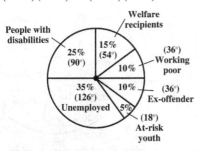

10.1 Section Exercises

1. A circle graph shows how a total amount is divided into <u>parts</u>.

3. Start with "Fish" and add in a clockwise direction.

$$\underbrace{150 + 90 + 75 + 18 + 16 + 11}_{\text{in millions}} = 360$$

The number of pets owned in the United States is

360 million or 360,000,000.

5. From Exercise 3, the total number of pets is 360 million.

$$\frac{\text{number of cats}}{\text{number of pets}} = \frac{90 \text{ million}}{360 \text{ million}}$$
$$= \frac{90 \div 90}{360 \div 90} = \frac{1}{4}$$

7. $\frac{\text{number of cats}}{\text{number of dogs}} = \frac{90 \text{ million}}{75 \text{ million}}$
$$= \frac{90 \div 15}{75 \div 15} = \frac{6}{5}$$

9. The largest sector represents "Sleeping." Americans spend 8 hours 40 minutes per day on this activity.

11. Total minutes per day
$$= (24 \text{ hours})(60 \text{ minutes per hour})$$
$$= \underline{1440} \text{ minutes}$$

Total minutes sleeping per day
$$= (8 \text{ hours})(60 \text{ minutes per hour}) + 40$$
$$= 480 + 40$$
$$= 520 \text{ minutes}$$

"Sleeping" to total:

$$\frac{520 \text{ minutes}}{1440 \text{ minutes}} = \frac{520 \div 40}{1440 \div 40} = \frac{13}{36}$$

13. "Educational/civic/religious activities" to "Eating and drinking":

$$\frac{50 \text{ minutes}}{1 \text{ hour 16 minutes}} = \frac{50 \text{ minutes}}{76 \text{ minutes}}$$
$$= \frac{50 \div 2}{76 \div 2} = \frac{25}{38}$$

15. Time working compared to household duties/caring for others:

$$\frac{3 \text{ hours 30 minutes}}{2 \text{ hours 30 minutes}} = \frac{210 \text{ minutes}}{150 \text{ minutes}}$$
$$= \frac{210 \div 30}{150 \div 30} = \frac{7}{5}$$

17. To determine how many people prefer onions for their hot dog topping, use the percent equation.

$$\text{part} = \text{percent} \cdot \text{whole}$$
$$x = (0.05)(3200)$$
$$= 160 \text{ people}$$

19. Sauerkraut = 3% of 3200
$$x = (0.03)(3200)$$
$$= 96 \text{ people}$$

21. Mustard = 30% of 3200
$$x = (0.30)(3200)$$
$$= 960 \text{ people}$$

23. Need the car to go 400 miles or more:

$$x = 34\% \text{ of } 3021$$
$$= (0.34)(3021)$$
$$= 1027 \text{ people (rounded)}$$

25. Need the car to go 50 miles or fewer:

$$x = 4\% \text{ of } 3021$$
$$= (0.04)(3021)$$
$$= 121 \text{ people (rounded)}$$

27. Need the car to go 100 miles:

$$x = 16\% \text{ of } 3021$$
$$= (0.16)(3021)$$
$$= 483 \text{ people (rounded)}$$

29. First find the percent of the total that is to be represented by each item. Next, multiply the percent by 360° to find the size of each sector. Finally, use a protractor to draw each sector.

31. 25% of the total is Adult sports.

$$\text{Degrees of a circle} = 25\% \text{ of } 360°$$
$$= (0.25)(360)$$
$$= 90°$$

Copyright © 2018 Pearson Education, Inc.

33. From the text, $5460 is the total amount spent during one month.

$$\text{Percent for Day camp} = \frac{\$546}{\$5460}$$
$$= 0.10 = 10\%$$

$$\text{Degrees of a circle} = 10\% \text{ of } 360°$$
$$= (0.10)(360°)$$
$$= 36°$$

35. 15% of the total is Annual egg hunt.

$$\text{Degrees of a circle} = 15\% \text{ of } 360°$$
$$= (0.15)(360°)$$
$$= 54°$$

37. Percent for Mommy and baby exercise
$$= \frac{54°}{360°}$$
$$= 0.15 = 15\%$$

39. **(a)** Total sales = $12,500 + $40,000 + $60,000 + $50,000 + $37,500 = $200,000

(b) Adventure classes = $12,500
$$\text{percent of total} = \frac{12,500}{200,000} = 0.0625 = 6.25\%$$
$$\text{number of degrees} = (0.0625)(360°) = 22.5°$$

Grocery and provision sales = $40,000
$$\text{percent of total} = \frac{40,000}{200,000} = 0.2 = 20\%$$
$$\text{number of degrees} = (0.2)(360°) = 72°$$

Equipment rentals = $60,000
$$\text{percent of total} = \frac{60,000}{200,000} = 0.3 = 30\%$$
$$\text{number of degrees} = (0.3)(360°) = 108°$$

Rafting tours = $50,000
$$\text{percent of total} = \frac{50,000}{200,000} = 0.25 = 25\%$$
$$\text{number of degrees} = (0.25)(360°) = 90°$$

Equipment sales = $37,500
$$\text{percent of total} = \frac{37,500}{200,000} = 0.1875 = 18.75\%$$
$$\text{number of degrees} = (0.1875)(360°) = 67.5°$$

(c)

41. The total number of adults surveyed is

$$2578 + 1813 + 1155 + 749 + 469 + 266 = 7030.$$

(a) $$\text{Percent for Work/school} = \frac{2578}{7030} \approx 0.37 = 37\%$$

$$\text{Degrees for Work/school} = (0.37)(360°) \approx 133°$$

$$\text{Percent for Place of Worship} = \frac{266}{7030} \approx 0.04 = 4\%$$

$$\text{Degrees for Place of Worship} = (0.04)(360°) \approx 14°$$

$$\text{Percent for Online dating site} = \frac{1155}{7030} \approx 0.16 = 16\%$$

$$\text{Degrees for Online dating site} = (0.16)(360°) \approx 58°$$

$$\text{Percent for Friend/family member} = \frac{1813}{7030} \approx 0.26 = 26\%$$

$$\text{Degrees for Friend/family member} = (0.26)(360°) \approx 94°$$

$$\text{Percent for Bar/club/social event} = \frac{749}{7030} \approx 0.11 = 11\%$$

$$\text{Degrees for Bar/club/social event} = (0.11)(360°) \approx 40°$$

$$\text{Percent for Other} = \frac{469}{7030} \approx 0.07 = 7\%$$

$$\text{Degrees for Other} = (0.07)(360°) \approx 25°$$

How They Met	Number in Survey
Work/school	2578
Place of Worship	266
Online dating site	1155
Friend/family member	1813
Bar/club/social event	749
Other	469

(continued)

Copyright © 2018 Pearson Education, Inc.

How They Met	Percent of Total	Number of Degrees
Work/school	37% (≈)	133° (≈)
Place of Worship	4% (≈)	14° (≈)
Online dating site	16% (≈)	58° (≈)
Friend/family member	26% (≈)	94° (≈)
Bar/club/social event	11% (≈)	40° (≈)
Other	7% (≈)	25° (≈)

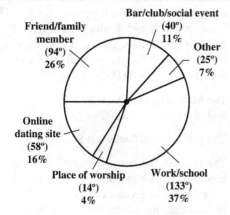

(b) Total percent =

$$37\% + 4\% + 16\% + 26\% + 11\% + 7\% = 101\%$$

No. The total is 101% due to rounding.

(c) Total degrees =

$$133° + 14° + 58° + 94° + 40° + 25° = 364°$$

No. The total is 364° due to rounding.

10.2   Bar Graphs and Line Graphs

10.2   Margin Exercises

1. **(a)** The bar for 2014 rises to 20, showing that the number of members was

$$20 \times 1000 = 20,000.$$

(b) 2013: $25 \times 1000 = 25,000$ members

(c) 2017: $35 \times 1000 = 35,000$ members

(d) 2016: $30 \times 1000 = 30,000$ members

2. **(a)** 1st quarter, 2016:

$400 \times 1000 = \underline{400,000 \text{ subscribers}}$

1st quarter, 2017:

$700 \times 1000 = \underline{700,000 \text{ subscribers}}$

(b) 3rd quarter, 2016:

$400 \times 1000 = 400,000$ subscribers

3rd quarter, 2017:

$500 \times 1000 = 500,000$ subscribers

(c) 4th quarter, 2016:

$700 \times 1000 = 700,000$ subscribers

4th quarter, 2017:

$800 \times 1000 = 800,000$ subscribers

(d) The tallest bar rose to 800, showing 800,000 subscribers for the 4th quarter of 2017.

3. **(a)** June: The dot is at 5.5 on the vertical axis.
$5.5 \times 10,000 = 55,000$ trout

(b) May: $3 \times 10,000 = 30,000$ trout

(c) April: $4 \times 10,000 = 40,000$ trout

(d) July: $6 \times 10,000 = 60,000$ trout

4. **(a)** The number of desktop computers sold:

2012: $30 \times 1000 = \underline{30,000}$
2014: $40 \times 1000 = \underline{40,000}$
2015: $20 \times 1000 = \underline{20,000}$
2016: $15 \times 1000 = \underline{15,000}$

(b) The number of laptop computers sold:

2012: $10 \times 1000 = 10,000$
2013: $20 \times 1000 = 20,000$
2014: $30 \times 1000 = 30,000$
2015: $50 \times 1000 = 50,000$

(c) The number of laptop computers sold was greater than the number of desktop computers sold when the red line was above the blue line. The first full year in which this happened was 2015.

10.2   Section Exercises

1. A double-bar graph *can* be used to compare two sets of data, so the statement is *false.*

3. The country which had the highest percent of income spent on food is Nigeria, which spent 56.6% of household income.

5. The countries in which less than 15% of household income is spent, on average, for food are the USA, Britain, Germany, Japan, and France.

7. What is 23.3% of 365 days?

$$x = (0.233)(365 \text{ days})$$
$$= \underline{85} \text{ days (rounded)}$$

9. The month of May in 2017 had the greatest number of plants shipped. The total was 10,000 plants shipped.

Copyright © 2018 Pearson Education, Inc.

11. There are two bars for February. One is for 2016 and one is for 2017. The bar for 2017 rises to 7. Multiply 7 by 1000 because the label on the left side of the graph says in thousands. The bar for 2016 rises to halfway between 5 and 6. So multiply 5.5 by 1000.

    Plants shipped in February 2017 = 7000
    Plants shipped in February 2016 = 5500

    $$7000 - 5500 = 1500$$

    There were 1500 more plants shipped in February 2017.

13. The number of plants shipped increased from 5500 in February 2016 to 9500 in April 2017. The increase was $9500 - 5500 = 4000$ plants shipped.

15. 150,000 gallons of supreme unleaded gasoline were sold in 2013.

17. The greatest difference occurred in 2013. The difference was
    $400,000 - 150,000 = 250,000$ gallons.

19. 700,000 gallons of supreme unleaded gasoline were sold in 2017. 150,000 gallons of supreme unleaded gasoline were sold in 2013.

    $$700,000 - 150,000 = 550,000$$

    There was an increase of 550,000 gallons of supreme unleaded gasoline sales.

21. Above the dot for 1990 is 24.1. Looking at the left edge of the graph, notice that this number is in millions. The number of PCs sold in 1990 was 24.1 million or 24,100,000.

23. The increase in the number of PCs sold in 2015 from the number sold in 2005 is
    $402.1 - 197.4 = 204.7$ million or 204,700,000.

25. Answers will vary. Some possibilities are: greater demand as a result of lower price; more uses and applications for the owner; improved technology; multiple computers in each location; more laptop computers and tablets sold.

27. Store A sold $300 \cdot 1000 = 300,000$ MP3 wireless headphones in 2017.

29. In 2015, the annual sales for Store A were $150 \cdot 1000 = 150,000$ wireless headphones.

31. Find the dot above 2016 on the line for store B. Use a ruler or straightedge to line up the dot with the numbers along the left edge. The dot is aligned with 350, so chain store B sold $350 \cdot 1000 = 350,000$ wireless headphones in 2016.

33. Probably Store B with greater sales. Predicted sales might be 450,000 wireless headphones to 500,000 wireless headphones in 2018.

35. A single bar or a single line must be used for each set of data. To show multiple sets of data, multiple sets of bars or lines must be used.

37. Find the dot above 2017 on the line for total sales. Use a ruler or straightedge to line up the dot with the numbers along the left edge. The dot is aligned with 400. Multiply 400 by 1000 because the label on the left side of the graph say *in thousands*. The total sales in 2017 were $400 \cdot \$1000 = \$400,000$.

39. The total sales in 2015 were $250 \cdot \$1000 = \$250,000$.

41. The profit in 2016 was $50 \cdot \$1000 = \$50,000$.

43. The answers will vary. Some possibilities are that the decrease in sales may have resulted from poor service or greater competition. The increase in sales may have been a result of more advertising or better service.

45. The five-flavor roll was invented in 1935. The Pep-O-Mint flavor was invented in 1912. $1935 - 1912 = 23$, so the five-flavor roll was invented 23 years after the first Life Savers flavor was invented.

46. There are a total of 25 flavors today. $25 - 5 = 20$, so there are 20 flavors in addition to the five of the five-flavor roll.

47. Each day 3,000,000 Life Savers candies are produced. Each roll contains 14 candies. Therefore, the number of rolls produced daily is

    $$\frac{3,000,000}{14} \approx 214,286.$$

48. The number of pounds of sugar used each day in the making of Life Savers equals 250,000. According to Exercise 47, 214,286 rolls are produced each day. So, the amount of sugar in one roll of Life Savers equals roughly

    $$\frac{250,000}{214,286} \approx 1.17 \text{ lb (rounded).}$$

49. The weight of one roll of Life Savers is much less than 1.17 lb. The answer is "correct" using the information given, but some of the data given must not be accurate. Perhaps 3 million **rolls** of Life Savers are produced daily.

Copyright © 2018 Pearson Education, Inc.

50. Answers will vary. Possible answers are: misprints or typographical errors; careless reporting of data; math errors.

Summary Exercises *Graphs*

1. Start with "Vanilla" and add in a clockwise direction.

$$660 + 380 + 140 + 100 + 80 + 640 = 2000$$

The number of people in the survey was 2000.

3. From Exercise 1, the total number of people in the survey was 2000.

$$\frac{\text{prefer chocolate}}{\text{total number}} = \frac{380}{2000} = \frac{380 \div 20}{2000 \div 20} = \frac{19}{100}$$

5.

$$\frac{\text{prefer all other flavors}}{\text{prefer strawberry}} = \frac{640}{80} = \frac{640 \div 80}{80 \div 80} = \frac{8}{1}$$

7. The longest bar corresponds to the year 2100, so that is when the world population is expected to reach its peak. The 11.2 is in billions, so the estimated population will be

11.2 billion or 11,200,000,000 people.

9. Subtract the 2050 value, 9.7, from the 2075 value, 10.5.

$$10.5 - 9.7 = 0.8$$

This number is in billions, so the increase is

0.8 billion or 800,000,000 people.

11. **Live in China:** 11% of 10.5 billion

$$\text{part} = \text{percent} \cdot \text{whole}$$
$$x = (0.11)(10.5 \text{ billion})$$
$$x = 1.155 \text{ billion}$$

1.089 billion or 1,089,000,000 in China

Live in the United States: 4.5% of 9.9 billion

$$\text{part} = \text{percent} \cdot \text{whole}$$
$$x = (0.045)(10.5 \text{ billion})$$
$$x = 0.4725 \text{ billion}$$

0.4725 billion or 472,500,000 in the United States

13. In 1950, the Divorced graph was about $2\frac{1}{2}\%$. Doubling this value would give us 5% and the Divorced graph is clearly above 5%, so yes, the percent of those divorced has more than doubled since 1950.

15. In 2010, about 6% of the people were living together.

10.3 Frequency Distributions and Histograms

10.3 Margin Exercises

1. **(a)** The least number of sales calls made in a week is 20 calls.

(b) The most common number of sales calls made in a week is 30 calls (frequency = 9).

(c) The number of weeks in which 35 calls were made is 4 weeks.

(d) The number of weeks in which 45 calls were made is 5 weeks.

2. **(a)** Add the class frequencies that correspond to the class intervals 20–29, 30–39, and 40–49. Fewer than 50 calls were made in

$$9 + 13 + \underline{13} = \underline{35} \text{ weeks.}$$

(b) Add the class frequencies that correspond to the class intervals 50–59, 60–69, and 70–79. 50 or more calls were made in

$$5 + 4 + 6 = 15 \text{ weeks.}$$

3. **(a)** Add the class frequencies that correspond to the class intervals 60–69 and 70–79. 60 or more calls were made in

$$4 + \underline{6} = \underline{10} \text{ weeks.}$$

(b) Add the class frequencies that correspond to the class intervals 20–29, 30–39, 40–49, and 50–59. Fewer than 60 calls were made in

$$9 + 13 + 13 + 5 = 40 \text{ weeks.}$$

10.3 Section Exercises

1. A large number of data can be easier to understand if they are in a special type of <u>table</u> called a frequency distribution.

3. Because the bar associated with 26–34 is the highest, the age group of 26–34 years old has the greatest number of users, which is 45 million users.

5. To find the number of users 35 years of age and over, add the number of users in the 35–44, 45–54, and 55–65 groups.

$$30 + 25 + 30 = 85 \text{ million users}$$
$$\text{or } 85,000,000 \text{ users}$$

7. The number of users in the 35–44 years age group is 30 million users or 30,000,000 users.

Copyright © 2018 Pearson Education, Inc.

9. The greatest number of employees is in the $4100 to $5000 group. There are <u>16</u> employees in that earnings group.

11. The bar for 31–40 rises to 11, so there are 11 employees who earn $3100 to $4000.

13. The number of employees earning $5000 or less is $6 + 10 + 6 + 11 + 16 = \underline{49}$ employees.

15. Class intervals are the result of combining data into groupings. Class frequency is the number of data items that fit in each class interval. These are used to group data and to have multiple responses (frequency) in a class interval—this makes the data easier to interpret.

17. 140–149: 6 tally marks and a class frequency of 6

19. 160–169: 1 tally mark and a class frequency of 1

21. 180–189: 2 tally marks and a class frequency of 2

23. 0–5: 4 tally marks and a class frequency of 4

25. 11–15: 5 tally marks and a class frequency of 5

27. 21–25: 3 tally marks and a class frequency of 3

29. 31–35: 7 tally mark and a class frequency of 7

31. 70–79: 3 tally marks and a class frequency of 3

33. 90–99: 26 tally marks and a class frequency of 26

35. 110–119: 17 tally marks and a class frequency of 17

37. 130–139: 1 tally mark and a class frequency of 1

10.4   Mean, Median, and Mode

10.4 Margin Exercises

1.
$$\text{Mean} = \frac{\text{sum of all values}}{\text{number of values}}$$
$$= \frac{96 + 98 + 84 + 88 + \underline{82} + \underline{92}}{6}$$
$$= \frac{540}{6}$$
$$= \underline{90}$$

2. **(a)** The sum of the 12 values is
$50.28 + \$85.16 + \$110.50 + \$78 + \$120.70$
$+ \$58.64 + \$73.80 + \$86.24 + \$67.85$
$+ \$96.56 + \$138.65 + \$48.90 = \$1015.28.$

$$\text{Mean} = \frac{\$1015.28}{12} \approx \$84.61 \text{ (rounded)}$$

(b) The sum of the 8 values is
$48,076 + 58,595 + 37,874 + 46,289 + 29,235$
$+ 59,311 + 45,675 + 39,721 = 364,776.$

$$\text{Mean} = \frac{364,776}{8} = 45,597 \text{ people}$$

3.

Value	Frequency	Product
$2	4	$8
$4	6	$24
$6	5	$30
$8	6	$48
$10	12	$120
$12	5	$60
$14	8	$112
$16	4	$64
Totals	50	$466

$$\text{Mean} = \frac{\$466}{50} = \$9.32$$

The mean daily amount spent for lunch was $9.32.

4.

Course	Credits	Grade	Credits · Grade
Mathematics	3	A (= 4)	$3 \cdot 4 = 12$
P.E.	1	C (= 2)	$1 \cdot 2 = 2$
English	3	C (= 2)	$3 \cdot 2 = 6$
Keyboarding	2	B (= 3)	$2 \cdot 3 = 6$
Biology	4	B (= 3)	$4 \cdot 3 = 12$
Totals	13		38

$$\text{GPA} = \frac{\text{sum of Credits} \cdot \text{Grade}}{\text{total number of Credits}} = \frac{38}{13} \approx 2.92$$

5. Arrange the numbers (weights) in numerical order from least to greatest.

$$10, 14, 15, 17, 18, 19, 20$$

The list has 7 numbers. The middle number is the 4th number, so the median is <u>17</u> lb.

6. Arrange the numbers in numerical order from least to greatest.

$$74, 87, 96, 108, 125, 136$$

$$\frac{96 \text{ m} + 108 \text{ m}}{2} = \frac{204 \text{ m}}{2} = 102 \text{ m}$$

7. **(a)** $20, 30, \underline{26}, 24, \underline{26}$

The only number that occurs more than once is 26.
The mode is 26 laps.

(b) $\underline{38}, \underline{45}, 52, 60, \underline{38}, \underline{45}$

Because both 38 minutes and 45 minutes occur two times, each is a mode. This list is *bimodal.*

Copyright © 2018 Pearson Education, Inc.

(c) $1706, $1289, $1653, $1892, $1301, $1782

No number occurs more than once. This list has *no mode*.

10.4 Section Exercises

1. "Another word for mean is average." is a *true* statement.

3. A common use of the weighted mean is to find a student's <u>grade point average</u>.

5. Mean $= \dfrac{\text{sum of all values}}{\text{number of values}}$

$= \dfrac{9 + 14 + 18 + 11 + 12 + 20}{6}$

$= \dfrac{84}{6}$

$= 14$ hours

The mean (average) time spent playing video games each week is 14 hours.

7. Mean $= \dfrac{3.1 + 1.5 + 2.8 + 0.8 + 4.1}{5}$

$= \dfrac{12.3}{5}$

≈ 2.5 in. of rain (rounded)

The mean (average) inches of rain per month is 2.5.

9. Mean

$= \dfrac{\$38{,}500 + \$39{,}720 + \$42{,}183 + \$21{,}982 + \$43{,}250}{5}$

$= \dfrac{\$185{,}635}{5}$

$= \$37{,}127$

The mean (average) annual salary is $37,127.

11. Mean $= \dfrac{\text{sum of all values}}{\text{number of values}}$

$= \dfrac{\begin{array}{c}(18.38 + 168.75 + 28.63 + 72.85 + 39.60 \\ + \ 183.74 + 15.82 + 33.18 + 87.45 \\ + \ 98.72 + 50.70)\end{array}}{11}$

$= \dfrac{797.82}{11} \approx \$72.53$

The average (mean) cost of the prescriptions was $72.53.

13.

Customers Each Hour	Frequency	Product
8	2	$(8 \cdot 2) = 16$
11	12	$(11 \cdot 12) = 132$
15	5	$(15 \cdot 5) = 75$
26	1	$(26 \cdot 1) = \underline{26}$
Totals	20	249

Weighted Mean $= \dfrac{\text{sum of products}}{\text{total number of customers}}$

$= \dfrac{249}{20} \approx 12.5$ customers (rounded)

15.

Fish per Boat	Frequency	Product
12	4	$(12 \cdot 4) = 48$
13	2	$(13 \cdot 2) = 26$
15	5	$(15 \cdot 5) = 75$
19	3	$(19 \cdot 3) = 57$
22	1	$(22 \cdot 1) = 22$
23	5	$(23 \cdot 5) = \underline{115}$
Totals	20	343

Weighted Mean $= \dfrac{\text{sum of products}}{\text{total number of fish}}$

$= \dfrac{343}{20} = 17.15$

≈ 17.2 fish (rounded)

17.

Policy Amount	Number of Policies Sold	Product ($)
$ 5000	5	25,000
$ 10,000	10	100,000
$ 25,000	7	175,000
$ 50,000	5	250,000
$ 100,000	12	1,200,000
$ 250,000	5	1,250,000
$ 500,000	5	2,500,000
$1,000,000	1	1,000,000
Totals	50	6,500,000

Weighted Mean $= \dfrac{\text{sum of products}}{\text{total number of policies}}$

$= \dfrac{6{,}500{,}000}{50} = 130{,}000$

The mean (average) amount of money won was $130,000.

19. Arrange the numbers in numerical order from least to greatest.

$$100, 114, 125, 135, 150$$

The list has 5 numbers. The middle number is the 3rd number, so the median is 125 guests.

Copyright © 2018 Pearson Education, Inc.

21. Arrange the numbers in numerical order from least to greatest.

$$298, 346, 412, \underbrace{501, 515}_{\text{middle two numbers}}, 521, 528, 621$$

The list has 8 numbers. The middle numbers are the 4th and 5th numbers, so the median is

$$\frac{501 + 515}{2} = \frac{1016}{2} = \underline{508} \text{ calories.}$$

23. $\underline{21\%}, 18\%, \underline{21\%}, 28\%, 22\%, \underline{21\%}, 25\%$

The mode is the value that occurs most often. The mode is 21%.

25. $\underline{74}, \underline{68}, \underline{68}, \underline{68}, 75, 75, \underline{74}, \underline{74}, 70$

Because both 68 and 74 occur three times, each is a mode. This list is *bimodal.*

27. The median is a better measure of central tendency when the list contains one or more extreme values. Examples will vary. One possibility is to find the mean and the median of the following home values:

$182,000; $164,000; $191,000; $115,000; $982,000

mean home value = $326,800;
median home value = $182,000

29.

Credits	Grade	Credits · Grade
4	B (= 3)	$4 \cdot 3 = 12$
2	C (= 2)	$2 \cdot 2 = 4$
2	A (= 4)	$2 \cdot 4 = 8$
1	C (= 2)	$1 \cdot 2 = 2$
3	D (= 1)	$3 \cdot 1 = 3$
12		29

$$\text{GPA} = \frac{\text{sum of Credits} \cdot \text{Grade}}{\text{total number of credits}}$$
$$= \frac{29}{12} \approx 2.42 \text{ (rounded)}$$

31.

Credits	Grade	Credits · Grade
4	B (= 3)	$4 \cdot 3 = 12$
2	A (= 4)	$2 \cdot 4 = 8$
5	C (= 2)	$5 \cdot 2 = 10$
1	F (= 0)	$1 \cdot 0 = 0$
3	B (= 3)	$3 \cdot 3 = 9$
15		39

$$\text{GPA} = \frac{\text{sum of Credits} \cdot \text{Grade}}{\text{total number of credits}}$$
$$= \frac{39}{15} = 2.60$$

33.

Credits	Grade	Credits · Grade
2	A (= 4)	$2 \cdot 4 = 8$
3	C (= 2)	$3 \cdot 2 = 6$
4	A (= 4)	$4 \cdot 4 = 16$
1	C (= 2)	$1 \cdot 2 = 2$
4	B (= 3)	$4 \cdot 3 = 12$
14		44

$$\text{GPA} = \frac{\text{sum of Credits} \cdot \text{Grade}}{\text{total number of credits}}$$
$$= \frac{44}{14} \approx 3.14 \text{ (rounded)}$$

35. Samuels total:
$$= 39 + 15 + 40 + 22 + 13 + 22 + 17 + 8$$
$$= 176 \text{ calls}$$

Stricker total:
$$= 21 + 22 + 20 + 23 + 19 + 24 + 25 + 22$$
$$= 176 \text{ calls}$$

Each made 176 sales calls.

36. Mean for Samuels $= \dfrac{176}{8} = 22$

Mean for Stricker $= \dfrac{176}{8} = 22$

37. Arrange the data in numerical order.

Samuels: $8, 13, 15, \underline{17}, \underline{22}, 22, 39, 40$

$$\text{Median} = \frac{17 + 22}{2} = \frac{39}{2} = 19.5$$

Stricker: $19, 20, 21, \underline{22}, \underline{22}, 23, 24, 25$

$$\text{Median} = \frac{22 + 22}{2} = \frac{44}{2} = 22$$

38. The mode is the value that occurs most often.

Mode for Samuels $= 22$

Mode for Stricker $= 22$

39. The mean and mode are identical for both sales representatives, and the medians are close.

40. The number of weekly sales calls made by Samuels varies greatly from week to week while the number of weekly sales calls made by Stricker remains fairly constant.

41. Samuels range $= 40 - 8 = 32$

42. Stricker range $= 25 - 19 = 6$

43. No, not with any certainty. There probably are additional questions that need to be answered, such as the dollar amount of sales, number of repeat customers, and so on.

Copyright © 2018 Pearson Education, Inc.

44. Answers will vary. Some possible answers are: He works hard one week, then takes it easy the next week; the characteristics of the sales territories vary greatly; illness or personal problems may be affecting performance.

Chapter 10 Review Exercises

1. The largest sector represents Basketball teams. There are 18,000 girls basketball teams.

2. The total number of teams is $18{,}000 + 16{,}000 + 15{,}000 + 14{,}000 + 13{,}000 = 76{,}000$.

$$\frac{\text{track and field}}{\text{total teams}} = \frac{16{,}000}{76{,}000} = \frac{16 \div 4}{76 \div 4} = \frac{4}{19}$$

3.
$$\frac{\text{softball teams}}{\text{total teams}} = \frac{15{,}000}{76{,}000} = \frac{15}{76}$$

4.
$$\frac{\text{volleyball teams}}{\text{total teams}} = \frac{14{,}000}{76{,}000} = \frac{14 \div 2}{76 \div 2} = \frac{7}{38}$$

5.
$$\frac{\text{basketball teams}}{\text{track and field}} = \frac{18{,}000}{16{,}000} = \frac{18 \div 2}{16 \div 2} = \frac{9}{8}$$

6.
$$\frac{\text{track and field}}{\text{volleyball teams}} = \frac{16{,}000}{14{,}000} = \frac{16 \div 2}{14 \div 2} = \frac{8}{7}$$

7. Casual dress
82% of 4800 companies
$$x = 0.82 \cdot 4800$$
$$= 3936 \text{ companies}$$

8. Free food/Beverages
36% of 4800 companies
$$x = 0.36 \cdot 4800$$
$$= 1728 \text{ companies}$$

9. Fitness centers
15% of 4800 companies
$$x = 0.15 \cdot 4800$$
$$= 720 \text{ companies}$$

10. Flexible hours
60.5% of 4800 companies
$$x = 0.605 \cdot 4800$$
$$= 2904 \text{ companies}$$

11. On-site child care and Fitness centers
Answers will vary. Perhaps employers feel that they are not needed or would not be used. Or, it may be that they would be too expensive for the benefit derived.

12. Flexible hours and Casual dress
Answers will vary. Perhaps employees request them and appreciate them. Or, it may be that neither of them cost the employer anything to offer.

13. In 2017, the greatest amount of water in the lake occurred in March when there were 8,000,000 acre-feet of water.

14. In 2016, the least amount of water in the lake occurred in June when there were 2,000,000 acre-feet of water.

15. In June 2017, there were 5,000,000 acre-feet of water in the lake.

16. In May 2016, there were 4,000,000 acre-feet of water in the lake.

17. From March 2016 to June 2016, the water went from 7 to 2 million acre-feet, a decrease of 5,000,000 acre-feet of water.

18. From April 2017 to June 2017 the water went from 7 to 5 million acre-feet, a decrease of 2,000,000 acre-feet of water.

19. In 2013, Center A sold
$$50 \cdot \$1{,}000{,}000 = \$50{,}000{,}000$$
worth of floor covering.

20. In 2015, Center A sold
$$20 \cdot \$1{,}000{,}000 = \$20{,}000{,}000$$
worth of floor covering.

21. In 2014, Center B sold
$$20 \cdot \$1{,}000{,}000 = \$20{,}000{,}000$$
worth of floor covering.

22. In 2016, Center B sold
$$40 \cdot \$1{,}000{,}000 = \$40{,}000{,}000$$
worth of floor covering.

23. The floor-covering sales decreased for 2 years and then moved up slightly. Answers will vary. Perhaps there is less new home construction, remodeling, and home improvement in the area near center A, or better product selection and service have reversed the decline in sales.

24. The floor-covering sales are increasing. Answers will vary. Perhaps new construction and home remodeling have increased in the area near center B, or greater advertising has attracted more customers.

Copyright © 2018 Pearson Education, Inc.

25. Mean

$$= \frac{18 + 12 + 15 + 24 + 9 + 42 + 54 + 87 + 21 + 3}{10}$$

$$= \frac{285}{10} = 28.5 \text{ purchases}$$

26. Mean

$$= \frac{31 + 9 + 8 + 22 + 46 + 51 + 48 + 42 + 53 + 42}{10}$$

$$= \frac{352}{10} = 35.2 \text{ matches}$$

27.

Dollar Value	Frequency	Product
$42	3	$126
$47	7	$329
$53	2	$106
$55	3	$165
$59	5	$295
	20	$1021

$$\text{Weighted Mean} = \frac{\$1021}{20} = \$51.05$$

28.

Total Points	Frequency	Product
243	1	243
247	3	741
251	5	1255
255	7	1785
263	4	1052
271	2	542
279	2	558
	24	6176

$$\text{Weighted Mean} = \frac{6176}{24}$$
$$\approx 257.33 \text{ points (rounded)}$$

29. Arrange the numbers in numerical order from least to greatest.

$$13, 28, 35, 37, 39, 43, 54, 68, 75$$

The list has 9 numbers. The middle number is the 5th number, so the median is 39 concerts.

30. Arrange the numbers in numerical order from least to greatest.

$151, $525, $542, $559, $565, $576, $578, $590

The list has 8 numbers. The middle numbers are the 4th and 5th numbers, so the median is

$$\frac{\$559 + \$565}{2} = \$562.$$

31. $79, $56, $110, $79, $72, $86, $79

The value $79 occurs three times, which is more often than any other value. Therefore, $79 is the mode.

32. 18, 25, 63, 32, 28, 37, 32, 26, 18

Because both 18 and 32 contestants occur two times, each is a mode. This list is *bimodal*.

33.

Credits	Grade	Credits · Grade
3	B (= 3)	3·3 = 9
4	A (= 4)	4·4 = 16
2	B (= 3)	2·3 = 6
4	C (= 2)	4·2 = 8
3	A (= 4)	3·4 = 12
16		51

$$\text{GPA} = \frac{\text{sum of Credits} \cdot \text{Grade}}{\text{total number of credits}}$$
$$= \frac{51}{16} \approx 3.19 \text{ (rounded)}$$

34.

Credits	Grade	Credits · Grade
3	A (= 4)	3·4 = 12
4	B (= 3)	4·3 = 12
3	B (= 3)	3·3 = 9
3	A (= 4)	3·4 = 12
1	C (= 2)	1·2 = 2
14		47

$$\text{GPA} = \frac{\text{sum of Credits} \cdot \text{Grade}}{\text{total number of credits}}$$
$$= \frac{47}{14} \approx 3.36 \text{ (rounded)}$$

35. The weighted mean is used to find a student's GPA. The number of credits determines how many times the grade is counted. Semester 1 and Semester 2 have different total credits and different credits for each grade.

Chapter 10 Mixed Review Exercises

1. Books and supplies

Degrees of Circle = 10% of 360°
$$= (0.10)(360°) = 36°$$

2. Rent

The total dollar amount is $17,920.

$$\text{Percent of Total} = \frac{\$6272}{\$17,920} = 0.35 = 35\%$$

We could also use $\frac{126°}{360°} = 0.35$.

3. Food

$$\text{Percent of Total} = \frac{\$3584}{\$17,920} = 0.20 = 20\%$$

Degrees of Circle = 20% of 360°
$$= (0.20)(360°) = 72°$$

Copyright © 2018 Pearson Education, Inc.

4. Tuition/fees

$$\text{Percent of Total} = \frac{\$4480}{\$17,920} = 0.25 = 25\%$$

$$\text{Degrees of Circle} = 25\% \text{ of } 360°$$
$$= (0.25)(360°) = 90°$$

5. Miscellaneous

$$\text{Percent of Total} = \frac{\$1792}{\$17,920} = 0.10 = 10\%$$

$$\text{Degrees of Circle} = 10\% \text{ of } 360°$$
$$= (0.10)(360°) = 36°$$

6.

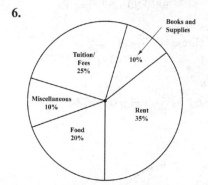

7. $\text{Mean} = \dfrac{48 + 72 + 52 + 148 + 180}{5}$

$$= \frac{500}{5} = 100 \text{ volunteers}$$

8. Mean

$$= \frac{122 + 135 + 146 + 159 + 128 + 147 + 168 + 139 + 158}{9}$$

$$= \frac{1302}{9} \approx 144.7 \text{ vaccinations (rounded)}$$

9. $\underline{48}, 43, 46, 47, \underline{48}, \underline{48}, 43$

The value 48 occurs three times, which is more often than any other value. Therefore, 48 applicants is the mode.

10. $26, \underline{31}, \underline{31}, 37, \underline{43}, 51, \underline{31}, \underline{43}, \underline{43}$

Because both 31 and 43 two-bedroom apartments occur three times, each is a mode. This list is *bimodal*.

11. Arrange the numbers in numerical order from least to greatest.

$$1.0, 2.9, 3.2, 4.7, 5.3, 7.1, 8.2, 9.4$$

The list has 8 numbers. The middle numbers are the 4th and 5th numbers, so the median is

$$\frac{4.7 + 5.3}{2} = 5.0 \text{ hours.}$$

12. Arrange the numbers in numerical order from least to greatest.

$$2, 3, 9, 12, 17, 19, 23, 27, 35, 39, 46, 51$$

The list has 12 numbers. The middle numbers are the 6th and 7th numbers, so the median is

$$\frac{19 + 23}{2} = 21 \text{ miles}$$

13. 30–39: 4 tally marks and a class frequency of 4

14. 40–49: 1 tally mark and a class frequency of 1

15. 50–59: 6 tally marks and a class frequency of 6

16. 60–69: 2 tally marks and a class frequency of 2

17. 70–79: 13 tally marks and a class frequency of 13

18 80–89: 7 tally marks and a class frequency of 7

19. 90–99: 7 tally marks and a class frequency of 7

20. Construct a histogram by using the data in Exercises 13–19.

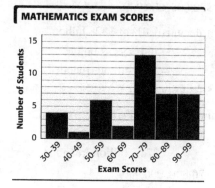

21.

Test Score	Frequency	Product
46	4	184
54	10	540
62	8	496
70	12	840
78	10	780
	44	2840

$$\text{Weighted Mean} = \frac{2840}{44}$$
$$\approx 64.5 \text{ (rounded)}$$

Copyright © 2018 Pearson Education, Inc.

22.

Units Sold	Frequency	Product
104	6	624
112	14	1 568
115	21	2 415
119	13	1 547
123	22	2 706
127	6	762
132	9	1 188
	91	10,810

$$\text{Weighted Mean} = \frac{10{,}810}{91}$$
$$\approx 118.8 \text{ units (rounded)}$$

Chapter 10 Test

1. Nuclear is 20.2% of $328 billion.
$$x = 0.202 \cdot \$328 \text{ billion}$$
$$= \$66.3 \text{ billion (rounded)}$$
or $66,300,000,000

2. Hydroelectric (conventional) is 6.9% of $328 billion.
$$x = 0.069 \cdot \$328 \text{ billion}$$
$$= \$22.6 \text{ billion (rounded)}$$
or $22,600,000,000

3. Petroleum is 1.0% of $328 billion.
$$x = 0.01 \cdot \$328 \text{ billion}$$
$$= \$3.3 \text{ billion (rounded)}$$
or $3,300,000,000

4. Natural gas is 23.4% of $328 billion.
$$x = 0.234 \cdot \$328 \text{ billion}$$
$$= \$76.8 \text{ billion (rounded)}$$
or $76,800,000,000

5. Coal is 44.9% of $328 billion.
$$x = 0.449 \cdot \$328 \text{ billion}$$
$$= \$147.3 \text{ billion (rounded)}$$
or $147,300,000,000

6. Other is 3.6% of $328 billion.
$$x = 0.036 \cdot \$328 \text{ billion}$$
$$= \$11.8 \text{ billion (rounded)}$$
or $11,800,000,000

Note: In Exercises 7–11, we can use either the degrees of a circle (360° total) or the dollar amounts ($1,440,000 total).

7. **Team sports**
Degrees of a Circle = 30% of 360°
$$= (0.30)(360°) = 108°$$

8. **Golf**
Degrees of a Circle = 10% of 360°
$$= (0.10)(360°) = 36°$$

9. **Hunting and fishing**
Degrees of a Circle = 20% of 360°
$$= (0.20)(360°) = 72°$$

10. **Athletic shoes**
Degrees of a Circle = 35% of 360°
$$= (0.35)(360°) = 126°$$

11. **Water sports**
$$\text{Percent of Total} = \frac{18°}{360°} = 0.05 = 5\%$$

12.

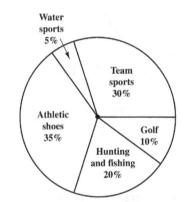

	Profit	Tally	Number of Weeks
13.	$120–129	|||	3
14.	$130–139	||	2
15.	$140–149	||||	4
16.	$150–159	|||	3
17.	$160–169	|||	3
18.	$170–179	|||||	5

Copyright © 2018 Pearson Education, Inc.

19. Construct a histogram by using the data in Exercises 13–18.

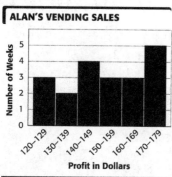

ALAN'S VENDING SALES

20. Mean
$$= \frac{52 + 61 + 68 + 69 + 73 + 75 + 79 + 84 + 91 + 98}{10}$$
$$= \frac{750}{10} = 75 \text{ mi}$$

21. Mean
$$= \frac{11 + 14 + 12 + 14 + 20 + 16 + 17 + 18}{8}$$
$$= \frac{122}{8} \approx 15.3 \text{ lb (rounded)}$$

22. Mean
$$= \frac{\begin{array}{c}(458 + 432 + 496 + 491 + 500 + 508 \\ + 512 + 396 + 492 + 504)\end{array}}{10}$$
$$= \frac{4789}{10} = 478.9 \text{ miles per hour}$$

23. The weighted mean is used because different classes are worth different numbers of credits. GPA problems will vary. One example is shown below.

Credits	Grade	Credits · Grade
3	A (= 4)	$3 \cdot 4 = 12$
2	C (= 2)	$2 \cdot 2 = 4$
4	B (= 3)	$4 \cdot 3 = 12$
9		28

$$\text{GPA} = \frac{\text{sum of Credits} \cdot \text{Grade}}{\text{total number of Credits}} = \frac{28}{9} \approx 3.11$$

24. Arrange the values in order, from least to greatest. When there is an odd number of values in a list, the median is the middle value. Students' problems will vary; one possibility is shown below.

$$8, 17, \underline{23}, 32, 64$$

The median is 23.

25.

Cost	Frequency	Product
$12	5	$ 60
$20	6	$120
$22	8	$176
$28	4	$112
$38	6	$228
$48	2	$ 96
	31	$792

$$\text{Weighted Mean} = \frac{\$792}{31}$$
$$\approx \$26 \text{ (rounded)}$$

26.

Points Scored	Frequency	Product
150	15	2250
160	17	2720
170	21	3570
180	28	5040
190	19	3610
200	7	1400
	107	18,590

$$\text{Weighted Mean} = \frac{18,590}{107}$$
$$\approx 174 \text{ points (rounded)}$$

27. Arrange the numbers in numerical order from least to greatest.

$$16, 28, 28, 31, 32, 35, 37, 41$$

The list has 8 numbers. The middle numbers are the 4th and 5th numbers, so the median is

$$\frac{31 + 32}{2} = 31.5 \text{ degrees.}$$

28. Arrange the numbers in numerical order from least to greatest.

$$6.2, 7.6, 9.1, 9.5, 10.0, 11.4, 12.5, 22.8, 31.7$$

The list has 9 numbers. The middle number is the 5th number, so the median is 10.0 meters.

29. $72, 46, \underline{52}, 37, 28, 18, \underline{52}, 61$

The value 52 occurs two times, which is more often than any other value. Therefore, 52 milliliters is the mode.

30. $96, \underline{104}, \underline{103}, \underline{104}, \underline{103}, \underline{104}, 91, 74, \underline{103}$

Because both 103 and 104 degrees occur three times, each is a mode. This list is *bimodal*.

Copyright © 2018 Pearson Education, Inc.

Cumulative Review Exercises (Chapters 1–10)

1. *Estimate:* *Exact:*

$$
\begin{array}{r}
900 \\
-\ 60 \\
\hline
840
\end{array}
\qquad
\begin{array}{r}
\overset{4\ 15\ 11\ 10}{87\cancel{5}.\cancel{6}\ \cancel{2}\ \cancel{0}} \\
-\ 63.7\ 5\ 7 \\
\hline
811.8\ 6\ 3
\end{array}
$$

2. *Estimate:* *Exact:*

$$
\begin{array}{r}
7000 \\
\times\ \ \ 600 \\
\hline
4{,}200{,}000
\end{array}
\qquad
\begin{array}{r}
7064 \\
\times\ \ \ \ 635 \\
\hline
35\,320 \\
211\,92\ \ \\
4\,238\,4\ \ \ \\
\hline
4{,}485{,}640
\end{array}
$$

3. *Estimate:* *Exact:*

$$
\begin{array}{r}
1\,5 \\
4\,\overline{)6\,0} \\
4\ \ \\
\hline
2\,0 \\
2\,0 \\
\hline
0
\end{array}
$$

$$
\begin{array}{r}
1\,4.\ 7\,2 \\
4.25_\wedge\overline{)6\,2.\,5\,6_\wedge0\,0} \\
4\,2\,5\ \ \ \ \ \ \\
\hline
2\,0\,0\,6\ \ \ \ \\
1\,7\,0\,0\ \ \ \\
\hline
3\,0\,6\,0\ \ \\
2\,9\,7\,5\ \ \\
\hline
8\,5\,0 \\
8\,5\,0 \\
\hline
0
\end{array}
$$

4.
$$
\begin{aligned}
4\tfrac{3}{5} &= 4\tfrac{9}{15} \\
+\ 5\tfrac{2}{3} &= 5\tfrac{10}{15} \\
\hline
9\tfrac{19}{15} &= 10\tfrac{4}{15}
\end{aligned}
$$

5.
$$
\begin{aligned}
6\tfrac{2}{3} &= 6\tfrac{8}{12} = 5\tfrac{20}{12} \\
-\ 4\tfrac{3}{4} &= 4\tfrac{9}{12} = 4\tfrac{9}{12} \\
\hline
& \qquad\qquad\ \ 1\tfrac{11}{12}
\end{aligned}
$$

6. $\left(9\tfrac{3}{5}\right)\cdot\left(4\tfrac{5}{8}\right) = \dfrac{\overset{6}{\cancel{48}}}{5}\cdot\dfrac{37}{\underset{1}{\cancel{8}}} = \dfrac{6\cdot37}{5\cdot1} = \dfrac{222}{5} = 44\tfrac{2}{5}$

7. $22\left(\dfrac{2}{5}\right) = \dfrac{22}{1}\cdot\dfrac{2}{5} = \dfrac{44}{5} = 8\tfrac{4}{5}$

8. $3\tfrac{1}{3} \div 8\tfrac{3}{4} = \dfrac{10}{3} \div \dfrac{35}{4} = \dfrac{\overset{2}{\cancel{10}}}{3}\cdot\dfrac{4}{\underset{7}{\cancel{35}}} = \dfrac{2\cdot4}{3\cdot7} = \dfrac{8}{21}$

9. $\dfrac{2}{3}\left(\dfrac{7}{8} - \dfrac{3}{4}\right) = \dfrac{2}{3}\left(\dfrac{7}{8} - \dfrac{6}{8}\right) = \dfrac{\overset{1}{\cancel{2}}}{3}\cdot\dfrac{1}{\underset{4}{\cancel{8}}} = \dfrac{1}{12}$

10. $4 + 10 \div 2 + 7(2)$
$= 4 + 5 + 7(2)$
$= 4 + 5 + 14$
$= 9 + 14$
$= 23$

11. $\sqrt{81} - 4(2) + 9$
$= 9 - 4(2) + 9$
$= 9 - 8 + 9$
$= 1 + 9$
$= 10$

12. $2^2 \cdot 3^3 = 4 \cdot 27 = 108$

13. $0.218 = 0.2180$
$0.22 = 0.2200$
$0.199 = 0.1990$
$0.207 = 0.2070$
$0.2215 = 0.2215$

From least to greatest:

$0.199, 0.207, 0.218, 0.22, 0.2215$

14. $0.6319 = 0.6319$
$\dfrac{5}{8} = 0.6250$
$0.608 = 0.6080$
$\dfrac{13}{20} = 0.6500$
$0.58 = 0.5800$

From least to greatest:

$0.58, 0.608, \tfrac{5}{8}, 0.6319, \tfrac{13}{20}$

15. $5\tfrac{1}{2}$ in. to 44 in.

$$\dfrac{\tfrac{11}{2}\text{ in.}}{44\text{ in.}} = \dfrac{11}{2} \div \dfrac{44}{1} = \dfrac{\overset{1}{\cancel{11}}}{2}\cdot\dfrac{1}{\underset{4}{\cancel{44}}} = \dfrac{1}{8}$$

16. 3 hr to 45 min

3 hr $= 3(60\text{ min}) = 180$ min

$$\dfrac{180\text{ min}}{45\text{ min}} = \dfrac{180 \div 45}{45 \div 45} = \dfrac{4}{1}$$

17. $\dfrac{1}{5} = \dfrac{x}{30}$
$5\cdot x = 1\cdot30$
$\dfrac{5\cdot x}{5} = \dfrac{30}{5}$
$x = 6$

Copyright © 2018 Pearson Education, Inc.

18. $\dfrac{15}{x} = \dfrac{390}{156}$

$x \cdot 390 = 15 \cdot 156$

$x \cdot 390 = 2340$

$\dfrac{x \cdot 390}{390} = \dfrac{2340}{390}$

$x = 6$

19. $\dfrac{200}{135} = \dfrac{24}{x}$

$200 \cdot x = 24 \cdot 135$

$200x = 3240$

$\dfrac{200x}{200} = \dfrac{3240}{200}$

$x = 16.2$

20. $\dfrac{x}{208} = \dfrac{6.5}{26}$

$26 \cdot x = 6.5 \cdot 208$

$26x = 1352$

$\dfrac{26x}{26} = \dfrac{1352}{26}$

$x = 52$

21. $x = 0.054 \cdot 6000$

$\quad = 324$

5.4% of 6000 homes is 324 homes.

22. Use the percent proportion.
part is 238; percent is 8.5; whole is unknown

$\dfrac{238}{x} = \dfrac{8.5}{100}$

$8.5 \cdot x = 238 \cdot 100$

$\dfrac{8.5x}{8.5} = \dfrac{23,800}{8.5}$

$x = 2800$

$8\frac{1}{2}$% of 2800 people is 238 people.

23. Use the percent proportion.
part is 1443; whole is 555; percent is unknown

$\dfrac{1443}{555} = \dfrac{x}{100}$

$555 \cdot x = 1443 \cdot 100$

$\dfrac{555 \cdot x}{555} = \dfrac{144,300}{555}$

$x = 260$

260% of $555 is $1443.

24. 28 qt = <u>7</u> gal

$\dfrac{28 \text{ qt}}{1} \cdot \dfrac{1 \text{ gal}}{4 \text{ qt}} = 7 \text{ gal}$

25. 400 mm to cm

Count 1 place to the *left* on the metric conversion line.

$\qquad$ 400 mm = 40.0 (or 40) cm

26. 230 g to kg

$\dfrac{230 \text{ g}}{1} \cdot \dfrac{1 \text{ kg}}{1000 \text{ g}} = \dfrac{230}{1000} \text{ kg} = 0.23 \text{ kg}$

27. <u>12,000</u> lb = 6 tons

$\dfrac{6 \text{ tons}}{1} \cdot \dfrac{2000 \text{ lb}}{1 \text{ ton}} = 12,000 \text{ lb}$

28. a rectangle 8.45 m by 5.6 m

$A = l \cdot w$

$\quad = 8.45 \text{ m} \cdot 5.6 \text{ m}$

$\quad = 47.32 \text{ m}^2$

$\quad \approx 47.3 \text{ m}^2 \text{ (rounded)}$

29. a triangle with base 4.25 ft and height 4.5 ft

$A = \frac{1}{2} \cdot b \cdot h$

$\quad = 0.5 \cdot 4.25 \text{ ft} \cdot 4.5 \text{ ft}$

$\quad = 9.5625 \text{ ft}^2$

$\quad \approx 9.6 \text{ ft}^2 \text{ (rounded)}$

30. a circle with diameter of 13 cm

radius $= \frac{13 \text{ cm}}{2} = 6.5$ cm

$A = \pi \cdot r^2$

$\quad \approx 3.14 \cdot 6.5 \text{ cm} \cdot 6.5 \text{ cm}$

$\quad \approx 132.7 \text{ cm}^2 \text{ (rounded)}$

31. a cylinder with radius 4.8 cm and height 7.6 cm

$V = \pi \cdot r^2 \cdot h$

$\quad \approx 3.14 \cdot 4.8 \text{ cm} \cdot 4.8 \text{ cm} \cdot 7.6 \text{ cm}$

$\quad \approx 549.8 \text{ cm}^3 \text{ (rounded)}$

32. a rectangular solid with length 9.5 m, width 3 m, and height 7 m

$V = l \cdot w \cdot h$

$\quad = 9.5 \text{ m} \cdot 3 \text{ m} \cdot 7 \text{ m}$

$\quad = 199.5 \text{ m}^3$

33. hypotenuse $= \sqrt{(\text{leg})^2 + (\text{leg})^2}$

$\qquad\qquad = \sqrt{(30)^2 + (16)^2}$

$\qquad\qquad = \sqrt{900 + 256}$

$\qquad\qquad = \sqrt{1156} = 34$

The hypotenuse is 34 cm.

Copyright © 2018 Pearson Education, Inc.

34. $-12 + (-10) = -22$

35. $-5.7 - (-12.6) = -5.7 + (+12.6) = 6.9$

36. $(-14.6)(-5.7) = 83.22$

37. $\dfrac{-34.04}{14.8} = -2.3$

38.
$$3x - 5 = 16$$
$$3x - 5 + 5 = 16 + 5$$
$$3x = 21$$
$$\frac{3x}{3} = \frac{21}{3}$$
$$x = 7$$

The solution is 7.

39.
$$-12 = 3(x + 2)$$
$$-12 = 3x + 6$$
$$-12 + (-6) = 3x + 6 + (-6)$$
$$-18 = 3x$$
$$\frac{-18}{3} = \frac{3x}{3}$$
$$-6 = x$$

The solution is -6.

40.
$$3.4x + 6 = 1.4x - 8$$
$$3.4x + 6 - 6 = 1.4x - 8 - 6$$
$$3.4x = 1.4x - 14$$
$$3.4x - 1.4x = 1.4x - 1.4x - 14$$
$$2x = -14$$
$$x = -7$$

The solution is -7.

41. Mean
$$= \frac{16 + 37 + 27 + 31 + 19 + 25 + 15 + 38 + 43 + 19}{10}$$
$$= \frac{270}{10} = 27 \text{ units}$$

Arrange the numbers in numerical order from least to greatest.

$$15, 16, 19, 19, 25, 27, 31, 37, 38, 43$$

The list has 10 numbers. The middle numbers are the 5th and 6th numbers, so the median is

$$\frac{25 + 27}{2} = 26 \text{ units}$$

The number 19 occurs two times, which is more often than any other number. Therefore, 19 units is the mode.

42. Mean
$$= \frac{10.3 + 4.3 + 1.65 + 2.85 + 5.3 + 5.7 + 2.3 + 4.35 + 2.85}{9}$$
$$= \frac{39.6}{9} = 4.4 \text{ acres}$$

Arrange the numbers in numerical order from least to greatest.

$$1.65, 2.3, 2.85, 2.85, 4.3, 4.35, 5.3, 5.7, 10.3$$

The list has 9 numbers. The middle number is the 5th number, so the median is 4.3 acres.

The number 2.85 occurs two times, which is more often than any other number. Therefore, 2.85 acres is the mode.

43. part is 874; whole is 2082; percent is unknown

$$\text{part} = \text{percent} \cdot \text{whole}$$
$$874 = x \cdot 2082$$
$$\frac{874}{2082} = \frac{x \cdot 2082}{2082}$$
$$x \approx 42.0$$

About 42.0% of the workers used all of their paid time-off.

44. **(a)** part is \$87.20; percent is 9.8; whole is unknown

$$\text{part} = \text{percent} \cdot \text{whole}$$
$$87.20 = (0.098)(x) \qquad 9.8\% = 0.098$$
$$\frac{87.20}{0.098} = \frac{(0.098)(x)}{0.098}$$
$$x \approx 889.80$$

The average heating bill before the increase was \$889.80 (rounded).

(b) The average heating bill after the increase was

$$\$889.80 + \$87.20 = \$977.00.$$

45. 12 nasal strips for $\$6.50 - \$2$ (coupon) $= \$4.50$

$$\frac{\$4.50}{12} = \$0.375 \text{ per strip}$$

24 nasal strips for \$7.50

$$\frac{\$7.50}{24} \approx \$0.313 \text{ per strip}$$

30 nasal strips for $\$8.95 - \$1$ (coupon) $= \$7.95$

$$\frac{\$7.95}{30} = \$0.265 \text{ per strip}$$

38 nasal strips for \$9.95

$$\frac{\$9.95}{38} \approx \$0.262 \text{ per strip } (*)$$

The best buy is 38 nasal strips for \$9.95.

Copyright © 2018 Pearson Education, Inc.

46. First find the total area.

$$A = l \cdot w$$
$$= 45 \text{ yd} \cdot 32 \text{ yd}$$
$$= 1440 \text{ yd}^2$$

Now find the area of the atrium.

$$A = \pi \cdot r^2$$
$$\approx 3.14 \cdot 5 \text{ yd} \cdot 5 \text{ yd}$$
$$= 78.5 \text{ yd}^2$$

Subtract the area of the atrium from the total area.

$$1440 \text{ yd}^2 - 78.5 \text{ yd}^2 = 1361.5 \text{ yd}^2$$

Multiply the cost of the carpeting by the number of yards needed.

$$\frac{\$43.50}{\text{yd}^2} \cdot \frac{1361.5 \text{ yd}^2}{1} = \$59,225.25$$

The cost is $59,225.25.

47. Change milliliters to liters.

$$\frac{1250 \text{ mL}}{1} \cdot \frac{1 \text{ L}}{1000 \text{ mL}} = \frac{1250}{1000} \text{ L} = 01.25 \text{ L}$$

Multiply the amount of muriatic acid needed for each pool by the number of pools to be serviced.

$$
\begin{array}{r}
01.25 \\
\times \ 140 \\
\hline
50\,00 \\
125 \ \ \ \\
\hline
175.00
\end{array}
$$

To service 140 pools, 175 L of muriatic acid will be needed.

48. $I = p \cdot r \cdot t \quad (1\frac{3}{4}\% = 0.0175)$
$$= \$11,400(0.0175)\left(\frac{8}{12}\right)$$
$$= \$133.00$$

The total amount in her account at maturity is

$$\$11,400 + \$133 = \$11,533.$$

Copyright © 2018 Pearson Education, Inc.

APPENDIX: INDUCTIVE AND DEDUCTIVE REASONING

Margin Exercises

1. $2, 8, 14, 20, \ldots$

   Find the difference between each pair of successive numbers.

   $$8 - 2 = 6$$
   $$14 - 8 = 6$$
   $$20 - 14 = 6$$

   Note that each number is 6 greater than the previous number. So the next number is $20 + 6 = 26$.

2. $6, 11, 7, 12, 8, 13, \ldots$

   The pattern involves addition and subtraction.

   $$6 + 5 = 11$$
   $$11 - 4 = 7$$
   $$7 + 5 = 12$$
   $$12 - 4 = 8$$
   $$8 + 5 = 13$$

   Note that we add 5, then subtract 4. To obtain the next number, 4 should be subtracted from 13, to get 9.

3. $2, 6, 18, 54, \ldots$

   To see the pattern, use division.

   $$6 \div 2 = 3$$
   $$18 \div 6 = 3$$
   $$54 \div 18 = 3$$

   To obtain the next number, multiply the previous number by 3. So the next number is $(54)(3) = 162$.

4. The next figure is obtained by rotating the previous figure clockwise the same amount of rotation ($\frac{1}{4}$ turn) as the second figure was from the first, and as the third figure was from the second.

5. All cars have four wheels.
   All Fords are cars.
   _____
   ∴ All Fords have four wheels.

   The statement "All cars have four wheels" is shown by a large circle that represents all items that have 4 wheels with a small circle inside that represents cars.

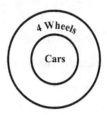

The statement "All Fords are cars" is represented by adding a third circle representing Fords inside the circle representing cars.

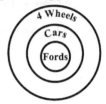

Since the circle representing Fords is completely inside the circle representing items with 4 wheels, it follows that:

   All Fords have four wheels.

The conclusion follows from the premises; it is valid.

6. **(a)** All animals are wild.
   All cats are animals.
   _____
   ∴ All cats are wild.

   "All animals are wild" is represented by a large circle representing wild creatures with a smaller circle inside representing animals.

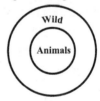

"All cats are animals" is represented by a small circle representing cats, inside the circle representing animals.

Since the circle representing cats is completely inside the circle representing wild creatures, it follows that:

   All cats are wild.

(*continued*)

Copyright © 2018 Pearson Education, Inc.

The conclusion follows from the premises; it is valid. (Note that correct deductive reasoning may lead to false conclusions if one of the premises is false, in this case, all animals are wild.)

(b) All students use math.
All adults use math.

∴ All adults are students.

A larger circle is used to represent people who use math. A small circle inside the larger circle represents students who use math.

Another small circle inside the larger circle represents adults. The circles should overlap since some students could be adults.

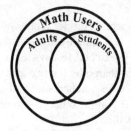

From the diagram, we can see that some adults are not students. Thus, the conclusion does *not* follow from the premises; it is invalid.

7. All 100 students in the class are represented by a large circle.

Students taking history are represented by a small circle inside and students taking math are also represented by a small circle inside. The small circles overlap since some students take both math and history.

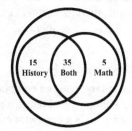

Since 35 students take history and math and 50 take history, $50 - 35 = 15$ students take history, but not math. Since 40 students take math, $40 - 35 = 5$ students take math, but not history. So $15 + 35 + 5 = 55$ students take history, math or both subjects. Therefore, $100 - 55 = 45$ students take neither math nor history.

8. A Chevy, BMW, Cadillac, and Ford are parked side by side.

1. The Ford is on the right side (fact a), so write "Ford" at the right of a line.

Ford

2. The Chevy is next to both the Ford and the Cadillac (fact c). So write "Chevy" between Ford and Cadillac.

Cadillac Chevy Ford

3. The BMW is beside the Cadillac (fact b), so the BMW must be on the other side of the Cadillac.

BMW Cadillac Chevy Ford

Therefore, the BMW is parked on the left end.

Appendix Exercises

1. $2, 9, 16, 23, 30, \ldots$

Inspect the sequence and note that 7 is added to a term to obtain the next term. So the term immediately following 30 is $30 + 7 = 37$.

3. $0, 10, 8, 18, 16, \ldots$

Inspect the sequence and note that 10 is added, then 2 is subtracted, then 10 is added, then 2 is subtracted, and so on. So the term immediately following 16 is $16 + 10 = 26$.

5. $1, 2, 4, 8, \ldots$

Inspect the sequence and note that 2 is multiplied times a term to obtain the next term. So, the term immediately following 8 is $(8)(2) = 16$.

7. $1, 3, 9, 27, 81, \ldots$

Inspect the sequence and note that 3 is multiplied times a term to obtain the next term. So, the term immediately following 81 is $(81)(3) = 243$.

9. $1, 4, 9, 16, 25, \ldots$

Inspect the sequence and note that the pattern is add 3, add 5, add 7, etc., or $1^1, 2^2, 3^2$, etc. So, the term immediately following 25 is $6^2 = 36$.

11. The first three figures are unique. The fourth figure is the same as the first figure except that it is reversed, and reversing the position of the second figure gives the fifth figure. So, the next figure will be the reverse of the third figure.

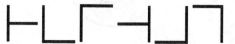

Copyright © 2018 Pearson Education, Inc.

13. The first three figures have the same shape with unique rotations. Since the third figure is the reverse of the first figure, the fourth figure should be a reverse of the second figure.

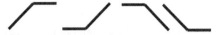

15. All animals are wild.
All lions are animals.
∴ All lions are wild.

The statement "All animals are wild" is shown by a large circle that represents all creatures that are wild with a smaller circle inside that represents animals.

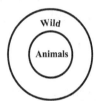

The statement "All lions are animals" is represented by adding a third circle representing lions inside the circle representing animals.

Since the circle representing lions is completely inside the circle representing creatures that are wild, it follows that:

All lions are wild.

The conclusion follows from the premises.

17. All teachers are serious.
All mathematicians are serious.
∴ All mathematicians are teachers.

The statement "All teachers are serious" is represented by a large circle representing serious and a smaller circle inside the large circle representing teachers.

The statement "All mathematicians are serious" is represented by another circle that represents mathematicians inside the larger circle that represents serious. (The circle for teachers would overlap the one for mathematicians.)

Since the circle representing mathematicians is not completely inside the circle representing teachers, the conclusion does *not* follow from the premises.

19. Two intersecting circles represent the days of television watching by the husband and wife. Since they watched 18 days together, place an 18 in the area shared by the two smaller circles. Since the wife watched a total of 25 days, place a $25 - 18 = 7$ in the other region of the circle labeled Wife.

Since the husband watched a total of 20 days, place a $20 - 18 = 2$ in the other region labeled Husband. The 2, 18, and 7 give a total of 27 days. This results in $30 - 27 = 3$ days for the region outside the intersecting circles. This represents 3 days of neither one watching television.

21. Tom, Dick, Mary, and Joan

One is a secretary, one is a computer operator, one is a receptionist, and one is a mail clerk. Find the one who is a computer operator.

1. Tom and Joan eat dinner with the computer operator (fact a), so neither Tom nor Joan is the computer operator.

2. Mary works on the same floor as the computer operator and the mail clerk (fact c), so Mary is not the computer operator.

We know that if Tom, Joan, and Mary are not the computer operator, then Dick must be the computer operator. Fact b is not needed.

Copyright © 2018 Pearson Education, Inc.